W9-CSP-500

NUCLEAR STRUCTURE AND DYNAMICS '09

To learn more about AIP Conference Proceedings,
including the Conference Proceedings Series, please visit the webpage
http://proceedings.aip.org/proceedings

NUCLEAR STRUCTURE AND DYNAMICS '09

Proceedings of the International Conference

Dubrovnik, Croatia   4–8 May 2009

EDITORS

Matko Milin
Tamara Nikšić
Dario Vretenar

University of Zagreb
Zagreb, Croatia

Suzana Szilner
Ruđer Bošković Institute
Zagreb, Croatia

SPONSORING ORGANIZATIONS
Physics Department, University of Zagreb, Croatia
Ruđer Bošković Institute, Zagreb, Croatia
Croatian Academy of Sciences and Arts
Ministry of Science, Education and Sports of the Republic of Croatia

AMERICAN INSTITUTE OF PHYSICS

Melville, New York, 2009
AIP CONFERENCE PROCEEDINGS ■ 1165

Editors

Matko Milin
Tamara Nikšić
Dario Vretenar

Physics Department
Faculty of Science
University of Zagreb
Bijenicka 32
10000 Zagreb, Croatia

E-mail: matko.milin@phy.hr
tniksic@phy.hr
vretenear@phy.hr

Suzana Szilner
Division of Experimental Physics
Ruđer Bošković Institute
Bijenicka 54
10000 Zagreb, Croatia

E-mail: szilner@irb.hr

L.C. Catalog Card No. 2009907355

ISBN 978-0-7354-0702-2
ISSN 0094-243X

Printed in the United States of America

CONTENTS

AB INITIO, CLUSTER MODEL AND SHELL MODEL APPROACHES

NUCLEI FAR FROM STABILITY

SYMMETRY DICTATED APPROACHES

DYNAMICS OF LIGHT-ION REACTIONS

WEAK INTERACTIONS AND ASTROPHYSICS

Preface

This volume collects the contributions to the international conference "Nuclear Structure and Dynamics", held in Dubrovnik, Croatia, from May 4 to 8, 2009. The conference was hosted by the Physics Department of the University of Zagreb and the Ruđer Bošković Institute.

Following the long tradition of nuclear physics meetings on the Adriatic coast, organized by our two institutions, the Conference presented a review of recent experimental and theoretical advances in the physics of nuclear structure and reactions, and provided a broad discussion forum on current and future research projects. The main focus was on the following topics:

- Nuclear structure and reactions far from stability
- Exotic modes of excitation and decays
- Collective phenomena and symmetries
- Ab initio, cluster model, and shell-model approaches
- Nuclear energy density functionals
- Heavy-ion reactions at near-barrier energies
- Dynamics of light-ion reactions
- Nuclear reactions of astrophysical interest
- Weak-interaction processes

The Conference was devoted to a discipline which has seen a strong revival of research activities in the last decade. More than 170 nuclear physicists from over 30 countries participated in "Nuclear Structure and Dynamics". The considerable interest in this meeting was reflected not only in the number of participants but also in the large number of received abstracts. Following the suggestions of the International Advisory Committee, only about half of the submitted abstracts could be selected for oral presentation. In addition to 24 invited talks, more than 80 oral contributions were presented. An evening poster session was organized as part of the conference. An effort was made to provide the opportunity for young researchers to present their work, and selected posters were accepted for very short oral presentation. Many interesting scientific discussions took place in the beautiful surroundings of Dubrovnik, with fruitful interaction between various communities from the fields of reaction mechanisms and nuclear structure.

We would like to express our gratitude to the Advisory Committee: Giacomo de Angelis (Legnaro), Thomas Aumann (GSI Darmstadt), Juha Äystö (Jyväskylä), Yorick Blumenfeld (CERN/Orsay), Rafal Broda (Krakow), Peter Butler (Liverpool), David Dean (Oak Ridge), Jacek Dobaczewski (Warsaw/Jyväskylä), Henning Esbensen (Argonne), Brian Fulton (York), Sydney Gales (GANIL), Florent Haas (Strasbourg), Sotirios Harissopulos (Athens), Rumiana Kalpakchieva (Dubna), Reiner Krücken (Munich), Karlheinz Langanke (Darmstadt/GSI), Marcello Lattuada (Catania), Jie Meng

(Beijing), Tohru Motobayashi (RIKEN), Wolfram von Oertzen (Berlin), Nigel A. Orr (Caen/GANIL), Takaharu Otsuka (Tokyo), Norbert Pietralla (Darmstadt), Karsten Riisager (CERN), Yukinori Sakuragi (Osaka), Peter Schuck (Orsay), Alan C. Shotter (TRIUMF/Edinburgh), Ivo Šlaus (IRB Zagreb), Michael Thoennessen (NSCL/MSU), Ian J. Thompson (LLNL), Piet Van Duppen (Leuven), Andrea Vitturi (Padova), Nicolae V. Zamfir (Bucharest), for helping us to prepare the Conference program and selecting the speakers.

We sincerely thank the local Organizing Committee: Zoran Basrak, Slobodan Brant, Roman Čaplar, Matko Milin, Đuro Miljanić, Tamara Nikšić, Nils Paar, Lovro Prepolec, Neven Soić, and the Conference Secretary Mrs. Ksenija Krsnik, for their diligent work in the preparation and coordination of this meeting.

We acknowledge the financial support of the organizing institutions, the Croatian Ministry of Science, Education and Sports, and the Croatian Academy of Sciences and Arts. Finally, we would like to thank all the participants whose enthusiasm, interest and scientific ideas present the main contribution to the success of the Conference.

Suzana Szilner and Dario Vretenar
Co-Chairs of "Nuclear Structure and Dynamics"

Nuclear Structure and Dynamics '09: opening address

Ivo Šlaus

*South-East European Division of the World Academy of Art and Science
and Ruđer Bošković Institute, Bijenička 54, Zagreb, Croatia*

ABOUT DUBROVNIK

There is a long tradition of physics conferences on our Adriatic coast and by choosing Dubrovnik the organizer selected our most beautiful city. Though the University of Dubrovnik has been established only in 2003, Dubrovnik has a rich history of education and research. Famous physicists Ruđer Bošković and Marin Getaldić are from Dubrovnik, and though Bošković, whom Leon Lederman [1] ranks among the very top physicists through the history, left the City when he was 16, he remained in the diplomatic service of the Republic of Dubrovnik throughout his entire life. The Republic of Dubrovnik can also boast with one of the oldest pharmacy in the world. It has introduced the first quarantine in Europe in 1377, and an obligatory vaccination of children against small pox. This action has been second only to that of the canton of Aargau in Switzerland. Since 1971 Dubrovnik is the seat of the Inter-University Centre including over 200 universities throughout the world, and several Croatian institutions. Thus it remains an important centre at the scientific map of Europe and world.

NUCLEAR PHYSICS CONFERENCES AT ADRIATIC COAST

As the part of development of nuclear physics in Croatia, the first summer schools have been organized at Island of Mali Lošinj in the northern Adriatic, then in the city of Hercegnovi, now Montenegro, and then in Brela, Split and Dubrovnik. Gradually these summers schools evolved into conferences, and the conference in 1967 in Brela is now listed as the second in a series of Few body conferences, actually initializing the series after it started in London in 1959 [2]. There was also a series of Adriatic International Conferences on Nuclear Physics held at the island of Hvar in 1977, 1979, 1981, 1984, Dubrovnik 1987 and island of Brioni 1991. Furthermore, Dubrovnik also hosted the Fast Neutron Physics Conference in 1986 and the 5th International Symposium on Radiation Physics in 1990. The 7th International Conference on Clustering Aspects of Nuclear Structure and Dynamics was held at the island of Rab in 1999, while the International Conference on Nuclear Many-body Problem took place at Brioni in 2001.

CP1165, *Nuclear Structure and Dynamics '09*
edited by M. Milin, T. Nikšić, D. Vretenar, and S. Szilner
© 2009 American Institute of Physics 978-0-7354-0702-2/09/$25.00

HISTORY OF NUCLEAR PHYSICS IN CROATIA

The biggest Croatian institute is named after Dubrovnik-born Ruđer Bošković. Immediately after the end of the World War II a group of professors from the University of Zagreb led by Professor Ivan Supek decided to establish a centre for physics, but rather quickly decided that it should be a multi- and interdisciplinary research institute. This is how the Ruđer Bošković Institute (RBI) was formally established in 1950. Obviously, the development of nuclear weapons and nuclear energy played an important role [3]. At the beginning, nuclear physics has been the central part of the RBI, wherefrom developed particle physics, nuclear chemistry, radiobiology and electronics departments. Since there have been no research in nuclear physics in Croatia till 1945, it has been decided to start it with two prongs: first, sending best graduate students to different research centres in Europe and in the USA, and second, initiating a series of summer schools (see previous section). International collaboration has been emphasized from the beginning. In 1957 the first graduate courses at the University of Zagreb have been introduced and they have been in nuclear physics and some of us just returning from our research abroad have been the first lecturers.

In 1960-ties Zagreb established itself on the nuclear physics world map. First, Alaga developed selection rules for beta- and gamma-transitions in collaboration with A. Bohr and B. Mottelson [4]. Second, Zagreb became a world renowned centre for nuclear data production. During 1960-ties RBI was one of the leading producers of 14 MeV neutron data [5,6]. Third, regarding a problem of nuclear interaction, half a century ago the Los Alamos group [7] and then Zagreb group did many few-nucleon studies culminating in a first successful attempt to determine neutron-neutron $^1S_0$ scattering length [8]. Study of nuclear interactions, two-body and three-body, and the study of few particle systems were and remain one of the outstanding research activities of the Zagreb group [9,10]. Fourth, study of nuclear dynamics and structure (models and spectroscopy) – the very subject of this conference – has been one of the focal points of Zagreb nuclear physics in 70-ties and 80-ties (e.g. studies of quasi-free reaction mechanisms) and remained so till these days. Among most important recent results in this area are those on the structure of very light nuclei [11], multinucleon transfer processes in heavy-ion reactions [12], relativistic Hartree-Bogoliubov theory [13], microscopic description of nuclear quantum phase transitions [14] and exotic modes of excitations far from stability [15].

Furthermore, one of the characteristics of Zagreb nuclear physicists has been their intensive collaboration with many research groups throughout the world, from all the continents. It was, therefore, natural that nuclear physicists from Zagreb were among the founding fathers of the European Physical Society (EPS) established on September 26, 1969, of Academia Europaea (AE) established in 1988, and of the Interacademy Panel, established in 1991. Finally, it is well-known that good basic research frequently gives remarkable applications. Zagreb nuclear physics group initiated in 70-ties the development of short-lived radio-pharmaceuticals including mathematical models [16] for their use and thereby strengthened the collaboration between physics and medicine.

It is equally important and even more instructive to analyze our failures and I am personally responsible for many of them. Anti-nuclear campaign in 80's resulted in significant reduction of interest for nuclear energy and technology; some countries were able to maintain high interest in nuclear physics research while Croatia did not. Second failure: after a very successful start in development of nuclear instruments, which are crucial for any scientific research, this activity was not continued. Third, the unique opportunity to seat the present-day ITER facility [17] in former Yugoslavia failed and that could have enormous consequences on R&D (and politics) at these areas.

FUTURE PROSPECTS

Nowadays we are witnessing the renaissance of nuclear physics. Scientific research can be represented in a 3-dimensional space of curiosity-driven research which is a prime generator of breakthroughs and new literacies, incremental research and instrument development. Let me attempt to describe this nuclear physics renaissance through these dimensions. First, excellent persons - as all participants and organizers of this conference are - are the best sign of this renaissance. Physics research always underlines the validity of the statement that "people are the true wealth of nations" [18]. Obviously, we are now witnessing that best young talents are once again coming to nuclear physics and your activity and you as role-models are what attracts them. Second, it became clear that nuclear physics has a central position between particle physics and astrophysics and that many essential problems require nuclear physics research directly or at least indirectly. Third, there are 17,500 accelerators in the world [19] and their number constantly increases as well as the size of the domain of their applications. Detectors are now as expensive as accelerators ranging in costs to several hundred million dollars. In addition to nuclear research contribution to basic research (including not only nuclear physics, but also particle physics, astronomy, chemistry, biology *etc.*), nuclear physics nowadays has very important applications in medicine, as a basis of femto-technology (that will add new dimension to the present nanotechnology), in fighting the climate change *etc.*

It seems to me that most important contribution of physics and particularly nuclear physics to society is through education and through introducing our "methodology" - a reasonable argument is based on the fact that physicists have not only obtained all Nobel prizes in physics, but they got Nobels in all other activities (chemistry, economics, medicine, biology). Even Nobel prize for peace has been several times awarded to physicists. We are witnessing a recent transition of many young talented physicists to financing and economics in general. The current economic crisis with accompanying ecological, energy and ethical crisis compounded by demographic transition and climate change call all of us to be involved and to try to contribute toward solving these obviously acute problems. It was pointed out [20] that present economists made some of their concepts sacred cows. Robert Nelson wrote a book entitled «Economics as Religion» and argued that economists have deified some of their concepts like market and individual property. Physicists, on the other hand, learned to be suspicious of their axioms and models. It is this heretical feature of the spirit of science that we physicists have to bring in when we develop scenarios

3

(scenarios are very different from predictions that we learned to do) and this is the core of the Galileo' goal that the arguments of a humble individual prevail over the authorities. This is the main reason why I will conclude by paraphrasing the answer of an English admiral to Peter the Great. When Peter the Great was in England he was so impressed by the English navy that he said to an English admiral "If I were not the Czar of all Russia, I would like to be admiral of the English fleet!" The admiral replied "Sire, if I were not admiral of the English fleet, I would like to be admiral of the English fleet." Similarly, I will tell you "If I were not a physicist, I would like to be a physicist!"

On behalf of *Croatian Academy of Sciences and Arts* I wish you a successful and fruitful conference!

REFERENCES

1. L. Lederman (with D. Teresi): *"The God Particle: If the Universe is the Answer, What is the Question?"*, Houghton Mifflin Company, 1993.
2. The Editors, Preface, *Nucl. Phys.* **A737**, xiii (2004).
3. W.C. Potter, Đ. Miljanić, I. Šlaus, *Bull. At. Scientists* **56**, 63 (2000).
4. G. Alaga, *Phys Rev* **100**, 432 (1955); G. Alaga, *Nucl. Phys.* **4**, 625 (1957).
5. I. Šlaus, Invited talk, Proc. Int. Conf. "Interactions of Neutrons with Nuclei" (Lowell, MA, USA, 1976) , Ed. E. Sheldon, Tech. Inf. Div. ERDA, 1976, p. 272.
6. N. Cindro, Proc. Intl. Conf. "Interactions of Neutrons with Nuclei"(Lowell, MA, USA, 1976) , Ed. E. Sheldon, Tech. Inf. Div. ERDA, 1976, p. 347.
7. J.D. Seagrave, in Few-Body Problem, Proc. Brela conference, eds. G. Paić and I. Šlaus, Gordon and Breach (1968), 787 and references therein.
8. K. Ilakovac, et al, *Phys. Rev. Lett* **6**, 356 (1961).
9. I. Šlaus, *Rev. Mod. Phys.* **39** (1967) 575; I. Šlaus, Few Nucleon Systems, Invited talk at the Inter. Nucl. Phys. Conf. Tokyo, 1977, J. Phys. Soc. Japan **44**, Suppl. (1978) 57
10. I. Šlaus, Summary talk of the 18th Inter. IUPAP Conf. on Few-body problems in phys., *Nucl. Phys.* **A 790,** 199c (2007), see also G.A. Miller, B.M.K. Nefkens, I. Šlaus, *Phys. Rep.* **194**,1 (1990).
11. M. Milin, M. Zadro, et al., *Nucl. Phys.* **A 753**, 263 (2005).
12. L. Corradi, G. Pollarolo, S. Szilner, *J. Phys.* G, invited topical review, to be published.
13. D. Vretenar, A. Afanasjev, G.A. Lalazissis, P. Ring, *Phys. Rep.* **409**, 101 (2005).
14. T. Nikšić, D. Vretenar, G.A. Lalazissis, P. Ring, *Phys. Rev. Lett.* **99**, 092502 (2007).
15. N. Paar, D. Vretenar, E. Khan, G. Colo, *Rep. Prog. Phys.* **70**, 691 (2007).
16. J. Nosil, Š. Spaventi, I. Šlaus, *Eur. J. Nucl. Med.* **2** (1977) 1.
17. C. Llewellyn-Smith and D. Ward, *Eur. Rev.* **13**, 337 (2005).
18. The State of Human Development, Human development indicators, 2004, p.127, UNDP.
19. U. Amaldi, *Nucl. Phys.* **A 751**, 494c (2005).
20. J.P. Bouchaud, arXiv:0810.5306 (June 1, 2009); J.P. Bouchaud, arXiv:0904.0805 (June 2, 2009); Dirk Helbing (ed.) «Managing Complexity: Insights, Concepts, Applications» (Book series: Understanding complex systems) Springer Berlin, Heidelberg (2008)

AB INITIO, CLUSTER MODEL AND SHELL MODEL APPROACHES

Clusters and Halos in Light Nuclei

Thomas Neff and Hans Feldmeier

GSI Helmholtzzentrum für Schwerionenforschung GmbH, Planckstraße 1, 64291 Darmstadt, Germany

Abstract. The structure of light nuclei in the p- and sd-shell features exotic phenomena like halos and clustering. In the Fermionic Molecular Dynamics (FMD) approach we aim at a consistent microscopic description of well bound nuclei and of loosely bound exotic systems. This is possible due to the flexibility of the single-particle basis states using Gaussian wave-packets localized in phase space. Many-body basis states are Slater determinants projected on parity, angular and total linear momentum.

The structure of ^{12}C is discussed. Here the ground state band can be well described within a shell model picture but excited states above the three-α threshold, including the famous Hoyle state, show a pronounced cluster structure. As another example we study the structure of the Neon isotopes $^{17-22}$Ne. In ^{17}Ne we find a large s^2 occupation related to a large charge radius. The charge radius decreases for ^{18}Ne but gets again very large for ^{19}Ne and ^{20}Ne which is explained by significant admixtures of ^{3}He and ^{4}He cluster components into to the ground state wave functions.

Keywords: exotic nuclei, clustering, microscopic cluster model, no-core shell model, Hoyle state, di-proton halo
PACS: 21.10.-k, 21.60.-n, 25.30.-c

INTRODUCTION

In recent years *ab initio* calculations starting from realistic nuclear interactions have become possible for example in the Green's Function Monte Carlo (GFMC) approach and the No-Core Shell Model (NCSM). Unfortunately these approaches are restricted to light nuclei because of the numerical effort. Furthermore exotic nuclei with clustering and halos are very difficult to describe in an harmonic oscillator basis as used in the NCSM approach. Cluster models have been used successfully but have to rely on simple effective interactions. The Fermionic Molecular Dynamics (FMD) approach aims at a consistent description of nuclei in the p- and sd-shell including well bound nuclei with shell model configurations as well as loosely bound nuclei featuring clusters and halos.

FERMIONIC MOLECULAR DYNAMICS

In the Fermionic Molecular Dynamics model [1, 2] Slater determinants are used as many-body basis states

$$|Q\rangle = \mathscr{A}\left\{|q_1\rangle \otimes \cdots \otimes |q_A\rangle\right\}. \tag{1}$$

CP1165, *Nuclear Structure and Dynamics '09*
edited by M. Milin, T. Nikšić, D. Vretenar, and S. Szilner
© 2009 American Institute of Physics 978-0-7354-0702-2/09/$25.00

The single-particle states $|q\rangle$ are given by a single or a superposition of two Gaussian wave packets localized in phase space

$$\langle \vec{x}|q\rangle = \sum_i c_i \exp\left\{-\frac{(\vec{x}-\vec{b}_i)^2}{2a_i}\right\} |\chi_i^\uparrow, \chi_i^\downarrow\rangle \otimes |\xi\rangle. \tag{2}$$

The complex parameter $\vec{b}$ encodes mean position and mean momentum of the wave packet. The width a is a variational parameter and can be different for each wave packet. The spin can assume any orientation, whereas the isospin assumes values of $\pm\frac{1}{2}$ describing either protons or neutrons. The wave packet basis is very flexible. Slater determinants are invariant under linear transformations of the single-particle states and harmonic oscillator single-particle states can be obtained by linear combinations of slightly shifted Gaussians. On the other hand Bloch-Brink type cluster states can be obtained by localizing groups of wave packets. The FMD model space therefore includes the harmonic oscillator shell model and the microscopic cluster model as limiting cases.

The FMD solution on the mean-field level is obtained by minimizing the intrinsic Hamiltonian with respect to all the parameters of the single-particle states.

$$\min_{\{q_i\}} \frac{\langle Q|H - T_{\text{cm}}|Q\rangle}{\langle Q|Q\rangle} \tag{3}$$

The symmetries of the Hamiltonian are restored by projecting the intrinsic state $|Q\rangle$ on parity, angular momentum and total linear momentum.

$$\left|Q; J^\pi M K, \vec{P} = 0\right\rangle = P^\pi P^J_{MK} P^{\vec{P}=0} |Q\rangle \tag{4}$$

The implementation of these projections is straightforward as a reflected, rotated and translated Slater determinant of Gaussian wave packets is a Slater determinant of the reflected, rotated and translated wave packets.

The correlation energies can be very large especially if we are dealing with well deformed or clustered states. Therefore instead of the simple projection after variation (PAV) approach a variation of projection (VAP) should be performed. Unfortunately this comes with a huge numerical effort and straightforward VAP calculations can only be performed for light nuclei. For heavier nuclei we perform VAP calculations in a generator coordinate picture. The energy of the intrinsic state $|Q\rangle$ is here minimized under constraints on the radius, dipole, quadrupole or octupole deformation and we can investigate the energy surface as a function of these constraints. With both methods we generate a set $|Q^{(a)}\rangle$ of intrinsic states that are used to construct our many-body states

$$\left|\Psi; J^\pi M \alpha\right\rangle = \sum_{Ka} P^\pi P^J_{MK} P^{\vec{P}=0} |Q^{(a)}\rangle c^{J^\pi \alpha}_{Ka}, \tag{5}$$

in a multi-configuration mixing calculation for which we have to solve the generalized eigenvalue problem

$$\sum_{K'b} \langle Q^{(a)} | (H - T_{cm}) P^\pi P^J_{KK'} P^{\vec{P}=0} | Q^{(b)} \rangle c^{J^\pi \alpha}_{K'b} =$$

$$E^{J^\pi \alpha} \sum_{K'b} \langle Q^{(a)} | P^\pi P^J_{KK'} P^{\vec{P}=0} | Q^{(b)} \rangle c^{J^\pi \alpha}_{K'b} . \quad (6)$$

Interaction

We use an effective interaction based on the realistic Argonne V18 interaction. By means of the Unitary Correlation Operator Method (UCOM) short-range central and tensor correlations are included explicitly, conserving the phase shifts of the bare interaction [3]. V_{UCOM} is an effective low-momentum interaction which can be used for *ab initio* calculations in many-body approaches like the no-core shell model [4, 5] . The FMD model space using a small set of many-body basis states is not able to describe the correlations due to medium-range tensor forces. To partly account for that a longer ranged tensor correlation operator is used. In addition a phenomenological momentum-dependent two-body correction term is added and fitted to binding energies and radii of the "closed-shell" nuclei ^{4}He, ^{16}O and ^{40}Ca. An additional spin-orbit correction is added and fitted to the binding energies of ^{24}O, ^{34}Si and ^{48}Ca. The correction terms contribute about 15% to the potential energy.

CLUSTER STATES IN ^{12}C

While NCSM calculations are perfectly able to describe the properties of the ground state band of ^{12}C they fail miserably in the description of the first excited 0^+ state, the famous Hoyle state located just above the three-α threshold [6]. This state is supposed to have a pronounced α-cluster structure and microscopic cluster model calculations within the RGM approach [7] were quite successful in describing its properties but use simple effective interactions. Based on the cluster model wave function an interpretation of this state as a Bose condensate of α-particles was proposed recently [8].

In the FMD approach [9] α-cluster configurations are a subset of the Hilbert space. Further configurations are obtained by VAP calculations with constraints on radius and quadrupole deformation. These additional configurations are necessary to describe properties of the ground state band where α-clusters are broken due to the spin-orbit force. In a Hilbert space that consists only of α-cluster configurations the FMD ground state is underbound by more than 10 MeV. Including all configurations we can reproduce the properties of the ground state band as well as that of the Hoyle state. The Hoyle state has an overlap of 85% with three-α configurations. Such an admixture is needed to explain for example the observed β-transition strength from ^{12}B and ^{12}N [10]. The Hoyle state has a very large radius of 3.38 fm. This spatially extended nature of the Hoyle state can also be tested by electron scattering data, measuring the transition from

9

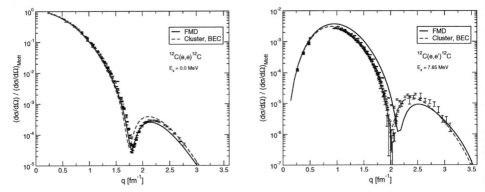

FIGURE 1. (Color online) Elastic and inelastic scattering cross sections calculated with FMD and the microscopic cluster model compared to experiment.

the ground state to the Hoyle state. In Fig. 1 elastic and inelastic scattering cross sections are compared with experimental data for the FMD and the microscopic cluster model.

To understand why the NCSM fails in the description of the Hoyle state we decompose the FMD wave function into $N\hbar\Omega$ shell model configurations. In Fig. 2 the results are presented for the ground state and for the Hoyle state. While the ground state has only components up to about $8\hbar\Omega$ the Hoyle is essentially a coherent excitation spanning many shells up to about $50\hbar\Omega$ which is way beyond the treatable model space size of the NCSM.

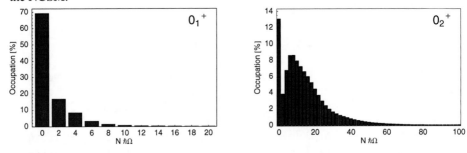

FIGURE 2. (Color online) Decomposition of the ^{12}C FMD ground and Hoyle state wave functions into N-$\hbar\Omega$ shell model components.

NEON ISOTOPES $^{17-22}$NE

^{17}Ne is located at the proton drip-line and is considered as the best candidate for a two-proton halo nucleus. In the simplest picture ^{17}Ne can be considered as an ^{15}O core in its ground state and two protons in either an s^2 or d^2 configurations. Interaction cross sections [11] and longitudinal momentum distributions [12] support the halo picture. Three-body cluster model calculations [13, 14] predict s^2 contributions of about 45%. On the other hand shell model calculations that focus on the Coulomb displacement energies between ^{17}Ne and ^{17}N predict rather small s^2 admixtures of only 20% [15].

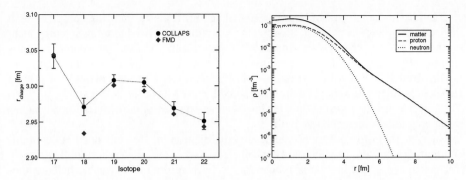

FIGURE 3. (Color online) Left: Charge radii for Neon isotopes measured by COLLAPS at ISOLDE and calculated in FMD. Right: Matter, proton and neutron distributions of FMD ground state for ^{17}Ne. The proton skin thickness is 0.45 fm.

Recent measurements of the charge radii by the COLLAPS group at ISOLDE now allow for a direct test of the wave function. In a joint paper the experimental results for the charge radii of $^{17-22}$Ne are compared with FMD calculations [16].

In the calculation we perform a variation after parity projection. In the case of ^{17}Ne we find two minima which correspond to s^2- and d^2-dominated configurations for the valence protons around an ^{15}O core. The situation in ^{18}Ne is similar. In ^{19}Ne the experimental $1/2^+$ and $1/2^-$ states are almost degenerate. We therefore include intrinsic states minimized for positive and minimized for negative parity. As the cluster thresholds are pretty low for most of the Neon isotopes we include explicit ^{3}He and ^{4}He cluster configurations. Additional configurations are created by cranking the strength of the spin-orbit force which changes the properties of the single-particle orbits.

The binding energies and radii are then calculated in a multi-configuration mixing

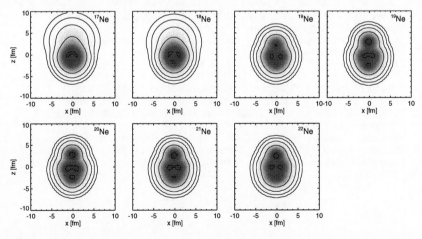

FIGURE 4. (Color online) Cuts through intrinsic density distributions of dominant FMD configurations for the Neon isotopes. For ^{19}Ne the densities of intrinsic states optimized for positive and negative parity are shown.

11

calculation. In the left part of Fig. 3 the calculated FMD charge radii are compared with the experimental data. We find a large charge radius for ^{17}Ne due to an s^2 component of about 42%. In ^{18}Ne the s^2 component is only about 15% and the charge radius is significantly smaller. The charge radius increases again for ^{19}Ne. Here the s-orbits are not involved but we have both strong α- and ^{3}He-cluster components in the $1/2^+$ ground state which enlarge the charge radius compared to the FMD mean-field result by 0.17 fm. This effect is also very strong for ^{20}Ne but becomes smaller for the heavier Neon isotopes explaining the dropping charge radii.

The right part of Fig. 3 shows the density distribution for ^{17}Ne obtained in the multi-configuration mixing calculation. One can clearly see the extended proton distribution. The calculated s^2 occupation is about 42%, ^{17}Ne therefore has a sizeable halo component.

REFERENCES

1. R. Roth, T. Neff, H. Hergert, and H. Feldmeier, *Nucl. Phys.* **A745**, 3 (2004).
2. T. Neff, and H. Feldmeier, *Eur. Phys. J Special Topics* **156**, 69 (2008).
3. T. Neff, and H. Feldmeier, *Nucl. Phys.* **A713**, 311 (2003).
4. R. Roth, H. Hergert, P. Papakonstantinou, T. Neff, and H. Feldmeier, *Phys. Rev.* **C72**, 034002 (2005).
5. R. Roth, P. Papakonstantinou, N. Paar, H. Hergert, T. Neff, and H. Feldmeier, *Phys. Rev.* **C73**, 044312 (2006).
6. P. Navrátil, V. G. Gueorguiev, J. P. Vary, W. E. Ormand, and A. Nogga, *Phys. Rev. Lett.* **99**, 042501 (2007).
7. M. Kamimura, *Nucl. Phys.* **A351**, 456 (1981).
8. Y. Funaki, A. Tohsaki, H. Horiuchi, P. Schuck, and G. Röpke, *Phys. Rev.* **A67**, 051306(R) (2003).
9. M. Chernykh, H. Feldmeier, T. Neff, P. von Neumann-Cosel, and A. Richter, *Phys. Rev. Lett.* **98**, 032501 (2007).
10. H. O. U. Fynbo, C. A. Diget, U. C. Bergmann, M. J. G. Borge, J. Cederkäll, P. Dendooven, L. M. Fraile, S. Franchoo, V. N. Fedosseev, B. R. Fulton, W. Huang, J. Huikari, H. B. Jeppesen, A. S. Jokinen, P. Jones, B. Jonson, U. Köster, K. Langanke, M. Meister, T. Nilsson, G. Nyman, Y. Prezado, K. Riisager, S. Rinta-Antila, O. Tengblad, M. Turrion, Y. Wang, L. Weissman, K. Wilhelmsen, J. Äystö, and T. I. Collaboration, *Nature* **433**, 136 (2005).
11. A. Ozawa, T. Kobayashi, H. Sato, D. Hirata, I. Tanihata, O. Yamakawa, K. Omata, K. Sugimoto, D. Olson, W. Christie, and H. Wieman, *Phys. Lett.* **B334**, 18 (1994).
12. R. Kanungo, M. Chiba, B. Abu-Ibrahim, S. Adhikari, D. Q. Fang, N. Iwasa, K. Kimura, K. Maeda, S. Nishimura, T. Ohnishi, A. Ozawa, C. Samanta, T. Suda, T. Suzuki, Q. Wang, C. Wu, Y. Yamaguchi, K. Yamada, A. Yoshida, T. Zheng, and I. Tannihata, *Eur. Phys. J.* **A25**, 327 (2005).
13. L. V. Grigorenko, I. G. Mukha, and M. V. Zhukov, *Nucl. Phys.* **A713**, 372 (2003).
14. E. Garrido, D. V. Fedorov, and A. S. Jensen, *Nucl. Phys.* **A733**, 85 (2004).
15. H. T. Fortune, and R. Sherr, *Phys. Lett.* **B503**, 70 (2001).
16. W. Geithner, T. Neff, G. Audi, K. Blaum, P. Delahaye, H. Feldmeier, S. George, C. Guénaut, F. Herfurth, A. Herlert, S. Kappertz, M. Keim, A. Kellerbauer, H.-J. Kluge, M. Kowalska, P. Lievens, D. Lunney, K. Marinova, R. Neugart, L. Schweikhard, S. Wilbert, and C. Yazidjian, *Phys. Rev. Lett.* **101**, 252502 (2008).

Cluster Structure of ^{12}C and ^{11}Be

M. Freer*, H. Fujita†, Z. Buthelezi**, J. Carter†, R. W. Fearick‡, S. V. Förtsch**, R. Neveling**, S. M. Perez**, P. Papka§, F. D. Smit**, J. A. Swartz§, I. Usman†, P. J. Haigh*, N. I. Ashwood*, T. Bloxham*, N. Curtis*, P. McEwan*, H. G. Bohlen¶, T. Dorsch¶, Tz. Kokalova¶, Ch. Schulz¶ and C. Wheldon¶

*School of Physics and Astronomy, University of Birmingham, Edgbaston, Birmingham, B15 2TT, United Kingdom
†School of Physics, University of the Witwatersrand, Johannesburg 2050, SA
**iThemba LABS, PO Box 722, Somerset West 7129, SA
‡Physics Department, University of Cape Town, Rondebosch 7700, SA
§Physics Department, University of Stellenbosch, Stellenbosch, SA
¶Helmholtz-Zentrum Berlin, Glienicker Strasse 100, D-14109 Berlin, Germany

Abstract. The structure of ^{12}C is discussed, in particular the spectrum of states above the α-decay threshold. A search for the 2^+ excitation of the Hoyle-state is reported. The structural link between halo-like states and molecular states is explored in the case of ^{11}Be.

Keywords: Nuclear Clustering, Light Nuclei, Experimental Studies
PACS: 25.60Je, 27.20+n, 21.60.Gx

THE STRUCTURE OF ^{12}C

Despite being one of the lightest, and most studied, nuclei in nature the structure of the nucleus ^{12}C remains far from being fixed. It lies at the extremes of the application of *ab initio* approaches to the calculation of the structure of nuclei from an understanding of the nucleon-nucleon interaction, including 3-body forces, e.g. the Greens Function Monte Carlo approach (GFMC) [1]. Thus, its experimental properties provide one of the best tests of our understanding of the nucleon-nucleon force. Intriguingly, the GFMC calculations show that the ground state of the lighter system, ^{8}Be possesses an $\alpha+\alpha$ cluster structure. The inert nature of the α particle (it possesses a high binding energy and a first excited state close to 20 MeV) suggests that it may have an important influence on the structure of other light systems. In the case of ^{12}C, the system can be constructed from a variety of geometric arrangements of three α-particles. Due to the much stronger interaction between three α-particles in an equilateral-triangle arrangement, it might be expected that this should be the lowest energy configuration. Such an arrangement possesses a D_{3h} point group symmetry. The corresponding rotational and vibrational spectrum is described by a form [2]

$$E(v_1, v_2^l, L, K, M) = E_0 + Av_1 + Bv_2 + CL(L+1) + D(K \pm 2l)^2 \qquad (1)$$

where $v_{1,2}$ are vibrational quantum numbers (v_2 is doubly degenerate); $l = v_2, v_2 - 2, ..., 1$ or 0, M is projection of an angular momentum L on a laboratory fixed and K on a body-

CP1165, *Nuclear Structure and Dynamics '09*
edited by M. Milin, T. Nikšić, D. Vretenar, and S. Szilner

fixed axis [2]. *A, B, C* and *D* are adjustable parameters; the spectrum got by the choice *A*=7.0, *B*=9.0, *C*=0.8 and *D*=0.0 MeV is shown in Fig. 1.

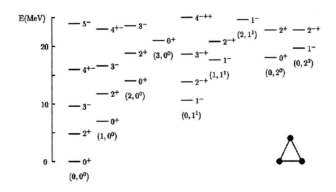

FIGURE 1. Spectrum of the energy levels of an equilateral triangle configuration. The bands are labelled by (v_1, v_2^l). See Ref. [2] for further details.

The ground state band, (v_1, v_2^l)=$(0,0^0)$, contains no vibrational modes and coincides well with the observed experimental spectrum. Here the states correspond to different values of K ($K = 3n$, n=0,1,2...) and L. For K=0, L=0, 2, 4 etc....., which is a rotation of the plane of the triangle about a line of symmetry, whereas for $K \neq 0$ $L = K$, $K + 1$, $K + 2$..... In the present case K=0 or 3 is plotted with the parity being given by $(-1)^K$. The K=0 states coincide well with the well-known 0^+ (ground-state), 2^+ (4.4 MeV) and 4^+ (14.1 MeV) states. The K=3 states correspond to a rotation about an axis which passes through the centre of the triangle, the first of which has spin and parity 3^- and coincides with the 9.6 MeV, 3^-, excited state. A prediction of this model is that there should be a 4^- state almost degenerate with the 4^+ state. A recent measurement involving studies of the α-decay correlations indicated that the 13.35 MeV unnatural-parity state possessed J^π=4^- [3]. The close degeneracy with the 14.1 MeV 4^+ state would appear to confirm the D_{3h} symmetry. In this picture the 0^+ state at 7.65 MeV would correspond to a vibrational mode (v_1=1). The coupling of rotational modes would then produce a corresponding 2^+ state at 4.4 MeV above 7.65 MeV, i.e. 12.05 MeV; there is no known 2^+ state at this energy.

Analysis of electron inelastic scattering data [4, 5] indicates that the Hoyle-state has a volume some 3.4 times larger than the ground state. This larger volume reduces the overlap of the α-particles and may allow them to obtain their quasi-free characteristics in something approaching an α-particle gas, or perhaps a bosonic condensate (BEC) [6]. This latter possibility is intriguing as it would correspond to a new form of nuclear matter where the bosonic nature of the α-particles would allow the constituents to all occupy the lowest energy level of the mutual interaction potential – unlike fermions. Fermionic Molecular Dynamics (FMD) calculations also find that the 7.65 MeV state has a similar structure [7].

From an experimental perspective the determination of the nature of the Hoyle-state has been something of a challenge. One approach might be to determine where a 2^+

excitation exists, if at all, as this would provide some information on the deformation, or in the condensate picture the nature of the interaction potential. What is relatively certain is that there is no 2^+ state below 9.64 MeV (the energy of the 3^- state). This excludes the possibility of a linear chain structure. Above this energy the situation is much more complex since the ^{12}C spectrum is dominated by the broad 0^+ state at 10.3 MeV, and no strong candidate has been observed for a 2^+ state in a reasonable energy window from the 7.65 MeV state. There is a possible 2^+ state at 11.16 MeV[8], which has only been seen in once in the ^{11}B(^{3}He,d)^{12}C reaction. The β-decay measurements of Fynbo et $al.$ [9] find no strong evidence for a narrow (few hundred keV) 2^+ state in the region of 9-13 MeV. In principle the β-decay measurements are the most sensitive studies as selection rules only allow 0^+, 1^+ and 2^+ states to be populated with appreciable strength. One possible reason for this lack of observation, aside from the possibility that the state is extremely broad, is that the matrix element for the β-decay may be very small owing to the decay from a shell model-like state to one which is strongly clustered. In this instance, if the 2^+ state is not associated with the 11.16 MeV state, then an alternate possibility for the lack of observation is because the 2^+ state lies below another state which is strongly populated in reactions. The analysis of the ^{12}C(α,α') reaction by Itoh et $al.$ [10] has suggested that a 2^+ state might lie at 9.7 MeV with a width of a few hundred keV. In this instance the spectrum is dominated by the very strong 9.64 MeV (3^-) excitation. This analysis is very difficult and an unambiguous result has not emerged.

In order to clarify the situation we have recently performed a high resolution, 24 keV, measurement of the ^{12}C(p,p') reaction. The measurements were performed with a 66 MeV proton beam (25 nA) provided by the SSC accelerator at iThemba LABS, SA. The beam was incident on a 1 mg/cm^2 natural carbon target. Measurements were performed at a number of spectrometer angles; $\theta_{lab} = 10°$, $16°$ and $28°$. In the excitation energy region of interest the spectrum is complicated. At 280 keV above the decay threshold is the narrow (8.5 eV) 7.65 MeV, 0^+, Hoyle-state; a 9.641 MeV, 3^- state which is listed as having a width of 34(5) keV [8]; a 1^- state at 10.844 MeV ($\Gamma = 315(25)$ keV) and the unnatural parity state listed as $J^\pi = 2^-$ at 11.828 MeV ($\Gamma = 260(25)$ keV) [8]. Underlying all of these is a very broad 0^+ state at $\sim$10.3 MeV with a width of 3000(700) keV. In particular, it is this very broad 0^+ state which masks all other contributions in this region and inhibits the search for any 2^+ excitation of the Hoyle-state.

In order to suppress the 0^+ contribution, a measurement has been performed at an angle which coincides with a diffractive minimum ($\theta_{lab} = 16°$) in the angular distributions measured for the 0^+ 7.65 MeV state [11, 12]. If a broad 2^+ component exists, then it should be emphasized at $16°$. The spectra at the three angles have been directly compared in Fig. 2. The yields have been normalized to the area of the 7.65 MeV 0^+ peak. In each case a flat background indicated by blank target measurements of 15, 25 counts and 140 counts at $28°$, $16°$ and $10°$, respectively, has been subtracted. The background subtraction yields spectra which have close to zero counts at 7.4 MeV (indicating it has been performed correctly). Figure 2 shows that between 9 and 11 MeV there is a significantly different spectral shape for the $16°$ data compared with the other two angles (note that the angular distribution for the 10.3 MeV state should follow that of the 7.65 MeV data). All three curves agree in amplitude close to 8.4 MeV indicating that the 0^+ strength is the dominant contribution at this excitation energy. The $10°$ and $28°$ data

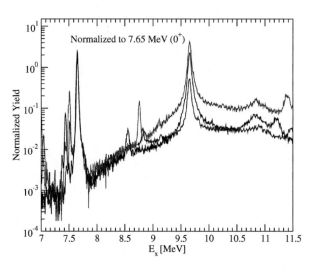

FIGURE 2. (color online) The three excitation energy spectra measured at $\theta_{lab} = 10°$ (blue), $16°$ (red) and $28°$ (black). Spectra have been normalized to the area of the 7.65 MeV, 0^+, peak.

agree across the range of excitation energies, indicating, as expected, that it is the broad 0^+ state which dominates. The $16°$ data show there to be an extra component above 9 MeV. This enhancement coincides with the location of the 2^+ strength found by Itoh *et al.* [10]. The present measurements indicate that the enhanced strength could be associated with a resonance at ~9.6 MeV with a width of 600(100) keV. If this resonance is associated with a 2^+ state then this would coincide with that predicted for a loose assembly of α-particles [7], i.e. an α-particle gas state.

TWO α-PARTICLES PLUS THREE NEUTRONS: ^{11}BE

The concepts of nuclear clustering and nuclear molecules are best illustrated with the beryllium isotopes. For example, ^{8}Be is an unbound nucleus which has a well established α-α cluster structure [13]. The ^{9}Be ground state has a similar two-centre structure but unlike ^{8}Be it is bound by 1.67 MeV. The binding energy which stabilizes the system is provided by the extra neutron which resides in delocalised orbits around the α-cores. The signature of such a molecular structure is a rotational band, two of which have been identified in ^{9}Be [14]; a band built on the $K^\pi = 3/2^-$ ground state in which the valence neutron is in a π-binding orbital and a band built on the $K^\pi = 1/2^+$ first excited state in which the valence neutron is in a σ-binding orbital.

The corresponding molecular structures in ^{10}Be are expected to be dominant just below the $2n + {}^8$Be decay threshold. Thus, the molecular structure of the ^{10}Be ground state, which is bound with respect to two neutron emission by 8.48 MeV and to single neutron decay by 6.81 MeV, is unlikely to be strongly developed. The 5.9583 MeV [2^+], 5.9599 MeV [1^-], 6.1793 MeV [0^+] and 6.2633 MeV [2^-] quartet of states lie much closer to the cluster decay threshold and studies suggest that they may have a molecular

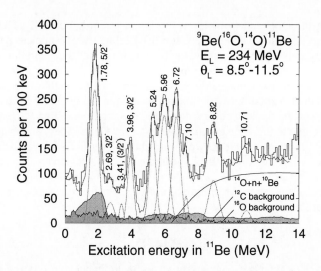

FIGURE 3. The ^{11}Be excitation energy spectrum from the two-neutron transfer reaction ^{9}Be(^{16}O,^{14}O)^{11}Be, fitted with Gaussian peaks (centroids indicated) and a smooth background related to three bodied processes (^{14}O + n + ^{10}Be). Background contributions from ^{12}C and ^{16}O target contaminations are labelled.

structure which is much more pronounced [15, 16]. The 1^- and 2^- states have been interpreted as being members of a $K^\pi = 1^-$ rotational band [14] with the two valence neutrons in a σ-π configuration.

These molecular configurations may also have a connection with halo-like structures, since both imply a core plus valence neutrons. It has been suggested that if the core of a halo nucleus such as ^{11}Li undergoes β-decay then it is possible for the halo wavefunction to retain its features after the β-decay. Sarazin *et al.* [17] suggested that the 8.82 MeV ^{11}Be state, populated in the β-decay of ^{11}Li, is a possible candidate for a two-neutron halo structure which can subsequently survive in a halo-like configuration after the neutron decay to ^{10}Be. In this instance the ^{9}Li core undergoes β-decay to ^{9}Be and the two valence neutrons remain in the $2s_{1/2}$ orbital. The rotational band of the ground state of ^{9}Be indicates a large deformation and is consistent with a structure based on a 2α+n cluster structure [14]. Thus, the two valence "halo" neutrons would orbit a very different core to ^{9}Li. This would strongly overlap with a $(1p_{3/2})(2s_{1/2})^2$ shell model configuration in ^{11}Be, or in the language of molecular orbitals two α-cores with two neutrons in σ-type molecular orbitals and one neutron in a π-orbital. In other words, the two-neutron halo state in ^{11}Be may in fact correspond to a molecular configuration with two delocalised covalent neutrons exchanged between the α-particles. The decay by emission of a valence neutron would be expected to populate similar molecular/halo states in ^{10}Be. The emission of a $2s_{1/2}$ (σ) neutron from ^{11}Be would be expected to populate the negative parity states states close to 6 MeV in ^{10}Be.

In order to determine the structure of the 8.82 MeV ^{11}Be excited state, we have measured its decay probabilities. i.e. the branching ratios for the neutron decay to states of ^{10}Be. The two-neutron transfer reaction ^{9}Be(^{16}O,^{14}O)^{11}Be*, ^{11}Be$^* \rightarrow$ ^{10}Be + n was measured using the Q3D spectrometer at the Ionen-Strahl-Labor (ISL) facility, at the Hahn-Meitner-Institut (now Helmholtz-Zentrum), Berlin. The ^{14}O nucleus has no bound excited states and thus its position along the focal plane yields the excitation of the ^{11}Be nucleus, as shown in Fig. 3. The decay products of the excited ^{11}Be recoil were detected, in coincidence with the oxygen ejectile, using an array of four double sided silicon strip detectors. More details about the experimental technique can be found in Ref. [18, 19].

The 8.82 MeV state is clearly observed in Fig. 3. As reported in Ref. [19], the state has a weak decay branch to the ^{10}Be ground state and first excited state, and instead decays to the group of ^{10}Be states close to 6 MeV (and above). The beta-delayed neutron decay measurements of Hirayama *et al.* [20] indicate that it is the negative parity states which are populated; these are the states with a $\sigma \otimes \pi$ molecular configuration. Such decay properties are consistent with the emission of a valent σ neutron leaving behind a ^{9}Be core with a single σ neutron.

SUMMARY

The search for α-particle gas/condensate states in light nuclear systems is fascinating, but experimentally challenging. The important question as to the nature of the Hoyle-state in ^{12}C via the determination of the energy of its 2^+ excitation remains to be resolved. We present some evidence for a possible 2^+ state. Measurements of the neutron decay widths of excited states in ^{11}Be have been performed using the ^{9}Be(^{16}O,^{14}O)^{11}Be reaction. These indicate that the 8.82 MeV excited state decays to states in ^{10}Be with a molecular character, which points to its molecular nature, suggested to also possess halo-like characteristics.

REFERENCES

1. R. B. Wiringa, et al., *Phys. Rev.* C **62**, 014001 (2000).
2. R. Bijker and F. Iachello, *Phys. Rev.* C **61**, 067305 (2000).
3. M. Freer, I. Boztosun, et al., *Phys. Rev.* C **76**, 034320 (2007).
4. I. Sick and J.S. McCarthy, *Nucl. Phys.* A **150**, 631 (1970); A. Nakada, Y. Torizuka and Y. Horikawa, *Phys. Rev. Lett.* **27**, 745 (1971); and 1102 (Erratum); P. Strehl and Th. H. Schucan *Phys. Lett.* **27B**, 641 (1967).
5. Y. Funaki, A. Tohsaki, H. Horiuchi, P. Schuck and G. Ropke, *Eur. Phys. J.* A **28**, 259 (2006).
6. A. Tohsaki, H. Horiuchi, P. Schuck, and G. Röpke, *Phys. Rev. Lett.* **87**, 192501 (2001).
7. M. Chernykh, et al., *Phys. Rev. Lett.* **98**, 032501 (2007).
8. F. Azjenberg-Selove, *Nucl. Phys.* A **506**, 1 (1990).
9. H.O.U. Fynbo, et al., *Nature* **433**, 136 (2005).
10. M. Itoh, et al., *Nucl. Phys.* A **738**, 268 (2004).
11. G.R. Satchler, *Nucl. Phys.* A **100**, 497 (1967).
12. S. Chiba, et al., *Nucl. Sci. Tech.* **37** 498 (2000).
13. B. Buck, H. Friedrich, and C. Wheatley, *Nucl. Phys.* A **275**, 246 (1977).
14. W. von Oertzen, *Z. Phys.* A **354**, 37 (1996); *Z. Phys.* A **357**, 355 (1997).
15. Y. Kanada-En'yo, H. Horiuchi and A. Dóte, *J. Phys.* B **24**, 1499 (1998).
16. Y. Kanada-En'yo, H. Horiuchi and A. Doté, Phys. Rev. C **60**, 064304 (1999).
17. F. Sarazin, et al., *Phys. Rev.* C **70**, 031302 (2004).
18. P. J. Haigh, et al., *Phys. Rev.* C **78**, 014319 (2008).
19. P. J. Haigh, M. Freer, et al., *Phys. Rev.* C **79**, 014302 (2009).
20. Y. Hirayama, T. Shimoda, et al., *Phys. Lett.* B **611**, 239 (2005).

The ^{14}C-Cluster and Molecular bands in the Oxygen Isotopes ^{18,20}O

W. von Oertzen[*][†], T. Dorsch[*] and H. G. Bohlen[*]

[*]Helmholtz-Zentrum Berlin, Glienicker Strasse 100, D-14109 Berlin, Germany
[†]Fachbereich Physik, Freie Universität Berlin

Abstract. We have studied states in ^{18}O and ^{20}O with the (^{7}Li,p) reaction on ^{12}C and ^{14}C targets at E_{lab}(^{7}Li) = 44 MeV, using the high resolution Q3D magnetic spectrometer at the Maier-Leibnitz-Laboratory in Munich[1]. The systematics of the excitation energies and cross sections were used to construct rotational bands with high moments of inertia. The bands observed are discussed in terms of underlying (^{14}C$\otimes^4$He)-cluster structure for ^{18}O, and for ^{20}O the cluster structures are (^{14}C$\otimes^6$He) and (^{14}C$\otimes 2n\otimes\alpha$). The intrinsically reflection asymmetric shapes give rise to molecular bands, which appear as parity inversion doublets.

Keywords: Nuclear clusters, molecular bands
PACS: 21.10.Hw, 21.10.Pc, 25.70.Hi, 27.30.+t

^{14}C-CLUSTERS AND NUCLEAR MOLECULES

In light nuclei clustering is observed as a general phenomenon at higher excitation energies, in particular close to the α-decay thresholds [1, 2, 3]. This observation has been summarized in the "Ikeda"-diagram for N=Z nuclei [1] in 1968 and for nuclei with extra neutrons in an extended diagram by von Oertzen [4]. With the additional neutrons specific molecular structures appear, with binding effects based on covalent molecular neutron orbitals. In these diagrams α-clusters and ^{16}O-clusters are the main ingredients. Actually, the ^{14}C-nucleus has equivalent properties as a cluster, as compared to ^{16}O. These are: (i) closed neutron p-shells (a better closure then in ^{16}O), (ii) the first excited states are above 6 MeV excitation energy, (iii) the binding energies of nucleons are very high ($E_B(p)$=20.83 MeV, $E_B(n)$= 8.17 MeV), (iv) the binding energy for α-particles is also high ($E_B(\alpha)$= 12.02 MeV). Therefore we must observe pronounced clustering and molecular configurations in the isotopes of oxygen [5], ^{18,19,20}O. At present the ($\alpha\otimes^{14}$C)-cluster structure for a number of low-energy states of ^{18}O is reasonably well established, the literature [6] shows a large variety of reaction studies (see ref. [7] for more recent work), as well as calculations, which established the level scheme up to 12 MeV and the main shell model structures. For ^{20}O results are less numerous, we expect structures defined by the strong binding of ^{14}C as (^{14}C$\otimes^6$He)-bands and the (^{14}C$\otimes 2n\otimes\alpha$)-structure, a covalently bound molecular band based again on the ^{14}C-cluster and valence neutrons. These cases are illustrated in Fig. 1, where

---

[1] In collaboration with: R. Hertenberger, Tz. Kokalova, R. Krücken, T. Kröll, M. Mahgoub, M. Milin, C. Wheldon, H.-F. Wirth

CP1165, *Nuclear Structure and Dynamics '09*
edited by M. Milin, T. Nikšić, D. Vretenar, and S. Szilner

also the corresponding shell model structures as multi-particle multi-hole excitations are given. These shapes actually correspond to octupole deformations. In the present work the multi-nucleon transfer reaction (^{7}Li,p) on ^{12}C is used with the aim to establish the higher-lying cluster states and find the parity inversion doublets and the high spin members of the rotational bands. Due to the angular momentum mismatch between

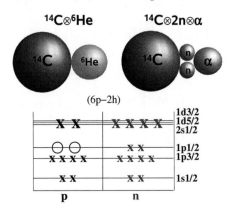

FIGURE 1. Schematic illustration of two possible cluster and shell model structures in ^{20}O.

the incoming and outgoing channels (particles ^{7}Li and p, the latter can carry only small amounts of angular momentum) the reaction is expected to populate strongly the states of higher spin of rotational bands. These are in addition expected to be strongly populated, if they belong to cluster structures.

EXPERIMENTAL RESULTS

Spectra were taken at three angles $\theta_L = 10°, 20°$ and $39°$. The solid angle was set to 13.85 msr, and further ten magnetic field settings (with regions of overlap) were needed per angle in order to cover an excitation energy range of 21 MeV. Further details of the detectors in the focal plane and the experimental set-up are given in [5]. A complete spectrum for ^{18}O is shown in Fig. 2, strong and rather narrow lines are observed well above the particle thresholds. The larger number of narrow states at high excitation energy points to their particular structure, and potentially high spins, their line width has been fitted using Breit-Wigner shapes. The energy calibration was obtained using the positions of the known levels below 12 MeV, the overall agreement was found to be better within this range than 5 keV. Due to the careful calibrations and the good resolution we were able to determine the position of a new state at 7.796 MeV(0^+). This is very important, because it can be identified as the "missing" band head of the molecular band in ^{18}O, shown in Fig. 3 (left side).

In the construction of bands we used: i) the excitation energy $E_x(J)$ systematics as function of $J(J+1)$, ii) the cross section dependence on $(2J+1)$. Further support is found from the results of calculations, i) the shell model using the code Oxbash [8], ii) the generator-coordinate-method of Descouvemont et al. [10, 11] and, iii) the Antisymmetrized Molecular Dynamics (AMD)-calculations of Furutachi, Kanada-Enyo et

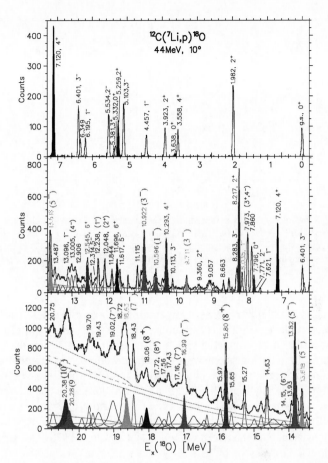

FIGURE 2. (Color online) Energy spectrum of protons obtained with the Q3D-spectrometer for the reaction ^{7}Li+^{12}C→ p+^{18}O. At higher E_x energies the three- and four-body backgrounds are included.

al. [7]. Slope parameters $\hbar^2/2\Theta$ obtained in a linear fit to the excitation energies data indicate the moments of inertia, Θ, of the rotational bands, as shown in Fig. 3. The instrinsic structure of the cluster bands is reflection asymmetric, the parity projection gives an energy splitting between the partner bands, which is an important quantity for the discussion of bands. Furutachi *et al.* [7] for example predict for ^{20}O a splitting energy for the $K=0_2^\pm$ doublet of about 6 MeV, the $K=0_2^-$ band head is then expected in the region of 10 MeV.

For ^{18}O we are able to identify two parity inversion doublets ($K=0_2^\pm$, the negative parity bands have to start with 1^-), for which we have extended tentatively the systematics up to 20 MeV of excitation energy and spins of 9^- and 10^+. The first band with the (^{14}C⊗α)-configuration starts at 3.64 MeV, i.e. 2.6 MeV below the (^{14}C⊗α)-threshold. The energy splitting with 4.2 MeV is slightly smaller than for $K=0^\pm$-bands in ^{20}Ne.

21

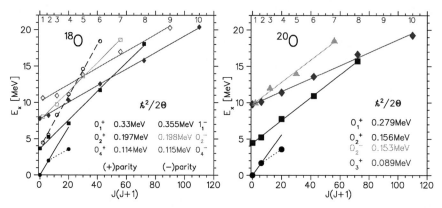

FIGURE 3. (Color online) Energy systematics for the members of the rotational bands forming parity inversion doublets with K=0, for ^{18}O (left) and ^{20}O (right) as a function of $J(J+1)$. The slope parameters $\hbar^2/2\Theta$ are indicated which contain the moments of inertia Θ. Symbols for ^{18}O (left): squares-cluster bands, diamonds-molecular bands; ^{20}O (right): squares/triangles-cluster bands, diamonds-molecular band.

The second parity doublet has a larger moment of inertia (see Fig. 3, left side) and is expected to represent the (^{12}C$\otimes 2n \otimes \alpha$)-structure. The smaller energy splitting of 2.5 MeV indicates that in the more elongated structure the exchange between the two reflected shapes is more difficult, consistent with the larger value of the moment of inertia Θ. This interpretation differs from that in the work of ref. [9], where large deformation was observed, but no inversion doublets (octupole deformation) were considered.

For ^{20}O we can compare the bands (Fig. 3) with those of ^{18}O in the same figure. The first doublet ($K = 0_2^{\pm}$) has a moment of inertia which is slightly larger (smaller slope parameter), consistent with the interpretation as (^{14}C$\otimes ^6$He or ^{16}C$\otimes ^4$He)-structures (they start well below the thresholds of 16.8 MeV and 12.32 MeV, respectively). The second band, for which the negative parity partner is yet to be determined, has a slope parameter slightly smaller as compared to ^{18}O, (the moment of inertia is 25 % larger). This is consistent with the interpretation as a (^{14}C$\otimes 2n \otimes \alpha$)-structure, the energy splitting is expected to be around 2.0 MeV.

REFERENCES

1. H. Horiuchi and K. Ikeda, *Prog. Theor. Phys.* (Japan) **40**, 277 (1968).
2. W. von Oertzen, M. Freer and Y. Kanada-En'yo, *Phys. Rep.* **432**, 43 (2006).
3. W. von Oertzen, *Z. Physik* A **354**, 37 (1996); A **357**, 355 (1997).
4. W. von Oertzen, *Eur. Phys. J.* A **11**, 403 (2001).
5. T. Dorsch, PhD Thesis , TU München (2008).
6. D.R. Tilley, H.R. Weller, C.M. Cheves and R.M. Chasteler, *Nucl. Phys.* A **595**, 1 (1995).
7. N. Furutachi, et al., *Prog. Theor. Phys.* (Japan) **119**, 403 (2008).
8. W. Rae, A. Etchegoyen, B. A. Brown, OXBASH code, Tech. Rep. 524, MSU Cycl. Lab. (1985).
9. M. Gai, et al., *Phys. Rev.* C **43**, 2127 (1992).
10. P. Descouvemont and D. Baye *Phys. Rev.* C **31**, 2274 (1985).
11. P. Descouvemont, *Phys. Lett.* **437B**, 7 (1998).

New Cluster States in ^{12}Be

M. Dufour*, P. Descouvemont† and F. Nowacki*

*IPHC Bat27, IN2P3-CNRS/Université de Strasbourg, BP28, F-67037 Strasbourg Cedex 2 - France
†Physique Nucléaire Théorique et Physique Mathématique, C.P. 229, Université Libre de Bruxelles (ULB), B 1050 Brussels, Belgium

Abstract. This work is devoted to new investigations of the ^{12}Be spectrum with No-core Shell Model (NCSM) calculations and with the Generator Coordinate Method (GCM). The NCSM calculations do not reproduce the breaking of the $N = 8$ shell closure. The NCSM and the GCM calculations do not support the existence of a band based on the ground state. The GCM calculations lead to a good description of the molecular band of Freer et al. and Saito et al.. The 0_2^+ isomeric state and the 2_2^+ state are also well described by the GCM. New bands of positive and negative parities are proposed.

Keywords: ^{12}Be, NCSM, GCM, cluster states
PACS: 21.60.Gx, 21.10.Jx, 24.10.Cn, 26.20.Fj

INTRODUCTION

The description of light exotic nuclei such as ^{12}Be is a very interesting challenge for experiments as well as for theoretical calculations. In Fig. 1, we represent experimental ^{12}Be states whose spin and parity are clearly assigned. Below the ^{11}Be+n threshold, four states are known: the 0_1^+ ground state, the 2_1^+ state at 2.10 MeV, the 0_2^+ state at 2.24 MeV, and the 1_1^- at 2.68 MeV. The 0_2^+ is believed to be an isomeric state [1]. Above this threshold, let us mention the 2_2^+ at 4.56 MeV, and a tentatively assigned $(3^-, 4^+)$ state at 5.70 MeV. Above the ^{8}He+α threshold and the ^{6}He+^{6}He threshold, 4^+, 6^+ and 8^+ states have been identified in the breakup of ^{12}Be into ^{6}He+^{6}He and ^{8}He+α channels by Freer et al. [2]. They are believed to be members of a molecular band. 0^+ and 2^+ members have been identified later at 10.9 MeV and 11.3 MeV, respectively, in the ^{6}He+^{6}He channel by Saito et al. [3].

In this paper, we briefly report new theoretical investigations of the ^{12}Be nucleus performed with two sophisticated microscopic models [4]. We begin with a study of the low-lying 0^+, 2^+ and 4^+ states with the NCSM approach (see Ref. [5] and References therein) and the GCM (see Ref. [6] and References therein). Our purpose is to clarify the structure of the ground state and to discuss the possible existence of a rotational ground-state band as it is proposed in the literature [7, 8]. Such an hypothesis could be justified only if a breaking of the $N = 8$ shell closure is confirmed [9]. We then pursue within the GCM framework which appears to be better adapted to describe the ^{12}Be spectrum.

The NCSM calculations are performed with the code ANTOINE developed in Strasbourg by Caurier et al. [10]. For the GCM, we have developed a two-cluster model which allows the inclusion of the ^{8}He(0^+, 2^+)+α, and ^{6}He(0^+, 2^+)+^{6}He(0^+, 2^+) channels simultaneously. It generalizes a previous GCM study of Descouvemont and Baye

CP1165, *Nuclear Structure and Dynamics '09*
edited by M. Milin, T. Nikšić, D. Vretenar, and S. Szilner
© 2009 American Institute of Physics 978-0-7354-0702-2/09/\$25.00

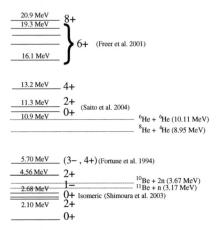

FIGURE 1. ^{12}Be experimental spectrum. Only states clearly assigned in spin and parity are shown. Thresholds are indicated with dotted lines. Experimental data are from Refs. [2, 3, 1]. The 6^+ states represented here correspond to states observed in the ^{6}He+^{6}He channel of Ref. [2].

[11] where only channels involving the ^{6}He and ^{8}He ground states are included. The addition of new channels involving the 2^+ excited states of ^{6}He and ^{8}He is expected to significantly improve the description of the ^{12}Be nucleus.

THEORETICAL FRAMEWORKS

In a NCSM calculation, the infinite dimensional many-body problem is replaced by a diagonalization in a finite model space whose dimension is expected sufficiently large to give converged results. A detailed description of the NCSM approach as it is implemented in this work can be found in Refs. [10] and References therein.

The size of the harmonic oscillator (HO) basis is controled by the parameter N_{max}. Our calculations are performed up to $N_{max} = 8$ with two different high-precision realistic NN interactions: the 'CD-Bonn 2000' (CDB2k) and the 'Inside Nonlocal Outside Yukawa' (INOY) potentials [12].

For the GCM, the Schrödinger equation of the 12-nucleon system is approximately solved with the Generator Coordinate Method (GCM) and the Microscopic R-Matrix-Method [6]. The hamiltonian is an effective interaction well adapted to cluster models [6]. It contains 3 free parameters. Two parameter sets have been selected. Both reproduce exactly the energy of the $(^6$He+^{6}He)-$(^8$He+$\alpha)$ threshold. In addition, one set reproduces exactly the binding energy of the 0_1^+ ground state and the other one, the energy of the 0^+ molecular band-head [3]. The first set is used in the calculation of the low-lying 0^+, 2^+, and 4^+ states. We keep the same set to study the negative parity band according to Ref. [11]. The second set is used for the other positive parity states.

In the present GCM study, the ^{12}Be wave functions are expressed as sums of antisymmetrized basis functions involving two identical clusters of nucleons. In this framework, a multi-channel wave function can be written

$$\Psi^{JM\pi}_{^{12}\text{Be}} = \sum_i \Psi^{JM\pi}_{^8\text{He}_{(i)}+\alpha} + \sum_{j,k} \Psi^{JM\pi}_{^6\text{He}_{(j)}+^6\text{He}_{(k)}},\tag{1}$$

where the sums run over all internal states equal to the 0^+ ground states and 2^+ excited states of ^{6}He and ^{8}He.

RESULTS

The 0_1^+, 2_1^+, and 4_1^+ states

NCSM calculations show that the 0_1^+ and 2_1^+ energies are converged [4]. The analysis of the corresponding wave functions show that they are similar and dominated by (s,p) configurations. In this condition, a breakdown of the $N = 8$ shell closure is not confirmed by the present NCSM calculations. On the other hand, we cannot precisely conclude that the 4_1^+ is converged. Nevertheless, we can notice that the corresponding wave function is dominated by (s,d) configurations [4]. Keeping in mind that a 4^+ state cannot be generated by s, p shells, we can conclude that the present NCSM calculations do not confirm an hypothetical band based on the ground state. Let us also notice that we get very similar results for both interactions.

However, our results are in disagreement with previous studies which predict a breaking of the $N = 8$ shell closure [9]. In this context, intend to test the accuracy of our results by including a $3N$ term in our hamiltonian [4]. Anyway, we can consider that the present NCSM calculation have established a baseline of results at the pure two-body realistic interaction level.

The GCM calculations confirm the non exixtence of a ground-state rotational band. Indeed, we do not find a GCM 4^+ state which could be interpreted as a possible member of a ground-state rotational band [7, 8, 4].

GCM cluster states of ^{12}Be

The present GCM calculations support the existence of a molecular band as it is proposed in the experiments of Freer *et al.* [2] and Saito *et al.* [3] (see Fig. 2). They confirm with better conditions of calculation the previous GCM investigation of Descouvemont and Baye [11, 4].

The 0_2^+ state and the 2_2^+ state are also well reproduced (see Fig. 2). According to Refs. [7, 8, 13], we confirm that these states belong to a same band. We also propose other 0_2^+ band members and a new $K = 2^+$ band with non-natural-parity states (see Fig. 2). All these results are new as compared to the previous GCM work [11].

The present GCM calculations confirm the negative parity band $K = 0_1^-$ found in Ref. [11] near the ^{8}He+α threshold (see Fig. 3). In addition, another negative-parity band $K = 1^-$ which could correspond to the tentatively assigned band seen in three-neutron stripping reaction on the ^{9}Be [13] is also obtained (see Fig. 3).

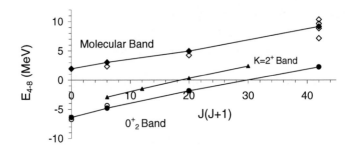

FIGURE 2. Positive-parity ^{12}Be states predicted by the GCM. Energies are expressed with respect to the ^{8}He+α threshold. Diamonds represent the molecular band. The full ones correspond to GCM calculations, the open ones to experimental data [2, 3]. Triangles are for the $K = 2^{+}$ band. Circles represent the 0_{2}^{+} band. The full ones correspond to the GCM calculations, the open ones to expated data [1].

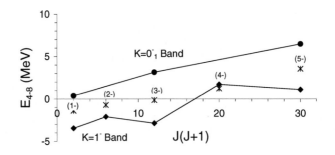

FIGURE 3. Negative-parity ^{12}Be states predicted by the GCM. Full circles and full diamonds correspond to GCM calculations. Crosses are experimental data from Ref. [13].

REFERENCES

1. S. Shimoura, et al., *Phys. Lett.* **654B**, 87 (2007).
2. M. Freer, et al., *Phys. Rev.* C **63**, 034301 (2001).
3. A. Saito, et al., *Nucl. Phys.* **A 738**, 337 (2004).
4. M. Dufour, P. Descouvemont and F. Nowacki, submitted to *Phys. Rev.* C.
5. P. Navrátil, et al., *Phys. Rev. Lett.* **99**, 042501 (2007).
6. M. Dufour and P. Descouvemont, *Phys. Rev.* C **78**, 015808 (2008).
7. Y. Kanada-En'yo and H Horiushi, *Phys. Rev.* C **68**, 014319 (2003).
8. M. Ito, N. Itagaki, H. Sakurai, and K. Ikeda, *Phys. Rev. Lett.* **100**, 182502 (2008).
9. S. D. Pain, et al., *Phys. Rev. Lett.* **96**, 032502 (2006).
10. E. Caurier, et al., *Phys. Rev.* C **66**, 024314 (2002).
11. P. Descouvemont and D. Baye, *Phys. Lett.* **505B**, 71 (2001).
12. C. Forssén, et al., *Phys. Rev.* C **71**, 044312 (2005).
13. H. G. Bohlen, et al., *Int. Jour. of Mod. Phys.* E, **2067** (2008).

Nuclear Structure of ^{12}C from Break-up Studies in Complete Kinematics

M. Alcorta*, M.J.G. Borge*, M. Cubero*, R. Dominguez-Reyes*, L. Fraile*, B. Fulton[†], H.O.U. Fynbo[†], D. Galaviz*, S.Hyldegaard[†], H. Jeppesen[†], B. Jonson**, O. Kirsebom[†], M. Madurga*, A. Maira*, A. Muñoz[‡], T. Nilsson**, G. Nyman**, D. Obradors*, A. Perea*, K. Riisager[†], O. Tengblad* and M. Turrión*

*Instituto de Estructura de la Materia, CSIC, Serrano 113 bis, Madrid E-28006, Spain
[†]Department of Physics and Astronomy, University of Aarhus, DK-8000 Aarhus C, Denmark
**Fundamental Physics, Chalmers Univ. of Technology, S-41296 Göteborg, Sweden
[‡]CMAM, Universidad Autónoma de Madrid, Cantoblanco, Spain

Abstract. A complete kinematics study of the ^{10}B(^{3}He,p$\alpha\alpha$) and ^{11}B(^{3}He,d$\alpha\alpha$) reactions has been performed to study the multi-particle break-up of ^{12}C resonances above the triple-alpha threshold. Four-particle coincidence detection gives us complete information on the direction and energy of the individual alpha particles from the decay of ^{12}C, allowing us to extract new information on the structure of ^{12}C which we shall present in this contribution. We have observed gamma de-excitation of the T=1 15.11 MeV resonance using charged particle detectors, and have constructed Dalitz plots of the individual resonances in ^{12}C using the complete kinematics information of the alpha particles which come from their break-up.

Keywords: Complete kinematics, Dalitz plot, Deduced gamma transition, ^{12}C
PACS: 21.10.-k,21.10.Hw,25.55.-e,27.20.+n

INTRODUCTION

The nuclear structure of ^{12}C has been the subject of many recent experiments. This is especially true for the region above the triple-alpha threshold, where the predicted 2^+ state around 9 MeV of the rotational band of the Hoyle state has yet to be confirmed [1]. There are many approaches to probe the region of ^{12}C just above the triple-alpha threshold, each with its advantages and disadvantages. We chose to study the excited states in ^{12}C using reaction experiments with high Q-values, with a setup designed to detect the outgoing particles in complete kinematics. By using two different reactions, we were able to get independent measurements of each resonance. The two reactions used were the ^{10}B(^{3}He,p$\alpha\alpha$) reaction with a beam energy of 4.9 MeV, and the ^{11}B(^{3}He,d$\alpha\alpha$) reaction with a beam energy of 8.5 MeV. An advantage of this method is that a reaction experiment will indiscriminately populate the resonances in ^{12}C, and we are therefore not limited from the selection rules imposed by e. g. in β-decay.

With the use of highly segmented detectors, we determine the position and energy of each outgoing product, thus determining both the energy and momentum of the four outgoing particles in the reactions. This does not only greatly reduce the contribution from other channels, as discussed in [2], but it enables us to indirectly determine γ

CP1165, *Nuclear Structure and Dynamics '09*
edited by M. Milin, T. Nikšić, D. Vretenar, and S. Szilner
© 2009 American Institute of Physics 978-0-7354-0702-2/09/$25.00

branches, as discussed below. In addition, detecting the alpha particles from the breakup of ^{12}C helps us determine the spin and parity assignments in unique ways.

EXPERIMENT

The study of the ^{10}B(^{3}He,p$\alpha\alpha$) and the ^{11}B(^{3}He,d$\alpha\alpha$) reactions were carried out at the Centro de Microanálisis de Materiales (CMAM), located in the campus of the Universidad Autónoma de Madrid. This center houses a 5 MV tandetron that uses the Cockroft-Walton power system to obtain the terminal voltage [3]. This device, due to the absence of mechanical moving parts, provides very stable beams, making the CMAM accelerator an ideal place for these types of reaction studies. The ^{3}He beam intensity in each case was of the order of 7.5×10^9 ions/s with a beam diameter of 3 mm. A Faraday cup was used throughout the experiment to monitor the beam intensity. The energy spread of the beam is negligible (tens of eV's) compared to that of our detectors (30 keV). An advantage of using stable beams is that statistics is not a problem, and only 10-20 hours of beam-time was used for each reaction.

The experimental setup covered 38% of 4π and consisted of four Double Sided Si Strip Detectors (DSSSD) [4], each 60 μm thick. Each DSSSD was backed by a non-segmented Si-PAD 1500 μm thick. Three of the DSSSDs have 16×16 perpendicular strips, while one of the DSSSDs has 32×32 perpendicular strips. The 16×16 DSSSDs have an active area of 50×50 mm^2, giving a pixel size of 3×3 mm^2, while the 32×32 DSSSD has an active area of 67×67 mm^2, giving a pixel size of 2×2 mm^2. The telescopes were arranged to maximize multi-particle detection, placing two DSSSDs as close to 0° as possible. The target thickness was minimized to reduce energy losses in the target foil. The targets used were 18.9 μg/cm^2 enriched ^{10}B (90%) and 22.0 μg/cm^2 ^{11}B, each with a 4 μg/cm^2 C backing.

RESULTS

By comparing the excitation spectra given by the invariant mass of the decay fragments versus the excitation energy given by the primary ejectile one can search for missing energy. This technique can be used to find γ de-excitations of resonances to states above the multi-particle breakup threshold by imposing only momentum conservation (the missing momentum from the low-energy γ is below the experimental uncertainty) and no condition on conservation of energy [2]. This technique is especially useful to detect γ branches to broad states which are difficult to identify using standard γ spectroscopy. Figure 1 shows an example of this method applied to the ^{10}B(^{3}He,p$\alpha\alpha$) reaction where the presence of decay via γ-emission to lower lying states is seen. The figure shows clear evidence of γ de-excitation of the 15.11 MeV 1^+ resonance (abscissa) to other resonances in ^{12}C above the triple-α threshold (ordinate). The highlighted vertical region corresponds to a fixed proton energy feeding the 15.11 MeV state in ^{12}C, which is known to decay by gamma-emission to states in ^{12}C above the triple-alpha threshold [5]. The fact that this line is below the diagonal indicates an energy mismatch between the initial ^{12}C state calculated from the proton energy and the final ^{12}C state calculated using the invariant mass of the three alpha particles. The missing energy is thus the energy of the gamma emitted from the 15 MeV state to lower-lying states. Gamma branches of the 15.11 MeV state to the 7.65 MeV, the broad 10 MeV, the 11.83 MeV, and the 12.71 MeV resonances can be seen in figure 1. Relative branching ratios of these states and further information can be found in [6].

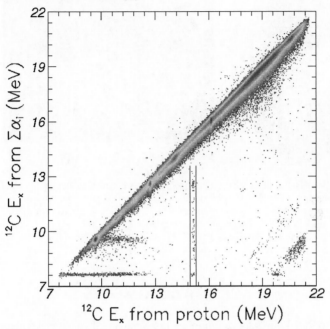

FIGURE 1. ^{12}C excitation energy calculated from the sum of the alpha energies versus the ^{12}C excitation energy calculated from the proton energy for the $^{10}B(^{3}He,p\alpha\alpha\alpha)$ reaction. The narrow region of the 15.11 MeV state is highlighted with vertical lines, see the text for details. The diagonal seen in the lower right are reaction events from the ^{11}B present in the target, $^{11}B(^{3}He,p\alpha\alpha\alpha)n$. The horizontal lines extending from the diagonal on the lower left are a result of punch-through protons.

As mentioned earlier, because we have the information of the alpha particles originating from the breakup of resonances in ^{12}C, we can determine properties of these resonances through the use of Dalitz plots [7]. The intensity distribution of the Dalitz plot gives insight into the decay mechanism and can be used in certain cases to determine the spin and parity of the initial three-body resonance [8]. In Figure 2 the 3α-decay of the 13.35 MeV state in ^{12}C is used as an example, though Dalitz plots have been constructed for each resonance in ^{12}C. In addition, specific structures due to interactions between the α particles give information on the decay mechanism. The experimentally observed distribution strongly supports a 4$^-$ assignment, dismissing a 2$^-$ assignment which was considered a possibility [9].

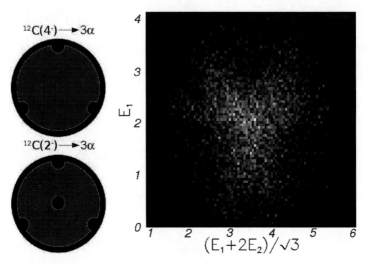

FIGURE 2. (left) Dalitz plots which show the regions of vanishing density in black if the initial resonance is assumed to have $J^\pi = 4^-$ (top) and $J^\pi = 2^-$ (bottom). These regions are due to spin and parity conservation and Bose statistics, and are independent of the reaction mechanism. (right) Experimantal Dalitz plot of the 13.35 MeV resonance in ^{12}C which decays through the broad 2^+ state at 3.03 MeV in ^8Be. The shape seen here is indicative of $J^\pi = 4^-$ break-up through a broad resonance, and rules out the possibility of a 2^- assignment.

ACKNOWLEDGMENTS

We would like to acknowledge support by the Spanish CICYT research grant FP2007-62170 and MICINN Consolider Project CSD2007-00042, as well as support by the European Union Sixth Framework through RII3- EURONS/ JRA4-DLEP (contract no. 506065). M. Alcorta acknowledges the support of the CSIC I3P program cofinanced by the European Social Fund.

REFERENCES

1. H. Morinaga, *Phys. Lett.* **21**, 78 (1966).
2. M. Alcorta, et al., *Nucl. Instr. Meth. A* **605**, 318 (2009).
3. O. Enguita, et al, *Symp. Northeastern Acc. Pers.* (2005).
4. O. Tengblad, et al, *Nucl. Inst. Meth. A* **525**, 458 (2004).
5. D. Alburger, and D. Wilkinson, *Phys. Rev. C* **5**, 384 (1972).
6. O. Kirsebom, et al, *Phys. Lett. B* (2009).
7. R. Dalitz, *Philisophical Magazine* **44**, 1068 (1953).
8. C. Zemach, et al, *Phys. Rev.* **133**, B1201 (1964).
9. F. Ajzenberg-Selove, *Nucl. Phys.* **A 506**, 1 (1990).

Analysis of $T = 1$ ^{10}B States Analogue to ^{10}Be Cluster States

M. Uroić*, Đ. Miljanić*, S. Blagus*, M. Bogovac*, L. Prepolec*,
N. Skukan*, N. Soić*, M. Majer†, M. Milin†, M. Lattuada**,
A. Musumarra** and L. Acosta‡

*Ruđer Bošković Institute, Bijenička 54, Zagreb, Croatia
†Department of Physics, University of Zagreb, Zagreb, Croatia
**INFN - Laboratori Nazionali del Sud, Catania, Italy
‡Departamento de Fisica Aplicada, Universidad de Huelva, Huelva, Spain

Abstract. Current status of the search for T=1 cluster states in ^{10}Be, ^{10}B and ^{10}C is presented. The best known of the three, ^{10}Be, has an established rotational band (6.18, 7.54 and 10.15 MeV) with unusually large moment of inertia. Search of their isobaric analogue in ^{10}B is presented, with emphasis on ^{3}He+^{11}B reaction.

Keywords: nuclear cluster, nuclear reaction, ^{3}He, ^{10}B, isobaric analogue
PACS: 25.55.-e, 27.20.+n

INTRODUCTION

The states at 6.18 MeV in ^{10}Be and 7.56 MeV in ^{10}B are confirmed to be "intruder" states in shell model, both having $(0^+,1)$ assignment. Their analogue in ^{10}C, expected above $\alpha+^6$Be treshold (6.10 MeV), has yet to be established [1]. In ^{10}Be, the state was found to be the head of a rotational band with unusually large moment of inertia [2, 3, 4]. Isospin symmetry predicts similar structures in ^{10}B (T=1) and ^{10}C, not yet observed.

^{11}B(^{3}He,α)^{10}B REACTION

A measurement was performed using 15 MeV ^{3}He beam from 6 MV EN Tandem Van de Graaff accelerator of the "Ruđer Bošković" Institute in Zagreb. Two detector telescopes, with 6.5 and 9.7 μm thin and 500 μm thick PSD detectors were used to reconstruct three-particle events, consisting of α-particle from reaction, and two ^{10}B decay products of which one was recorded and the other kinematically reconstructed. Details of experimental setup can be found in [5].

Figure 1 presents the excitation energy spectra for ^{10}B decaying into the most probable particle emission, i.e. by proton and α-particle emission. In the spectra on the left side, the ^{6}Li in the second excited state was reconstructed(up) and detected(down). As this state is analogue to ^{6}He ground state (thus having T=1), this guarantees the T=1 isospin assignment for the decaying ^{10}B nucleus. A peak emerges arround 11.3 MeV, with T=1 isospin and natural parity assignment. Aside from α-decay channals, it is visible in the proton-decay with the remaining ^{9}Be in the excited state at 2.43 MeV.

CP1165, *Nuclear Structure and Dynamics '09*
edited by M. Milin, T. Nikšić, D. Vretenar, and S. Szilner
© 2009 American Institute of Physics 978-0-7354-0702-2/09/$25.00

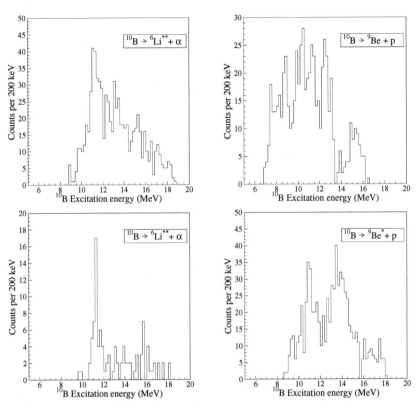

FIGURE 1. Excitation energy spectra of ^{10}B for α and p particle decay. The α-paticle from the ^{11}B(^{3}He,α)^{10}B reaction was detected at 30 ± 5^0, and the decay product at 67 ± 5^0 on the opposite side.

FUTURE PROSPECTS

With the cluster structure of the entrance chanel of the reaction in mind, the most promising reactions for investigating $A = 10$ $T = 1$ triplet cluster states should be obtained using ^{7}Li and ^{7}Be beam on ^{7}Li and ^{6}Li targets. Both measurements were recently performed, and the following reactions are expected to populate cluster states: ^{7}Li(^{7}Li,α)^{10}Be, ^{7}Be(^{7}Li,α)^{10}B, ^{7}Be(^{6}Li,t)^{10}C, ^{7}Be(^{6}Li,^{3}He)^{10}B, ^{7}Li(^{6}Li,t)^{10}B and ^{7}Li(^{6}Li,^{3}He)^{10}Be.

REFERENCES

1. H. T. Fortune and R. Sherr, *Phys. Rev. C*, **73**, 064302 (2006).
2. N. Curtis, et al, *J. Phys. G*, **36**, 015108 (2009).
3. N. Soić, et al., *Europhys. Lett.*, **34**, 7 (1996).
4. M. Freer, et al., *Phys. Rev. Lett.*, **96**, 042501 (2006).
5. M. Uroić, et al., *Int. Journ. Mod. Phys. E*, **17**, 2345 (2008).

Shape Isomers and Cluster Structure

J. Cseh[*], J. Darai[†] and A. Lépine-Szily[**]

*Institute of Nuclear Research, Hungarian Academy of Sciences, Debrecen, Pf. 51, Hungary-4001
†Institute of Experimental Physics, University of Debrecen, Debrecen, Pf. 105, Hungary-4010
**Instituto de Fisica, Universidade de Sao Paolo, CP66318, 05389-970, Sao Paolo, Brazil

Abstract. We discuss how shape-isomers can be determined from the quasi-dynamical $SU(3)$ symmetry based on the Nilsson-model. The possible clusterization of superdeformed and hyperdeformed states are also investigated. The allowed cluster-configurations give a hint, for the favourable reaction channels for populating these states. An illustrative example is presented, in which the theoretical prediction is justified by the experimental observation.

Keywords: super- and hyperdeformed states, cluster-configurations, molecular resonances, quasi-dynamical $SU(3)$ symmetry
PACS: 21.60.Cs, 21.60.Fw, 21.60.Gx, 25.55.Ci, 25.70.Ef, 27.30.+t

INTRODUCTION

The exotic shapes of atomic nuclei attract much attention recently both from the experimental and from the theoretical sides. Theoretically they are usually obtained from shell-model or mean-field calculations as the minima of the energy-surface. Here we discuss a possibility of determining these states from the Nilsson-model, via their quasi-dynamical (or effective) $SU(3)$ symmetry [1, 2]. The study of the possible clusterizations of these exotic states are important not only for the better understanding of their structure, but also also for finding the preferred reaction channels, in which they can be populated.

Some superdeformed bands in $N = Z$ nuclei were observed experimentally during the last decade. As for their hyperdeformed states are concerned, practically they belong to a domain for theoretical studies.

A remarkable example is that of the ^{36}Ar nucleus. Its superdeformed shape was detected in 2000 [3], and some theoretical predictions were done on the hyperdeformed state from alpha-cluster [4], and binary cluster studies [5]. A recent study on molecular resonances in heavy-ion reactions seems to justify these predictions [6] not only from the respect of the moment of inertia of the band, but also concerning its clusterization, i.e. the preferred reaction channels for its population. Furthermore, the above-mentioned shape-calculation from Nilsson-model + quasi-dynamical symmetry gives a hyperdeformed state completely in line with the cluster models and with the experimental finding [7].

SHAPE ISOMERS

The shape isomers can be obtained from symmetry-considerations based on a self-consistency argument, as follows. The $U(3)$ symmetry, which is an approximately good

CP1165, *Nuclear Structure and Dynamics '09*
edited by M. Milin, T. Nikšić, D. Vretenar, and S. Szilner
© 2009 American Institute of Physics 978-0-7354-0702-2/09/$25.00

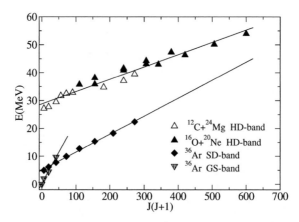

FIGURE 1. Excitation energy versus the angular momentum for the ground, superdeformed, and the recently observed hyperdeformed bands in the ^{36}Ar Nucleus.

symmetry of light nuclei [8], is known to be uniquely related to the quadrupole shape [9]. Furthermore, when the real $U(3)$ symmetry breaks down (due to some symmetry breaking interactions, like spin-orbit, pairing, etc), a generalized version of it, called quasi-dynamical or effective $U(3)$ symmetry still survives [1]. The quasi-dynamical $U(3)$ quantum numbers can be obtained from Nilsson-calculations [10, 11]. Thus, the self-consistency calculation consists in the continuous variation of the quadrupole deformation, as an input for a Nilsson-calculation, and determination of the effective $U(3)$ quantum numbers or, from them, the corresponding β_{out} quadrupole deformation. For light nuclei, like ^{24}Mg and ^{28}Si, where detailed comparison could be made, the results of this self-consistency calculation are in very good agreement with that of the energy-minima calculations [12, 13].

For the ^{36}Ar nucleus this kind of calculation gives the ground state, of course, as the first stable shape. When proceeding towards larger deformation, the next one corresponds to the superdeformed shape, representing 4 nucleon excitation, being very much in line with the more recent Nilsson, as well as with the shell-model calculations [14, 3].

With increasing β-values two further stable plateaus appear, a less-pronounced one around 0.8 (β_{out}), and a very big one at cca 1.1. The first one coincides with the prolate state of [14]. The second one is in complete agreement with the prediction of the alpha-cluster model for the hyperdeformed state [4], as well as with the experimentally observed largely prolate band, as shown by Figure 1.

CLUSTER STRUCTURE

When studying the possible cluster structure of a specific nuclear state, like the ones of the previous section, one has to take into account the effects of two basic natural laws: the energy-minimum principle and the Pauli-exclusion principle. Fully microscopic cluster models are able to handle both of them in a completely satisfactory manner, however, their range of applicability is limited due to the large-scale calculations. We are interested in applying a semimicroscopic approach, which can be extended even to heavy nuclei and exotic clusterization [15]. Symmetry-considerations can be helpful along this line, and exclude the Pauli-forbidden states without carrying out the antisymmetrization explicitly. For light nuclei even the real $SU(3)$ can lead us to useful results, but for the heavy ones we need to change to the more general quasi-dynamical symmetry. (This is applicable also for light nuclei, of course, as illustrated e.g. by the previous section. When the real $SU(3)$ is approximately valid, then it coincides with the quasi-dynamical one [11].)

There are two simple recipes, which are based on the microscopic picture, yet they are easy to apply systematically. These are the $U(3)$ selection rule [16], and Harvey's prescription [17]. Both of them applies the harmonic oscillator basis, thus there is a considerable similarities between them. However, they are not identical, rather, they are complementary to each other in a sense. Therefore, they should be applied in a combined way [13].

Since the $U(3)$ symmetry is related to the quadrupole deformation of the nucleus, the $U(3)$ selection rule can be interpreted as a consistency check of the quadrupole deformation, between the shell model (or collective model) states and the corresponding cluster states. It should also be mentioned that the $U(3)$ selection rule, which deals with the space-symmetry of the states, is always accompanied by a similar $U^{ST}(4)$ [18] selection rule for the spin-isospin degrees of freedom.

As for the energetic stability of the cluster-configuration is concerned, either binding-energy arguments, or double-folding calculation can be applied [19]. Both methods indicate that the alpha-like (i.e. $N = Z = even$) configurations are energetically preferred. Thus we concentrate here on alpha-like binary clusterizations.

The superdeformed state of the ^{36}Ar nucleus shows a little shape-uncertainty. This is a joined conclusion of several theoretical studies. In our method this uncertainty is reflected by the appearance of close-lying, but not completely identical $U(3)$ representations. Therefore, we have considered the shapes corresponding to the [32,14,10] effective quantum numbers, and to the [32,16,8], which correspond to a simple harmonic oscillator state. The latter one allows (^{4}He, ^{8}Be, ^{12}C, ^{16}O)+core clusterizations, while the former one allows only the last two configurations (though the other ones are not very strongly forbidden either).

The hyperdeformed state on the other hand seems to have a very well-defined symmetry, and it allows only ^{24}Mg+^{12}C and ^{20}Ne+^{16}O binary configurations. (In these considerations the clusters are supposed to be in their ground intrinsic states. E.g. a ^{28}Si+^{8}Be configuration is allowed, too with the prolate shape of the ^{28}Si, but it is thought to be not the dominant component of the ground-state wavefunction.)

It is also remarkable that the ^{24}Mg+^{12}C clusterization is possible both in the ground-sate, and in the superdeformed, as well as in the hyperdeformed state. The difference between these configurations is the relative orientation of the two deformed clusters with respect to the molecular axis. This observation could be made due to the facts that i) the Pauli-principle was taken into account, and ii) no oversimplifying model assumptions (e.g. spherically or cylindrically symmetric cluster shapes) were made.

CONCLUSIONS

With this contribution we wished to demonstrate that symmetry-considerations can be helpful in studying both the shape isomers and the possible clusterizations of atomic nuclei. They are able to shed some light also on the interrelations of these two phenomena. In this respect they can be helpful in predicting the favourable reactions for the population of superdeformed and hyperdeformed states. The hyperdeformed state of the ^{36}Ar nucleus serves as a very interesting example. Its existence was first foreseen from alpha-cluster model [4], then binary-cluster studies predicted their population in ^{24}Mg+^{12}C and ^{20}Ne+^{16}O reactions [5]. Recent reaction studies [6] seem to justify these predictions, and structure-calculations [7] show that the same state is obtained from the Nilsson-model via the quasi-dynamical $SU(3)$ symmetry.

ACKNOWLEDGMENTS

This work was supported by the OTKA (Grant No. K72357), as well as by Fundação de Apoio a Pesquisa do Estado de Sao Paulo (FAPESP).

REFERENCES

1. P. Rochford, D. J. Rowe, *Phys. Lett.* **B210**, 5 (1988);
   D. J. Rowe, P. Rochford, J. Repka, *J. Math. Phys.* **29**, 572 (1988).
2. J. Cseh, *Proc. IV Int. Symp. on Quantum Theory and Symmetries (Varna)* (Sofia: Heron Press) p. 918 (2006).
3. C. E. Svensson, A. O. Machiavelli, A. Juodagalvis, A. Poves, I. Ragnarsson, S. Aberg, D. E. Appelbe, R. A. E. Austin, C. Baktash, G. C. Ball et al., *Phys. Rev. Lett.* **85**, 2693 (2000).
4. W. D. M. Rae, A. C. Merchant, *Phys. Lett.* **B279**, 207 (1992).
5. J. Cseh, A. Algora, J. Darai, P. O. Hess, *Phys. Rev.* **C70**, 034311 (2004).
6. W. Sciani, Y. Otani, A. Lepine-Szily, E.A. Benjamin, L.C. Chamon, R. Lichtenthaler, J. Darai, J. Cseh, *Phys. Rev. C*, in press.
7. J. Cseh, J. Darai, W. Sciani, Y. Otani, A. Lepine-Szily, E.A. Benjamin, L.C. Chamon, R. Lichtenthaler, in preparation.
8. J. P. Elliott, *Proc. Roy. Soc.* **A245**, 128 and 562 (1958).
9. D. J. Rowe, *Rep. Prog. Phys.* **48**, 1419 (1985).
10. M. Jarrio, J. L. Wood, and D. J. Rowe, *Nucl. Phys.* **A528**, 409 (1991).
11. P. O. Hess, A. Algora, M. Hunyadi, J. Cseh, *Eur. Phys. J.* **A15**, 449 (2002).
12. J. Cseh, J. Darai, A. Algora, H. Yepez-Martinez, P. O. Hess, *Rev. Mex. Fis. S.* **54** (3) 30.
13. J. Cseh, J. Darai, *AIP Conf. Proc. 1098: Fusion08*, 225 (2008).
14. G. Leander, S. E. Larsson, *Nucl. Phys.* **A239**, 93 (1975).
15. J. Darai, J. Cseh, N.V. Antonenko, A. Algora, P.O. Hess, R.V. Jolos, W. Scheid, in the present volume.
16. J. Cseh, *J. Phys.* **G19**, L97 (1993), and references therein.
17. M.Harvey, *Proc. 2nd Int. Conf. on Clustering Phenomena in Nuclei, (College Park) USDERA report ORO-4856-26*, 549 (1975).
18. E.P. Wigner *Phys. Rev.* **51**, 106 (1937).
19. J. Cseh, J. Darai, N.V. Antonenko, A. Algora, P.O. Hess, R.V. Jolos, W. Scheid, *Rev. Mex. Fis. S.* **52** (4) 11, and references therein.

A Semi-microscopic Approach to Clusterization in Heavy Nuclei

J. Darai[*], J. Cseh[†], N.V. Antonenko[**], A. Algora[‡], P.O. Hess[§], R.V. Jolos[**] and W. Scheid[¶]

[*]Institute of Experimental Physics, University of Debrecen, Debrecen, Pf. 105, Hungary-4010
[†]Institute of Nuclear Research, Hungarian Academy of Sciences, Debrecen, Pf. 51, Hungary-4001
[**]Joint Insitute for Nuclear Research, 141980 Dubna, Russia
[‡]IFIC, CSIC-Universidad de Valencia, A. C. 22085, E 46071, Valencia, Spain
[§]Instituto de Sciencias Nucleares, UNAM, A. P. 70-543, 04510 Mexico D. F.
[¶]Insitute of Theoretical Physics, Justus Liebig University, D-35392 Giessen, Germany

Abstract. We present a semimicroscopic approach to clusterization in heavy nuclei. The method is largely based on symmetry-considerations. As an example we determine the possible binary clusterizations of the shape isomers of the ^{56}Ni nucleus. We combine our structure-considerations with energy-calculations.

Keywords: clusterization, selection rules, symmetries
PACS: 21.60.Cs, 21.60.Fw, 21.60.Gx, 27.40.+z

Clusterization seems to be an important phenomenon in many different regions of the isotopic table. However, its fully microscopic description, which deals with the Pauli-principle appropriately, is available only for light nuclei, or for some special simple cases of the heavy ones. Phenomenological models are applied systematically both for light and for heavy nuclei. But they do not really contain the exclusion principle, thus we may very well miss something important, and do not even know what it is. Therefore, a semimicroscopic approach, which is systematically applicable on the one hand, and incorporates the consequences of the exclusion principle in a well-controlled way, on the other hand, is highly desirable. Here we present an attempt along this line. It is largely based on symmetry-consideration. In particular we follow the scenario: i) based on Nilsson-calculations we determine quasidynamical (or effective) $U(3)$ quantum numbers for a given state [1, 2]; ii) we apply the Harvey prescription and a $U(3)$ selection rule in order to determine if a cluster-configuration is allowed or forbidden [3]. The structure considerations, as described above, should be combined with energy-calculations. For this latter problem we use both simple binding-energy arguments [4], and double-folding calculations, according to the dinuclear system model [5].

The quasi-dynamical $U(3)$ quantum numbers characterizing the shape isomers of ^{56}Ni nucleus (especially the ground state, a superdeformed state and a hyperdeformed state) were determined from self-consistent shape calculations in the Nilsson-model. For more details see [6].

Once the shape isomers have been found, the next question is how they are related to cluster configurations. To find their connection we use the Harvey prescription and the

CP1165, *Nuclear Structure and Dynamics '09*
edited by M. Milin, T. Nikšić, D. Vretenar, and S. Szilner
© 2009 American Institute of Physics 978-0-7354-0702-2/09/$25.00

$U(3)$ selection rule. In the case of a binary clusterization the $U(3)$ selection rule reads:

$$[n_1, n_2, n_3] = [n_1^{(1)}, n_2^{(1)}, n_3^{(1)}] \otimes [n_1^{(2)}, n_2^{(2)}, n_3^{(2)}] \otimes [n^{(R)}, 0, 0], \qquad (1)$$

where $[n_1, n_2, n_3]$ is the set of $U(3)$ quantum numbers of the parent nucleus, the superscripts (i) stand for the ith cluster, (R) indicates relative motion. A given clusterization is allowed if there is a matching between $[n_1, n_2, n_3]$ and the product representations. When a cluster configuration is forbidden, we can characterize its forbiddenness quantitatively [7].

The selection rule can incorporate the effects of the exclusion-principle, only in an approximate way, of course. But it can be checked by making a comparison with the results of the fully microscopic description, where they are available [1]. In geometrical terms it can be illustrated by the similarity of the quadrupole-deformation of the cluster-configuration and the shell-model (or collective model) state.

Energetic preference calculations represent a complementary viewpoint for the selection of clusterization. The criterium of maximal stability [4] requires the largest value of the summed differences of the measured binding energies and the corresponding liquid drop values. In the dinuclear system model the potential energy U is minimized with respect to the mass asymmetry for each fixed charge asymmetry [5].

Based on the $U(3)$ selection rule method we determined the possible binary alpha-like cluster-configurations for the shape isomers of ^{56}Ni. As a result we got, that in the ground state of the nucleus only the ^{4}He+^{52}Fe clusterization is allowed, the ^{8}Be+^{48}Cr is not very strongly forbidden, the others are forbidden. In the superdeformed state more clusterizations are allowed: ^{8}Be+^{48}Cr, ^{12}C+^{44}Ti, ^{20}Ne+^{36}Ar, ^{24}Mg+^{32}S and ^{28}Si+^{28}Si. Concerning the hyperdeformed state the ^{24}Mg+^{32}S and ^{28}Si+^{28}Si clusterizations turned out to be allowed. As for the energetic preferences, from both energy-calculations the ^{4}He is the most favoured, much ahead of the ^{8}Be, which is followed by the group of ^{12}C, ^{28}Si, and ^{16}O. The ^{24}Mg and ^{20}Ne turn out to be the least-preferred alpha-like clusters.

As a summary, we can say that the quasi-dynamical $SU(3)$ symmetry seems to be able to extend some of the useful algebraic methods to the study of the heavy nuclei, where real symmetry is not valid. In particular, the selection of the allowed cluster-configurations in this way can be interesting.

This work was supported by the OTKA (Grant No K72357), by the NKTH (project no. ES-26/2008), by the MTA-CONACyT exchange programme, and by the A. von Humboldt-Foundation.

REFERENCES

1. P. O. Hess, A. Algora, M. Hunyadi, J. Cseh, *Eur. Phys. J.* **A15**, 449 (2002).
2. M. Jarrio, J. L. Wood, and D. J. Rowe, *Nucl. Phys.* **A 528**, 409 (1991).
3. J. Cseh, J. Darai, *AIP Conf. Proc. 1098: Fusion08*, 225 (2008), and references therein.
4. B. Buck, A. C. Merchant, S. M. Perez, *Few-Body Systems* **29**, 53 (2000).
5. T. M. Schneidman et al., *Phys. Lett.* **B526**, 322 (2002).
6. J. Cseh, J. Darai, A. Lépine-Szily, in the present volume.
7. A. Algora, J. Cseh, *J. Phys. G: Nucl. Part. Phys.* **22**, L39 (1996).

α-Particle Condensation in ^{16}O

P. Schuck[*], Y. Funaki[†], T. Yamada[**], H. Horiuchi[‡], G. Röpke[§] and A. Tohsaki[‡]

[*]Institut de Physique Nucléaire, CNRS, UMR 8608, Orsay, F-91406, France
Université Paris-Sud, Orsay, F-91505, France
[†]Institute of Physics, University of Tsukuba, Tsukuba 305-8571, Japan
[**]Laboratory of Physics, Kanto Gakuin University, Yokohama 236-8501, Japan
[‡]Research Center for Nuclear Physics (RCNP), Osaka University, Osaka 567-0047, Japan
[§]Institut für Physik, Universität Rostock, D-18051 Rostock, Germany

Abstract. In order to explore the 4α-particle condensate state in ^{16}O, we solve a full four-body equation based on the 4α OCM (Orthogonality Condition Model) in a large 4α model space spanned by Gaussian basis functions. A full spectrum up to the 0_6^+ state is reproduced consistently with the lowest six 0^+ states of the experimental spectrum. It is suggested that the 0_6^+ state is the analog to the Hoyle state of ^{12}C, to be identified with the well known 0_6^+ state at 15.1 MeV in ^{16}O.

Keywords: Bose-Einstein condensation; alpha-particle clusters; Ikeda energy in ^{16}O
PACS: 21.10.Dr,21.10.Gv,21.60.Gx,03.75Hh

It is well established that α-clustering plays a very important role for the structure of lighter nuclei [1, 2]. The importance of α-cluster formation also has been discussed in infinite nuclear matter, where α-particle type condensation is expected at low density [3], quite in analogy to the recently realised Bose-Einstein condensation of bosonic atoms in magneto-optical traps [4]. On the other hand, for trapped fermions, quartet condensation also is an emerging subject, discussed, so far, only theoretically [5]. In nuclei the bosonic constituents always are only very few in number, nevertheless possibly giving rise to clear condensation characteristics, as is well known from nuclear pairing [6]. Concerning α-particle condensation, only the Hoyle state, i.e. the 0_2^+ state in ^{12}C has clearly been established, so far.

The 4α-particle condensate state was first investigated in Ref. [7] and its existence was predicted around the 4α threshold with a new type of microscopic wave function of α-particle condensate character. While that so-called THSR wave function can well describe the dilute α cluster states as well as shell model like ground states, other structures such as $\alpha+^{12}$C clustering are smeared out and only incorporated in an average way.

We here explore the 4α condensate state by solving a full OCM four-body equation of motion without any assumption with respect to the structure of the 4α system. We take the 4α OCM with Gaussian basis functions, the model space of which is large enough to cover the 4α gas, the $\alpha+^{12}$C cluster, as well as the shell-model configurations. The OCM is extensively described in Ref. [8]. Many successful applications of OCM are reported in Ref. [2]. The presentation is largely based on Ref. [9]. The 4α OCM Hamiltonian is given as follows:

CP1165, *Nuclear Structure and Dynamics '09*
edited by M. Milin, T. Nikšić, D. Vretenar, and S. Szilner
© 2009 American Institute of Physics 978-0-7354-0702-2/09/$25.00

$$\mathcal{H} = \sum_{i}^{4} T_i - T_{cm} + \sum_{i<j}^{4} \left[V_{2\alpha}^{(N)}(i,j) + V_{2\alpha}^{(C)}(i,j) + V_{2\alpha}^{(P)}(i,j) \right]$$

$$+ \sum_{i<j<k}^{4} V_{3\alpha}(i,j,k) + V_{4\alpha}(1,2,3,4), \tag{1}$$

where T_i, $V_{2\alpha}^{(N)}(i,j)$, $V_{2\alpha}^{(C)}(i,j)$, $V_{3\alpha}(i,j,k)$ and $V_{4\alpha}(1,2,3,4)$ stand for the operators of kinetic energy for the i-th α particle, two-body, Coulomb, three-body and four-body forces between α particles, respectively. The center-of-mass kinetic energy T_{cm} is subtracted from the Hamiltonian. $V_{2\alpha}^{(P)}(i,j)$ is the Pauli exclusion operator [10], by which the Pauli forbidden states between two α-particles in $0S$, $0D$ and $1S$ states are eliminated, so that the ground state with the shell-model-like configuration can be described correctly. The effective α-α interaction $V_{2\alpha}^{(N)}$ is constructed by the folding procedure from two kinds of effective two-nucleon forces. One is the Modified Hasegawa-Nagata (MHN) force [11] and the other is the Schmidt-Wildermuth (SW) force [12], see Refs. [13] and [14] for applications, respectively. We should note that the folded α-α potentials reproduce the α-α scattering phase shifts and energies of the ^{8}Be ground state and of the Hoyle state. The three-body force $V_{3\alpha}$ is as in Refs. [13] and [14] where it was phenomenologically introduced, so as to fit the ground state energy of ^{12}C. In addition, the phenomenological four-body force $V_{4\alpha}$ which is taken to be a Gaussian is adjusted to the ground state energy of ^{16}O, where the range is simply chosen to be the same as that of the three-body force. The origin of the three-body and four-body forces is considered to derive from the state dependence of the effective nucleon-nucleon interaction and the additional Pauli repulsion between more than two α-particles. However, they are short-range, and hence only act in compact configurations. The expectation values of those forces do not exceed 7 percent of the one of the corresponding two-body term, even for the ground state with the most compact structure, i.e. being the most sensitive to those forces.

The total wave function Ψ of the 4α system is expanded in terms of Gaussian basis functions [15]. Figure 1 shows the energy spectrum with $J^\pi = 0^+$, which is obtained by diagonalizing the Hamiltonian, Eq. (1), in a model space as large as given by 5120 Gaussian basis functions. All levels are well converged. With the above mentioned effective α-α forces, we can reproduce the full spectrum of 0^+ states, and tentatively make a one-to-one correspondence of those states with the six lowest 0^+ states of the experimental spectrum. In view of the complexity of the situation, the agreement is considered to be very satisfactory. We show in TABLE 1 the calculated rms radii and monopole matrix elements to the ground state, together with the corresponding experimental values. The $M(E0)$ values for the 0_2^+ and 0_5^+ states are consistent with the corresponding experimental values. The consistency for the 0_3^+ state is within a factor of two. As mentioned above, the structures of the 0_2^+ and 0_3^+ states are well established as having the $\alpha + ^{12}$C(0_1^+) and $\alpha + ^{12}$C(2_1^+) cluster structures, respectively. These structures of the 0_2^+ and 0_3^+ states are confirmed in the present calculation. We also mention that the ground state is described as having a shell-model configuration within the present framework, the calculated rms value agreeing with the observed one (2.71 fm).

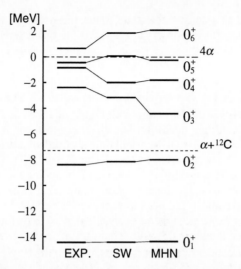

FIGURE 1. Comparison of energy spectra between experiment and the present calculation. Two kinds of effective two-body nucleon-nucleon forces MHN and SW are adopted (see text). Dotted and dash-dotted lines denote the $\alpha+{}^{12}C$ and 4α thresholds, respectively. Experimental data are taken from Ref. [16], and from Ref. [17] for the 0_4^+ state. The assignments with experiment are tentative, see, however, detailed discussion in the text.

On the contrary, the structures of the observed 0_4^+, 0_5^+ and 0_6^+ states in Fig. 1 have, in the past, not clearly been understood, since they have never been discussed with the previous cluster model calculations [18, 19, 20]. Although Ref. [7] predicts the 4α condensate state around the 4α threshold, it is not clear to which of those states it corresponds to.

As shown in Fig. 1, the present calculation succeeded, for the first time, to reproduce the 0_4^+, 0_5^+ and 0_6^+ states, together with the 0_1^+, 0_2^+ and 0_3^+ states. In Table 1, the largest rms value of about 5 fm is found for the 0_6^+ state. The α decay width constitutes a very important information to identify the 0_6^+ state from the experimental point of view. It can be estimated, based on R-matrix theory [21]. We find that the total α decay width of the 0_6^+ state is as small as 50 keV (experimental value: 166 keV). This means that the

TABLE 1. The rms radii R and monopole transition matrix elements to the ground state $M(E0)$ in units of fm and fm^2, respectively. $R_{\text{exp.}}$ and $M(E0)_{\text{exp.}}$ are the corresponding experimental data. The finite-size effect of α particle is taken into account in R and $M(E0)$ (see Ref. [13] for details).

	R		$M(E0)$		$R_{\text{exp.}}$	$M(E0)_{\text{exp.}}$
	SW	MHN	SW	MHN		
0_1^+	2.7	2.7			2.71 ± 0.02	
0_2^+	3.0	3.0	4.1	3.9		3.55 ± 0.21
0_3^+	2.9	3.1	2.6	2.4		4.03 ± 0.09
0_4^+	4.0	4.0	3.0	2.4		no data
0_5^+	3.1	3.1	3.0	2.6		3.3 ± 0.7
0_6^+	5.0	5.6	0.5	1.0		no data

state can be observed as a quasi-stable state. Thus, the width, as well as the excitation energy, are consistent with the observed data. All the characteristics found from our OCM calculation, including the large radius, therefore, indicate that the calculated 6th 0^+ state with 4 alpha condensate nature can probably be identified with the experimental 0_6^+ state at 15.1 MeV.

Concerning the 0_4^+ and 0_5^+ states, our present calculations show that the 0_4^+ and 0_5^+ states mainly have $\alpha + {}^{12}C(0_1^+)$ structure with higher nodal behaviour and $\alpha + {}^{12}C(1^-)$ structure, respectively. Further details will be given in a forthcoming extended paper.

In conclusion, the present 4α OCM calculation succeeded in describing the structure of the full observed 0^+ spectrum up to the 0_6^+ state in ${}^{16}O$. The 0^+ spectrum of ${}^{16}O$ up to about 15 MeV is thus essentially understood, including the 4α condensate state. This is remarkable improvement concerning our knowledge of the structure of ${}^{16}O$. We found that the 0_6^+ state above the 4α threshold has a very large rms radius of about 5 fm and has a rather large occupation probability of 61% of four α particles sitting in a spatially extended single-α $0S$ orbit. These results are strong evidence of the 0_6^+ state for being the 4α condensate state, i.e. the analog to the Hoyle state in ${}^{12}C$. Further experimental information is very much requested to confirm the structure of this state. Also independent theoretical calculations are strongly needed for confirmation of our results.

REFERENCES

1. K. Wildermuth and Y. C. Tang, *A Unified Theory of the Nucleus* (Vieweg, Braunschweig, 1977).
2. K. Ikeda et al., *Prog. Theor. Phys. Suppl.* No. **68**, 1 (1980).
3. G. Röpke et al., *Phys. Rev. Lett.* **80**, 3177 (1998).
4. F. Dalfovo et al., *Rev. Mod. Phys.* **71**, 463 (1999).
5. A. S. Stepanenko et al., arXiv: cond-mat/9901317; B. Doucot et al., *Phys. Rev. Lett.* **88**, 227005 (2002); H. Kamei et al., *J. Phys. Soc. Jpn.* **74**, 1911 (2005); S. Capponi et al., *Phys. Rev.* **A77**, 013624 (2008).
6. P. Ring and P. Schuck, *The Nuclear Many-Body Problem* (Springer-Verlag, Berlin, 1980).
7. A. Tohsaki et al., *Phys. Rev. Let.* **87**, 192501 (2001).
8. S. Saito, *Prog. Theor. Phys.* **40**, 893 (1968); **41**, 705 (1969); *Prog. Theor. Phys. Suppl.* No. **62**, 11 (1977).
9. Y. Funaki et al., *Phys. Rev. Lett.* **101**, 082502 (2008).
10. V. I. Kukulin et al., *Nucl. Phys.* **A417**, 128 (1984).
11. A. Hasegawa et al., *Prog. Theor. Phys.* **45**, 1786 (1971); F. Tanabe et al., ibid. **53**, 677 (1975).
12. E. W. Schmid et al., *Nucl. Phys.* **26**, 463 (1961).
13. T. Yamada et al., *Euro. Phys. J.* **A26**, 185 (2005).
14. C. Kurokawa et al., *Phys. Rev.* **C71**, 021301 (2005); *Nucl. Phys.* **A792**, 87 (2007).
15. M. Kamimura, *Phys. Rev.* **A38**, 621 (1988); E. Hiyama et al., *Prog. Part. Nucl. Phys.* **51**, 223 (2003).
16. F. Ajzenberg-Selove, *Nucl. Phys.* **A46**, 1 (1986).
17. T. Wakasa et al., *Phys. Lett.* **B653**, 173 (2007).
18. Y. Suzuki, *Prog. Theor. Phys.* **55**, 1751 (1976); **56**, 111 (1976).
19. M. Libert-Heinemann et al., *Nucl. Phys.* **A339**, 429 (1980).
20. K. Fukatsu et al., *Prog. Theor. Phys.* **87**, 151 (1992).
21. A. M. Lane et al., *Rev. Mod. Phys.* **30**, 257 (1958).

Dominance of Low Spin and High Deformation in Ab Initio Approaches to the Structure of Light Nuclei

T. Dytrych[*], J. P. Draayer[*], K. D. Sviratcheva[*], C. Bahri[*] and J. P. Vary[†]

[*]Department of Physics and Astronomy, Louisiana State University, Baton Rouge, LA 70803, USA
[†]Department of Physics and Astronomy, Iowa State University, Ames, IA 50011, USA

Abstract. Ab initio no-core shell-model solutions for the structure of light nuclei are shown to be dominated by low-spin and high-deformation configurations. This implies that only a small fraction of the full model space is important for a description of bound-state properties of light nuclei. It further points to the fact that the coupling scheme of choice for carrying out calculations for light nuclear systems is an algebraic-based, no-core shell-model scheme that builds upon an LS coupling [SO(3) ⊗ SU(2)] foundation with the spatial part of the model space further organized into its symplectic [SO(3) ⊂ SU(3) ⊂ Sp(3,R)] structure. Results for ^{12}C and ^{16}O are presented with the cluster nature of the excited 0^+ states in ^{16}O analyzed within this framework. The results of the analysis encourages the development of a no-core shell model code that takes advantage of algebraic methods as well as modern computational techniques. Indeed, although it is often a very challenging task to cast complex algebraic constructs into simple logical ones that execute efficiently on modern computational systems, the construction of such a next-generation code is currently underway.

Keywords: symplectic symmetry, nuclear shell model, nuclei with mass number 6 to 19
PACS: 21.60.Cs; 21.10.Re; 21.60.Fw; 27.20.+n

The no-core shell model (NCSM) [1] is a prominent ab initio method that has achieved a good description of low-lying states in few-nucleon systems as well as in more complex p-shell nuclei. The main limitation of this method is inherently coupled with its use of an m-scheme basis, which grows combinatorially in size with increasing nucleon number and increasing many-body basis cutoff N_{max}. To extend the scope of the NCSM to heavier nuclei and to incorporate configurations spanning larger model spaces, we propose to carry out NCSM calculations in a symmetry-adapted basis built upon an LS coupling scheme with the spatial part of the basis organized according to its transformation properties with respect to the physically relevant subgroups of the symplectic $Sp(3, \mathbb{R})$ symmetry group, which underpins a microscopic description of the nuclear collective motion [2].

We expressed well-converged NCSM eigenstates in ^{12}C as a superposition of wave functions carrying intrinsic spin of protons and neutrons, S_π and S_ν, coupled to an overall intrinsic spin S, and calculated corresponding mixing amplitudes. The NCSM eigenstates employed in this study, the ground state band and $J = 1_1^+$ states, were obtained with effective interactions derived from the realistic JISP16 and N3LO NN potentials in the $N_{max} = 6$ model space. We found that only 4 configurations with the lowest intrinsic spins in the proton and neutron sector, namely $\{S_\pi S_\nu\}=\{0\ 0\}$, $\{1\ 0\}$, $\{0\ 1\}$, and $\{1\ 1\}$, overlap at a 98% level with the NCSM ground state band for both the JISP16 and N3LO results (see Fig. 1). The remaining 44 spin combinations cover

CP1165, *Nuclear Structure and Dynamics '09*
edited by M. Milin, T. Nikšić, D. Vretenar, and S. Szilner
© 2009 American Institute of Physics 978-0-7354-0702-2/09/$25.00

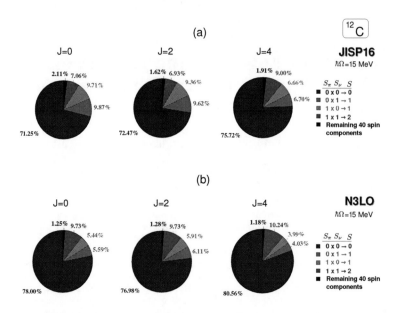

FIGURE 1. (Color online) Intrinsic spin structure of $J = 0^+_{gs}, 2^+_1$, and 4^+_1 NCSM states in ^{12}C obtained using: (a) JISP16, (b) N3LO, effective interactions in the $N_{max} = 6$ model space with $\hbar\Omega = 15$ MeV.

typically less than 2% of the NCSM wave functions. Qualitatively similar results were obtained for 1^+_1 state, with the lowest 4 spin configurations describing about 95% of the wave function, (see Fig. 2). If the $N_{max} = 6$ model space is restricted to the wave functions with good total angular momentum J and $S_\pi \leq 1$ and $S_\nu \leq 1$, the size of the basis drops by a factor of three relative to the basis which does not impose any restrictions on intrinsic spins. This reduction further improves, albeit slowly, for heavier nuclei and larger model spaces.

We also identified the most dominant spatial configurations within well-converged NCSM eigenstates for ^{12}C and ^{16}O by their projection on a subspace spanned by irreducible representations (irreps) of the $Sp(3,\mathbb{R})$ symmetry group of the symplectic shell model [2]. This microscopic model describes the monopole-quadrupole vibrational and rotational modes of nuclear collective motion, and also partially incorporates α-cluster correlations. Since Elliott's $SU(3)$ group is a subgroup of $Sp(3,\mathbb{R})$, all symplectic states can be classified by the two $SU(3)$ quantum numbers $(\lambda\mu)$, which are related to the quadrupole deformation and the asymmetry variables β and γ of the geometric collective model.

The NCSM wave functions for the lowest 0^+, 2^+ and 4^+ states in ^{12}C and the ground state in ^{16}O project at the 85-90% level onto a few spurious center-of-mass free symplectic irreps built on the most deformed $0\hbar\Omega$ and $2\hbar\Omega$ $Sp(3,\mathbb{R})$ bandhead configurations. The symplectic configurations with oblate shapes clearly dominate the ^{12}C ground state rotational band (Fig. 3). In addition, the three dominant symplectic

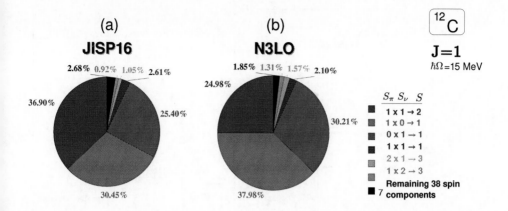

FIGURE 2. (Color online) Intrinsic spin structure of $J = 1_1^+$ NCSM wave function in ^{12}C obtained using: (a) JISP16, (b) N3LO, effective interactions in the $N_{max} = 6$ model space with $\hbar\Omega = 15$ MeV.

irreps closely reproduce the NCSM B(E2) estimates. About 80% of the ground state in ^{16}O fall within a subspace spanned by only seven $J = 0$ basis states of the symplectic irrep that is constructed from the $0\hbar\Omega$ spherical bandhead. While the ground state in ^{16}O is dominated by a mixture of spherical and slightly prolate symplectic configurations (see Fig. 4a), the projection of the second 0_2^+ state in ^{16}O on the symplectic model space reveals the presence of larger deformations, and in particular, a significant contribution of the 2p-2h Sp(3, $\mathbb{R}$) irrep (4 2) (see Fig. 4b), whose bandhead configuration describes

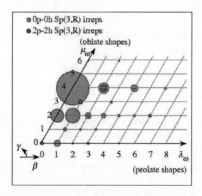

FIGURE 3. (Color online) Probabilities (specified by the area of the circles) for the symplectic states which make up the most important 0p-0h (blue) and $2\hbar\Omega$ 2p-2h (red) symplectic irreps, within the NCSM ground state in ^{12}C, $\hbar\Omega = 15$ MeV. The Sp(3, $\mathbb{R}$) states are grouped according to their $(\lambda\mu)$ SU(3) symmetry, which is mapped onto the $(\beta\gamma)$ shape variables of the collective model.

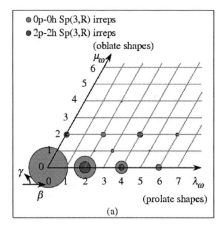
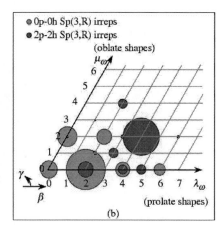

FIGURE 4. (Color online) Probabilities (specified by the area of the circles) for the symplectic states, which make up the most dominant 0p-0h (blue) and $2\hbar\Omega$ 2p-2h (red) symplectic irreps, within (a) 0^+_{gs} and (b) 0^+_2 state in ^{16}O calculated by NCSM, $\hbar\Omega = 15$ MeV. The Sp$(3,\mathbb{R})$ states are grouped according to their $(\lambda\,\mu)$ SU(3) symmetry, which is mapped onto the $(\beta\,\gamma)$ shape variables of the collective model.

the $\alpha + {}^{12}$C cluster structure of ^{16}O. Nevertheless, the $N_{max} = 6$ model space falls short of incorporating correlations needed to describe more pronounced cluster structures and deformations.

In short, we demonstrated that ^{12}C and ^{16}O NCSM wave functions are reproduced remarkably well in terms of a small number of low-spin and high-deformation many-body states classified according to the $[\text{SO}(3) \supset \text{SU}(3) \supset \text{Sp}(3,\mathbb{R})] \otimes \text{SU}(2)$ coupling scheme. A space spanned by these states is computationally manageable even when high-$\hbar\omega$ configurations are included. NCSM calculations in the $\text{SU}(3) \supset \text{Sp}(3,\mathbb{R})$ symmetry-adapted basis thus holds promise to resolve the scale explosion problem in ab initio nuclear structure calculations.

ACKNOWLEDGMENTS

This work was supported in part by the US NSF 0500291, the Southeastern Universities Research Association, and by the US DOE DE-FG02-87ER40371. We acknowledge the Center for Computation and Technology at Louisiana State University and the Louisiana Optical Network Initiative (LONI) for providing HPC resources.

REFERENCES

1. P. Navrátil, J. P. Vary, and B. R. Barret, *Phys. Rev. Lett.* **84**, 5728 (2000).
2. G. Rosensteel and D. J. Rowe, *Phys. Rev. Lett.* **38**, 10 (1977); D. J. Rowe, *Rep. Progr. Phys.* **48**, 1419 (1985).
3. T. Dytrych, K. D. Sviratcheva, C. Bahri, J. P. Draayer, and J. P. Vary, *Phys. Rev. Lett.* **98**, 162503 (2007).

Hadronic Interaction and Exotic Nuclei

Takaharu Otsuka*,†, Naofumi Tsunoda*, Koshiroh Tsukiyama*, Toshio Suzuki**, Michio Honma‡, Yutaka Utsuno§, Morten Hjorth-Jensen¶, Jason D. Holt‖ and Achim Schwenk‖

*Department of Physics, University of Tokyo, Hongo, Tokyo, 113-0033, Japan
†Center for Nuclear Study, University of Tokyo, Hongo, Tokyo, 113-0033, Japan, National Superconducting Cyclotron Laboratory, Michigan State University, East Lansing, Michigan, 48824, U.S.A
**Department of Physics, College of Humanities and Sciences, Nihon University, Sakurajosui 3, Tokyo 156-8550, Japan
‡Center for Mathematical Sciences, University of Aizu, Tsuruga, Ikki-machi, Aizu-Wakamatsu, Fukushima 965-8580, Japan
§Japan Atomic Energy Agency, Tokai, Ibaraki, 319-1195, Japan
¶Department of Physics and Center of Mathematics for Applications, University of Oslo, N-0316 Oslo, Norway
‖TRIUMF, 4004 Wesbrook Mall, Vancouver, BC, V6T 2A3, Canada

Abstract. The shell evolution and drip line are discussed with links to the nuclear two- and three-body forces.

Keywords: shell evolution, two- and three-body forces, exotic nuclei, shell model
PACS: 21.60.Cs,24.10.Cn,24.30.Gd,27.30.+t

Exotic nuclei provide us with new phenomena which are not found in stable nuclei. One of them is the evolution of the shell structure over a wide change of the proton number (Z) or the neutron number (N) [1, 2]. While the conventional magic numbers are valid for stable nuclei [3], the evolution indeed ends up, at extreme cases, with the appearance of new magic numbers and/or the disappearance of conventional ones. The nuclear force plays crucial roles in this evolution. As Z increases, there are more exotic isotopes between the β-stability line and the drip line, creating a wider frontier to be challenged. Most of such exotic nuclei are far inside the drip line, being well bound [4]. The driving force of changing structure should be the combination of the unbalanced Z/N ratio and the nuclear force mainly of two- and three-body nature. Thus, it is crucial to see the basic robust features of the nuclear force in exotic nuclei. We shall present, in this talk, such features governing the shell evolution. The tensor force has been shown to change the spin-orbit splitting of exotic nuclei, resulting in shifts of magic numbers [5, 6]. The two-body nuclear force is comprised also of its central part. We shall extract basic but novel features of effective nucleon-nucleon (NN) interactions, and suggest that the central and tensor forces of the effective NN interaction seem to have simple structures with their own characteristic effects. We further discuss the basic nature of the effects of the three-body force.

We start with the shell-model NN interactions which are successful in reproducing experimental data. These interactions have been obtained from so-called microscopic interactions based on NN scattering data and by including a treatment of the short-range

CP1165, *Nuclear Structure and Dynamics '09*
edited by M. Milin, T. Nikšić, D. Vretenar, and S. Szilner
© 2009 American Institute of Physics 978-0-7354-0702-2/09/$25.00

repulsion and core polarization effects. A good example of such microscopic interactions is G-matrix [7, 8]. For successful shell-model calculations, the microscopic interaction has to be modified. This modification has been carried out for the families of the KB interaction [9] and GXPF1 interaction [10, 11].

The change of the shell structure, or the shell evolution, may have different origins. We first focus upon the shell evolution by the tensor force. It is well-known that the one-pion exchange process is the major origin of the tensor force, which is written as

$$V_T = (\vec{\tau}_1 \cdot \vec{\tau}_2)\left([\vec{s}_1\vec{s}_2]^{(2)} \cdot Y^{(2)}\right)f(r), \tag{1}$$

where $\vec{\tau}_{1,2}$ ($\vec{s}_{1,2}$) denotes the isospin (spin) of nucleons 1 and 2, $[\ ]^{(K)}$ means the coupling of two operators in the brackets to an angular momentum (or rank) K, Y denotes the spherical harmonics for the Euler angles of the relative coordinate, and the symbol $(\cdot)$ means a scalar product. Here, $f(r)$ is a function of the relative distance, r. Eq. (1) is equivalent to the usual expression containing the S_{12} function. Because the spins $\vec{s}_1$ and $\vec{s}_2$ are dipole operators and are coupled to rank 2, the total spin S of two interacting nucleons must be $S=1$. If both of the bra and ket states of V_T have $L=0$, with L being the relative orbital angular momentum, their matrix element vanishes because of the $Y^{(2)}$ coupling. These properties are used later.

The (spherical) bare single-particle energy of an orbit j is given by its kinetic energy and the effects from the inert core (closed shell) on the orbit j. As some nucleons are added to another orbit j', the single-particle energy of the orbit j is changed. The nucleons on j' can form various many-body states, but we are interested in monopole effects independent of details of such many-body states. The monopole component of an interaction, V, is [12]:

$$V_{j,j'}^T = \frac{\sum_J (2J+1) < jj'|V|jj' >_{JT}}{\sum_J (2J+1)}, \tag{2}$$

where $< jj'|V|jj' >_{JT}$ stands for the (diagonal) matrix element of a state where two nucleons are coupled to an angular momentum J and an isospin T. In the summation in eq. (2), J takes values satisfying antisymmetrization. We then construct a two-body interaction, called V_M, consisting of two-body matrix elements $V_{j,j'}^T$ in eq. (2). Because the J-dependence is averaged out in eq. (2), the monopole interaction, V_M, represents the angular-free, i.e., monopole property of the original interaction, V, while it still depends on the isospin. If neutrons occupy j' and one looks into the orbit j ($\neq j'$) as a proton orbit, the shift of the single-particle energy of j is given by

$$\Delta\varepsilon_p(j) = \frac{1}{2}\{V_{j,j'}^{T=0} + V_{j,j'}^{T=1}\} n_n(j'), \tag{3}$$

where $n_n(j')$ is (the expectation value of) the number of neutrons in the orbit j'. The same is true for $\Delta\varepsilon_n(j)$ as a function of $n_p(j')$. The monopole effects from orbits j', j'', ... are added as these orbits are filled. The single-particle energy, including this monopole effect, is called the effective single-particle energy (ESPE), and it depends on the configurations. We shall discuss, in this talk, how the ESPE of an orbit j varies due to the tensor force as an orbit j' is filled.

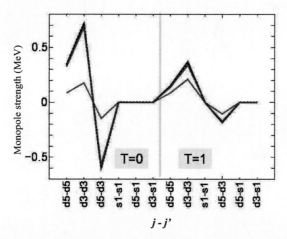

FIGURE 1. (Color online) Monopole interaction of the tensor component obtained by the spin-tensor decomposition of V_{lowk} interaction from AV8' interaction [13]. The green and red solid line indicate the results of the cut-off 1.0 and 2.1 (fm^{-1}), respectively. The black dashed line is for the cut-off 4.0 (fm^{-1}), while the blue solid line denotes the bare AV8' potential. Left (Right) panel shows the results for isospin T=0 (1).

If the orbit j' is fully occupied by neutrons in eq. (3), only the monopole effect remains over the other multipoles and eq. (3) gives the shift of the bare single-particle energy for this shell closure. If protons and neutrons are occupying the same orbit, the change of ESPE becomes slightly more complicated due to isospin symmetry [9, 12].

We begin with cases with orbital angular momenta l or l', protons are in either $j_> = l + 1/2$ or $j_< = l - 1/2$, while neutrons are in either $j'_> = l' + 1/2$ or $j'_< = l' - 1/2$. In results to be presented, the radial wave functions are given by the harmonic oscillator potential for simplicity.

From now on, V is the tensor force. For the orbits j and j', the monopole component in eq. (2) is calculated. We begin with the effect of the treatment of the short-range correlation. For this purpose, we use the Argonne V8' (AV8') potential [13], while the following argument is not specific to a particular choice of the interaction. The AV8' potential contains the tensor part explicitly, and therefore is suitable for the present purpose. The AV8' potential in fact include both isoscalar and isovector tensor components, and we keep them. This is not a problem, because the following discussion is made in the isospin formalism.

We obtain a low-momentum potential V_{lowk} based on Ref. [14] from the AV8' potential [13]. By transforming a given NN interaction this way, the short-range correlation effect is renormalized, and only low-momentum properties remain. Thus, the interaction becomes free from difficulties of the short-range strong repulsion and suitable for studies of the ground and low-lying excited states, *i.e.*, states of the shell model.

We then carry out the spin-tensor decomposition [15] for the obtained potential. We thus obtain the tensor part of the V_{lowk} potential. Figure 1 indicates monopole component of the tensor part of such V_{lowk} potentials in comparison to the original AV8'. The cut-off parameter is taken to be 1.0, 2.1 and 4.0 (fm^{-1}).

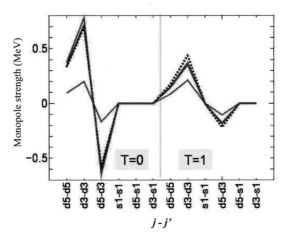

FIGURE 2. (Color online) Monopole interaction of the tensor component obtained by the spin-tensor decomposition of the second order Q-box treatment of the interaction discussed in Fig. 1. See the caption of Fig. 1 for details.

The cut-off 1.0 (fm^{-1}) is too small in the momentum space, or, too large in the coordinate space. We loose much of the tensor force effect, and the monopole component is quite small. This is, however, taken for the sake of comparison. The cut-off 2.1 (fm^{-1}) is taken from a typical low-momentum range. We can find a good agreement to the result of the AV8'. We also note that cut-off 4.0 (fm^{-1}) produces the basically identical result to the one by the cut-off 2.1 (fm^{-1}). We here conclude that the V$_{lowk}$ treatment of the NN interaction conserves the tensor component to a great extent. This feature should be found in results of UCOM formalism, where the short-range contributions are treated by unitary transformations [16].

The Q-box formalism is applied to the V$_{lowk}$ potential, in order to take into account effects of the inert core, including its polarization effect [17]. Figure 2 indicates monopole component of the tensor part of such effective interactions in comparison to the original AV8'. Here, the second order Q-box calculation has been carried out. The cut-off parameter is taken to be 1.0, 2.1 and 4.0 (fm^{-1}). It is remarkable that the results are very close to each other. except for the case with cut-off parameter 1.0 (fm^{-1}). We have already discussed for Fig. 1 that the cut-off parameter 1.0 (fm^{-1}) is extreme, and it is included in Fig. 2 for comparison. We thus find that the renormalization of various effects of the core does not change the monopole part of the tensor force in the valence shell. The V$_{lowk}$ formalism is basically for the short-range correlation, while the monopole properties are of long-range within a shell. As for the Q-box, the particle-hole excitation due to the tensor force changes mainly the central part. The present consequences can be verified for the pf shell.

We investigate the monopole properties of phenomenologically improved shell-model effective interactions, the SDPF-M for the sd shell [18] and the GXPF1A for the pf shell [11]. The results cannot be shown here due to the space limitation, but one can see that the orbital variation of the monopole strength comes almost entirely from the tensor force. The rest can be explained by a simple central force with a Gaussian dependence on the relative distance. The tensor force here is the one obtained from $\pi + \rho$ meson exchange potential [19], which is the bare tensor force between free nucleons. Namely, the tensor force in the shell model effective interactions are nothing but the bare tensor force to a good extent. This remarkable conclusion is consistent with the observations made with V_{lowk} and Q box calculations, presented above.

The combination of simple Gaussian central forces and the bare tensor forces is somewhat similar to what Weinberg proposed for the interaction between free nucleons[20], which led to the Chiral Perturbation theory. In this model, the NN interaction is comprised of the zero-range central force and the one-pion exchange tensor force. In the shell-model effective interaction, the former becomes Gaussian central force with the range of about 1 fm, while the latter is the $\pi + \rho$ meson exchange potential. This similarity between two simple modelings is of certain interest.

The second part of this talk is about effects of the three-body force. The three-body force has been discussed, for instance, in Refs. [21, 22, 23]. We, however, present a robust repulsive effect of the three-body force. The dominant contribution to this effect originates in the Fujita-Miyazawa force [24], which incorporates the virtual excitation from a nucleon to a Delta particle. In fact, it has been argued that although the shell-model effective interaction should be repulsive in its $T=1$ monopole part except for the contribution from the pairing force, this property cannot be explained if one includes NN forces only [25].

We show [26] that the three-body force explains the drip line of oxygen isotopes at the right place, whereas the calculations including only microscopic NN forces, G-matrix type [8] or more modern Chiral EFT type [27, 28], predict the drip line too far. In other words, these forces are too attractive as a common general feature. This is because the Fujita-Miyazawa-type three-body force produces effective two-body interaction between valence neutrons which is robustly repulsive. This can be seen in terms of single-particle energies and also in terms of the energies of the ground states, calculated by conventional pion-nucleon-Delta coupling [29] and by Chiral EFT [30]. As the mechanism for repulsive valence-shell effective interaction is robust and general, we can predict similar effects on other regions of the nuclear chart.

On the other hand, by adding a proton to oxygen, one can create fluorine isotopes which have the drip line much further away. This is because the added proton is mainly in the $0d_{5/2}$ orbit, and produces strong attraction with neutrons in the $0d_{3/2}$ orbit, making it bound [31]. As the tensor force plays a crucial role here also, the drastic change between oxygen and fluorine isotopes is one of the prominent examples of the importance of the nuclear forces in exotic nuclei.

In summary, the shell evolutions due to the tensor and three-body forces are presented with the underlying mechanisms. The tensor and three-body forces produce general and robust effects on the shell and (sub-)magic structures from the p-shell to the superheavy regions. The significant role of the tensor force as a direct consequence of π and ρ meson exchange is discussed in the comparison to the well-established phenomenologically improved shell-model interactions, and is analyzed in the light of V_{lowk} and Q box theories. We end up with a simple Ansatz that the shell-model effective interaction is comprised of simple Gaussian central and bare tensor forces [32]. This picture can be related to the Chiral Perturbation idea of Weinberg [20].

The shell evolution is also due to the three-body force. The three-body force is dominated by Fujita-Miyazawa force in its long-range part, and indeed produces decisive repulsive effects on the structure of exotic nuclei. As an example, we presented the case of oxygen isotopes [26].

This work was supported in part by a Grant-in-Aid for Scientific Research (A) (20244022) from the MEXT, and also by the JSPS core-to-core program "International Research Network for Exotic Femto Systems (EFES)". KT thanks JSPS for a fellowship. This work was supported in part also by the Natural Sciences and Engineering Research Council of Canada (NSERC). TRIUMF receives funding via a contribution through the National Research Council Canada.

REFERENCES

1. A. Gade and T. Glasmacher, *Prog. Part. Nucl. Phys.* **60**, 161 (2008).
2. O. Sorlin and M.-G. Porquet, *Prog. Part. Nucl. Phys.* **61**, 602 (2008).
3. M.G. Mayer, *Phys. Rev.* **75** 1969 (1949); O. Haxel, J.H.D. Jensen and H.E. Suess, *Phys. Rev.* **75** 1766 (1949).
4. T. Otsuka, T. Suzuki, and Y. Utsuno, *Nucl. Phys.* **A805**, 127c (2008).
5. T. Otsuka et al., *Phys. Rev. Lett.* **87**, 082502 (2001).
6. T. Otsuka, T. Suzuki, R. Fujimoto, H. Grawe and Y. Akaishi, *Phys. Rev. Lett.* **95**, 232502 (2005).
7. T.T.S. Kuo and G.E. Brown, *Nucl. Phys.* **A114**, 241 (1968).
8. M. Hjorth-Jensen, T.T.S. Kuo and E. Osnes, *Phys. Rep.* **261**, 125 (1995).
9. A. Poves and A. Zuker, *Phys. Rep.* **70**, 235 (1981).
10. M. Honma, T. Otsuka, B.A. Brown, and T. Mizusaki, *Phys. Rev.* **C65**, 061301(R) (2002); *Phys. Rev.* **C69**, 034335 (2004).
11. M. Honma, T. Otsuka, B.A. Brown, and T. Mizusaki, *Eur. Phys. J.* **A25**, s01, 499 (2005).
12. R.K. Bansal and J.B. French, *Phys. Lett.* **11**, 145 (1964).
13. B.S. Pudliner, et al., *Phys. Rev.* **C56**, 1720 (1997).
14. S.K. Bogner, T.T.S. Kuo and A. Schwenk, *Phys. Rep.* **386**; S.K. Bogner et al., *Nucl. Phys.* **A784**, 79 (2007).
15. B.A. Brown, W.A. Richter, R.E. Julies and H.B. Wildenthal, *Ann. Phys. (N.Y.)* **182**, 191 (1988).
16. H. Feldmeier, T. Neff, R. Roth and J. Schnack, *Nucl. Phys.* **A632**, 61 (1998); T. Neff and H. Feldmeier, *Nucl. Phys.* **A713**, 311 (2003).
17. T. Morita, *Prog. Theor. Phys.*, **29**, 351 (1963); T.T.S. Kuo E. Osnes, *Lecture notes in physics*, **364**, 1 (1990).
18. Y. Utsuno, et al., *Phys. Rev.* **C60**, 054315 (1999); *Phys. Rev.* **C70**, 044307 (2004).
19. F. Osterfeld, *Rev. Mod. Phys.* **64**, 491(1992).
20. S. Weinberg, *Phys. Lett.* **B251**, 288 (1990).
21. S.C. Pieper and R.B. Wiringa, *Ann. Rev. Nucl. Part. Sci.* **51**, 53 (2001); S.C. Pieper, *Nucl. Phys.* **A751**, 516 (2005).
22. P. Navratil et al., *Phys. Rev. Lett.* **99**, 042501 (2007).
23. A. Schwenk and J.D. Holt, *AIP Conf. Proc.* **1011**, 159 (2008).
24. J. Fujita and H. Miyazawa, *Prog. Theor. Phys.* **17**, 360 (1957).
25. A.P. Zuker, *Phys. Rev. Lett.* **90**, 042502 (2003).
26. T. Otsuka, T. Suzuki, J.D. Holt, A. Schwenk, Y. Akaishi, to be submitted.
27. E. Epelbaum, *Prog. Part. Nucl. Phys.* **57**, 654 (2006); E. Epelbaum, H.-W. Hammer and U.-G. Meißner, arXiv:0811.1338.
28. D.R. Entem and R. Machleidt, *Phys. Rev.* **C68**, 041001(R) (2003).
29. A.M. Green, *Rep. Prog. Phys.* **39**, 1109 (1976).
30. U. van Kolck, *Phys. Rev.* **C49**, 2932 (1994); E. Epelbaum et al., *Phys. Rev.* **C66**, 064001 (2002).
31. Y. Utsuno, T. Otsuka, T. Mizusaki and M. Honma, *Phys. Rev.* **C64**, 011301 (2001).
32. T. Otsuka, T. Suzuki, M. Honma, Y. Utsuno, N. Tsunoda, K. Tsukiyama, M. H.-Jensen, to be submitted.

Structure of Light Neutron-Rich Nuclei and Important Roles of Tensor Interaction

Toshio Suzuki[*][†] and Takaharu Otsuka[**][‡]

[*]*Department of Physics, College of Humanities and Sciences, Nihon University, Sakurajosui 3-25-40, Setagaya-ku, Tokyo 156-8550, Japan*
[†] *Center for Nuclear Study, University of Tokyo, Hirosawa, Wako-shi, Saitama, 351-0198, Japan*
[**]*Department of Physics and Center for Nuclear Study, University of Tokyo, Hongo, Bunkyo-ku, Tokyo 113-0033, Japan*
[‡] *RIKEN, Hirosawa, Wako-shi, Saitama 351-0198, Japan*

Abstract. Structure of light neutron-rich nuclei are investigated by shell model calculations with the use of our new shell model Hamiltonians. The new Hamiltonians take into account important roles of the tensor interaction, and can explain well the Gamow-Teller (GT) transitions in nuclei with mass 14 and 12. A new version with repulsive corrections in $T=1$ monopole terms is applied to study structure of neutron-rich carbon isotopes. Characteristics of effective neutron single particle enrgies and shell evolutions toward the drip-line are discussed. Ground state energies and low lying energy levels of the isotopes are found to be well reproduced. The anomalous quenching of the magnetic dipole (M1) transition strength in ^{17}C is also found to be well explained. GT transitions in ^{19}C are studied. The halo nature of ^{19}C is investigated from a retarted GT transition. Contributions from three-nucleon forces induced by Δ excitations are shown to be the main origin of the repulsive monopole corrections in the $T=1$ channel.

Keywords: shell model, neutrino-nucleus reactions, nucleosynthesis
PACS: 21.60.Cs, 21.30.Fe, 23.20.Lv, 23.40.-s

SHELL EVOLUTIONS IN CARBON ISOTOPES

Structure of light neuton-rich nuclei are studied by shell model calculations. Our shell model Hamiltonian for p-shell, SFO [1], properly takes into account important roles of the tensor interaction. The tensor components of the interaction have proper sign character, that is, attractive between $j_> = \ell + \frac{1}{2}$ and $j_< = \ell - \frac{1}{2}$ orbits while repulsive between $j_>$ and $j_>$ ($j_<$ and $j_<$) orbits. This property is quite important for the shell evolution toward drip-lines [2, 3]. The configuration space with $2 \sim 3\hbar\omega$ excitations is used for the Hamiltonians. Systematic improvements in the descriptions of the magnetic moments of the p-shell nuclei as well as Gamow-Teller (GT) transitions, for example, in ^{12}C and ^{14}N have been obtained [1]. The $B(GT)$ for the transition to the ground state of ^{14}C vanishes consistent with the observation. The tensor interacion involved in the transition is found to be consistent with that of the $\pi+\rho$ meson exchang model, and no need to be reduced contrary to the result obtained in Ref. [4]. The SFO also gives rise to the observed splitting of the strength for the 2^+ states in ^{14}C [5].

The modified version of SFO, SFO-tls [6], is applied to study structure of neutron-rich carbon isotopes. Important roles of the tensor interaction is fully taken into account in the SFO-tls. The tensor components of the p-sd cross shell matrix elements are taken to be those of the $\pi+\rho$ meson exchanges. Their two-body spin-orbit components are also

CP1165, *Nuclear Structure and Dynamics '09*
edited by M. Milin, T. Nikšić, D. Vretenar, and S. Szilner
© 2009 American Institute of Physics 978-0-7354-0702-2/09/$25.00

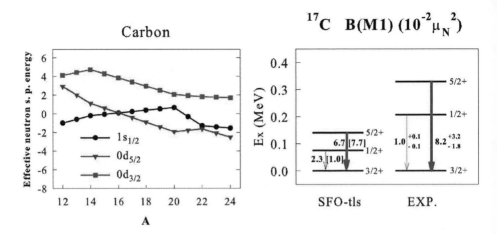

FIGURE 1. (Color online) (a) Effective neutron single particle energies for carbon isotopes obtained by the SFO-tls Hamiltonian (left). (b) Calculated and experimental energy levels and $B(M1)$ values in ^{17}C (right). Numbers in the parentheses [ ] are obtained with the effects of loosely bound nature of ^{17}C. Experimental values are taken from Refs. [8, 9].

replaced by those of the $\sigma+\omega+\rho$ meson exchanges. The repulsive corrections are made for the monopole terms with isospin $T=1$ in the sd-shell, in the $0d_{5/2}$-$1s_{1/2}$ and $0d_{5/2}$-$0d_{3/2}$ orbit pairs [6]. Note that the sd shell part of the two-body interactions in the SFO Hamiltonian are Kuo's G-matrix elements. The effective neutron single particle energies for $0d_{5/2}$, $1s_{1/2}$ and $0d_{3/2}$ orbits are shown in Fig. 1(a). The $1s_{1/2}$ orbit is below the $0d_{5/2}$ orbit for A < 16, while the $0d_{5/2}$ orbit becomes lower than the $1s_{1/2}$ orbit for A > 16. They are almost degenerate around A =16. This behavior is important to reproduce the low-lying energy levels as well as the magnetic dipole (M1) transition strengths in ^{17}C. The $3/2^+$ and $5/2^+$ states made of three $0d_{5/2}$'s come down and the $3/2^+$ state becomes the ground state. The observed $B(M1)$ for the $1/2^+ \rightarrow 3/2^+$ transition is considerably hindered compared to that for the $5/2^+ \rightarrow 3/2^+$ transition [8]. This anomaly can be well described by the SFO-tls as shown in Fig. 1(b). Both the strong tensor components in the p-sd cross shell matrix elements and the repulsive monopole corrections in the $T=1$ channel are important to reproduce the anomaly.

Energies of the ground states of carbon isotopes are shown in Fig. 2(a). The SFO-tls reproduces the experimental values rather well up to A=22. A considerable improvement from SFO is obtained for A>19. When further monopole corrections in the $T=1$ channel due to the three body force induced by the Δ excitations are made in the orbit pairs except for the $0d_{5/2}$-$1s_{1/2}$ and $0d_{5/2}$-$0d_{3/2}$ (denoted as 'SFO-tls+3Nmono'), the energies above A=22 increse and become similar to those of WBT [10] obtained within $0\hbar\omega$ space.

Near degeneracy between the $0d_{5/2}$ and $1s_{1/2}$ orbits around A=16 leads to nearly constant excited energies for the 2^+ states in ^{16}C, ^{18}C and ^{20}C as shown in Fig. 2(b). As for ^{19}C, excitation energies are calculated to be higher compared to experiments. When the halo nature of ^{19}C is taken into account by lowering the single particle energy of the $1s_{1/2}$ orbit, the observed energy levels are well reproduced. When the $1s_{1/2}$ orbit comes

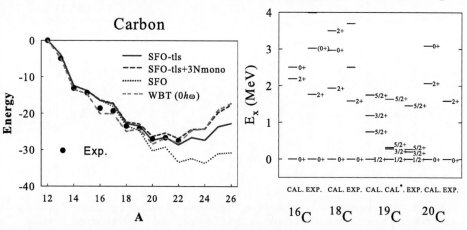

FIGURE 2. (Color online) (a) Ground state energies of carbon isotopes obtained by SFO-tls, SFO-tls+3Nmono (see text), SFO and WBT Hamiltonians. Experimental value [7] are shown by filled circles (left side). (b) Energy levels of carbon isotopes. CAL. and EXP. denote calculated energies obtained by SFO-tls and experimetal ones, respectively. CAL.* include the halo effects in ^{19}C. Experimental values are taken from Refs. [11, 12, 9, 13] (right side) .

close to the $0d_{5/2}$ orbit, two neutrons are excited into the $1s_{1/2}$ orbit and the remaining three $0d_{5/2}$ neutrons can form $3/2^+$ and $5/2^+$ states more easily, resulting in the lowering of these $3/2^+$ and $5/2^+$ states.

GT transitions in ^{19}C are studied by taking into account the halo effects. A retarded GT transition to the $1/2^+$ (2.14 MeV) state in ^{19}N is used to investigate the nature of the $1s_{1/2}$ halo wave function. The observed $\log ft$ value [14] is reproduced for halo wave functions which have overlaps of about 0.7 with harmonic oscillator wave functions. This corresponds to one neutron separation energy around 600 keV, that is consistent with the experimental values obtained from the Coulomb dissociation of ^{19}C [15, 7]. Relatively large GT strength of $\log ft$ = 4.6~4.8 found in the energy region as low as E_x = 6~7 MeV in ^{19}N [14] are found to be rather well explained by SFO-tls.

MONOPOLE CORRECTIONS FROM THREE-BODY FORCES

We now discuss possible origins of the repulsive monopole corrections in the T=1 channel. Three body forces induced by Δ excitations can give rise to repulsive contributions to the T=1 monopole terms [16]. Calculated monopole terms for the sd shell are shown in Fig. 3 (a) for π and $\pi+\rho$ meson exchanges. They are about 100~400 keV depending on the orbits. Those that involve 1s orbit are relatively larger. We stress here that main contributions come from the tensor components of the meson exchange potentials. Their effects become sizable near drip-lines in the energies of carbon and oxygen isotopes as shown in Fig. 2(a) and Fig. 3(b). In case of oxygen isotopes, the $0d_{3/2}$ orbit remains unbound near the drip-line due to the repulsive monopole terms from the three-nucleon

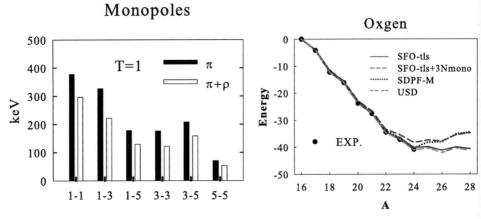

FIGURE 3. (Color online) (a) Monopole terms due to three-nucleon forces induced by Δ excitations by π or $\pi+\rho$ meson exchanges in the T=1 channel. The $1s_{1/2}$, $0d_{3/2}$ and $0d_{5/2}$ orbits are denoted by 1,3 and 5, respectively. (left) (b) Ground states energies of oxygen isotopes obtained by SFO-tls, SFO-tls+3Nmono, SDPF-M and USD Hamiltonians. Experimental values [7] are denoted by filled circles. (right)

forces, and ^{24}O becomes the drip-line nucleus consistent with the observation.

Thus, we have shown the important roles of the tensor interaction induced by π meson exchanges in the shell evolutions of carbon and oxgen isotopes toward the drip-lines. Both the full inclusion of the tensor components in the two-body matrix elements and the repulsive monopole corrections from the tensor components of the π meson exchange in the Fujita-Miyazawa Δ excitation processes are pointed out to be important.

REFERENCES

1. T. Suzuki, R. Fujimoto and T. Otsuka, *Phys. Rev.* **C67**, 044302 (2003).
2. T. Otsuka, T. Suzuki, R. Fujimoto, R. Grawe and Y Akaishi, *Phys. Rev. Lett* **95**, 232502 (2005).
3. T. Suzuki, S. Chiba, T. Yoshida, T. Kajino and T. Otsuka, *Phys. Rev.* **C74**, 034307 (2006).
4. J. W. Holt et al., *Phys. Rev. Lett.* **100**, 062501 (2008).
5. A. Negret et al., *Phys. Rev. Lett.* **97**, 062502 (2006).
6. T. Suzuki and T. Otsuka, *Phys. Rev.* **C78**, 061301(R) (2008).
7. G. Audi, A. H. Wapstra and C. Thibault, *Nucl. Phys.* **A729**, 337 (2003).
8. D. Suzuki et al., *Phys. Lett.* **B666**, 222 (2008).
9. Z. E. Elekes et al., *Phys. Lett.* **B614**, 174 (2005); H. G. Bohlen et al., *Eur. Phys. J* **A31**, 279 (2007).
10. S. E. Warburton and B. A. Brown, *Phys. Rev.* **C46**, 923 (1992).
11. H. J. Ong et al., *Phys. Rev.* **C78**, 014308 (2008).
12. M. Stanoiu et al., *Phys. Rev.* **C78**, 034315 (2008).
13. Y. Satou et al., *Phys. Lett.* **B660**, 320 (2008).
14. T. Onishi, Doctor Thesis (University of Tokyo) (2008); H. Sakurai, private communication.
15. T. Nakamura et al., *Phys. Rev. Lett.* **83**, 1112 (1999).
16. T. Otsuka, T. Suzuki, J. D. Holt, A. Schwenk and Y. Akaishi, to be published; T. Otsuka, in this *Proceedings*.

Monopole Interaction and Single-Particle Energies

A. Umeya[*], G. Kaneko[†] and K. Muto[†]

[*]Nishina Center for Accelerator-Based Science, RIKEN, Wako, Saitama 351-0198, Japan
[†]Department of Physics, Tokyo Institute of Technology, Meguro, Tokyo 152-8551, Japan

Abstract. We derived an expression of single-particle energies, and it is expressed in terms of monopole strengths and number operators. The monopole component of NN interaction, which is defined as the lowest-order terms of multipole expansion, is well renormalized into the single-particle energies.

Keywords: monopole interaction, single-particle energy
PACS: 21.60 Cs, 21.10 Dr, 21.30 Fe

We derived an expression of single-particle (s.p.) energies by employing a shell-model sum rule technique [1], according to the prescription of Baranger [2]. The s.p. energies are expressed as

$$\varepsilon_j = \varepsilon_j^{\text{core}} + \langle \Psi | \sum_{j'} \Delta\varepsilon_{jj'} \widehat{N}_{j'} | \Psi \rangle, \tag{1}$$

where $\varepsilon_j^{\text{core}}$ is the s.p. energy with respect to the core nucleus, and the second term arises from interaction between valence nucleons. The monopole strengths $\Delta\varepsilon_{jj'}$ are defined by

$$\Delta\varepsilon_{jj'} = \frac{\sum_J \left(1 - \delta_{jj'}(-1)^{2j-J}\right)(2J+1) \langle jj'|V|jj'\rangle_J}{(2j+1)(2j'+1)}, \tag{2}$$

and $\widehat{N}_{j'}$ are number operators. The above expression of s.p. energies can be applied to any shell-model eigenstate $|\Psi\rangle$ of any nucleus, and thus allows, for the first time, to discuss quantitatively changes of s.p. energies from a nucleus to another due to the monopole interaction. Some results of successful applications of the present formulation are shown below.

Figure 1 (a) shows the behavior of neutron s.p. energies in $N = 7$ nuclei, calculated in the p-sd shell model with SFO effective interaction [3]. The shell gap at $N = 8$, which is large enough to lead to a good magic number in $N \approx Z$ nuclei, is reduced with the decrease of proton number. In the neutron-rich nucleus ^{11}Be, the ground state has $1/2^+$ and a $1/2^-$ state appears just above it, implying the vanishing of the shell gap [4]. However, there still is a gap of about $\varepsilon(s_{1/2}) - \varepsilon(p_{1/2}) \approx 3$ MeV, and configuration mixing causes fragmentation of the $s_{1/2}$ strength, resulting in the $1/2^+$ state to be the ground state of ^{11}Be. We also showed, by using the spin-tensor decomposition, that the shell evolution in those nuclei is due mainly to attractive proton-neutron interaction in the triplet-even channel.

CP1165, *Nuclear Structure and Dynamics '09*
edited by M. Milin, T. Nikšić, D. Vretenar, and S. Szilner
© 2009 American Institute of Physics 978-0-7354-0702-2/09/$25.00

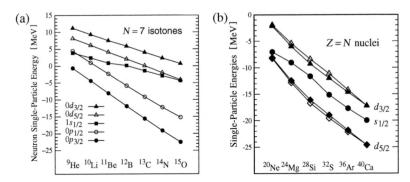

FIGURE 1. (a) Neutron s.p. energies of the p- and sd-shell orbits calculated for $N = 7$ isotones with $Z = 2\text{--}8$. (b) The s.p. energies of $Z = N$ sd-shell nuclei, calculated with (filled) and without (open) tensor component.

Figure 1 (b) shows the s.p. energies of sd-shell nuclei calculated with the USD interaction [5] and the tensor component of monopole has been separated, which is considered to play a role in the shell-structure evolution [6]. It is revealed that the central triplet-even attraction plays the major role in the changes of s.p. energies, and the tensor force reduces the energy gap between the spin-orbit partners.

We have studied the monopole interaction in detail [7]. It is the lowest order term in the multipole expansion of NN interactions. It is expressed by number operators multiplied by the monopole strength,

$$V_{pn}^{(0)} = \sum_{j_p j_n} \Delta \varepsilon_{j_p j_n} \widehat{N}_{j_p} \widehat{N}_{j_n},$$

$$V_{pp/nn}^{(0)} = \sum_{[jj']} \frac{1}{(1+\delta_{jj'})^2} \Delta \varepsilon_{jj'} \widehat{N}_j \widehat{N}_{j'} + \frac{1}{4} \sum_j (2j+1) \Delta \varepsilon_{jj} \widehat{N}_j.$$

We have also shown that the two-body monopole interaction is renormalized, as a very good approximation, to the one-body s.p. energies. The monopole, therefore, dominates the nuclear binding due to NN interactions. This indicates that the above formulation of s.p. energies is consistent with the two-body interaction, and thus a reasonable extension of the Hartree-Fock method to the shell model which uses a number of configurations. On the other hand, however, since the monopole interaction is expressed by number operators, it is diagonal with respect to shell-model configurations, and further binding is gained mainly by the quadrupole component.

REFERENCES

1. A. Umeya and K. Muto, *Phys. Rev.* **C69**, 024306 (2004); **C74**, 034330 (2006).
2. M. Baranger, *Nucl. Phys.* **A149**, 225 (1970).
3. T. Suzuki, R. Fujimoto and T. Otsuka, *Phys. Rev.* **C67**, 044302 (2003).
4. I. Talmi and I. Unna, *Phys. Rev. Lett.* **4**, 469 (1960).
5. B. H. Wildenthal, *Prog. Part. Nucl. Phys.* **11**, 5 (1984).
6. T. Otsuka et al., *Phys. Rev. Lett.* **87**, 082502 (2001); **95**, 232502 (2005).
7. A. Umeya, S. Nagai, G. Kaneko and K. Muto, *Phys. Rev.* **C77**, 034318 (2008).

Low-lying Continuum States in Oxygen Isotopes

K. Tsukiyama[*], T. Otsuka[*,†] and R. Fujimoto[**]

[*]*Department of Physics, University of Tokyo, Hongo, Tokyo, 113-0033, Japan*
[†]*Center for Nuclear Study, University of Tokyo, Hongo, Tokyo, 113-0033, Japan*
RIKEN Nishina Center, Hirosawa, Wako-shi, Saitama 351-0198, Japan
[**] *Energy and Environmental Systems Laboratory, Hitachi, Ltd., Hitachi-shi, Ibaraki, Japan*

Abstract. Low-lying continuum states of exotic oxygen isotopes are studied, by introducing the Continuum-Coupled Shell Model (CCSM) and an interaction for continuum coupling constructed in a close relation to realistic shell-model interaction. Neutron emission spectra from exotic oxygen isotopes are calculated. The results agree with experiment remarkably well, as an evidence that the continuum effects are stronger than ~ 1 MeV.

Keywords: Weakly bound, shell model
PACS: 21.60.Cs,24.10.Cn,24.30.Gd,27.30.+t

Exotic nuclei far from the β-stability line provide us with new interesting features. An important example can be the evolution of shell structure, including the change of magic numbers, due to characteristic properties of the nuclear force, as being increasingly observed. Although there are many bound nuclei between β-stability and drip lines showing the shell evolution [1], one eventually approaches the drip line, by adding more neutrons (or protons). Low-lying states are then in the continuum even if the ground state is still bound. Thus, the physics of continuum and that of shell evolution should meet. We shall study continuum spectra of neutron-rich O isotopes. The neutron drip line is quite different between O and F isotopes, and this can be explained by unbound (bound) neutron $0d_{3/2}$ orbit for O (F) isotopes.

We extend the normal shell model (SM) so as to include continuum states as a part of SM basis. As the $0d_{3/2}$ orbit remains unbound in oxygen isotopes, we generate discretized continuum single-particle basis states of $d_{3/2}$, by introducing an infinite wall on top of the Woods-Saxon potential. The infinite wall is placed at 1000 fm from the center of the nucleus so as to obtain a sufficiently high density of discretized states The final results do not depend on the position of the infinite wall if the wall is far enough.

The inert ^{22}O core is assumed in the following calculations. The CCSM is proposed to treat correlations due to nuclear forces in continuum states in a manner consistent with the SM for bound states. We introduce a simple $\hat{V}(r) = \sum_{i=1,2} g_i(1 + a_i\sigma \cdot \sigma)e^{-r^2/d_i^2}$, as a modeling of a realistic interaction within this small model space. The r is the inter-nucleon distance, and σ implies spin of a nucleon. Here, $d_{1,2}$=1.4, 0.7 fm are prefixed for simplicity. The other parameters are determined so that the low-lying levels calculated by SDPF-M interaction [2] can be reproduced by the CCSM with bound approximation (*i.e.*, the calculation with Harmonic Oscillator) . The CCSM Hamiltonian is then written as

$$H = H_0 + \hat{V} = \sum_j \tilde{\varepsilon}_j n_j + \hat{V}, \tag{1}$$

CP1165, *Nuclear Structure and Dynamics '09*
edited by M. Milin, T. Nikšić, D. Vretenar, and S. Szilner
© 2009 American Institute of Physics 978-0-7354-0702-2/09/$25.00

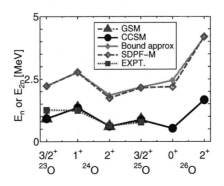

FIGURE 1. (Color online) Energies from the particle-emmision thresholds [MeV].

where $\tilde{\varepsilon}_j$ and n_j are, respectively, the single-particle energy (SPE) and the occupation number operator, and j denotes single-particle states including discretized continuum ones. The spectra of emitted neutrons from knockout reactions are shown in Figure 1. The results of "bound approx." are almost identical to those of "SDPF-M". This is the constraint for the construction of CCSM Hamiltonian. Therefore, the difference between "bound approx." and "CCSM" is the pure effect from the continuum coupling. The CCSM reproduces the experimentally observed levels and their spacing [4]. It should be noted that the CCSM treats properly the asymptotic behavior of the correlated states. We also calculate the low-lying states by Gamow Shell Model (GSM) [5] with the same Hamiltonian Eq. (1) following the prescription [6]. The GSM shows the same results as CCSM (only the real parts of the energies are shown in Figure 1).

The CCSM Hamiltonian is designed to include correlations due to a two-body nuclear force in the continuum, in a manner consistent with the realistic SM interaction for bound states. Spectra of emitted neutrons from knockout reactions are calculated, in good agreements to recent experiments without adjustment to final results. The continuum effects are significant; the difference from bound-state calculation is more than 1 MeV for the cases studied. This suggests that the appropriate treatment of the nuclear force is still important in the continuum. The present work seems to mean that the SM SPE for the bound $0d_{3/2}$ should be really high in O isotopes, and there should be a strong shell evolution mechanism so that this orbit comes down in F isotopes [7]. As the CCSM idea naturally extends the successful SM interaction to continuum states (or open quantum systems), its future developments into full configurations and more applications to exotic nuclei are intriguing issues.

REFERENCES

1. T. Otsuka, T. Suzuki, and Y. Utsuno, *Nucl. Phys.* **A805**, 127c (2008)
2. Y. Utsuno, et al., *Phys. Rev.* **C60**, 054315 (1999); *Phys. Rev.* **C70**, 044307 (2004).
3. C. Hoffman et al., *Phys. Rev. Lett.* **100**, 152502 (2008).
4. C. Hoffman et al., *Phys. Lett.* **B672**, 17 (2009).
5. N. Michel, W. Nazarewicz, M. Płoszajczak, and J. Rotureau, *Phys. Rev.* **C74**, 54305 (2006).
6. G. Hagen, M. H-Jensen, and N. Michel, *Phys. Rev.* **C73**, 64307 (2006).
7. Y. Utsuno, T. Otsuka, T. Mizusaki and M. Honma, *Phys. Rev.* **C64**, 011301 (2001).

Shell Model Description of Negative Parity Intruder States in ^{36}S

M. Bouhelal[*, **], F. Haas[*], E. Caurier[*], and F. Nowacki[*]

[*] IPHC, CNRS/IN2P3, Université de Strasbourg, F-67037 Strasbourg Cedex 2, France
[**] Département des Sciences de la Matière, Université de Tébessa, Tébessa 12002, Algérie

Abstract. A new $1\hbar\omega$ interaction called PSDPF which spans three major shells has been developed to calculate the properties of the intruder negative parity states of sd shell nuclei. PSDPF has been applied to the N=20 neutron rich ^{36}S nucleus and a very good description of the first negative parity states with $J^{\pi}=0^{-}$ to 5^{-} has been obtained at excitation energies below 6 MeV.

Keywords: Shell Model, $1\hbar\omega$ interaction, negative parity states in ^{36}S.
PACS: 21.60.Cs, 21.10.Hw, 21.10.Tg, 27.30.+t

A NEW PSDPF INTERACTION

The structure of sd shell nuclei has been the subject of numerous experimental and theoretical investigations. The normal $0\hbar\omega$ states of these nuclei have positive parity and are well described by the shell model using the USD interaction [1]. In this case, the sd valence space is used and an inert ^{16}O core is assumed. In sd shell nuclei, intruder negative parity states are known throughout the shell but for them no consistent description exists. We have thus developed a new PSDPF interaction where the model space is now composed of the three major shells p, sd, pf and a ^{4}He core. PSDPF is based on existing interactions for the major shells: the $0\hbar\omega$ CK interaction for the p shell [2], the $0\hbar\omega$ USDB interaction for the sd shell [1], the $0\hbar\omega$ SDPF-U interaction for the sd-pf valence space [3]. However for the description of the $1\hbar\omega$ states of interest the cross shell parts of the interaction are essential. They have been adjusted through a fitting procedure involving the negative parity states at the beginning and at the end of the sd shell. More details concerning the construction of the interaction and its application to calculate the properties of negative parity states in N=Z and N=Z+1 sd nuclei throughout the shell can be found in Refs. [4] and [5].

NEGATIVE PARITY STATES IN ^{36}S

In the N=20 neutron rich ^{36}S nucleus, six excited states of established negative parity and at an excitation energy E* below 6 MeV are reported in the NNDC database [6]. We have calculated such states with PSDPF, seven of them were found which have essentially a neutron 1p-1h structure with one jump from sd to pf. A one to one

CP1165, *Nuclear Structure and Dynamics '09*
edited by M. Milin, T. Nikšić, D. Vretenar, and S. Szilner
© 2009 American Institute of Physics 978-0-7354-0702-2/09/$25.00

correspondence can be established between the experimental levels with all energies given in keV at E*= 4193, 5021, 5206, 5573 and the calculated ones at 4293, 5153, 5379, 5397 with J^π= 3⁻,4⁻,5⁻,1⁻, respectively. No experimental partners with established J^π exist for the calculated levels at 5398 and 5519 with J^π= 0⁻ and 2⁻. We thus propose that the state reported in NNDC at 5338 corresponds to the calculated 0⁻ at 5398. This state decays by a 816 keV γ transition which has been, in our opinion, erroneously attributed in the past to a level reported in NNDC at 5391 [7, 8]. Based on our shell model calculation we propose that there is a doublet of states in the ~ 5.35 MeV excitation region: one at 5378 ± 3 with J^π= 2⁺ which decays mainly to the ground state and one at 5339 ± 1 with J^π = 0⁻ which decays mainly through an E1 transition to the J^π = 1⁺ state at 4523. The calculated lifetime of the 0⁻ state is 43.5 ps, the corresponding value reported in NNDC is $\tau > 0.3$ ps.

All six J^π = 0⁻ to 5⁻ states have thus been identified except the 2⁻ state predicted at 5519. We propose that it corresponds to the state at 5509 with J= 2, 3 or 4 in NNDC, this assignment is not only based on the excitation energy of the level but also on its experimental γ-decay in agreement with our decay calculations.

The Yrast negative parity states at 4193 (3⁻), 5021 (4⁻) and 5206 (5⁻) are strongly fed in deep inelastic reactions induced by a ³⁶S beam on a heavy target [9]. It is planned to measure the lifetimes of these states with the Doppler recoil distance method. The calculated lifetimes for the 3⁻, 4⁻ and 5⁻ states are 0.25, 1.3 and 34.1 ps, respectively.

In conclusion, our new PSDPF interaction gives a very good 1p-1h description of the ³⁶S negative parity states with J^π = 0⁻ to 5⁻ at an excitation energy below 6 MeV. The next step will be to compare experimental and calculated electromagnetic transition strengths which will be an additional but probably more stringent test of the interaction.

REFERENCES

1. B. A. Brown and W.A. Richter, *Phys Rev.* **C 74**, 034315 (2006).
2. S. Cohen and D. Kurath, *Nucl. Phys.* **73**,1 (1965); *Nucl. Phys.* **A 101**, 1 (1967).
3. F. Nowacki and A. Poves, *Phys Rev.* **C 79** , 014310 (2009).
4. M. Bouhelal et al., *Acta Phys. Pol.* **B 40**, 639 (2009).
5. M. Bouhelal et al., *Eur. Phys. J.* **A 40** (2009) in press.
6. http://www.nndc.bnl.gov
7. A. Hogenbirk et al., *Nucl Phys.* **A 516**, 205 (1990).
8. E. A. Samworth and J. W. Olness, *Phys. Rev.* **C 5**, 1238 (1973).
9. X. Liang et al., *Phys. Rev.* **C 66**, 014302 (2002).

NUCLEI FAR FROM STABILITY

Discovery of the Most Neutron-rich Nuclei With $12 < Z < 25$ and Future Prospects With the FRIB project

D.J. Morrissey

*National Superconducting Cyclotron Laboratory
and Dept. of Chemistry, Michigan State University, East Lansing, MI, USA 48824*

Abstract. The production of the most neutron-rich nuclei by the fragmentation of ^{48}Ca and ^{76}Ge beams at Michigan State are presented. The cross sections were measured for a large range of nuclei including fifteen new isotopes that are the most neutron-rich nuclides of magnesium, aluminum, silicon, and the elements from chlorine to manganese. The observation of ^{42}Al was itself surprising. The cross sections of several new nuclei are enhanced relative to a simple thermal evaporation framework, previously shown to describe similar production cross sections, indicates that precursor excited nuclei in the region around ^{62}Ti that decay to the observed nuclei may be more stable than predicted by current mass models. This may be evidence for a new island of inversion similar to that centered on ^{31}Na. It was recently announced that Michigan State will be the site of the next generation radioactive beam facility, FRIB, for the United States. A brief overview of the proposed facility are presented.

Keywords: projectile fragmentation dripline
PACS: 27.50.+e,25.70.Mn

INTRODUCTION

One of the important challenges facing nuclear physics is to push the study of neutron-rich isotopes to higher atomic number. The benchmark in this exploration is to find the maximum number of neutrons that can be bound for each atomic number. Often the delineation of the heavy limit of stability, called the neutron drip-line, itself has yielded surprises. Some years ago the heaviest bound oxygen isotope was shown to be ^{24}O while many theories predicted that it should be the much heavier doubly-magic ^{28}O. Recently we found that heavy isotopes of aluminum are likely more bound than predicted [6]. Continuing this work, we report here the next step in this challenging exploration in the region of calcium.

The neutron drip line is only confirmed by experiment at present up to $Z = 8$ (^{24}O$_{16}$) through years of work at various projectile fragmentation facilities. The neutron drip line has been found to rapidly shift to higher neutron numbers at $Z = 9$, i.e., ^{31}F$_{22}$ has been observed several times, see e.g. [6]. The nuclide ^{30}F$_{21}$ has been shown not to exist and ^{32}F$_{23}$ is thought to be unbound based on systematics while the particle stability of ^{33}F$_{24}$ is an open question. The shift is predicted to continue at higher masses [7, 8, 9] and makes the search for the neutron drip line in this region even more challenging. The fragmentation of ^{48}Ca$_{28}$ projectiles has produced a number of heavier nuclei in this region including ^{40}Mg$_{28}$ and ^{42}Al$_{29}$, but no clear limit has been established yet. [10] On

CP1165, *Nuclear Structure and Dynamics '09*
edited by M. Milin, T. Nikšić, D. Vretenar, and S. Szilner
© 2009 American Institute of Physics 978-0-7354-0702-2/09/$25.00

the other hand, all nuclei up to $Z = 12$ with $A = 3Z + 3$ have been shown to be unbound. In the short term, the fragmentation of heavier stable beams such as ^{76}Ge in which ^{52}Ar was observed [12] is necessary in order to go beyond the previous work.

In the longer term, the production of the most neutron-rich nuclei will be an important component of the broad scientific programs at the next generation facilities such as **FAIR** in Germany, **RIBF** in Japan, and the recently announced Facility for Rare-Ion Beams, **FRIB**, in the United States. Michigan State University was recently selected by the US Department of Energy to be the site of the FRIB. The proposed facility will provide primary beams that will be up to 400 kW in power, compared to approximately 4 kW at the present National Superconducting Cyclotron Laboratory (NSCL) facility. In addition, all stable beams will be available at this high power level and a new higher acceptance separator to provide significantly higher secondary beam rates. An overview of some of these features are given below.

PRODUCTION OF NEUTRON-RICH NUCLEI

In the present work a primary beam of ^{76}Ge was fragmented and a search for new neutron-rich isotopes above ^{40}Mg was carried out using the recently developed tandem fragment separator technique [6]. A 132 MeV/u ^{76}Ge primary beam from the NSCL was used to irradiate a series of ^{9}Be targets and finally a tungsten target located at the target position of the A1900 fragment separator [14]. The average beam intensity for the measurements of the most exotic fragments was 32 pnA. The A1900 fragment separator was combined with the S800 analysis beam line to form a two-stage separator system as described in Ref. [6]. Such a two-stage separator provides a high degree of rejection of unwanted reaction products and allows the identification of each fragment of interest. During the search for the most exotic fragments, a Kapton wedge (20.2 mg/cm^2) was used at the center of the A1900 to reject less exotic fragments at the A1900 focal plane by an 8 mm aperture. The transmitted fragments were identified by event-by-event momentum analysis and particle identification. The momentum acceptance of the A1900 was set to $\Delta p/p = \pm 0.05\%$, $\pm 0.5\%$, $\pm 1\%$ and $\pm 2.5\%$ as the production rate of the increasingly exotic nuclei decreased, always with an angular acceptance of 8.2 msr.

The particles stopped in a stack of eight silicon PIN diodes (50×50 mm^2) with a total thickness of 8.0 mm that provided multiple measurements of the energy-loss and thus made redundant determinations of the nuclear charge of each fragment along with the total kinetic energy. The time of flight (TOF) of each particle that reached the detector stack was measured in four ways: (1) over the 46.04 m flight path between a thin plastic scintillator located at A1900 focal plane and the second PIN detector, (2) over the 20.97 m flight path between another thin plastic scintillator at the object point of the S800 analysis beam line and the third PIN detector, (3) over the entire 81.51 m flight path by measuring the arrival time relative to the phase of the cyclotron rf-signal and the third PIN detector, and (4) over the 25.07 m path between the scintillators at the object point and the A1900 focal plane. This simultaneous measurement of multiple ΔE signals, the magnetic rigidity, the total energy, and the TOF's of each particle provided an unambiguous identification of the atomic number, charge state and mass number of each isotope. The detection system and particle identification was calibrated with

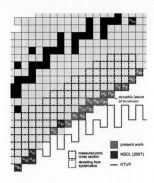

FIGURE 1. The region of the chart of nuclides under investigation. The solid line is the limit of bound nuclei from the KTUY mass model [9]. Nuclei in the green squares were recently discovered [10, 6, 12], those in red squares are new in this work, and some of the cross sections for those in dashed boxes are shown in Fig. 2. The center of the new proposed island of inversion at ^{62}Ti [13] is highlighted.

the primary beam and by the locations of gaps corresponding to unbound nuclei in the particle identification spectrum. The location of the observed fragments is shown in the chart of nuclides in Fig.1.

Tarasov et al. [10] have shown that the cross sections for projectile fragments in this region have an exponential dependance on Q_g (the difference in mass-excess of the beam particle and the observed fragment that is independent of the target in contrast to the older Q_{gg} four-body analysis applied to low energy reactions) and deviations from the predicted yield may be used to identify anomalies in the mass surface such as the new island of inversion near ^{62}Ti predicted by Brown [13] similar to the original island of inversion near ^{31}Na.

Figure 2 shows that the Q_g function using masses from from Ref. [9] and Ref. [19] which represent other mass models that give similar results as the differences among mass models are small on this scale. The figure showns that the logarithm of the cross sections for each isotopic chain falling on an approximately straight line except for the heaviest isotopes. On the other hand, most of isotopic chains with $15 \leq Z \leq 24$ can be fit with a single exponential slope of $1/1.8$ MeV [11]. The heaviest members of the isotopic chains with $Z=19$, 20, 21 and 22 break away from the uniform fit and the heaviest four or five isotopes have a shallower slope or enhanced cross sections. Recall that the masses of these most neutron-rich nuclei are not known but rather represent extrapolations. The observed increase in cross section might indicate that the excited precursors that give rise to the observed nuclei are more bound (i.e., less negative Q_g) than current mass models predict. One reason for a stronger binding can be deformation. In a shell-model framework, the wave functions of the ground and low-lying excited states of nuclei in the new island of inversion around ^{62}Ti would be dominated by neutron particle-hole intruder excitations across the $N = 40$ sub-shell gap, leading to deformation and shape coexistence.

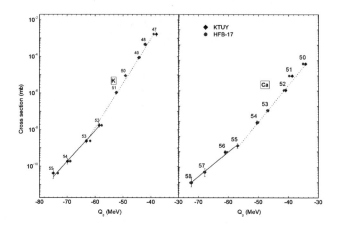

FIGURE 2. The cross sections for the potassium and calcium isotopes are shown as a function of Q_g using the [9] and [19] mass models.

FACILITY FOR RARE-ION BEAMS

The Department of Energy announced that Michigan State University has been selected to design and establish the Facility for Rare Isotope Beams (FRIB), a cutting-edge research facility to advance understanding of rare nuclear isotopes and the evolution of the cosmos. The new facilityÑexpected to take about a decade to design and build, and to cost an estimated \$550 millionÑwill provide a broad range of research opportunities for an international community. The facility is basd on a superconducting-RF driver linear accelerator that will provide a maximum beam power of 400 kW for all beams ranging from uranium accelerated to 200 MeV/nucleon, to lighter ions with increasing energy, down to protons at 600 MeV/nucleon. In addition, space will be provided in the linac tunnel and shielding in the production area to allow a future upgrade of the driver linac energy to 400 MeV/u for uranium and 1000 MeV for protons without significant interruption of the future science program. Space will also be provided in the production target area for other potential upgrades such as ISOL or even a second fragment separator. A schematic layout of the proposed FRIB facility and its connection to the existing NSCL facility is shown in Fig. 3. The heart of the FRIB facility will be a next-generation three-stage projectile fragment separator specifically designed to handle the very intense primary and secondary beams. This layout has several important advantages, for example it allows the continued operation of the Coupled Cyclotron Facility (CCF) using fast, stopped and reaccelerated beams, it allows a relatively rapid connection of the new accelerator complex to the existing facility, and reuse of the experimental equipment.

The opportunity for a pre-FRIB science program using the existing in-flight separated beams from the existing CCF and the new ReA3 reaccelerator presently under construction. Users will be able to mount and test equipment and techniques and push nuclear

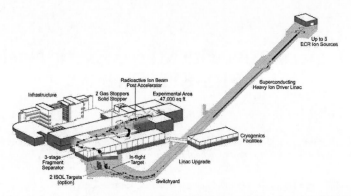

FIGURE 3. A schematic view of the major components of the FRIB facility in relation to the existing NSCL facility.

science forward with a variety of thermalized and reaccelerated fragmentation beams so that the science program will seamlessly continue on when FRIB is complete. This transition will allow for a continually evolving science program during the time FRIB is under construction that will form the basis of the research program at FRIB.

It is important to note that the work reported here was carried out by a large group of people at the National Superconducting Cyclotron Laboratory. The reader is referred to the author list in Ref.[11] for the names of the participants in the projectile fragmentation study. The FRIB project is made up from an even larger group headed by C.K. Gelbke (director), T. Glasmacher (project manager), with G. Bollen, B.M. Sherrill, and R.C. York.

REFERENCES

1. I. Tanihata, *J. Phys.* G **22**, 157 (1996).
2. T. Otsuka, T. Matsuo, and D. Abe, *Phys. Rev. Lett.* **97**, 162501 (2006).
3. C. Thibault, et al., *Phys. Rev.* C **75**, 644 (1975).
4. X. Campi, et al., *Nucl. Phys.* A **251**, 193 (1975).
5. E.K. Warburton, J.A. Becker, and B.A. Brown, *Phys. Rev.* C **41**, 1147 (1990).
6. T. Baumann, et al., *Nature* **449**, 1022 (2007).
7. P.-H. Heenen, *Nature* **449**, 992 (2007).
8. M. Samyn, S. Goriely, M. Bender, and J.M. Pearson, *Phys. Rev.* C **70**, 044309 (2004).
9. H. Koura, et al., *Prog. Theo. Phys.* **113**, 305 (2005).
10. O.B. Tarasov, et al., *Phys. Rev.* C **75**, 064613 (2007).
11. O.B. Tarasov, et al., *Phys. Rev. Lett.* **102**, 142501 (2009).
12. P.F. Mantica, et al., *Bull. Am. Phys. Soc.* **53**, 64 (2008).
13. B.A. Brown, *Prog. Part. Nucl. Phys.* **47**, 517 (2001).
14. D.J. Morrissey, et al., *Nucl. Instrum. Meth. Phys. Res.* B **204**, 90 (2003).
15. M. Lewitowicz, et al., *Z. Phys.* A **335**, 117 (1990).
16. R.J. Charity, *Phys. Rev.* C **58**, 1073 (1998).
17. K. Sümmerer and B. Blank, *Phys. Rev.* C **61**, 034607 (2000).
18. O.B. Tarasov, et al., *Phys. Rev.* C , accepted (2009).
19. S. Goriely, N. Chamel, and J. M. Pearson, *Phys. Rev. Lett.* **102**, 152503 (2009).

New Experiments with Stored Exotic Nuclei at the FRS-ESR Facility

H. Geissel

GSI Helmholtzzentrum für Schwerionenforschung,Planckstraße 1, 64291 Darmstadt, Germany and Justus-Liebig Universität Gießen, Heinrich-Buff-Ring 16, 35392 Gießen

Abstract. High accuracy mass and novel nuclear lifetime measurements have been performed with bare and few-electron ions produced via projectile fragmentation and fission, separated in flight and stored at relativistic energies. Characteristic experimental results and new developments are reviewed. A new generation of studies with exotic nuclei will be possible with the advent of the proposed international Facility for Antiproton and Ion Research (FAIR).

Keywords: Masses, Lifetimes, Cooled Exotic Nuclei
PACS: 21.10.-k, 21.10.Tg, 29.38.Gj

INTRODUCTIONS

A new experimental dimension has been opened up at GSI with the combination of in-flight separators with high-resolution systems like storage rings [1], traps [2] and new gamma-ray [3] and particle detector arrays. Here, we will report on new experimental developments and results obtained with the in-flight separator FRS [4] and the storage-cooler ring ESR [5]. Exotic nuclei are produced via projectile fragmentation and fission and separated in flight at about 70% light velocity before they are injected in the storage-cooler ring. At these high velocities the nuclear reaction products emerge from the target as bare and few-electron ions. The opportunity of these high ionic charge states is that the in-flight separation can be very efficient and pure. Furthermore, for the first time nuclear decay and reaction studies can be performed in the laboratory under conditions which prevail in hot stellar matter. New nuclear decay properties have been observed for bare and highly-ionized atoms. Furthermore, decay channels which occur in neutral atoms can be suppressed. In the recent years we have steadily refined the accuracy and sensitivity of our experimental methods such that along with time-resolved Schottky Mass Spectrometry new neutron-rich isotopes and isomers have been discovered in the element range of Tl to U.

The intensity of the stored exotic nuclear beams at GSI can be substantially improved by roughly a factor of 10^4 with the planned new accelerator system coupled to a new high-acceptance fragment separator followed by a dedicated collector and cooler ring system [6]. The main features of this next-generation facility will also be presented in this contribution.

CP1165, *Nuclear Structure and Dynamics '09*
edited by M. Milin, T. Nikšić, D. Vretenar, and S. Szilner
© 2009 American Institute of Physics 978-0-7354-0702-2/09/$25.00

Schottky Mass Spectrometry (SMS)

The formation of projectile fragments in nuclear collisions cause an inevitable velocity spread on the order of several percent depending on the mass and charge difference of the selected projectile and fragment which would aggravate precision measurements in flight. This disadvantageous property can be elegantly eliminated by applying novel experimental methods like high-resolution spectrometers operated in special ion-optical modes or storage-cooler devices. In the storage-cooler ring ESR the large phase space is reduced by electron cooling which enforces the stored ions to an identical mean velocity and reduces the velocity spread to roughly $3 \cdot 10^{-7}$ at low intensities. This is the basis for Schottky Mass Spectroscopy (SMS) [7, 8]. Presently, the electron cooling takes (5-10) s depending on the initial velocity spread. Employing stochastic precooling [9] we have extended the access to fragments down to half-lives of about (1-2) s.

With the measured time correlation the isotope peaks in the Schottky spectra are traced down to a single stored ion. In this way ground or isomeric states can be assigned even for very small excitation energies which cannot be resolved when both states are simultaneously present (Single-Particle Method) [8]. With time-resolved SMS we have achieved an improved mass accuracy of 30 keV and a resolution of $2 \cdot 10^6$ [8, 10]. As many as 285 new and more than 300 improved mass values of neutron-deficient isotopes in the range $36 \leq Z \leq 85$ have been contributed to the present knowledge of the mass surface by SMS. The research potential of SMS has been extended in our recent experiments with neutron-rich uranium fragments. In this run the power of time-resolved SMS has been demonstrated with the discovery of 5 new neutron-rich isotopes in the element range between Tl and U, i.e., along with the unambiguous identification their mass and lifetimes have been measured [14, 15]. In the same run also 5 new relatively long-lived isomers have been observed for the first time.

SMS is the most effective method for accurate large scale mass measurements. Unknown and reference nuclei are all included in one single Schottky spectrum. This condition has many advantages, especially with respect to possible systematic errors. Comparisons of SMS data with theoretical models over a large part of the mass surface have been often presented and illustrate clearly local and global needs for model improvements [14, 11]. However, it might also be of interest that only a small range of isotopes is investigated by SMS, then the accuracy can be boosted to the 10^{-8} range, completely dominated by the reference nuclides in the narrow window [14].

What will be the next experimental developments for SMS? A major goal is to access with SMS more exotic nuclei with shorter half-lives. Stochastic precooling certainly would bring us close to 1 s half-lives but this cooling process significantly narrows our isotope range in the recorded spectra, i.e., with this precooling mode we have to scan differently as we did with pure electron cooling to cover a large mass surface. In this case we have to match with the selected isotope range the fixed operating velocity of about 400 MeV/u.

Another direction of new developments is to make the Schottky probes more sensitive. Reducing the thermal and picked-up noise of the present probes with improved electronics and cooled plates could be a next approach. Another development is a resonant Schottky probe which was successfully employed with antiprotons at CERN [12].

Studying nuclei very far from the valley of stability with SMS will eventually suffer from the small number of adequate reference nuclei in the spectrum. A possible solution of this problem could be to produce with different primary beams from SIS a set of quite different fragments and different atomic charge-state combinations for additional injection. In such measurements the Bρ of the ESR is fixed throughout the experiment and from pulse to pulse the source of fragment beams can be changed. Already the different atomic charge states of one fragment setting can provide calibration points close to bare unknown neutron-rich nuclei., i.e. fragments with several electrons attached will come close to the unknown bare neutron-rich nuclei in the spectrum. In addition to this hardware consideration our established correlation-network evaluation [8] and the grid of "own" reference masses will be powerful tools as well.

Isochronous Mass Spectrometry (IMS)

The duration of the cooling process is a limitation for the access of very short-lived stored nuclei. This restriction can be circumvented by tuning the storage ring to its transition energy (isochronous mode). In this special ion-optical mode the revolution time is independent of the velocity spread because the faster ions of each isotope are guided on longer trajectories to preserve a constant revolution time [13]. The isochronicity condition is strictly fulfilled only for one mass-to-charge (m/q) ratio in a narrow Bρ range. However, recording an isochronous time-of-flight (ToF) spectrum for uranium fission fragments represents an accepted mass-to-charge ratio of more than 10%. Therefore, the high resolution of the FRS is used to determine the Bρ of the injected fragments within $1.5 \cdot 10^{-4}$ at the second dispersive focal plane via special slits. The mass resolution achieved with Bρ tagging is about $5 \cdot 10^5$ (σ value) which significantly improves the identification of isomers compared to our previous IMS measurements. Indeed, the accuracy over the full m/q range has been improved by a factor of more than 2. This pilot experiment with ToF measurement in the isochronous ESR and additional Bρ determination in the FRS has clearly demonstrated the rich potential of IMS.

SMS involves only the stability of the electro-magnetic fields in the ESR including the performance of the electron cooler whereas for the IMS also the conditions from SIS and the FRS can cause drifts in a long-time run. Of course the most abundant fragments in the spectra can serve as an indicator for off-line drift corrections. However, in regions of the most exotic nuclei such "pulser lines" will not be frequently enough available, thus another calibration-stabilization method has to be employed. A solution could be to take advantage of the universal technical performance of the SIS and its control system. SIS can deliver different projectile beams with different energies and intensities in each cycle [20]. In this way one is very flexible and can find suitable fragments or primary beams matching the selected Bρ range of the FRS and ESR.

Lifetime Measurements of Stored Fragments

Storage and cooling of exotic nuclei at relativistic energies present a unique opportunity to study nuclear decay properties as a function of the number of bound atomic

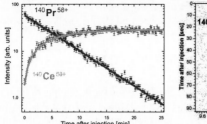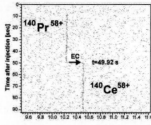

FIGURE 1. Lifetime studies applying Schottky analysis with many circulating ions (left panel) and with single particle analysis (right panel) [17, 18, 27]. The single particle measurements represent basically a counting run. In the later method the correlated traces of mother and daughter ions are easy to recognize.

electrons. Under these conditions decay channels can be blocked, or on the contrary, new decay modes can be opened up with respect to neutral atoms [21]. These basic nuclear physics studies have also great astrophysical relevance because in the stellar environment of nucleosynthesis the temperatures can reach 100 keV and more and thus cause a high degree of ionization for the ions in the plasma.

Lifetime measurements in the storage ring can be performed with different experimental methods depending on the magnetic rigidity difference $\Delta B\rho$ of the mother and daughter nuclei. For small $\Delta B\rho$, below 2.5 %, both species stay on closed storage orbits and thus can be detected with the tools described in the sections for mass measurements. In this case, it is especially attractive to perform high-resolution Schottky analysis and to observe the mass-resolved intensities of the circulating ions. In the Schottky spectra the mass and charge of the stored ions are unambiguously defined. Furthermore, the mass difference represents directly the Q-value of the decay. When $\Delta B\rho$ exceeds the acceptance of the storage ring the daughter nuclei will leave the closed orbits and will be detected in particle counters equipped with full identification capacity. Both methods have been employed already in our pioneering experiments [1, 22, 23, 24, 25].

Meanwhile, we have observed that the time-correlated intensity measurements via monitoring the area of a Schottky peak representing many ions is rather complex and can cause systematic errors. Therefore, we have recently introduced a solution which employs single-particle decay measurements of stored and cooled ions, i.e. one counts simultaneously single mother and daughter ions in their decay and appearance, respectively [16]. Many-particle analysis and single-particle Schottky analysis have been applied in recent electron capture decay studies [17, 18].

In our pioneer experiments [23, 29] we have demonstrated the new research potential for lifetime studies of bare and few-electron fragments. Recently, the β^- decay of bare Tl ions have been measured, where the ratio of bound-state beta decay to the "common" emission into the continuum have been studied [26, 25]. In the course of experimental half-live studies of neutron-deficient ions we have measured decay and appearance curves and the decay branching ratio of bare, H-like and He-like ^{140}Pr and ^{142}Pm ions [27]. The focus in this experimental series has been the nuclear electron-capture study. In both cases the EC rate of the H-like ions is by about a factor of 3/2 larger than the He-like system under the same condition. The observation can be explained by the hyperfine interaction and the conservation of angular momenta [19, 14, 17].

FUTURE PERSPECTIVES

The present experimental program at the SIS-FRS-ESR facilities has been quite success-ful and has led to several basic discoveries. An extended discovery potential for future long-range research programs will be presented with the future international Facility for Antiprotons and Ion Research FAIR [6].

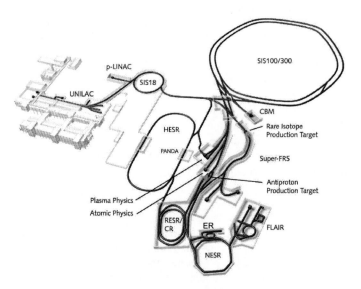

FIGURE 2. Layout of the planned FAIR facility. The present GSI accelerator system and an additional proton linear accelerator serves as an injector system for FAIR [6].The present multi-tasking storage-cooler ring ESR will be replaced by dedicated storage rings. The hot fragment beams will be efficiently collected by the so-called Collector Ring (CR). The CR is equipped with stochastic cooling and can also be operated in the isochronous mode for mass measurements of short-lived nuclei. For nuclei with lifetimes longer than 1 s the fragments can be precooled and transferred to the NESR where electron cooling can achieve the most brilliant fragment beams for mass and lifetime measurements or for nuclear reaction studies with the internal target. A new tool for studies of exotic nuclei via electron scattering will be the Electron-Collider Ring (ER). The RESR is a transfer and a slowing-down ring to provide the optimum energies for reactions with fragment beams in the NESR.

A new double-ring synchrotron system (100/300 Tm) will accelerate ions up to uranium with intensities of 10^{12} /s. The beam of stable isotopes will be converted to rare isotopes with a large-acceptance SUPERconducting FRagment Separator (Super-FRS) [31] which will efficiently handle also the large phase space of the fission fragments. A dedicated storage-ring system [32] will collect, store and cool the fragment beams with minor beam losses. These new facilities will allow us to substantially extend the research of nuclear structure physics and also the major set of astrophysical relevant nuclei in the nucleosynthesis paths can be studied for the first time. The ILIMA project

(Isomeric beams, LIfetimes, and MAsses), the continuation of the present mass and lifetime program, will explore the exotic nuclei at the outskirts of the chart of nuclei including the astrophysical nucleosynthesis paths. The present layout of the storage-cooler ring scenario of FAIR is illustrated in Fig. 2.

ACKNOWLEDGMENTS

The work reported here was carried out by a large collaboration at GSI, see e.g. the author list in Ref. [17]. Special acknowledgement is attributed to the PhD students involved.

REFERENCES

1. H. Geissel, et al., *Phys. Rev. Lett.* **68**, 3412 (1992).
2. M. Block, et al., *Eur. Phys. J. A* **25 Sup. 1**, 49 (2005).
3. H.J. Wollersheim, et al., *Nucl. Instr. and Meth. A* **537**, 637 (2005).
4. H. Geissel, et al., *Nucl. Instr. and Meth. B* **70**, 286 (1992).
5. B. Franzke, *Nucl. Instr. and Meth. B* **24/25**, 18 (1987).
6. An International Accelerator Facility for Beams of Ions and Antiprotons, Conceptual Design Report, GSI (2001), *http://www.gsi.de/GSI-Future/cdr/*.
7. T. Radon, et al., *Nucl. Phys. A* **677**, 75 (2000).
8. Yu.A.Litvinov, et al., *Nucl. Phys. A* **756**, 3 (2005).
9. F. Nolden, et al., *Nucl. Instr. and Meth. A* **441** 219, (2000).
10. H.Geissel, Yu.A. Litvinov, *J. Phys. G* **31**, S1779 (2005).
11. F. Bosch et al., *Int. J. Mass Spectrometry* **251**, 212 (2006).
12. F. Kaspers priv. communication
13. H. Wollnik, et al., SIS Experiment Proposal 1987 and M. Hausmann, et al., *Nucl. Instr. and Meth. A* **446**, 569 (2000).
14. H. Geissel, et al., *Eur. Phys. J. Special Topics* **150**, 109 (2007).
15. L. Chen, Doctoral Thesis, Universität Giessen 2008.
16. H.Geissel, et al., AIP Conference Proceedings Vol **831** (2006) 108, New York, Edts. S. Harissopulos, P. Demetriou, R. Julin
17. Yu. A. Litvinov, et al., *Phys. Rev Lett.* **99**, 262501 (2007).
18. Yu. A. Litvinov, et al., *Phys. Lett.* **664B**, 162 (2008).
19. Z. Patyk, et al., *Phys. Rev. C* **77**, 014306 (2008).
20. K. Blasche, B. Franczak, in Proc.: 3rd Eur. Part. Acc. Conf., Berlin, (1992), 9.
21. K. Takahashi, et al., *Phys. Rev. C* **36**, 1522 (1987).
22. M. Jung, et al., *Phys. Rev. Lett.* **69**, 2164 (1992).
23. H. Irnich, et al., *Phys. Rev. Lett.* **75**, 4182 (1995).
24. F. Bosch, et al., *Phys. Rev. Lett.* **77**, 5190 (1996).
25. T. Ohtsubo, et al., *Phys. Rev. Lett.* **95**, 052501 (2005).
26. D. Boutin, Doctoral Thesis, Universität Giessen 2005.
27. N. Winckler, Doctoral Thesis, Universität Giessen in preparation.
28. Yu.A. Litvinov, et al., *Nucl. Phys. A* **734**, 473 (2004).
29. Yu.A. Litvinov, et al., *Phys. Lett.* **573B**, 80 (2003).
30. Yu.A. Litvinov, et al., *Phys. Rev. Lett.* **95**, 042501 (2005).
31. H. Geissel, et al., *Nucl. Instr. and Meth. B* **204**, 71 (2003).
32. P. Beller et al., in Proc.: 9th Eur. Part. Acc. Conf., Lucerne, Switzerland, 2004, 1174.

In Beam γ-ray Spectroscopy at RI Beam Factory

Nori Aoi

RIKEN Nishina Center , 2-1 Hirosawa Wako Saitama 351-0198, Japan

Abstract. A variety γ-ray spectroscopy experiments are successfully performed at RIKEN. From the recent topics of the experiments performed at RIPS, the study of the the collectivity in the neutron-rich Cr isotopes are reported. The experiments recently performed at the new-generation facility RIBF are also introduced.

Keywords: Unstable nuclei, proton inelastic scattering, RI beam facility
PACS: 21.10.-k, 27.50.+e, 25.60.-t, 25.40.Ep

INTRODUCTION

Experiments using γ-ray spectroscopy technique have accomplished a variety of fruitful outcome in the structure study of unstable nuclei using fast RI beams. Systematic studies on the low-lying states have revealed various peculiar structures in the high isospin nuclei, such as rearrangement of shell structure [1, 2]. At RIKEN, these studies have been thus far performed mostly in the region with the mass number smaller than 60 at the present RI beam facility, RIPS [3]. This research field is now about to be extended toward heavier and/or more neutron-rich region due to the advent of the new-generation RI beam facility RI Beam Factory (RIBF) [4]. In this paper, we report on the recent result from the experiments at RIPS and the news from the DayOne experiments recently performed at RIBF.

DEFORMATION IN THE NEUTRON RICH CR ISOTOPES

Neutron rich pf-shell nuclei are of particular interest because of the rearrangement of shell structure in this region. The emergence of a new magic number $N=32$ in the neutron rich region was established by the observation of the high excitation energies of 2^+ states ($E_x(2^+)$) [5–9] and the small B(E2) values [10, 11] in ^{56}Cr, ^{54}Ti, and ^{52}Ca. In more neutron rich region, development of deformation is suggested in the Cr isotopes near $N=40$ [12].

In the Cr isotopes beyond $N=32$, $E_x(2^+)$ decreases monotonically with neutron number until $N=38$ (^{62}Cr) [12–14]. The lowering of the 2^+ state was interpreted as the indication of deformation due to the increasing occupancy of the $g_{9/2}$ neutrons. In order to further investigate the nature of collectivity in this region, we have studied inverse-kinematics proton inelastic scatterings on ^{60}Cr and ^{62}Cr with the technique of in-beam γ-spectroscopy [15].

The experiment was performed at RI Beam Factory operated by RIKEN Nishina Center and Center for Nuclear Study, University of Tokyo. The cocktail beam including

CP1165, *Nuclear Structure and Dynamics '09*
edited by M. Milin, T. Nikšić, D. Vretenar, and S. Szilner
© 2009 American Institute of Physics 978-0-7354-0702-2/09/$25.00

^{60}Cr and ^{62}Cr was produced using the RIKEN projectile-fragment separator (RIPS) [3] from the fragmentation reaction of a ^{70}Zn primary beam and bombarded the liquid hydrogen cell of the "Cryogenic proton and alpha target system" (CRYPTA) [16]. The levels populated by the secondary reactions were identified by measuring the de-

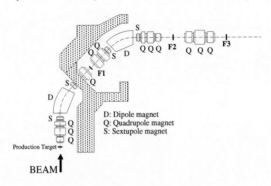

FIGURE 1. RIKEN Projectile-fragment Separator, RIPS.

excitation γ rays using the NaI(Tl) scintillator array DALI2 [17]. The scattered particles were identified using the "TOF mass analyzer for RI beam experiments" (TOMBEE) (Fig. 2) by measuring their time-of-flight (TOF), energy loss (ΔE) and total kinetic energy (E).

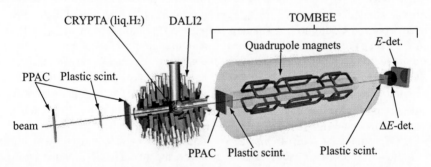

FIGURE 2. Schematic view of the TOF spectrometer.

The Doppler-shift corrected γ-ray spectra obtained in coincidence with inelastically scattered ^{60}Cr and ^{62}Cr are shown in Fig. 3 (a) and (b). The known transitions from the 2^+ states to the ground states [12–14] are clearly seen at 645(4) keV and 449(4) keV for ^{60}Cr and ^{62}Cr, respectively. In the γ spectra gated on the $2^+ \rightarrow 0^+$ peaks [Fig. 3 (c) and (d)], weak peaks are identified at 819(12) keV and 734(10) keV for ^{60}Cr and ^{62}Cr, respectively. The 819 keV peak for ^{62}Cr corresponds to the transition from the 4^+ state at 1462 keV to the 2^+ state observed in the previous work [14]. The 734 keV peak in the ^{62}Cr spectrum was observed for the first time in this work and is assigned to be the γ-rays corresponding to the transition from a level at 1180(10) keV to the 2^+ state. The

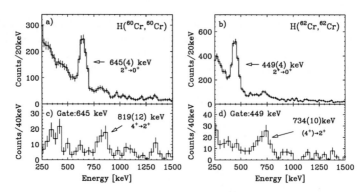

FIGURE 3. Doppler-shift corrected γ-ray spectra for the proton inelastic scattering on ^{60}Cr (a) and ^{62}Cr (b). The γ-ray spectra gated on the 2+→0+ transitions are shown in (c) and (d) for ^{60}Cr and ^{62}Cr, respectively.

1180-keV state is most probably a 4^+ state because of the similarity between ^{60}Cr and ^{62}Cr in the excitation energies, the population strength and the decay schemes.

The angle integrated cross sections of the 2^+ state excitations were obtained from the γ-ray yields after considering the contribution of the cascade transition to be 25(7) mb and 38(6) mb for ^{60}Cr and ^{62}Cr, respectively. The deformation lengths $\delta_{p,p'}$ were then extracted to be 1.12(16) fm and 1.36(14) fm, for ^{60}Cr and ^{62}Cr, respectively, by adopting the distorted wave Born approximation.

The extracted $\delta_{p,p'}$ values are plotted in Fig. 4(a) (filled circles) along with the ones for the stable Cr isotopes, ^{50,52,54}Cr (open circles) [18]. The triangles show $\delta_{p,p'}$ estimated from the Coulomb deformation lengths (δ_C) obtained from $B(E2)$ [11, 19] using Bernstein's prescription [20]. In the estimation, the ratio of the neutron to proton quadrupole matrix elements was taken from the shell model calculation within the pf-shell using the GXPF1A effective interaction [21, 22]. The $\delta_{p,p'}$ value is the smallest at $N=32$ and gradually increases with neutron number until ^{62}Cr. The increase of $\delta_{p,p'}$ is a direct evidence for the enhancement of collectivity from ^{56}Cr to ^{62}Cr, which was suggested by the decreasing $E_x(2^+)$ energies [12].

The $E_x(4^+)$ and $R_{4/2}$ ($=E_x(4^+)/E_x(2^+)$) values in ^{62}Cr are compared with those of the lighter even-even Cr isotopes in Fig. 4 (b) and (c). The $R_{4/2}$ value is close to the vibrational limit of 2.0 in the neutron sub-shell magic nucleus ^{56}Cr. In ^{60}Cr, $R_{4/2}$ increases slightly but still remains close to 2.0 despite the enhancement of the collectivity. This is consistent with the interpretation that its large collectivity originates from surface vibration. [14]. In ^{62}Cr, in contrast, the $R_{4/2}$ value rises rapidly to 2.7. This is an indication that the nature of collectivity starts to change from vibrational to rotational, signifying the development of deformation in ^{62}Cr.

The solid lines in Fig. 4 show the shell model results within the pf-shell using the GXPF1A interaction [21, 22]. The calculation well reproduces the experimental $\delta_{p,p'}$ values up to ^{58}Cr. However, it deviates from the observed large $\delta_{p,p'}$ values of ^{60}Cr and ^{62}Cr, suggesting that the pf-shell is not sufficient to describe the structure of ^{60}Cr and

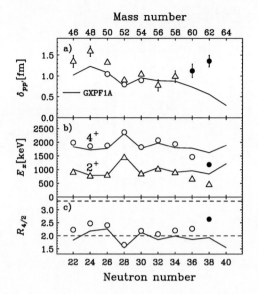

FIGURE 4. (a) Plot of $\delta_{p,p'}$ for Cr isotopes as a function of the neutron number. The filled circles are the result from the present work and the open circles are those on the stable Cr isotopes [18]. The triangles show the ones estimated from the $B(E2)$ values (see text for details). (b) Excitation energies of the 2^+ states (triangles) and 4^+ states (circles). (c) The $E_x(4^+)/E_x(2^+)$ ratio ($R_{4/2}$). Solid lines are the results of the shell model calculation using the GXPF1A interaction.

^{62}Cr. This is consistent with the conclusion of Ref. [12] based on the $E_x(2^+)$ energies. The large deformation in ^{62}Cr may be the consequence of the valence neutrons being located at the middle of the strongly-mixed pf- and $g_{9/2}$-shells.

DAYONE EXPERIMENTS AT RIBF

RI Beam Factory (RIBF) [4] is a new generation radioactive isotope beam facility recently commissioned, which will extend the capability of the RI beam experiments toward more neutron rich and heavier region. The accelerator complex consisting of an injector liniac and three ring cyclotrons provides intense heavy-ion beams, such as, ^{48}Ca, ^{86}Kr, ^{136}Xe or ^{238}U with a typical energy of 345A MeV. From these heavy-ion beams, RI beams are produced using the in-flight fragment separator BigRIPS [23, 24], which was designed to have large angular and momentum acceptance to achieve efficient collection of fragments produced not only by projectile fragmentation reactions but also by in-flight fissions. The commissioning of BigRIPS was performed in 2007 followed by the identification of new isotopes of ^{125}Pd and ^{126}Pd [25] in the in-flight fission of an U beam with a very weak intensity of 4×10^7 particles/s, which demonstrates the performance of BigRIPS.

The ZeroDegree spectrometer is a multi-purpose spectrometer constructed downstream of BigRIPS to analyze the reaction residues of secondary reactions. Various type

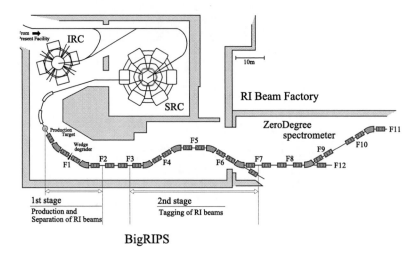

FIGURE 5. Layout of BigRIPS and the ZeroDegree spectrometer.

of detectors can be placed at the target position of ZeroDegree (F8) so that a variety of experiments can be performed. For example, γ-ray detectors, such as the NaI array DALI2 [17], or the Ge array GRAPE [26] will be placed for γ-ray spectroscopy experiments.

The first series of experiments utilizing the secondary reaction at RIBF were performed in December 2008. In this so-called "DayOne" campaign, three experiments were performed: in beam γ-ray spectroscopy of ^{32}Ne[27], exclusive Coulomb breakup of neutron rich oxygen and neon isotopes[28], and reaction cross section measurement for neutron rich neon isotopes[29]. The DayOne experiments utilized an intense ^{48}Ca beam with a maximum intensity of 170pnA. Secondary beams of neutron-rich carbon, neon and silicon isotopes were then produced in the ^{48}Ca+Be reaction. The obtained production yields are summarized in Table 1 together with the thickness of the Be target. The predicted production yields are also shown, which were calculated using the semi-empirical cross section formula, EPAX2[30]. The transmission of BigRIPS was obtained by the LISE++ code [31]. The ratios of the measured to the calculated yields are shown as R_{EPAX2}.

The R_{EPAX2} ratio is distributed around one, showing the good agreement between the measured cross sections at 345A MeV and the EPAX2 prediction. This is in contrast to the disagreement at around 100A MeV, where EPAX2 is known to overestimate the production yields of very neutron-rich nuclei significantly [32, 33]. This suggests that the production cross section at 345A MeV is larger than that at 63A MeV and closer to the EPAX2 formula. Another possible reason for the increased production yield is the multiple reactions inside the target. The high beam energy enables us to use very thick targets, which increases the contribution of the multi-step reaction. In either cases, the increase in the production yields at BigRIPS promises to extend the research opportunity toward very exotic regions with very neutron-rich or proton-rich regions.

TABLE 1. Measured and calculated production yields of neutron rich nuclei from the ^{48}Ca+Be reaction at 345AMeV. R_{EPAX2} shows the ratios of the measured to the calculated yields.

	Target Thickness [g/cm^2]	Yield (*meas.*) [cps/100pnA]	Yield *EPAX*2) [cps/100pnA]	R_{EPAX2}
^{20}C	3.6	2×10^4	3×10^2	50
^{22}C	3.6	1×10^1	2	7
^{29}Ne	2.8	5×10^2	5×10^3	0.1
^{30}Ne	2.8	3×10^2	6×10^2	0.6
^{31}Ne	2.8	1×10^1	6×10^1	0.2
^{32}Ne	2.8	5	2	3
^{40}Si	0.93	5×10^2	3×10^3	0.2
^{42}Si	3.6	2×10^1	1×10^2	0.1

REFERENCES

1. A. Gade, and T. Glasmacher, *Prog. Part. Nucl. Phys.* **60**, 161 (2008).
2. O. Sorlin, and M. G. Porquet, *Prog. Part. Nucl. Phys.* **61**, 602 (2008).
3. T. Kubo et al., *Nucl. Instrum. Methods in Phys. Res.* B **70**, 309 (1992).
4. Y. Yano, *Nucl. Instrum. Methods in Phys. Res.* B **261**, 1009 (2007).
5. A. Huck, et al., *Phys. Rev.* C **31**, 2226 (1985).
6. J. I. Prisciandaro, et al., *Phys. Lett.* B **510**, 17 (2001).
7. R.V.F. Janssens, et al., *Phys. Lett.* B **546**, 55 (2002).
8. S. N. Liddick, et al., *Phys. Rev. Lett.* **92**, 072502 (2004).
9. S. N. Liddick, et al., *Phys. Rev.* C **70**, 064303 (2004).
10. D.-C. Dinca, et al., *Phys. Rev.* C **71**, 041302 (2005).
11. A. Bürger, et al., *Phys. Lett.* B **622**, 29 (2005).
12. O. Sorlin, et al., *Eur. Phys. J.* A **16**, 55 (2003).
13. N. Mărginean, et al., *Phys. Lett.* B **633**, 696 (2006).
14. S. Zhu, et al., *Phys. Rev.* C **74**, 064315 (2006).
15. N. Aoi, et al., *Phys. Rev. Lett.* **102**, 012502 (2009).
16. H. Ryuto, et al., *Nucl. Instrum. Methods in Phys. Res.* A **555**, 1 (2005).
17. S. Takeuchi, et al., *RIKEN Accel. Prog. Rep.* **36**, 148 (2003).
18. E. Fabrici, et al., *Phys. Rev.* C **21**, 844 (1980).
19. S. Raman, et al., *At. Data Nucl. Data Tables* **78**, 1 (2001).
20. A. Bernstein, et al., *Phys. Rev. Lett.* **42**, 425 (1979).
21. M. Honma, et al., *Phys. Rev.* C **69**, 034335 (2004).
22. M. Honma, et al., *Eur. Phys. J.* A **25**, s01, 499 (2005).
23. T. Kubo, et al., *Nucl. Instrum. Methods in Phys. Res.* B **2004**, 97 (2003).
24. T. Kubo, et al., *IEEE Trans. Appl. Supercond.* **17**, 1069 (2007).
25. T. Ohnishi, et al., *J. Phys. Soc. Jpn.* **77**, 083201 (2008).
26. S. Shimoura, et al., *CNS Annu. Rep.* p. 5 (2001).
27. P. Doornenbal, et al. (2009), Phys. Rev. Lett., in press.
28. T. Nakamura, et al., in preperation.
29. T. Ohtsubo, et al., in preparation.
30. K. Summerer, and B. Blank, *Phys. Rev.* C **61**, 034607 (2000).
31. O. Tarasov, and D. Bazin, *Nucl. Phys.* **A746**, 411 (2004).
32. P. T. Hosmer, et al., *Phys. Rev. Lett.* **94**, 112501 (2005).
33. M. Mocko, et al., *Phys.Rev.* C **76**, 014609 (2007).

Exploring the Southern Boundaries of the "Island of Inversion" at the RIBF

P. Doornenbal*, H. Scheit*,†, N. Aoi*, S. Takeuchi*, K. Li*,†, E. Takeshita*,
H. Wang*,†, H. Baba*, S. Deguchi**, N. Fukuda*, H. Geissel‡,
R. Gernhäuser§, J. Gibelin¶, I. Hachiuma‖, Y. Hara††, C. Hinke§, N. Inabe*,
K. Itahashi*, S. Itoh‡‡, D. Kameda*, S. Kanno*, Y. Kawada**,
N. Kobayashi**, Y. Kondo*, R. Krücken§, T. Kubo*, T. Kuboki‖,
K. Kusaka*, M. Lantz*, S. Michimasa§§, T. Motobayashi*, T. Nakamura**,
T. Nakao‡‡, K. Namihira‖, S. Nishimura*, T. Ohnishi*, M. Ohtake*,
N.A. Orr¶, H. Otsu*, K. Ozeki*, Y. Satou**, S. Shimoura§§, T. Sumikama¶¶,
M. Takechi*, H. Takeda*, K. N. Tanaka**, K. Tanaka*, Y. Togano*,
M. Winkler‡, Y. Yanagisawa*, K. Yoneda*, A. Yoshida*, K. Yoshida* and
H. Sakurai*

*RIKEN Nishina Center, Wako, Saitama 351-0198, Japan
†Peking University, Beijing 100871, P.R.China
**Department of Physics, Tokyo Institute of Technology, Meguro, Tokyo 152-8551, Japan
‡GSI Helmholtzzentrum für Schwerionenforschung GmbH, 64291 Darmstadt, Germany
§Physik Department E12, Technische Universität München, 85748 Garching, Germany
¶LPC-Caen, ENSICAEN, Université de Caen, CNRS/IN2P3, 14050 Caen cedex, France
‖Department of Physics, Saitama University, Saitama 338-8570, Japan
††Department of Physics, Rikkyo University, Toshima, Tokyo 172-8501, Japan
‡‡Department of Physics, University of Tokyo, Bunkyo, Tokyo 113-0033, Japan
§§Center for Nuclear Study, The University of Tokyo, RIKEN Campus, Wako, Saitama 351-0198,
Japan
¶¶Department of Physics, Tokyo University of Science, Noda, Chiba 278-8510, Japan

Abstract. The nuclear structure of neutron rich Ne isotopes has been investigated at the recently commissioned Radioactive Ion Beam Factory (RIBF) at the RIKEN Nishina Center by means of inelastic scattering and knockout reactions. A cocktail beam of unstable nuclei around ^{32}Ne was provided by the BigRIPS fragment separator and incident on a natural carbon target at energies $\approx$ 220 A MeV. Reaction products were identified with the Zero Degree Spectrometer and de-excitation γ-rays detected by the DALI2 array. The reported results demonstrate the potential for in-beam γ-ray experiments at the RIBF.

Keywords: In-beam γ-ray spectroscopy, "Island of Inversion" , RIBF
PACS: 29.38.Db, 23.20.Lv, 27.30.+t

INTRODUCTION

When moving from stable nuclei to regions of extreme proton to neutron ratios, the standard ordering of nuclear shells may be abandoned in favor of new configurations. The most famous example is the so-called "Island of Inversion" , where $\nu f_{7/2}$ states intrude into the sd-shell. Despite the "Island of Inversion" being already discovered several decades ago by means of mass measurements [1, 2] and intense efforts on the experimental and theoretical side, the exact borders of the "Island of Inversion" remain unknown. In the pioneering shell-model study by Warburton *et al.* [3] $\nu(sd)^{-2}(fp)^2$ ($2\hbar\omega$) intruder configurations were predicted to become so low in energy that they form the ground states for $Z = 10$–12 and $N = 20$–22. The existing experimental data on the Mg isotopes shows, however, that also ^{31}Mg and ^{36}Mg with $N = 19$ and 24, respectively, are dominated by intruder configurations in the ground state [4, 5].

In the Ne isotopes, the experimental data is much scarcer. For ^{30}Ne, with $N = 20$, only the first excited state is known [6], while no spectroscopic information exists for the heavier isotopes due to the limitations imposed by the experimental facilities producing radioactive beams. With the Radioactive Ion Beam Factory (RIBF) [7] being recently commissioned, the previously inaccessible Ne nuclei are now within spectroscopic range. Here, we report on the first in-beam γ-ray experiment performed at the RIBF, which was aiming for the identification of the first excited state in ^{32}Ne.

EXPERIMENTAL SETUP

A primary ^{48}Ca beam with an energy of 345 MeV/u and a average intensity of 120 pnA was provided by the superconducting ring cyclotron SRC and incident on a 3.7 g/cm^2 rotating Be target [8]. In the first stage of the BigRIPS spectrometer, from focus F0 to focus F2, ^{32}Ne was separated and selected with a momentum acceptance of $\pm$ 3 % by the $B\rho$–ΔE–$B\rho$ method with a wedge shaped degrader of 4.0 g/cm^2 at the dispersive focal point F1. [1] In the second stage of the BigRIPS spectrometer (from focus F3 to focus F7) the secondary beam was identified on an event-by-event basis using the ΔE–$B\rho$–velocity method. The energy loss, which provided Z information, was measured by an ion chamber at focus F7. The $B\rho$ was deduced from a position measurement at the dispersive focal point F5. The velocity was obtained by a time-of-flight measurement

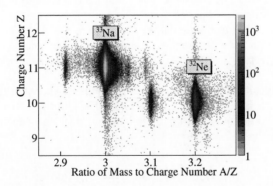

FIGURE 1. Particle identification before the secondary target. See text for details.

---

[1] See [9] for the location of all mentioned focal points.

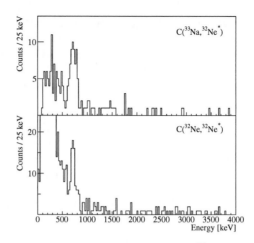

FIGURE 2. Doppler corrected γ-ray spectra in coincidence with ^{32}Ne detected in the Zero Degree Spectrometer. The top panel shows the 1p knockout reaction, the lower panel the inelastic scattering channel.

between two plastic scintillator mounted at the foci F3 and F7. Combining the latter two values unfolded the A/Z identification of the secondary beam. The particle identification plot is shown in Fig. 1. Main components were ^{33}Na (26 pps) and ^{32}Ne (6 pps). We would like to emphasize that the secondary beam rates obtained with BigRIPS mark an improvement of two orders of magnitude with respect to the RIPS separator.

The secondary beams were transported to the focus F8 and incident on a 2.54 g/cm^2 carbon secondary target with energies of 245 MeV/u for ^{33}Na and 226 MeV/u for ^{32}Ne at the mid-target position. The secondary target was surrounded by the DALI2 array [10], a γ-ray spectrometer consisting of 180 NaI(Tl) detectors and covering laboratory angles from from 11° to 147°. The γ-ray efficiency and energy resolution at 1.3 MeV were measured to 15 % and 6 %, respectively.

Reaction residues exiting the secondary target were selected and identified by the Zero Degree Spectrometer with momentum and angular acceptances of $\sim 80 \times 60$ mrad^2 and $\pm 4\%$, respectively. Again the ΔE–$B\rho$–velocity method was applied for particle identification.

A single γ-ray transition was observed for ^{32}Ne in the 1p-knockout channel from ^{33}Na as well as inelastic scattering on the carbon target. The Doppler corrected γ-ray spectra are shown in Fig. 2. The sum spectrum of both channels yields an energy of 722(9) keV, which we assign to the $2^+_1 \rightarrow 0^+_{gs}$ transition.

RESULTS AND DISCUSSION

Our result for ^{32}Ne is shown in Fig. 3 together with known $E(2^+_1)$ values from even-A Ne isotopes [6, 11]. The two low values for $N = 20$ and 22 indicate a shell quenching starting from ^{28}Ne. The figure includes shell-model calculations by Utsuno *et al.* [12] using

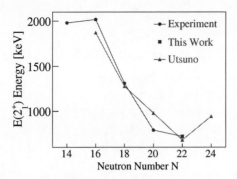

FIGURE 3. Comparison of experimental $E(2_1^+)$ energies [11, 6] in the Ne isotopes with shell model calculations by Utsuno *et al.* [12]. The lines are drawn to guide the eye.

the Monte Carlo shell model with the SDPF-M interaction, which allows unrestricted mixing of the *sd* and *pf* configurations. Their result for the $E(2_1^+)$ energy of ^{32}Ne is in agreement with our experimental findings. Furthermore, Utsuno *et al.* predict that an additional two neutrons occupy the *pf* shell with respect to normal filling, placing ^{32}Ne inside the "Island of Inversion" .

ACKNOWLEDGMENTS

We would like to thank the accelerator staff of the RIKEN Nishina Center for providing a high intensity ^{48}Ca beam. Moreover, we are grateful to T. Ohtsubo for letting us perform our measurements in the "Yakitori" mode. P.D. and M.L. are grateful for the financial support of the Japan Society for the Promotion of Science. This work was partially supported by the Deutsche Forschungsgemeinschaft Cluster of Excellence *Origin and Structure of the Universe*.

REFERENCES

1. R. Klapisch, et al., *Phys. Rev. Lett.* **23**, 652 (1969).
2. C. Thibault, et al., *Phys. Rev. C* **12**, 644 (1975).
3. E. K. Warburton, et al., *Phys. Rev. C* **41**, 1147 (1990).
4. G. Neyens, et al., *Phys. Rev. Lett.* **94**, 022501 (2005).
5. A. Gade, et al., *Phys. Rev. Lett.* **99**, 072502 (2007).
6. Y. Yanagisawa, et al., *Phys. Lett. B* **566**, 84 (2003).
7. Y. Yano, *Nucl. Instr. Meth. B* **261**, 1009–1013 (2007).
8. A. Yoshida, et al., *Nucl. Instr. Meth. A* **590**, 204–212 (2008).
9. H. Sakurai, *Nucl. Instr. Meth. B* **266**, 4080 – 4085 (2008).
10. S. Takeuchi, et al., *RIKEN Acc. Prog. R.* **36**, 148 (2003).
11. S. Raman, et al., *Atom. Nucl. Data Tab.* **78**, 1 (2001).
12. Y. Utsuno, et al., *Phys. Rev. C* **60**, 054315 (1999).

Structure of Neutron-rich N=126 Closed Shell Nuclei

Zs. Podolyák[1] *, S.J. Steer*, G. Farrelly*, M. Górska[†], H. Grawe[†],
H.K. Maier**, P.H. Regan*, J. Benlliure[‡], S. Pietri[†], J. Gerl[†] and
H.J. Wollersheim[†]

Department of Physics, University of Surrey, Guildford GU2 7XH, UK
[†]*GSI, Planckstrasse 1, D-64291, Darmstadt, Germany*
**The Henryk Niewodniczański Institute of Nuclear Physics, PL-31-342, Kraków, Poland*
[‡]*Universidad de Santiago de Compostela, E-15706, Santiago de Compostela, Spain*

Abstract.
   A series of experiments devoted to the study of the neutron-rich N~126 region have been performed at GSI, Darmstadt, within the Rare Isotopes Investigations at GSI (RISING) project. The highlights of the experimental results from these highly successful experiments include the first observation of excited states in three neutron-rich N=126 closed shell nuclei: (i) In ^{205}Au$_{79}$ conversion electron decay of the $\pi h_{11/2}^{-1}$ seconds lived isomeric state into the $\pi d_{3/2}^{-1}$ ground-state has been observed. In addition the yrast structure has been established up to spin-parity $(19/2^{+})$ via the observation of the decay of an isomeric state with configuration $\pi(h_{11/2}^{-1})_{10}(s_{1/2}^{-1})$; (ii) In ^{204}Pt$_{78}$, the yrast sequence has been observed following the internal decay of $I^{\pi}=(5^{-})$, (7^{-}) and (10^{+}) isomeric states; (iii) In ^{203}Ir$_{77}$ excited states have been observed following the decay of an isomeric state with structure 'similar' to that in its ^{205}Au isotone. Shell model calculations have been performed in order to get a deeper understanding of the structure of these N=126 nuclei. It was found that in order to get a good description for all available information on the N=126 isotones below lead, both on excitation energies and transition strengths, small modifications of the standard two-body matrix elements were required.

Keywords: gamma-ray spectroscopy
PACS: 21.10.-k, 21.60.Cs, 25.70.Mn

INTRODUCTION

Studies of the structure of single magic nuclei are of fundamental importance in our understanding of nuclear structure since they allow direct tests of the purity of shell model wave functions. Information on the single-particle energies and two-body residual interactions can be derived from the experimental observables such as energies of the excited states and transition probabilities.

The doubly magic ^{208}Pb nucleus, with 82 protons and 126 neutrons, provides the heaviest classic shell model core. Experimental information on the neutron-rich, $N = 126$ nuclei is very scarce. Prior to our investigations, information has been obtained on excited states for only two isotones with $Z < 82$: ^{207}Tl [1] and ^{206}Hg [2]. In the case of

---

[1] for the RISING collaboration

CP1165, *Nuclear Structure and Dynamics '09*
edited by M. Milin, T. Nikšić, D. Vretenar, and S. Szilner
© 2009 American Institute of Physics 978-0-7354-0702-2/09/$25.00

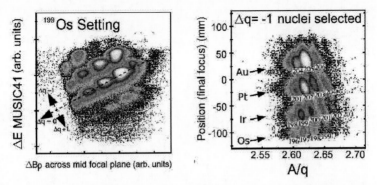

FIGURE 1. Identification plot showing the H-like nuclei arriving to the final focal plane of the fragment separator. This setting was optimised for the transmission of fully-stripped ^{199}Os ions.

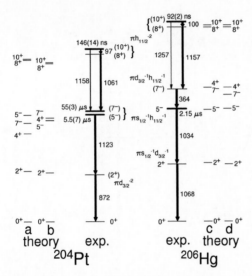

FIGURE 2. Experimental and calculated partial level schemes of the $N = 126$ ^{204}Pt [7] and ^{206}Hg [2] nuclei. Arrow widths denote relative intensities of parallel decay branches. The dominant state configurations are indicated. (a) and (d) are calculations using the Rydsrtöm matrix elements, while (b) and (c) are with the modified ones, as described in the text.

^{205}Au only the ground state was known [3].

EXPERIMENTAL TECHNIQUE

A number of experiments were performed with the aim to study the structure of N=126 nuclei. The nuclei of interest were populated via relativistic energy projectile fragmen-

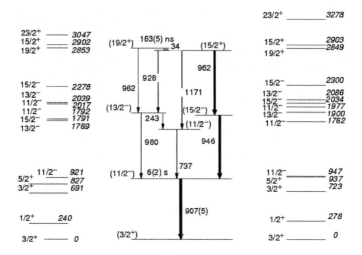

FIGURE 3. Experimental (middle) and calculated level schemes of ^{205}Au. Calculations using the Rydsrtöm matrix elements are shown on the left, while calculations with the modified TBMEs are on the right hand side.

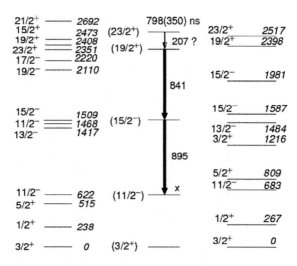

FIGURE 4. Same as figure 3, but for ^{203}Ir. The experimental level scheme is preliminary.

tation of a ^{208}Pb beam. The SIS-18 accelerator at GSI provided the primary beams at $E/A = 1$ GeV. The nuclei of interest were selected and identified in flight by the FRagment Separator (FRS) [4]. The identified ions were implanted in a stopper, positioned at the final focal plane of the FRS. Both passive and active stopper were used. A passive stopper was employed in internal isomeric decay studies. A passive, Silicon based stopper [5] was used in the case of charged-particle decays (in our case: conversion electron decay). The stopper was surrounded by the RISING array in the "Stopped Beam" configuration [6]. For details on the experimental setup, identification procedure and analysis see [7, 8, 9, 10].

RESULTS

An identification plot showing the H-like nuclei arriving to final focal plane of the fragment separator is shown in figure 1.

Isomeric decays were observed in three N=126 nuclei: ^{205}Au [8, 9, 10], ^{204}Pt [7], ^{203}Ir [8]. The nature of the isomeric states are explained in the abstract.

In order to obtain a quantitative understanding of the underlying single-particle structure of the excited states of the observed N=126 nuclei, shell-model calculations have been performed employing the OXBASH code [11]. The standard interaction two-body matrix elements (TBMEs) were used as taken from ref. [12]. The proton-hole energies were taken from the experimental level scheme of ^{207}Tl [1].

In order to get a good description for ^{204}Pt and ^{206}Hg, both on excitation energies and transition strengths, small modifications of the standard TBMEs were required [7]. Calculations with these modified matrix elements were also performed.

The experimental level schemes are compared with those from the shell-model calculations in figures 2-4.

ACKNOWLEDGMENTS

The present manuscript contains results form a series of experiments performed within the RISING collaboration. The work of all collaborators is acknowledged. The referenced individual papers contain the name of all those involved in the respective works. The excellent work of the GSI accelerator staff is acknowledged.

REFERENCES

1. D. Eccleshall, M.J.L.Yates, *Phys. Lett.* **19**, 301 (1965); M.J. Martin, *Nucl. Data Sheets* **70**, 315 (1993).
2. B. Fornal, et al., *Phys. Rev. Lett.* **87**, 212501 (2001).
3. Ch. Wennemann, et al., *Z. Phys.* A **347**, 185 (1994).
4. H. Geissel, et al., *Nucl. Instrum. Meth.* B **70**, 286 (1992).
5. R. Kumar, et al., *Nucl. Instrum. Meth.* A **598**, 754 (2009).
6. S. Pietri, et al., *Nucl. Instrum. Meth.* B **261**, 1079 (2007).
7. S. Steer, et al., *Phys. Rev.* C **78**, 061302(R) (2008).
8. S. Steer, et al., *Int. J. Mod. Phys.* E **18**, 1002 (2009).
9. Zs. Podolyák, et al., *Phys. Lett.* **672B**, 116 (2009).
10. Zs. Podolyák, et al., *Eur. Phys. J.* A (2009) DOI: 10.1140/epja/i2009-10794-5.
11. B.A. Brown A. Etchegoyen, W.D.M. Rae, The computer code OXBASH, MSU-NSCL rep. no. 524.
12. L. Rydström, et al., *Nucl. Phys.* A **512**, 217 (1990).

Shell Closure N=16 in ^{24}O

C. Nociforo[a], R. Kanungo[b], A. Prochazka[a,c], T. Aumann[a], D. Boutin[c],
D.Cortina-Gil[d], B. Davids[e], M. Diakaki[f], F. Farinon[a,c], H. Geissel[a],
R.Gernhäuser[g], J. Gerl[a], R. Janik[h], B. Jonson[i], B. Kindler[a], R. Knöbel[a,c],
R.Krücken[g], M. Lantz[i], H. Lenske[c], Yu.A. Litvinov[a], B. Lommel[a],
K.Mahata[a], P. Maierbeck[g], A. Musumarra[j], T. Nilsson[i], T. Otsuka[k], C.
Perro[e], C.Scheidenberger[a], B. Sitar[h], P. Strmen[h], B. Sun[a], I. Szarka[h],
I.Tanihata[l], Y. Utsuno[m], H. Weick[a] and M. Winkler[a]

[a]GSI, Darmstadt, Germany
[b]Saint Mary's University, Halifax, Canada
[c]Justus-Liebig Universität, Gießen, Germany
[d]Universidad de Satiago de Compostela, Santiago de Compostela, Spain
[e]TRIUMF, Vancouver, Canada
[f]National Technical University, Athens, Greece
[g]Technische Universität München, Garching, Germany
[h]Comenius University, Bratislava, Slovakia
[i]Chalmers University of Technology, Göteborg, Sweden
[j]Università di Catania, Catania, Italy
[k]University of Tokio, Saitama, Japan
[l]Osaka University, Osaka, Japan
[m]Japan Atomic Energy Agency, Ibaraki, Japan

Abstract. Advanced nuclear structure models predict the presence of the shell closures N=14, 16 in neutron-rich O isotopes rather than N=20. Spectroscopic investigations performed at the neutron drip line have recently confirmed such predictions showing that the ^{24}O is a doubly magic nucleus (Z=8, N=16). Predictions within the shell model calculation for the ^{23,24}O ground state have been confirmed measuring their spectroscopic factors. Results obtained in one-neutron removal reactions performed by using in-flight radioactive ion beams produced at the Fragment Separator FRS of GSI are reported.

Keywords: Nuclear structure, radioactive ion beams.
PACS: 21.10.Hw; 21.10.Jx; 21.60.Cs; 25.60.-t; 25.60.Gc; 27.30.+t; 29.38.Db.

N=14,16 MAGIC NUMBERS

The evolution of magic numbers and shell closures in exotic nuclei is one of the hottest topics in contemporary nuclear structure physics. Whether a usual magic number is preserved while approaching the drip lines is indeed crucial for testing single particle models. The disappearance of the N=20 magic number in the so called Island of Inversion is well known since long time [1]. On the other hand, the appearance of new shell gaps like N=14 have been proposed theoretically and could be

CP1165, *Nuclear Structure and Dynamics '09*
edited by M. Milin, T. Nikšić, D. Vretenar, and S. Szilner
© 2009 American Institute of Physics 978-0-7354-0702-2/09/$25.00

confirmed in some systematic studies of the first 2^+ states of the even-mass O isotopes [2]. In the O isotopic chain the minimum measured $B(E2)$ quadrupole strength (maximum $E(2^+)$ excitation energy) was found in correspondence of N=14. Evidence of the presence of N=14 (sub)shell closure in ^{22}O has been also found in Coloumb dissociation studies of the ^{23}O nucleus performed at the LAND-GSI experimental setup. There, the differential e.m. cross section of the outgoing channel $^{22}O(0^+)$ +n is well described only assuming the presence of a s-wave removed nucleon [3]. The associated spectroscopic factor close to unity establishes that the ground state of the ^{23}O can be considered to be a pure single-particle state. Similar predictions are obtained for the ^{24}O ground state configuration within the shell model calculations. Here, the valence neutron fills the $2s_{1/2}$ orbital making the N=16 shell closure possible. In a recent experiment performed at NSCL-MSU an $l=2$ resonance assumed as $d_{5/2}$ state has been found in ^{25}O continuum, providing a measurement of the shell gap $s_{1/2}$-$d_{5/2}$ equal to 4.86 MeV [4]. This value is high enough to confirm a shell closure in ^{24}O [5]. The existence of a spherical shell closure can be proven definitely measuring the neutron occupancy in ^{24}O. The first direct measurement of the spectroscopic factor of the ^{24}O ground state has been performed at the GSI using a single neutron removal reaction at 920 MeV/u [6].

EXPERIMENTAL RESULTS OF 1n REMOVAL REACTIONS

One neutron (1n) removal reactions at relativistic energies are known to be reliable methods in particular studying the structure of light weakly-bound nuclei. Applying the in-flight technique to radioactive ion beams at high energies in such reactions has several advantages, among them the possibility to make use of thick targets and small scattering angles. The key role is played by a magnetic separator working in a dispersion-matched mode, in our case the Fragment Separator FRS at GSI (see Fig. 1) [7]. The ^{24}O beam is produced and separated in the first half of the separator (mid-focal plane F2). There, the secondary reaction takes place and the outgoing ^{23}O fragments are analyzed by the second half of the separator. Full particle identification (see PID plots in Fig. 1) is performed on an event-by-event basis for the ion beams and the residual nuclei. The momentum after the 1n removal can be reconstructed in the laboratory frame at the last achromatic focal plane (F4) by means of the Eq. (1)

$$p_{lab} = B\rho\left(1 + \frac{x_{F2}M_{F2-F4} - x_{F4}}{D_{F2-F4}}\right)Z. \tag{1}$$

in the order of 10^{-4} due to precise $x_{F2,F4}$ position measurements and the known optical properties M (magnification), D (dispersion), and Bρ (magnetic rigidity) of the separator. The measured momentum distribution made independent on the momentum spread of the incoming beam, provides the momentum distribution of the removed neutron in the nucleus. The shape of the longitudinal momentum distribution dσ/d$p_{//}$ calculated in eikonal approximations is very sensitive to the orbital angular momentum of the removed neutron. In the framework of the Glauber theory the bound and scattering wave functions of initial and final states are represented by the system ^{23}O +n. The final d$\sigma_{//}$/dp distribution of $^{24}O \rightarrow {}^{23}O$ in the projectile frame is plotted in Fig.1, bottom right side (full points). It can be reproduced by a calculation which assumes an s-wave *(l=0)* configuration of the removal neutron (red solid line in Fig. 1). The calculation assuming a d-wave *(l=2)* component is also shown for a comparison (green dot-dashed line in Fig. 1).

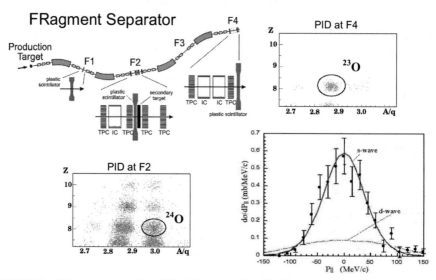

FIGURE 1. Schematic view of the FRS with particle identification plots (PID) at F2 and F4. On the bottom right side, $^{24}O \rightarrow {}^{23}O$ $d\sigma_{///}/dp$ distribution in the projectile frame (full points); the red solid (green dot-dashed) curve represents theoretical calculation within eikonal model.

Comparing the experimental $1n$ removal cross section σ_{-1n} with the theoretical value obtained from the eikonal calculation which assumes the neutron in an s-wave configuration and a spin value $1/2^+$ for the ^{23}O, a spectroscopic factor can be extracted. The obtained value is $S=1.74\pm0.19$. Spectroscopic factors obtained by shell model calculations assuming different effective nucleon-nucleon interaction (USDB [8] and SDPF-M [9]) are in good agreement with this value. The presence of the N=16 shell closure in the ^{24}O is theoretically justified by the lack of protons in the $d_{5/2}$ orbital which results in a weaker bound neutron in the $d_{3/2}$ orbital. This weakness of the so called tensor force is responsible of the upward shift of the $d_{3/2}$ levels and origin of the $s_{1/2}$-$d_{3/2}$ shell gap [5]. The results of the $d\sigma/dp_{//}$ measurements obtained for ^{24}O and $^{22,23}O$, respectively, are compared in Fig. 2 with previous GSI measurements [10] performed at similar energies (~900 MeV/u). The recent data are represented in Fig. 2 by red full points while the old data are the black squares. Looking into the neutron-rich O systematics, the change from a d- to s-wave character is more clearly pronounced in the decrease of the values of the width (FWHM) of the ^{23}O inclusive momentum distributions (Fig. 2, left panel) rather than in the $1n$ removal cross sections (Fig. 2, right panel). Access to inclusive cross sections is, in fact, not always sufficient to extract quantitatively spectroscopic results, like in the ^{24}O case.

CONCLUSIONS AND PERSPECTIVES

Studies of the neutron-rich isotopes done at in-flight facilities all over the world have shown that the neutron drip line in O isotopes is reached at N=16, where the presence of a new shell closure is foreseen. The latter has been confirmed in the direct measurement of the occupation probability of the ^{24}O ground state performed at GSI by means of $1n$ removal reaction at 920 MeV/u. The evolutions of magic numbers is of highest interest also in the C isotopes, where the N=16 shell closure is also predicted.

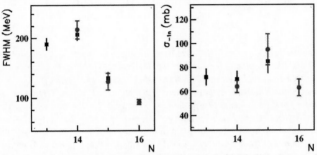

FIGURE 2. Systematics of experimental widths FWHM (left panel) and cross section measurements (right panel) of $1n$ removal reactions from $^{21\text{-}24}$O isotopes. The black square represent data from Ref. [10], the red full points are our measurements.

Comparing systematics in O and C isotopes can provide more understanding of the tensor component of the nuclear force at the drip lines [2]. Interaction cross section measurements have already provided estimations of C and O matter radii [11]. It is nowadays feasible to have access to the evolution of the neutron skins in both isotopic chains.

ACKNOWLEDGMENTS

The authors would like to thank K.-H. Behr, A. Brünle, C. Karagiannis and N. Kurz for their precious help in preparing the experiment.

REFERENCES

1. C. Thibault et al., *Phys. Rev.* **C12**, 644 (1975).
2. O. Sorlin and M.-G.Porquet, *Progr. in Part. and Nucl. Physics* **61**, 602 (2008).
3. C. Nociforo et al., *Phys. Lett.* **B605**, 79 (2005).
4. C. R. Hoffman et al., *Phys. Rev. Lett.* **100**, 152502 (2008).
5. T. Otsuka, D. Abe, *Progr. in Part. and Nucl. Physics* **59**, 425 (2007).
6. R. Kanungo et al., *Phys. Rev. Lett.* **102**, 152509 (2009).
7. H. Geissel et al., *Nucl. Instrum. Methods* **B70**, 286 (1992).
8. B. A. Brown and W. A. Richter, *Phys. Rev.* **C74**, 034307 (2004).
9. T. Otsuka et al. *Phys. Rev. Lett.* **87**, 082502 (2001).
10. D. Cortina-Gil et al., *Phys.Rev. Lett.* **93**, 062501 (2004).
11. M. Ozawa et al., *Nucl. Phys.* **A691**, 599-617 (2001).

Neutron Drip-Line Topography

E. Minaya Ramirez[a], G. Audi[a], D. Beck[b], K. Blaum[c], Ch. Böhm[d],
C. Borgmann[c], M. Breitenfeldt[e], N. Chamel[f], S. George[c], S. Goriely[f],
F. Herfurth[b], A. Herlert[g], A. Kellerbauer[c], M. Kowalska[g], D. Lunney,[a]*
S. Naimi[a], D. Neidherr[d], J.M. Pearson[h], M. Rosenbusch[e], S. Schwarz[i],
L. Schweikhard[e]

[a] Centre de Spectrométrie Nucléaire et de Spectrométrie de Masse, Université Paris Sud, Orsay, France
[b] GSI Helmholtzzentrum für Schwerionenforschung, Darmstadt, Germany
[c] Max-Planck-Institut für Kernphysik, Heidelberg, Germany
[d] Institut für Physik, Johannes-Gutenberg-Universität, Mainz, Germany
[e] Institut für Physik, Ernst-Moritz-Arndt-Universität, Greifswald, Germany
[f] Institut d'Astronomie et d'Astrophysique, Université Libre de Bruxelles, Belgium
[g] Physics Department, Centre Européan de Recherche Nucléaire, Geneva, Switzerland
[h] Department of Physics, Université de Montréal, Montréal, Québec, Canada
[i] National Superconducting Cyclotron Laboratory, Michigan State University, East Lansing MI, USA

Abstract. The development of microscopic mass models is a crucial ingredient for the understanding of how most of the elements of our world were fabricated. Confidence in drip-line predictions of such models requires their comparison with new mass data for nuclides far from stability. We combine theory and experiment using results that are state of the art: the latest mass measurements from the Penning-trap spectrometer ISOLTRAP at CERN-ISOLDE are used to confront the predictions of the latest Skyrme-Hartree-Fock-Bogoliubov (HFB) microscopic mass models. In addition, we compare the new data to predictions of other types of mass models and the extrapolative behavior of the various models is analyzed to highlight topographical trends along the shores of the nuclear chart.

Keywords: binding energies and masses, nuclear density functional theory, r process
PACS: 21.10.Dr, 21.60.Jz, 26.30.Hj

INTRODUCTION

For heavy elements, the neutron drip-line will (most likely) never be reached by radioactive beam facilities. However, its location is of great importance for modeling the rapid neutron-capture process [1]. Therefore, the development of microscopic mass models is the only hope in our quest to understand how most of the elements of our world were fabricated. The predictions of such models require detailed comparison with new masses of nuclides far from stability.

A comprehensive review of mass models was given in [2]. There are several ways of comparing models, notably by root-mean-square deviation with known masses. It is also interesting to examine the mass surfaces that different models provide. It is not

CP1165, *Nuclear Structure and Dynamics '09*
edited by M. Milin, T. Nikšić, D. Vretenar, and S. Szilner
© 2009 American Institute of Physics 978-0-7354-0702-2/09/$25.00

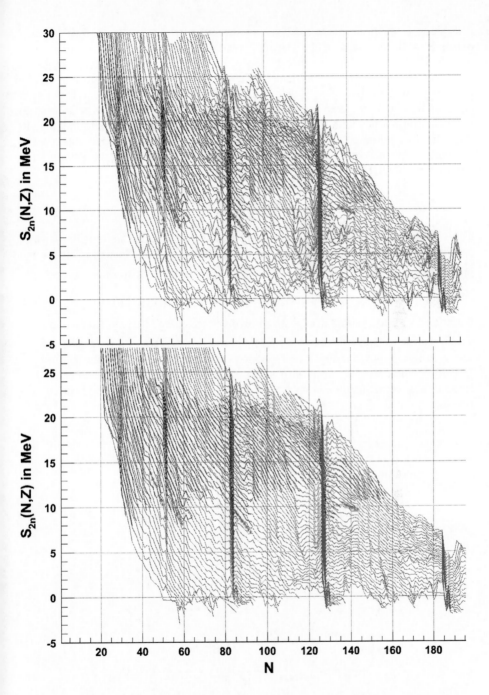

FIGURE 1. HFB-17 S_{2n} values before (top) and after (bottom) smoothing the deformation and collective-energy contributions. The dark (blue) lines are data from the AME2003 [6]. The (green) crosses are from recent ISOLTRAP measurements [7,8].

possible here to perform an analysis of several different models so we concern ourselves with the latest of the Hartree-Fock-Bogoliubov (HFB) microscopic mass models HFB-17 [3]. Shown in Fig. 1 are the two-neutron separation energies as calculated by HFB-17. Considerable fluctuations are visible, especially in the extremely neutron-rich region below 10 MeV, for $140 < N < 180$. This is in contradiction with the experimental picture and is due, among other things, to numerical difficulties with the energy minimizations with repect to deformation. In this work we explore possibilities for smoothing these fluctuations and to obtain better agreement with new measurements of masses from stability using the Penning-trap spectrometer ISOLTRAP [5] located at CERN-ISOLDE.

DISCUSSION

The strategy adopted for the present study was to examine the deformation-energy contributions to the binding energy. These are: the deformation energy related to the mean field [4] $E_{def}^{HFB} = E^{HFB}(\beta_2 = 0) - E^{HFB}(\beta_2)$ (where β_2 is the deformation parameter minimizing the total energy) and a collective-energy correction, E_{coll}. The form for E_{coll} is expressed in Fig. 2 and involves the energy derived from a cranking model that depends on β_2 (the other terms in the expression are free fit parameters). These terms are shown in Fig. 2 as a function of neutron number N for the Rn isotopes ($Z = 86$). These energy terms also show a lack of continuity so we attempted to smooth these terms using a simple averaging algorithm. The average is traced as a fine line in Fig. 2.

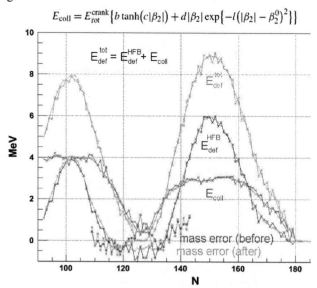

$$E_{coll} = E_{rot}^{crank}\left\{ b \tanh(c|\beta_2|) + d|\beta_2| \exp\left\{-l(|\beta_2| - \beta_2^0)^2\right\}\right\}$$

FIGURE 2. HFB-17 deformation and collective-energy contributions for the isotopes of Rn. Also shown are differences of calculated masses and experiment (mass error) before and after smoothing.

96

RESULTS

Figure 1 shows the comparison of the HFB-17 S_{2n} values calculated before and after applying the smoothing of the deformation-energy contributions as illustrated in Fig. 2. Also visible in the S_{2n} plots are the values from the recent ISOLTRAP measurements. Two chains of isotopes are used with new masses particularly far from what was known previously: $^{223-229}$Rn [7] and $^{144-146}$Xe [8].

The smoothed deformation energies have made the S_{2n} plot considerably smoother, at the same time even slightly reducing the root-mean-square (rms) deviation from the experimental mass values (from 581 to 574 keV). The rms deviation of the smoothed S_{2n} HFB-17 values compared to experiment is improved by this procedure, from 486 keV to 410 keV. The rms deviation of the smoothed HFB-17 masses with respect to the unsmoothed values is only 304 keV for the 8500 nuclides in the HFB-17 mass table. This shows that while better continuity is obtained on the mass surface, the values of the masses themselves, and consequently the extrapolations, are practically unchanged. In the case of Rn, the seven new masses were not predicted more accurately with the smoothing (720 keV compared to 664 keV before smoothing). However, the three new xenon masses were better with smoothing (666 keV), compared with 715 keV before. These first results will encourage us to continue our efforts.

ACKNOWLEDGMENTS

This work was supported by MPI and BMBF (Germany), FNRS (Belgium), CNRS (France) and NSERC (Canada)

REFERENCES

* e-mail: david.lunney@csnsm.in2p3.fr

1. M. Arnould, S. Goriely, K. Takahashi, *Phys. Rep.* **450**, 97 (2007).
2. D. Lunney, J.M. Pearson, C. Thibault, *Rev. Mod. Phys.* **75**, 1121 (2003).
3. S. Goriely, N. Chamel, J.M. Pearson, *Phys. Rev. Lett.* **102**, 152503(2009).
4. N. Chamel, S. Goriely, J.M. Pearson, *Nucl. Phys.* **A 812**, 72 (2008).
5. M. Mukherjee et al., *Eur. Phys. J.* A **35**,1 (2008).
6. G. Audi, A.H. Wapstra, C. Thibault, *Nucl. Phys.* **A 729**, 337(2003).
7. D. Neidherr et al., *Phys. Rev. Lett.* **102**, 112509 (2009).
8. D. Neidherr et al., *Phys. Rev.* C (2009) submitted for publication.

Investigation into behavior of weakly-bound proton via B(GT) measurement for the β decay of ^{24}Si

Y. Ichikawa[*,†], T. K. Onishi[*], D. Suzuki[*], H. Iwasaki[*], V. Banerjee[**],
T. Kubo[†], A. Chakrabarti[**], N. Aoi[†], B. A. Brown[‡], N. Fukuda[†],
S. Kubono[§], T. Motobayashi[†], T. Nakabayashi[¶], T. Nakamura[¶], T. Nakao[*],
T. Okumura[¶], H. J. Ong[||], H. Suzuki[*], M. K. Suzuki[*], T. Teranishi[††],
K. N. Yamada[†], H. Yamaguchi[§] and H. Sakurai[†]

[*]*Department of Physics, University of Tokyo, 7-3-1 Hongo, Bunkyo, Tokyo 113-0033, Japan*
[†]*RIKEN Nishina Center, RIKEN, 2-1 Hirosawa, Wako, Saitama 351-0198, Japan*
[**]*Variable Energy Cyclotron Centre, 1/AF, Bidhan, Nagar, Kolkata-700 064, India*
[‡]*Department of Physics and Astronomy, and National Superconducting Cyclotron Laboratory,
Michigan State University, East Lansing, Michigan 48824-1321, USA*
[§]*Center for Nuclear Study, University of Tokyo, 7-3-1 Hongo, Bunkyo, Tokyo 113-0033, Japan*
[¶]*Department of Physics, Tokyo Institute of Technology, 2-12-1 Oh-okayama, Meguro, Tokyo
152-8551, Japan*
[||]*Research Center for Nuclear Physics, Osaka University, 10-1 Mihogaoka, Ibaraki, Osaka
567-0047, Japan*
[††]*Department of Physics, Kyushu University, 6-10-1 Hakozaki, Higashi, Fukuoka 812-8581, Japan*

Abstract. The β-decay spectroscopy on ^{24}Si was carried out in order to investigate the behavior of a weakly-bound s-wave proton. We observed two β transitions to low-lying bound states in ^{24}Al for the first time. The Gamow-Teller transition strengths B(GT) to the 1_1^+ and 1_2^+ states were deduced to be 0.13(2) and 0.14(1), which were smaller than that of mirror nucleus ^{24}Ne by 22 % and 10 %, respectively. These mirror asymmetries of B(GT) indicate the configuration change in the wave function. By comparing with theoretical shell-model calculations, it is implied that the configuration change is attributed to the lowering of the $1s_{1/2}$ orbital.

Keywords: proton rich, weakly bound, beta decay, Gamow-Teller
PACS: 21.10.Hw, 23.20.Lv, 23.40.-s, 27.30.+t

INTRODUCTION

The behavior of a weakly-bound proton in a proton-rich nucleus has attracted much interest for exploring exotic nuclear structures such as proton halo [1]. The weakly-bound proton, especially in the s orbital, causes mirror "asymmetry" on nuclear structure. Thomas-Ehrman (TE) shift [2, 3] is one of the mirror asymmetries related to the weakly-bound proton. The TE shift is reduction of the Coulomb energy due to the spatial expansion of an s-wave proton. The TE shift is important for the spectra of nuclei near the proton-drip line.

There is a TE-type energy asymmetry in the sd-shell region. By comparing the low-lying 1^+ levels between mirror nuclei of ^{24}Al [5, 6, 7] and ^{24}Na [8], the energy difference

CP1165, *Nuclear Structure and Dynamics '09*
edited by M. Milin, T. Nikšić, D. Vretenar, and S. Szilner
© 2009 American Institute of Physics 978-0-7354-0702-2/09/$25.00

is only 46 keV for the 1^+_1 states, while it is 235 keV for the 1^+_2 states [1]. The significant energy asymmetry in the 1^+_2 state is attributed to a proton occupying the $1s_{1/2}$ orbital.

The TE shift in the lower sd-shell region is regarded as lowering of the single particle energy of the $1s_{1/2}$ orbital [4]. Then the $1s_{1/2}$ and the $0d_{5/2}$ orbitals are more likely to be mixed. Consequently the TE shift induces the change of configuration in the wave function. There also may be a configuration change in the bound states of ^{24}Al where the significant TE shift is observed.

The Gamow-Teller (GT) transition strength B(GT) allows to find out the configuration change. Since B(GT) reflects the space overlap between the initial and final wave functions, if there is a configuration change due to the TE shift, the configuration change would appear as a mirror asymmetry of B(GT). So far, no experimental work has been reported for the B(GT) of ^{24}Si to the low-lying bound 1^+ states in ^{24}Al where the TE shift is observed. Therefore in order to determine the B(GT) to the low-lying 1^+ states, we have performed β-decay spectroscopy on ^{24}Si with a measurement of the β-delayed γ rays from the bound 1^+ states.

β-DECAY SPECTROSCOPY ON ^{24}SI

The experiment was performed at the RIKEN Projectile Fragment Separator (RIPS) facility [9]. The secondary beam of ^{24}Si was produced by the projectile fragmentation of a 100-MeV/nucleon ^{28}Si beam with a primary target of a 0.72 mm-thick ^{nat}Ni. The secondary beam was identified event-by-event based on time-of-flight and energy-loss information. The beam was pulsed with cycles of 500ms/500ms for beam-on/off to measure the half-lives through detections of delayed γ rays.

The secondary beam was implanted into an active stopper located at the final focal plane of RIPS. The stopper which was a plastic scintillator with a thickness of 5 mm and an area of 100×100 mm^2 was tilted at an angle of 45 degrees with respect to the beam axis. The active stopper detected implanted ions as well as β rays. The delayed γ rays were detected using a clover-type Ge detector placed at 6.0 cm apart from the center of the stopper. The Ge detector was surrounded by eight BGO counters to suppress continuum backgrounds deriving from Compton scatterings. Each BGO counter had a size of $30 \times 80 \times 250$ mm^3. In front of the Ge detector, a plastic β-veto counter with a thickness of 1 mm and an area of 100×100 mm^2 was placed to reject β rays incident to the detector. The detailed description about the experiment is seen in Refs. [10, 11].

In this experiment, we observed two β branches to the bound states in ^{24}Al for the first time. Figure 1 shows the delayed γ rays of 0.426 and 0.664 MeV. The 0.426-MeV γ ray corresponds to de-excitation from the 1^+_1 isomeric state [5] to the ground state in ^{24}Al. The 0.664-MeV γ ray was identified to be a delayed γ ray of ^{24}Si, because its half life was determined to be 140.1(26) ms which was in good agreement with that of ^{24}Si [12, 13]. The 0.664-MeV γ ray was assigned to be a de-excitation γ ray from a level at 1.090(1) MeV ($= 0.426 + 0.664$) to the 1^+_1 state in ^{24}Al. The branching ratios to the

---

[1] Although the spin-parity of the "1^+_2 state" at 1.111-MeV in ^{24}Al is only tentatively assigned to be $(1^+, 2^+, 3^+)$ [6, 7], here it is assumed to be 1^+ based on the existence of a 1^+ level at 1.346 keV in ^{24}Na.

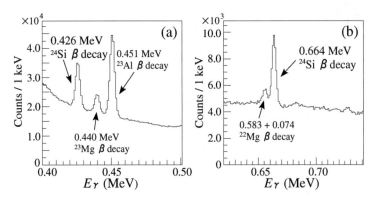

FIGURE 1. New observation of β-delayed γ rays of ^{24}Si. Figures (a) and (b) show the 0.426-MeV and 0.664-MeV lines, respectively.

1_1^+ and the 1.090-MeV states were determined to be $b_1 = 0.31(4)$ and $b_2 = 0.239(15)$, respectively, using the observables of the number of implanted ^{24}Si, the yields of the γ rays and the detection efficiency. In deduction of the b_1 value, a reference value of the electro-magnetic transition ratio $I_\gamma(^{24}\text{Al}^m) = 0.82(3)$ [14] on the 1_1^+ isomeric state was used.

For the observed β branches, log ft values were deduced to be 4.49(6) and 4.45(3) for the 1_1^+ and the 1.090-MeV states, respectively. The log ft value of 4.45 to the 1.090-MeV state is appropriate for an allowed transition. Thus it is possible to establish the spin-parity assignment for this state firmly to be 1^+. The GT transition strengths $B(\text{GT})$ were deduced to be 0.13(2) and 0.14(1) for the 1_1^+ and the 1_2^+ states, respectively.

DISCUSSION

The deduced $B(\text{GT})$ of ^{24}Si were compared with that of the mirror nucleus, ^{24}Ne [8]. The $B(\text{GT})$ asymmetry, defined as a ratio of $B(\text{GT}^+)/B(\text{GT}^-)$, were derived to be 0.78(11) and 0.90(8) for the 1_1^+ and 1_2^+ states, respectively. The $B(\text{GT})$ asymmetries appear both in the 1_1^+ and 1_2^+ states, though the large energy asymmetry is only seen in the 1_2^+ state. Rather the asymmetry is larger in the 1_1^+ state than that in the 1_2^+ state. The appearances of $B(\text{GT})$ asymmetry in both states indicate the configuration change, because the energy asymmetry is mainly sensitive to the s-orbital component, while the $B(\text{GT})$ asymmetry is sensitive to the change of configuration mixing between the s- and d-orbital components.

In order to clarify the origin of the $B(\text{GT})$ asymmetry microscopically, the experimental $B(\text{GT})$ were compared with theoretical ones, as shown in Table 1. The theoretical $B(\text{GT})$ were calculated in the sd-shell model space with USD[15] and more recent USDA and USDB [16] Hamiltonians, including the empirical quenching factor of 0.6 [17]. These Hamiltonians are isospin invariant and the $B(\text{GT})$ values for the mirror decays are equal. In addition to USD, calculations for the mirror asymmetry were carried out with the Hamiltonian [18] called USD+C in Table 1, by considering the Coulomb interaction together with charge-independence breaking and charge-asymmetry breaking of strong interactions. Further an effect of weak binding is taken into account by

TABLE 1. Comparison of experimental and theoretical B(GT) for mirror decays of ^{24}Si and ^{24}Ne. Experimental data for ^{24}Ne were taken from Ref. [8].

	1_1^+			1_2^+		
	$B(GT^+)$	$B(GT^-)$	ratio	$B(GT^+)$	$B(GT^-)$	ratio
Exp.	0.13(2)	0.167(4)	0.78(11)	0.14(1)	0.155(9)	0.90(8)
USD	0.1077	0.1077	1	0.1462	0.1462	1
USDA	0.0665	0.0665	1	0.1526	0.1526	1
USDB	0.1231	0.1231	1	0.1112	0.1112	1
USD+C	0.1051	0.1116	0.94	0.1353	0.1450	0.93
USD+C*	0.0872	0.1116	0.78	0.1418	0.1450	0.98

lowering the single-particle energy of the $1s_{1/2}$ proton orbital by 500 keV relative to the neutron single-particle energy. This lowers the energy of the 1_2^+ state in ^{24}Al by 202 keV compared to its experimental shift of 256 keV. The ratio of the mirror B(GT) values is in good agreement with experiment, as shown in the line denoted by USD+C* in Table 1. Thus, the origin of the B(GT) asymmetry in the 1_1^+ is attributed to the lowering of the $1s_{1/2}$ orbital which causes small differences in the various amplitudes of the one-body transition density. For the 1_2^+ states, the changes of some amplitudes are accidentally canceled, then the B(GT) asymmetry is not significant.

ACKNOWLEDGMENTS

The experiment was carried out at RIKEN Accelerator Research Facility operated by RIKEN Nishina Center and CNS, University of Tokyo. This work is supported by Grant-in-Aid for Scientific Research No.15204017 in Japan, NSF grant PHY-0758099 in USA and the Japan-US Theory Institute for Physics with Exotic Nuclei (JUSTIPEN). One of the authors (Y. I.) acknowledges Special Postdoctoral Researchers Program in RIKEN.

REFERENCES

1. X. Z. Cai et al., *Phys. Rev.* **C65**, 024610 (2002).
2. R. G. Thomas, *Phys. Rev.* **88**, 1109 (1952).
3. J. B. Ehrman, *Phys. Rev.* **81**, 412 (1951).
4. L. V. Grigorenko, et al., *Phys. Rev. Lett.* **88**, 042502 (2002).
5. A. J. Armini et al., *Phys. Lett.* **21**, 335 (1966).
6. G. C. Kiang et al., *Nucl. Phys.* **A499**, 339 (1989).
7. S. Kubono et al., *Nucl. Phys.* **A588**, 521 (1995).
8. R. E. McDonald et al., *Phys. Rev.* **181**, 1631 (1969).
9. T. Kubo et al., *Nucl. Instrum. Meth.* **B70**, 309 (1992).
10. Y. Ichikawa et al., *Eur. Phys. J.* **A**, published online: 29 March (2009).
11. Y. Ichikawa et al., *Phys. Rev.* **C**, to be published.
12. V. Banerjee et al., *Phys. Rev.* **C63**, 024307 (2001).
13. S. Czajkowski et al., *Nucl. Phys.* **A628**, 537 (1998).
14. J. Honkanen et al., *Phys. Scr.* **19**, 239 (1979).
15. B. A. Brown and B. H. Wildenthal, *Ann. Rev. of Nucl. Part. Sci.* **38**, 29 (1988).
16. B. A. Brown and W. A. Richter, *Phys. Rev.* **C74**, 034315 (2006).
17. B. A. Brown and B. H. Wildenthal, *Atomic Data Nuclear Data Tables* **33**, 347 (1985).
18. W. E. Ormand and B. A. Brown, *Nucl. Phys.* **A491**, 1 (1989).

Two-proton Decays from Light to Heavy Nuclei. Comparison of Theory and Experiment.

M. V. Zhukov[*] and L. V. Grigorenko[†]

[*]Fundamental Physics, Chalmers University of Technology, S-41296 Göteborg, Sweden
[†]Flerov Laboratory of Nuclear Reactions, JINR, RU-141980 Dubna, Russia

Abstract. Recently the complete three-body correlation pictures were, for the first time, experimentally reconstructed for the two-proton decays of the ^{6}Be and ^{45}Fe ground states. We are able to see qualitative similarities and differences between these decays. They demonstrate very good agreement with the predictions of a theoretical three-body cluster model.

Keywords: Two-proton decay, Three-body Coulomb problem, Hyperspherical harmonics method
PACS: 23.50.+z, 23.20.En, 21.60.Gx

INTRODUCTION

The ground state two-proton ($2p$) radioactivity was predicted by V.I. Goldansky in 1960 [1] as an exclusively quantum-mechanical phenomenon. True three-body decay, in his terms, is a situation where the sequential emission of the particles is energetically prohibited and all the final-state fragments are emitted simultaneously, which cause distinct energy systematics and specific correlation pattern of such decays. Since the experimental discovery of the ^{45}Fe two-proton radioactivity in 2002 [2, 3], the recent progress of this field is very fast. New cases of $2p$ radioactivity were found for ^{54}Zn [4], ^{19}Mg [5], and, maybe, ^{48}Ni [6].

The quantum-mechanical theory of $2p$ radioactivity and three-body decay was developed in Refs. [7, 8]. This is a three-body cluster approach utilizing the hyperspherical harmonics method with approximate boundary conditions for the three-body Coulomb problem. Exploratory studies of correlations performed in [9] predicted complex correlation patterns which are sensitive (in s/d shell nuclei) and very sensitive (in p/f shell nuclei) to the structure of the $2p$ emitter. Confirmation of the predictions [9] was obtained in paper [10] for inclusive correlation spectra of the ^{45}Fe decay. However, it should be mentioned that that the integrated distributions can give only a limited representation of the complete correlation picture.

^{6}BE AND ^{45}FE - COMPLETE CORRELATIONS

It should be noted, that the ^{6}Be nucleus is the lightest true two-proton emitter in the sense offered by Goldansky and is an isobaric partner of the famous Borromean halo nucleus ^{6}He. However, theoretical studies of ^{6}Be were limited, so far, mainly to the studies of energies and widths of the states (the complete correlations have never been calculated). The precise experimental data dedicated to correlations in ^{6}Be were also not existing

CP1165, *Nuclear Structure and Dynamics '09*
edited by M. Milin, T. Nikšić, D. Vretenar, and S. Szilner

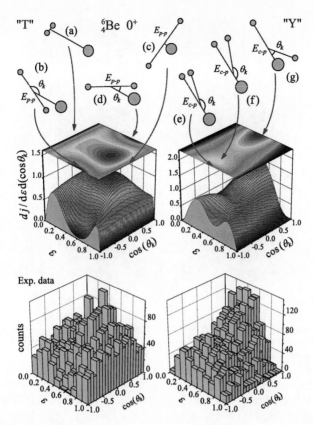

FIGURE 1. (Colour online) Complete correlation picture for ^{6}Be g.s. decay, presented in "T" and "Y" Jacobi systems (left and right columns respectively). The upper row is theory, lower is experimental data.

(there were only two papers [11, 12] where the inclusive α spectra from the ^{6}Be ground state decay were presented).

The ^{6}Be ground state decay was studied very recently in the new experiment [13], where ^{6}Be fragments were produced from the α-decay of ^{10}C projectiles excited via inelastic-scattering interactions on Be and C targets. The complete kinematics was reconstructed for the ^{6}Be events and they were separated from the other decay channels. Simultaneously theoretical calculations for the ^{6}Be ground state decay, were performed. More details of the experiment and theory can be found in Refs. [13, 14].

Since publication [10] the data treatment of the ^{45}Fe decay events has been improved, including the reconstruction of complete kinematics. Some details of the experiment are given in Ref. [10], the full description will be given in Ref. [15].

Joint experimental and theoretical studies complete correlations from two-proton decays of the ground states ^{6}Be and ^{45}Fe nuclei have been published recently in Ref. [16]. In this presentation we will discuss these results.

For fixed three-body decay energy E_T we end up with two parameters representing the

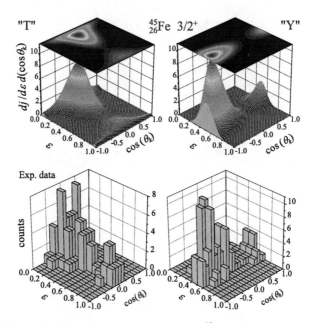

FIGURE 2. (Colour online) Complete correlation picture for ^{45}Fe g.s. decay, presented in "T" and "Y" Jacobi systems. The upper row is theory, lower is experimental data. Theoretical predictions are shown for the case of configuration mixing with $W(p^2) = 24\%$.

complete correlation picture. It is convenient to choose an energy distribution parameter $\varepsilon = E_x/E_T$ and $\cos(\theta_k) = (\mathbf{k_x} \cdot \mathbf{k_y})/(k_x k_y)$ where θ_k is an angle between the Jacobi momenta $\mathbf{k_x}, \mathbf{k_y}$. For two-proton emitters (protons are indistinguishable) there are two "irreducible" Jacobi systems, called "T"and "Y" (see top of Fig. 1). In the "T" Jacobi system, the parameter ε describes the energy distribution between the two protons and in the "Y" Jacobi system, ε corresponds to core-proton subsystem. Distributions in the different Jacobi systems are just different representations of the same physical picture. Each Jacobi system can reveal different aspects of the correlation picture.

Complete correlation pictures [16] for the ^{6}Be and ^{45}Fe g.s. decays are shown in Figs. 1 and 2. Schematic figures are included in Fig. 1 to help in visualizing the correlations associated with different regions of the Jacobi plots. Differences and similarities between the heavy and light $2p$ emitters can easily be seen in these plots. The general features of the experimental correlations are reproduced by the theory. This can be confidently stated even for the relatively low ^{45}Fe statistics.

Despite the strong differences in Figs. 1 and 2 there are certain kinematical regions (illustrated on top of Fig. 1) where the decay dynamics is governed by the same physics.

Strong Coulomb suppressions due to the core-p repulsion occur in regions (b) and (d) and a smaller effect for the p-p channel is found in (e). The magnitude of the suppressions in (b) and (d) is significantly larger for the heavier ^{45}Fe case. On the other hand, the p-p final-state interaction gives rise to enhancements in region (f). Although the g.s. core-p resonance is not energetically accessible for decay, the enhancement in region (g) is a hint of its presence.

It is clear that in the limits $\varepsilon \to 0$ and $\varepsilon \to 1$, the dependence on the relative orientation of $\mathbf{k}_x$ and $\mathbf{k}_y$ in the "T" system should become degenerate. The angular dependence practically vanishes for situations (a) and (c) but it can be seen in the theoretical plots that the scale of the phenomenon is very different for ^{6}Be and ^{45}Fe. At intermediate values of ε the angular dependence of distributions is very pronounced.

There is one more interesting aspect connected to the two-proton decay of ^{6}Be. In the experimental paper [11] the authors stressed that "... the largest cross section occurs when two protons are emitted ~ 180 degrees from one another with nearly equal energies". It is important that both new experiment and theory confirm this observation. It can easily be seen in Figs. 1 in the "Y" Jacobi system in the region (g) $(\cos(\theta_k) \sim 1.0$ and $\varepsilon \sim 0.5)$. Note that this effect is practically not seen in the integrated distributions.

CONCLUSION

The complete correlation pictures of $2p$ decays for ^{6}Be and ^{45}Fe are measured (and for ^{6}Be also calculated) for the first time. These correlations have common features (e.g., connected with significant weight of $[p^2]$ component in the WF) as well as differences (e.g., connected with the larger Coulomb repulsion in heavy $2p$ emitter and admixture of $[f^2]$ configuration).

Features of the three-body correlations are well reproduced by the quantum-mechanical model in a broad range of times and masses: from typical nuclear times $(\sim 10^{-20}$ s in ^{6}Be) to typical radioactivity times (some milliseconds in ^{45}Fe).

Theoretical models are further constrained when consistent descriptions of the widths and correlations are made simultaneously. This provides a cross check of the extracted structure information which has never been accessible before.

REFERENCES

1. V.I. Goldansky, Nucl. Phys. **19** (1960) 482.
2. M. Pfützner, et al., *Eur. Phys. J.* A **14**, 279 (2002).
3. J. Giovinazzo, et al., *Phys. Rev. Lett.* **89**, 102501 (2002).
4. B. Blank, et al., *Phys. Rev. Lett.* **94**, 232501 (2005).
5. I. Mukha, et al., *Phys. Rev. Lett.* **99**, 182501 (2007).
6. C. Dossat, et al., *Phys. Rev.* C **72**, 054315 (2005).
7. L.V. Grigorenko, et al., *Phys. Rev. Lett.* **85**, 22 (2000).
8. L.V. Grigorenko, R.C. Johnson, I.J. Thompson, and M.V. Zhukov, *Phys. Rev.* C **65**, 044612 (2002).
9. L.V. Grigorenko and M.V. Zhukov, *Phys. Rev.* C **68**, 054005 (2003).
10. K. Miernik, et al., *Phys. Rev. Lett.* **99**, 192501 (2007).
11. D.F. Geesaman, et al., *Phys. Rev.* C **15**, 1835 (1977).
12. O.V. Bochkarev, et al., *Nucl. Phys.* A **505**, 215 (1989).
13. K. Mercurio , et al., *Phys. Rev.* C **78**, 031602(R) (2008).
14. L.V. Grigorenko, et al., *Phys. Rev.* C (2009), submitted; arXiv:0812.4065.
15. K. Miernik, et al., *Eur. Phys. J.* A (2009), in press.
16. L.V. Grigorenko, et al., Phys. Lett. B **677**, 30 (2009).

Two-proton Emission from ^{29}S and ^{17}Ne Excited States via Coulomb Excitation

X.X. Xu[a], C.J. Lin[a], H.M. Jia[a], F. Yang[a], F. Jia[a], S.T. Zhang[a], Z.H. Liu[a], H.Q. Zhang[a], H.S. Xu[b], Z.Y. Sun[b], J.S. Wang[b], Z.G. Hu[b], M. Wang[b], R.F. Chen[b], X.Y. Zhang[b], C. Li[b], X.G. Lei[b], Z.G. Xu[b], G.Q. Xiao[b], and W.L. Zhan[b]

[a]*China Institute of Atomic Energy, Beijing 102413, P. R. China*
[b]*Institute of Modern Physics, The Chinese Academy of Science, Lanzhou 73000, P. R. China*

Abstract. The experiments of two-proton emission from ^{29}S and ^{17}Ne excited levels were performed at HIRFL-RIBLL facility of Institute of Modern Physics, Lanzhou. Complete kinematics measurements were achieved by the detectors of silicon strip and CsI+PIN array after the ^{197}Au target. The preliminary results show the strong correlation of two protons according to the spectrum of correlated proton-proton energies in the laboratory reference frame.

Key words: two-proton emission, complete kinematics measurements
PACS numbers: 23.50.+z, 25.60.-t, 25.70.-z, 29.40.Gx

INTRODUCTION

Two-proton (2p) radioactivity predicted by Goldansky[1] almost 50 years ago is one of the exotic decay modes near the proton drip line. Experimentally, the ground-state 2p radioactivity for ^{45}Fe [2-5], ^{54}Zn [6], and possibly ^{48}Ni [7] has been observed. Besides, 2p emission from excited states of ^{14}O [8], ^{17}Ne [9-11], ^{18}Ne [12, 13] and ^{94}Ag [14] was also reported. Recently, with the Optical Time Projection Chamber (OTPC) [15], the energy and angular correlations of 2p emitted from the ground state of ^{45}Fe have been investigated for the first time [5], which exhibit three-body decay for the mixtures of p^2 and f^2 configurations. For the 6.15 MeV state of ^{18}Ne, 2p decays proceeding through a ^{2}He diproton resonance (31%) and democratic/virtual-sequential decay, respectively, have been demonstrated in a very recent work [13]. However, the Final State Interaction (FSI) was neglected for the three-body decay of ^{18}Ne in the Monte Carlo (MC) simulation.

Study of 2p emission from excited states in our group, the Nuclear Reaction Group at China Institute of Atomic Energy, has been started in 2005, and until now, the complete kinematics measurements of ^{29}S+^{12}C [16], ^{17,18}Ne+^{197}Au, ^{28,29}S+^{197}Au have been achieved with the silicon strip and CsI+PIN detector array. In this presentation, we will show the recent results of 2p emission from ^{29}S and ^{17}Ne excited states.

EXPERIMENTS

The experiments were performed at HIRFL-RIBLL spectrometer [18] of the Institute of Modern Physics (Lanzhou, China) by using the ^{29}S Radioactive Ion Beam (RIB) at the energy of 49.2MeV/u and ^{17}Ne at 49.92MeV/u. The secondary ions were produced by the projectile fragmentation of a primary beam ^{32}S or ^{20}Ne bombarding on a ^{9}Be

CP1165, *Nuclear Structure and Dynamics '09*
edited by M. Milin, T. Nikšić, D. Vretenar, and S. Szilner
© 2009 American Institute of Physics 978-0-7354-0702-2/09/$25.00

target. Subsequently, the radioactive beams were separated by the combined Bρ-ΔE-Bρ method of ^{27}Al degrader for the isotope purification process and a Bρ setting optimized for the ^{29}S or ^{17}Ne transmission. About 10^4 ions/sec on the secondary target were observed with the purity of ^{29}S or ^{17}Ne at 3% of the total RIBs mixture. The Time-of-Flight (ToF) of the RIBs was measured by two scintillator detectors on the second and fourth focus plane of the RIBLL facility.

The schematic drawing of the detector array using in the ^{29}S experiment is shown in Fig. 1. The setup has the capability of the complete-kinematics measurement for all the decay products from excited states of ^{29}S by Coulomb excitation. The secondary beams were identified event by event through the ΔE-ToF measurements. Total of $2.5×10^7$ ^{29}S events were recorded in the experiment. Two parallel plate avalanche counters (PPACs) were placed in front of the target for beam tracking. The secondary reaction target was ^{197}Au with the thickness of 100 μm. The reaction products including heavy fragments and light particles were identified and tracked by a multiple stack telescope of particles (Silicon Strip Detectors，6×6 CsI detectors etc.). The maximum of the opening angle for the detector array was ±13°.

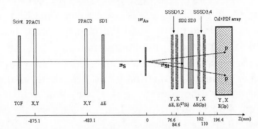

FIGURE 1. The schematic diagram of the experiment setup designed for the complete-kinematics measurement of all the reaction products following the Coulomb excitation of ^{29}S.

The detector array of the ^{17}Ne experiment was similar to the setup in Fig. 1. Figure 2(a) displays the ΔE versus ToF matrix for the particle identification. $5.05×10^8$ ^{17}Ne events were accumulated in total. Figures 2(b) and (c) exhibit the sum of the energy losses in the silicon detectors (ΔE) versus the rest energy deposited in the CsI array (E) for events in coincidence with ^{17}Ne projectiles. Five bands can be seen, in which the most intense one corresponds to the one-proton events and the 2p events appear at the double value of total energy and energy loss of the single proton band. For the band of d events, there are two interesting peaks.

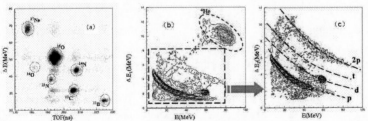

FIGURE 2. Particle identification: (a) Identification of the projectile fragmentation of a primary beam ^{20}Ne; (b) and (c) Light particle identification emitted from ^{17}Ne projectiles.

MC SIMULATIONS AND PRELIMINARY RESULTS

Three schematic pictures are used to describe 2p emission from ^{29}S excited states in the simulation. For the ^{2}He decay, the process consists of two steps: (1) the ^{2}He cluster is formed in the potential well of ^{27}Si. The total available energy of the ^{2}He is E_{av}= $[E_{ex}(^{29}$S)-5.35- $E_{ex}(^{27}$Si)] MeV, where E_{ex} is the excited energy of the ^{29}S or ^{27}Si. The fraction of E_{av}, ε, is converted to the ground state mass of ^{2}He, which is also recognized as the resonance energy of the two protons. So the mass of ^{2}He is defined as $M_{^2He} = \varepsilon + 2M_p$. The reminder E= E_{av}-ε is the energy available for the diproton to tunnel through the Coulomb barrier. (2) After the diproton penetrates the barrier it breakups into two protons with the energy ε that changes into the relative energy of the proton-proton (p-p) pair. As follows,

$$^{29}S^* \rightarrow {}^{27}Si^* + {}^2He \quad (Q=E_{ex}(^{29}S)-5.35-\varepsilon-E_{ex}(^{27}Si)) ; \quad {}^2He \rightarrow p+p \quad (Q=\varepsilon)$$

Sequential emission from excited states in ^{29}S can decay via intermediate states in ^{28}P and the two protons are unbound by the total available energy, E_{av}, which is divided into two steps as follows:

$$^{29}S^* \rightarrow {}^{28}P^* + p \quad (Q= E_{ex}(^{29}S)-3.29- E_{ex}(^{28}P))$$
$$^{28}P^* \rightarrow {}^{27}Si^* + p \quad (Q= E_{ex}(^{28}P)-2.06- E_{ex}(^{27}Si))$$

The third mechanism, three body decay, occurs when the two protons leaves the mother nucleus with no correlations beyond the phase-space constraints and FSI. In the simulation, FSI was neglected as the configurations (J^π) of excited levels in ^{29}S were still unknown. Fig. 3 shows the simulation results of correlated p-p energies in the laboratory reference frame. The distribution of ^{2}He decay is quite different from the others. The spectrum for 3-body decay with FSI may be similar to the ^{2}He decay. The experimental data were shown in Fig. 4. It indicates strong correlations of two protons which may originate from ^{2}He cluster or FSI.

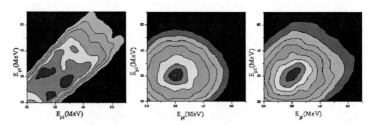

FIGURE 3. The simulated spectrum of correlated p-p energies in the laboratory frame:
(a) ^{2}He decay; (b) Phase Space; (c) Sequential two-body decay

The relative momentum (q_{pp}=|$\mathbf{P}_1$-$\mathbf{P}_2$|/2), the opening angle ($\theta_{pp}^{c.m.}$), and the invariant mass or excitation energy ($E^* = \sqrt{(\sum E_i)^2 - (\sum P_i c)^2} - M_{gr}$) have been deduced by relativistic-kinematics reconstruction in order to study 2p correlation. In the experiment of ^{29}S, results show the enhanced peaks at q_{pp}=20 MeV/c and $\theta_{pp}^{c.m.}$=35° for the 10.0 MeV (9.6<E^*<10.4 MeV) state. According to the MC simulation (no FSI), the branching ratio of ^{2}He emission is 29$^{+10}_{-11}$ %. While for the other states such as the 7.4 MeV (7.0<E^*<7.8 MeV) state, the experiment data show the maximum at q_{pp}=35 MeV/c and the opening angles show the feature of sinθ, indicating the branching ratio of ^{2}He emission less than 10%. The enhancement at 0.45 MeV for the relative energy of 2p was also observed for the 10.0 MeV state of ^{29}S. Details of ^{29}S experiment and its results can be found in Ref. [18]. In the ^{17}Ne experiment, preliminary results show the similar distribution of opening angle as in Ref [11]. Moreover, high statistics of 2p emission, about 2000 events, was picked out under strict conditions. The decay dynamics of 2p correlation for ^{17}Ne excited states can be studied under the precise theoretical framework which takes FSI into account.

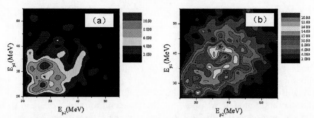

FIGURE 4. The spectrum of p1 energy versus p2 energy in the LAB;
2p emitted from excited states of (a) ^{29}S and (b) ^{17}Ne.

SUMMARY

2p emission from excited states of ^{17,18}Ne, ^{28,29}S has been experimentally investigated by complete kinematics measurements. Some preliminary results are presented to show the strong p-p correlation. For the sake of understanding decay dynamics of 2p emission from excited states, the modern theories, such as 3-body democratic decay model, the extended R-matrix theory, and the Faddeev equations *etc.* are urgently required.

ACKNOWLEDGEMENTS

This work was supported by the National Natural Science Foundation of China under Grants No. 10675169, 10735100, 10727505, and the Major State Basic Research Developing Program under Grant No. 2007CB815003.

REFERENCES

[1] V. Goldansky, *Nucl. Phys.* **19**, 482 (1960); ibid **27**, 648 (1961).
[2] M. Pfutzner, et al., *Eur. Phys. J. A* **14**, 279 (2002).
[3] J. Giovinazzo, et al., *Phys. Rev. Lett.* **89**, 102501 (2002); ibid **99**, 102501 (2007).
[5] K. Miernik, et al., *Phys. Rev. Lett.* **99**, 192501 (2007).
[6] B. Blank, A. Bey, G. Canche, et al., *Phys. Rev. Lett.* **94**, 232501 (2005).
[7] C. Dossat, et al., *Phys. Rev. C* **72**, 054315 (2005).
[8] C. Bain, et al., *Phys. Lett. B* **373**, 35(1996).
[9] M. J. Chromik, et al., *Phys. Rev. C* **55**, 1676 (1997).
[10] M. J. Chromik, et al., *Phys. Rev. C* **66**, 024313 (2002).
[11] T. Zeguerras, et al., *Eur. Phys. J. A* **20**, 389 (2004).
[12] J. Gomez del Campo, et al., *Phys. Rev. Lett.* **86**, 43 (2001).
[13] G. Raciti, et al., *Phys. Rev. Lett.* **100**, 192503 (2008).
[14] I. Mukha, et al., *Nature (London)* **439**, 298 (2006).
[15] K. Miernik, et al., *Nucl. Instrum. Meth. Phys. Res. A* **581**, 194 (2007).
[16] C. J. Lin, et al., *AIP Conf. Proc.* **961**, 117 (2007).
[17] Z. Y. Sun, et al., *Nucl. Instrum. Meth. Phys. Res A* **503**, 496 (2003).
[18] C. J. Lin, et al., submitted to *Phys. Rev. C*.

Two Neutron Correlations in Exotic Nuclei

H. Sagawa[*,†] and K. Hagino[**]

[*] Center for Mathematics and Physics, University of Aizu, Aizu-Wakamatsu, 965-8580 Fukushima,
Japan
[†] RIKEN Nishina Center, RIKEN, 2-1 Hirosawa, Wako, Saitama 351-0198, Japan
[**] Department of Physics, Tohoku University, Sendai, 980-8578, Japan

Abstract. We study the correlations between two neutrons in borromian nuclei ^{11}Li and ^{6}He by using a three-body model with a density-dependent contact two-body interaction. It is shown that the two neutrons show a compact bound feature at the nuclear surface due to the mixing of single particle states of different parity. We study the Coulomb breakup cross sections of ^{11}Li and ^{6}He using the same three-body model. We show that the concentration of the B(E1) strength near the threshold can be well reproduced with this model as a typical nature of the halo nuclei. The energy distributions of two emitted neutrons from dipole excitations are also studied using the correlated wave functions of dipole excitations.

Keywords: Di-neutron correlations, borromian nuclei
PACS: 21.30.Fe,21.60.Jz,21.65.-f,26.50.+x

It has been well recognized by now that the borromian nuclei such as ^{11}Li and ^{6}He show a strong di-neutron correlations in the ground states and also in the excited states. Recently, Nakamura *et al.* have remeasured the low-lying dipole excitations in ^{11}Li nucleus and have confirmed the strong concentration of the dipole strength near the threshold in the 2-neutron (2n) halo nucleus [1]. The low-lying dipole strength for another 2n halo nucleus, ^{6}He, has also been measured by Aumann *et al.* [2]. The two neutron correlations are further measured very recently in the Coulomb breakup process of dipole excitaitons in ^{11}Li [3].

The aim of this paper is to study the correlations between di-neutrons and also neutron-core correlations in borromian nuclei ^{6}He and ^{11}Li by using a three-body model [4, 5, 6]. In Ref. [7], the behavior of the two valence neutrons in ^{11}Li is studied at various positions from the center to the surface of the nucleus. It was found that the two-neutron wave function oscillates near the center whereas it becomes similar to that for a compact bound state around the nuclear surface, and the mean distance between the valence neutrons has a well pronounced minimum around the nuclear surface. We have pointed out that these are qualitatively the same behaviors as found in neutron matter [8]. To elucidate these points, we show in Fig. 1 the mean distance of the valence neutrons in ^{11}Li as a function of the nuclear radius R (the distance between the core and the center of two neutrons) obtained with and without the neutron-neutron (nn) interaction. For the uncorrelated calculations, we consider both the $[(1p_{1/2})^2]$ and $[(2s_{1/2})^2]$ configurations. One can see that, in the non-interacting case, the neutron pair almost monotonously expands, as it gets further away from the center of the nucleus. On the other hand, in the interacting case it first becomes smaller going from inside to the surface before expanding again into the free space configuration. These results confirm the strong and predominant influence of the pairing force in the nuclear surface of^{11}Li. We also show

CP1165, *Nuclear Structure and Dynamics '09*
edited by M. Milin, T. Nikšić, D. Vretenar, and S. Szilner
© 2009 American Institute of Physics 978-0-7354-0702-2/09/$25.00

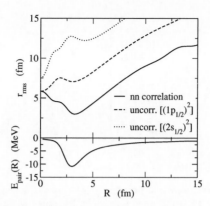

FIGURE 1. Upper panel: the root mean square distance r_{rms} between the valence neutrons in ^{11}Li as a function of the distance R between the core and the center of two neutrons. The solid line is obtained by taking into account the neutron-neutron correlations, while the dashed and the dotted lines are obtained by switching off the neutron-neutron interaction and assuming the $[(1p_{1/2})^2]$ and $[(2s_{1/2})^2]$ configurations, respectively. Lower panel: the neutron-neutron correlation energy as a function of the distance R.

the local neutron-neutron correlation energy as a function of the radius R in the lower panel of Fig. 1. It is clearly seen that the energy gain is the maximum at the surface where the correlaiton length is the minimum. The two panels in Fig. 1 confirm that the kink of the correlation length is induced by the strong pairing correlations at the surface.

Two particle densities of the correlated pair and the uncorrelated $[(1p_{1/2})^2]$ configuration are shown in Fig. 2. The reference particle is located at $(z,x) = (3.4,0)$fm. As can be seen in the right panel, the distribution has a symmetric two peaks in (z,x) plane with respect to the center of the core nucleus at $(z,x) = (0,0)$fm. This is due to the absence of mixing of opposite parity wave functions into the $[(1p_{1/2})^2]$ configuration. On the contrary, the peak appears only around the position of the reference particle when the two neutron correlations are taken into account in the wave functions. To compare two panels in Fig. 2, we can see a clear manifestation of the strong two neutron correlations in the wave function of the borromian nucleus ^{11}Li.

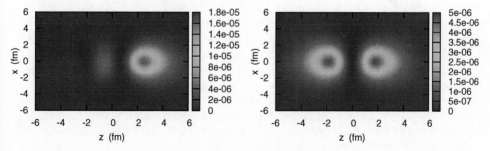

FIGURE 2. (Color online) Two dimensitional (2D) plots for the two particle density of the correlated pair (left panel) and uncorrelated $[(1p_{1/2})^2]$ configuration (right panel) in ^{11}Li. It represents the probability distributions for the spin-up neutron placing the spin-down neutron at $(z,x)=(3.4,0)$fm. The core nucleus is located at the origin $(z,x)=(0,0)$fm.

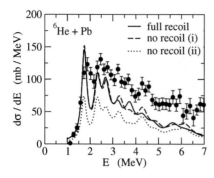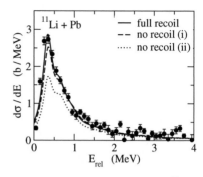

FIGURE 3. Coulomb breakup cross sections for ^{6}He+Pb at 240 MeV /nucleon and for the ^{11}Li +Pb at 70 MeV/nucleon. The solid line is the result of the full calculations, while the dashed line is obtained by neglecting the off-diagonal component of the recoil kinetic energy in the excited states. The dotted line is obtained by neglecting the off-diagonal recoil term both in the ground and the excited states. These results are smeared with an energy dependent width of $\Gamma = 0.15 \cdot \sqrt{E_{rel}}$ MeV for ^{6}He and $\Gamma = 0.25 \cdot \sqrt{E_{rel}}$ MeV for ^{11}Li. The experimental data are taken from Refs. [2] and [1] for ^{6}He and ^{11}Li, respectively.

Figs. 3 compare the Coulomb breakup cross sections calculated by taking into account the recoil term exactly (the solid curves) with those calculated approximately (the dashed and dotted curves). For the dashed curves, the off-diagonal component of the recoil kinetic energy is neglected in the excited $J^\pi = 1^-$ states, while it is fully taken into account in the ground state. It is interesting to notice that these calculations lead to similar results to the one in which the recoil term is treated exactly (the solid curves). The dotted curves, on the other hand, are obtained by neglecting the off-diagonal part of the recoil term both for the ground and the $J^\pi = 1^-$ states. By neglecting the recoil term in the ground state, the value for $\langle r^2_{c-2n} \rangle$ decreases, from 13.2 fm^2 to 9.46 fm^2 for ^{6}He and from 26.3 fm^2 to 20.58 fm^2 for ^{11}Li. Consequently, the B(E1) distribution as well as the breakup cross sections are largely underestimated. These results clearly indicate that the recoil term is important for the ground state, while it has a rather small effect on the excited states.

Figures 4 show the dipole strength distribution, $d^2B(E1)/de_1de_2$, as a function of the energies of the two emitted neutrons for the ^{11}Li and ^{6}He nuclei, respectively [9]. One immediately notices that the strength distribution is considerably different between ^{11}Li and ^{6}He. For ^{11}Li, a large concentration of the strength appears at about e_1=0.375 MeV and e_2=0.075 MeV (and at e_1=0.075 MeV and e_2=0.375 MeV), with a small ridge at an energy of about 0.5 MeV. On the other hand, for ^{6}He, the strength is largely concentrated at one peak around $e_1 = e_2 = 0.7$ MeV and only a large ridge at about 0.7 MeV appears. This difference between ^{11}Li and ^{6}He is due to the existence of virtual s-state in the residual ^{10}Li, but not in ^{5}He. Thus, if the interaction between the neutron and the core nucleus is switched off, the distribution become similar between the two nuclei because the virtual s-state is diappeared in ^{10}Li.

In summary, we have studied the di-neutron correlations and the neutron-core correlations in the borromian nuclei ^{6}He and ^{11}Li by using the three-body model with a density dependent contact interaction. It is shown that the two neutron wave functions show a strong di-nuetron correlation at the nuclear surface due to the mixing of differ-

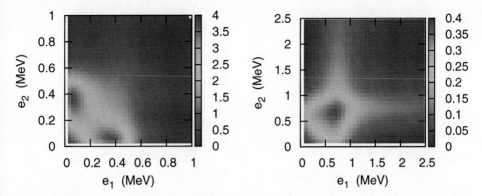

FIGURE 4. (Color online) The dipole strength distributions, $d^2B(E1)/de_1de_2$, of ^{11}Li (left panel) and ^{6}He (right panel) as a function of the energies of the two emitted neutrons relative to the core nucleus. They are plotted in units of e^2fm^2/MeV^2. Figures show the correlated response, which fully takes into account the ground state and final state interactions between the two neutrons.

ent parity single particle states. The same model is used to analyze the dipole strength distributions as well as the Coulomb breakup cross sections of the ^{6}He and ^{11}Li nuclei. We have shown that the strong concentration of the B(E1) strength near the continuum threshold can be well reproduced as a nature of the halo nuclei with the present model. It is shown that the recoil effect plays an important role in the ground state while it may be neglected in the excited states. We have carried out the calculations of the energy and angular distributions of the two emitted neutron from E1 excitations in ^{11}Li and ^{6}He nuclei. We have shown that these distributions are strongly affected by the existence of the virtual s-state in the residual ^{10}Li nucleus. Thus, the properties of the neutron-core potential is crucial to describe the energy distributions of two emitted neutrons, rather than the two neutron correlaions in the excited states.

ACKNOWLEDGMENTS

This work was supported by the Japanese Ministry of Education, Culture, Sports, Science and Technology by Grant-in-Aid for Scientific Research under the program numbers (C) 19740115 and 20540277.

REFERENCES

1. T. Nakamura et al., *Phys. Rev. Lett.* **96**, 252502 (2006).
2. T. Aumann et al., *Phys. Rev.* **C59**, 1252 (1999).
3. T. Nakamura et al., to be published.
4. G.F. Bertsch and H. Esbensen, *Ann. Phys. (N.Y.)* **209**, 327 (1991).
5. H. Esbensen and G.F. Bertsch, *Nucl. Phys.* **A542**, 310 (1992).
6. K. Hagino and H. Sagawa, *Phys. Rev.* **C72**, 044321 (2005).
7. K. Hagino, H. Sagawa, J. Carbonell, and P. Schuck, *Phys. Rev. Lett.* **99**, 022506 (2007).
8. M. Matsuo, *Phys. Rev.* **C73**, 044309 (2006).
9. K. Hagino, H. Sagawa, T. Nakamura and S. Shimoura, arXiv:0904.4775 [nucl-th].

Exploring Nuclear Radii from Total Interaction Cross Sections of Medium Mass Nuclei

D. Dragosavac[a,b], H. Àlvarez-Pol[a], J. Benlliure[a], B. Blank[c], E. Casarejos[a], V. Föhr[d], M. Gascón[a], W. Gawlikowicz[e], A. Heinz[f], K. Helariutta[g], A. Kelić[d], S. Lukić[d], F. Montes[d], D. Pérez-Loureiro[a], L. Pieńkowski[e], K-H. Schmidt[d], M. Staniou[d], K. Subotić[b], K. Sümmerer[d], J. Taieb[h] and A. Trzcińska[e]

[a]*Universidade de Santiago de Compostela, E-15782 Santiago de Compostela,Spain*
[b]*Institute of Nuclear Sciences Vinča, 11001 Belgrade, Serbia*
[c]*Centre d'Etudes Nucleaires, F-33175 Bordeaux-Gradignan, France*
[d]*Gesellschaft für Schwerionenforschung, D-64291 Darmstadt, Germany*
[e]*Heavy Ion Laboratory, Warsaw University,PL-02-093 Warsaw, Poland*
[f]*WNSL, Yale University, New Haven, CT-06520, USA*
[g]*University of Helsinki,FI-00014 Helsinki, Finland*
[h]*CEA DAM, DPTA/SPN, BP. 12, 91680 Bruyeres-le-Chatel, France*

Abstract. Experiments with radioactive nuclear beams have enabled us to investigate nuclear properties far from stability. Glauber model analysis of total interaction cross sections measurements provides a useful method to determine effective matter radii. In this work we have taken advantage of the relativistic heavy ion facility at GSI for producing and investigate total interaction cross sections of more than hundred medium-mass nuclei along long isotopic chains.

Keywords: Nuclear reactions, measured interaction σ
PACS: 25.60.-t, 27.60.+j

INTRODUCTION

Nuclear radii are one of the basic observables used in mean-field calculations for constraining the nuclear equation of state of asymmetric matter. Total interaction cross section measurements provide a useful method to determine effective matter radii. Moreover, experiments with radioactive nuclear beams have enabled us to investigate these observables far from stability. Up to now, total interaction cross sections have been extensively measured for light nuclei up to Kr [1-3].

In this work we have taken advantage of the relativistic heavy ion facility at GSI for producing and investigate total interaction cross sections of more than hundred medium mass nuclei along long isotopic chains.

CP1165, *Nuclear Structure and Dynamics '09*
edited by M. Milin, T. Nikšić, D. Vretenar, and S. Szilner
© 2009 American Institute of Physics 978-0-7354-0702-2/09/$25.00

EXPERIMENTAL PROCEDURE AND RESULTS

The experiment was performed at the GSI facilities in Darmstadt, Germany. Secondary beams of medium-mass nuclei were produced through the projectile fragmentation of a ^{132}Xe primary beam onto a Be target at 1200 A MeV and fission of a ^{238}U beam onto a Pb target at 950 A MeV. Many different isotopes of Ag, Cd, In, Sn, Sb, Te, I and Xe were separated and identified using the first stage of the high-resolving power magnetic spectrometer FRagment Separator (FRS) [4] as shown in Figure 1. All these nuclei impinged onto a secondary Be target placed at the intermediate image plane of the spectrometer. The further identification of the reaction residues using the second stage of the spectrometer allowed us to determine their total interaction cross sections with high accuracy. Details on the separation and identification method can be found in [5].

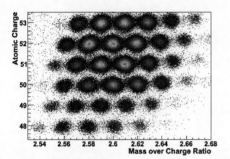

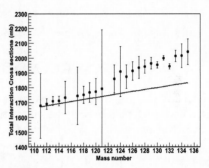

FIGURE 1. Part of a typical identification matrix of fission residues obtained at the intermediate image plane of the FRS.

FIGURE 2. Total interaction cross sections of tin isotopes measured in this work. See text for details.

Figure 2 depicts the measured total interaction cross sections for tin isotopes. As can be seen, the combination of two reaction mechanisms, fission and fragmentation allowed us to produce and measure their total interaction cross sections of a large number of tin isotopes. The measured cross sections are compared with simple estimations (solid line) using a geometrical assumption $\sigma_I \sim r_0 A^{1/3}$ (r_0=1.1 fm). The measured cross sections increase monotonically with mass number, but for the more neutron rich tin isotopes the measured values deviates significantly from the geometrical cross sections $A^{1/3}$. A detailed Glauber analysis of these cross sections combined with the large number of measurements performed in this work will make possible to obtained effective matter radii and investigate their evolution with the neutron excess.

REFERENCES

1. A. Ozawa et al., *Nucl. Phys.* **A693**, 32 (2001).
2. A. Ozawa et al., *Nucl. Phys.* **A709**, 60 (2001).
3. G.F. Lima et al., *Nucl. Phys.* **A735**, 303 (2004).
4. H. Geissel et al., *Nucl. Instr. and Meth.* B **70**, 286 (1992).
5. J. Benlliure et al., *Phys. Rev.* C **78**, 054605 (2008).

Improved Garvey-Kelson Local Relations

William J M F Collis

Strada Sottopiazzo 18, 14055 Boglietto(AT), Italy.   mr.collis @ physics.org

Abstract. We discuss methods of estimating atomic masses using Garvey - Kelson[1,2] like local relations. We show that both the longitudinal and transverse Garvey - Kelson and all other relations can be derived from a simpler mass relationship between just 4 nuclides, C4.

Keywords: Atomic mass, Garvey Kelson, local relations

PACS: 21.10.Dr Binding energies and masses

LOCAL MASS RELATIONS BASED ON 4 NUCLIDES

The difference in masses of the diagonal nuclides in the C4 quadruplet is close to zero. Two such patterns may be subtracted to form the transverse and longitudinal Garvey Kelson relations GKT[1] and GKL[2] as shown in Fig 1.

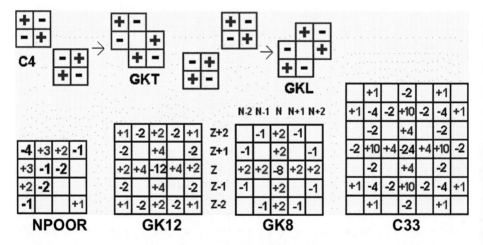

FIGURE 1. shows some how new local relations may be constructed, which consist of sums and / or differences of appropriately placed C4 units on a plane indexed by Z and N. In the simpler examples C4, GKT, GKL the coefficients of M are +1 or -1, and only the sign is shown. Other schemes have some non unity coefficients which are specified. The NRICH pattern (not shown) is similar to NPOOR but with the coefficients symmetrically moved to the lower right neutron rich corner. Note that coefficients in all columns and rows add up to 0.

$$C4(N,Z)=M(N+1,Z)-M(N,Z)+ M(N,Z+1)-M(N+1,Z+1) \approx 0$$

The residual r, on calculating mass using C4 shows a marked odd / even effect.[3,4] This systematic dependence can be used to improve the mass estimate. Empirically we found a rather arbitrary but adequate formula to do this: $r = k/A^{2/3}$ for constant k. If A (of the lightest nuclide) is odd then k is -16.1 otherwise -2.3 MeV. Other schemes appear to have no identifiable residual. Table 1 summarizes how well these local relations work. For comparison purposes, the last column of Table 1 should be examined as it calculates the RMS error on the same C33 subset of 588 nuclei.

TABLE 1. Mass Predictions by Local Relations for 263 > A > 15, N > Z or even N=Z

	No. of Samples	Number of Neighbours	Mean Abs. Error keV	RMS Error keV	Model Error keV	Model Error on C33 subset
GKL	1477	5	144	200	194	168
GKT	1358	5	149	205	199	166
C4	1598	3	411	560	559	450
C4r	1598	3	134	181	176	140
NRICH	1130	9	106	153	150	111
NPOOR	1102	9	100	141	136	129
GK12	847	20	53	73	70	70
GK8	970	16	50	71	68	64
C33	588	32	43	66	62	--

The GK12[3] and GK8 patterns shown in Fig. 1 estimate mass by applying 12 and 8 different GKT and GKL relations and taking the average. Such approaches and C33 are useful to verify the accuracy of the experimental data and models. Unknown masses of interest are found towards the neutron and proton drip lines where suitable surrounding neighboring masses may be unknown. Relations NPOOR and NRICH allow extrapolation of mass to proton and neutron rich nuclides. C4r is suited to calculating nucleon separation energies as only 2 neighbors are needed. C33 is particularly accurate when estimating even-even masses where for A>100, its model error is only 15 keV (and vanishes completely for A>160).

REFERENCES

1. G. T. Garvey and I. Kelson, *Phys. Rev. Lett.* **16**, 197 (1966).
2. G. T. Garvey et al., *Rev. Mod. Phys.* **41, S** 1 (1969).
3. J. G. Hirsch, V. Velázquez, J.Barea, A. Frank, P. van Isacker, *Eur. Phys. J. A* **25** s1.75 (2005) .
4. G. Zao-Chun, C. Yong-Shou and M. Jie, *Chin. Phys.Lett.* **18**, No. 9, 1186 (2001).
5. G.Audi, A.H.Wapstra and C.Thibault, *Nuclear Physics* **A729** 337 (2003).

The Ion Circus

E. Minaya Ramirez,[a] S. Cabaret,[a] R. Savreux,[a] D. Lunney[a] *

[a] *Centre de Spectrométrie Nucléaire et de Spectrométrie de Masse (CSNSM),*
Institut National de la Physique Nucléaire et la Physique des Particules (IN2P3),
Centre National de la Recherche Scientifique (CNRS),
Université de Paris Sud (Univ. Paris Sud-11), Orsay, France

Abstract. The study of exotic nuclides is chronically hampered by the overwhelming abundance of contamination, requiring the development of efficient instruments with good capacity for purification. We report on the first results of a novel circular ion trap designed to dynamically increase mass resolving power as the orbiting ions are cooled to avoid losses.

Keywords: binding energies and masses, cyclic accelerators and storage rings, charged-particle spectrometers electric and magnetic, beam handling; beam transport, ion trapping, ion cooling
PACS: 21.10.Dr, 29.20.D-, 29.30.Aj, 29.27.Eg, 37.10.Rs, 37.10.Ty

INTRODUCTION

The ability to prepare radioactive beams for experiments in nuclear structure has seen important developments in recent years. The use of ion traps and buffer-gas cooling is now routine used for the accumulation and purification of even short-lived nuclides [1]. This is a key point for future installations e.g. SPIRAL2 [http://www.ganil.fr/spiral2] and EURISOL [http://www.eurisol.org/] since higher intensity also brings increased isobaric contamination which can be disastrous for background and even radioprotection.

Until now, the development of beam cooler/bunchers has relied on linear (radiofrequency quadrupole) Paul traps. Due to practical constraints of limited space and manufacturing complications, linear devices have typically beem limited in length to less than one meter. Even slowing the incoming ions to less than 10 eV still requires the use of relatively high buffer-gas pressures (typically 10^{-2} mbar) which causes elevated pressures along the beamline and incurs high pumping costs. One way of increasing the stopping power, i.e. the product of interaction length and pressure is to increase the length by making the ions circulate. This has the added advantage of allowing the circular trap to operate in mass selective mode as the ions are cooled. The circular trap, called the ion circus for its resemblance to a traffic roundabout (as well as the fact that ions can be stopped and made to jump in any direction), has been described in a previous publication [2]. In this contribution we report results of the first tests and the future program.

Figure 1 illustrates the concept of the ion circus as well as the first test, performed to verify the injection and trapping of ions into the ring structure.

CP1165, *Nuclear Structure and Dynamics '09*
edited by M. Milin, T. Nikšić, D. Vretenar, and S. Szilner
© 2009 American Institute of Physics 978-0-7354-0702-2/09/$25.00

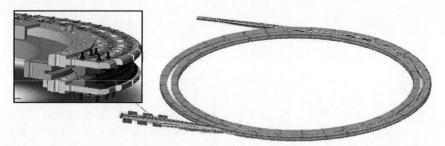

FIGURE 1. SIMION trajectory calculation of ions injected in the mid-plane of the ion circus. The RF trapping voltage is applied only to half the ring so the ions exit after half a turn and are detected. Inset left shows a cutaway view of the electrode configuration.

RESULTS

Figure 2 shows the results of the first injection tests with the ion circus. A $^{14}N^+$ beam, was injected through the midplane of the circus (see Fig. 1) at energies of 25 eV and 100 eV. The RF trapping voltage (0-280 V_{pp} at $\Omega/2\pi = 1.28$ MHz) was applied to one fraction of the electrodes such that the ions exited the circus tangentially and drifted towards a secondary-electron multiplier. The number of detected ions was recorded as a function of RF voltage amplitude (transformed into Mathieu parameter $q = 2eV/mr^2\Omega^2$ where $2r = 10$ mm is the inter-electrode separation).

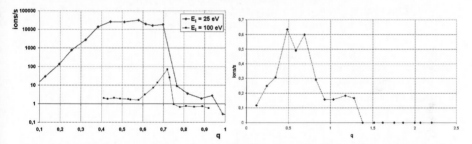

FIGURE 2. Transmission of an injected $^{14}N^+$ beam as a function of the Mathieu q parameter, varied using RF voltage amplitude after (left) one-quarter turn and (right) three-quarter turns.

ACKNOWLEDGMENTS

This work was supported by France's *Agence National de Recherche* (ANR) under contract number ANR-05-BLAN-0083-01 and by the European Community under the EURISOL Design Study, contract number 515768 RIDS.

REFERENCES

* e-mail: david.lunney@csnsm.in2p3.fr

1. F. Herfurth, *Nucl. Instr. Meth.* B **204**, 587 (2003).
2. E. Minaya Ramirez, S. Cabaret, D. Lunney, *Nucl. Instr. Meth.* B **266**, 4460 (2008).

Probing High-Velocity Transient-Field Strength Using Heavy-ions Traversing Fe and Gd

E. Fiori*, G. Georgiev*, A. E. Stuchbery[†], A. Jungclaus**,
D. L. Balabanski[‡], A. Blazhev[§], S. Cabaret*, E. Clement[¶], M. Danchev[‖],
J. M. Daugas[††], S. Grevy[¶], M. Hass[‡‡], V. Kumar[‡‡], J. Leske[§§], R. Lozeva*,
S. Lukyanov[¶¶], T. J. Mertzimekis***, V. Modamio**, B. Mouginot[†††],
Yu. E. Penionzhkevich[¶¶], L. Perrot[†††], N. Pietralla[§§], K. H. Speidel[‡‡‡],
I. Stefan[†††], C. Stodel[¶], J. C. Thomas[¶] and J. Walker**

*CSNSM, CNRS/IN2P3; Université Paris-Sud 11, UMR8609, F-91405 ORSAY-Campus, France
[†]Department of Nuclear Physics, Australian National University, Canberra, Australia
**Instituto de Estructura de la Materia, CSIC, Madrid, Spain
[‡]INRNE-BAS, Sofia, Bulgaria
[§]IKP, Cologne, Germany
[¶]GANIL, Caen, France
[‖]University of Sofia, Bulgaria
[††]CEA, DAM, DIF, 91297 Arpajon cedex, France
[‡‡]The Weizmann Institute, Rehovot, Israel
[§§]TU Darmstadt, Darmstadt, Germany
[¶¶]JINR, Dubna, Russia
***University of Ioannina, Department of Physics, Greece
[†††]IPN, CNRS/IN2P3; Université Paris-Sud 11, UMR8608, F-91405 ORSAY-Campus, France
[‡‡‡]Helmholtz-Institut für Strahlen- und Kernphysik, Bonn, Germany

Abstract. The transient field strength for ^{76}Ge ions, passing through iron and gadolinium layers at velocities $\sim Zv_0$, has been measured. Although a sizeable value has been obtained for Gd, a vanishing strength has been observed in Fe.

Keywords: Transient Field
PACS: 21.10.Ky, 21.60.Cs, 23.20.En

The Transient Field (TF) technique has been used for magnetic moment measurements of nuclear states with picosecond lifetimes throughout the nuclear chart. Generally the method has been applied for ion velocities well below the K-shell electron velocities $v_K = Zv_0$[1], where $v_0 = c/137$ is the Bohr velocity. With the exception of several light nuclei with atomic numbers between 6 and 24 (see ref. [1] and ref. therein) the behavior of the TF strength has been studied at velocities $v_{ion} \ll Zv_0$. For the above mentioned region of atomic numbers studies have been carried out up to $v_{ion} \sim Zv_0$, showing a steady increase of the TF strength, approaching its maximum value at Zv_0 [1, 2, 3]. The use of high velocity projectiles at the maximum TF strength, expected to occur at v_K, should allow measurements of very short lived nuclear states and give access to exotic nuclei produced as low-intensity radioactive beams.

In the low-velocity regime Gd is usually preferred as ferromagnetic layer compared to Fe. Larger precession angles can be achieved since Gd has a lower stopping power,

CP1165, *Nuclear Structure and Dynamics '09*
edited by M. Milin, T. Nikšić, D. Vretenar, and S. Szilner
© 2009 American Institute of Physics 978-0-7354-0702-2/09/$25.00

TABLE 1. The photopeak counts are the sums for all detectors, independently on their positions and field directions, $\langle v_{i(e)}/Zv_0 \rangle$ is the velocity range of the ions interacting with the TF, t_{eff} is the effective interaction time, G_2 is the de-orientation coefficient due to RIV and $\Delta\theta$ is the precession angle.

target	photopeak counts	$\langle v_i/Zv_0 \rangle$	$\langle v_e/Zv_0 \rangle$	t_{eff} [ps]	G_2	$\Delta\theta$ [mrad]
Gd	$8.3 \cdot 10^4$	1.05	0.68	1.9	0.68(2)	22(7)
PbFe	$3.6 \cdot 10^5$	1.07	0.64	2.4	0.56(4)	6(6)

enabling larger layer thicknesses and giving rise to longer interaction times of the ions with the TF. At much higher ion velocities the use of Fe might appear more advantageous, especially when looking for a significant decrease of the speed of the ions in order to reach values similar to v_K.

The present experiment was performed at the GANIL facility using a ^{76}Ge beam at 37.89 MeV/u. Its 2^+ state at 563 keV was Coulomb excited using two different targets: i) Pb-In-Fe-C with respective thickness of 91 mg/cm^2, 0.3 mg/cm^2, 94.2 mg/cm^2 and 1 mg/cm^2 and ii) Gd-C 204.2 mg/cm^2 and 0.6 mg/cm^2. The indium layer was used to improve adhesion and the carbon backing served as a stripper for the outgoing projectiles in order to minimize the de-orientation of the angular correlations due to recoil-in-vacuum (RIV) [1]. The targets were magnetized using using an external magnetic field of ≈ 0.1 T field in vertical direction. The direction of the field was reversed every 200 s to minimize systematic effects. Eight EXOGAM detectors, positioned at specific angles in a horizontal plane, were used to measure the de-exciting γ-rays in coincidence with the scattered ^{76}Ge nuclei. These were detected in a plastic scintillator, covering angles between $3°$ and $5.5°$ around the beam axis, guaranteeing safe Coulomb interaction.

The precession of the anisotropic angular distribution was determined in the conventional way [1, 2]. The preliminary results are shown in Table 1. The measured precession angle for Fe is consistent with zero, while sizeable effect was observed in Gd, in spite of factor of 5 lower statistical level of the measurement.

The present results show a distinctive difference between the high-velocity TF strength in Fe and Gd. More detailed studies are certainly needed for obtaining a working knowledge of the TF at $v \sim Zv_0$. Wide applications of this technique are envisaged in particular using high-energy radioactive beams.

We thank the GANIL staff for providing the beam for the experiment. This work was supported by the EU FP6 contract EURONS N° RII3-CT-2004-506065, by the IAP Program, P6/23 BriX, Belgian Science Policy, by the Bulgarian Science Fund VUF06/05, by the Spanish Ministerio de Ciencia e Innovación under contract FPA2007-66069, and by the Australian Research Council Discovery Grant, DP0773273.

REFERENCES

1. K.-H. Speidel, O. Kenn, and F. Nowacki, *Prog. Part. Nucl. Phys.* **49**, 91 (2002).
2. A.E. Stuchbery, et al., *Phys. Lett.* **611B**, 81 (2005).
3. U. Grabowy, et al., *Z. Phys. A* **359**, 377 (1997).

Low-lying Level Structure of Light Neutron-rich Nuclei Beyond the Dripline: ^{9}He and ^{10}Li

H. Al Falou

LPC-Caen, ENSICAEN, Université de Caen and IN2P3-CNRS, 14050 Caen Cedex, France [1]

Keywords: NUCLEAR REACTIONS C(^{11}Be/^{14}B, X)^{9}He/^{10}Li, 35 MeV/nucleon; measured (fragment)(neutron)-coin, invariant masses. ^{9}He, ^{10}Li Deduced energy of resonances.
PACS: 27.20.+n; 29.30.Hs; 21.10.Dr; 24.30.Gd ; 25.70.Mn

The very neutron-rich, light nuclei provide a fertile testing ground for our understanding of nuclear structure. From an experimental point of view this region is the only one for which nuclei lying at and beyond the neutron dripline may be accessed. Theoretically a wide range of models, including various shell model approaches and more ab initio type models are now capable of providing predictions. In addition, the structure of some unbound systems, such as ^{10}Li, is a key to constructing three-body descriptions of two-neutron halo nuclei ^{11}Li [1]. There is a special interest in the lightest N=7 isotones where the $1s_{1/2}$ orbital from the *sd*-shell is found to intrude into the *p*-shell, a phenomena long known in ^{11}Be [2]. There is evidence that this inversion occurs in ^{11}Li and ^{9}He [3]. In the following we describe briefly a new experimental investigation of the low-lying level structure of ^{9}He and ^{10}Li.

The experiments were performed at GANIL, using 35 MeV/u ^{11}Be and ^{14}B beams. The charged fragments arising from reactions on a C target were detected using a position sensitive Si-Si-CsI telescope. The neutrons were detected using the DEMON liquid scintillator array and the energy deduced from the time-of-flight. As all the beam velocity reaction products were identified and the momenta measured event-by-event, the ^{9}Li-n and ^{8}He-n invariant mass (IM) spectra could be reconstructed (Fig. 1).

Within the context of the sudden approximation, the removal at high energy of proton(s) from ^{11}Be should, to a reasonable approximation, leave the neutron configuration of the projectile unperturbed ($\Delta \ell_n$=0). Since the ground state of ^{11}Be is dominated by a $\nu s_{1/2}$ configuration [2], the final states populated in ^{10}Li and ^{9}He should be dominated by such a neutron configuration.

The results from the C(^{11}Be,^{9}Li+n) reaction are shown in Fig. 1a. The data is well described using two components: an *s*-wave virtual state and a broad uncorrelated continuum obtained by event mixing. The best adjustment was obtained for a scattering length $a_s = -17 \, ^{+7}_{-5} \, fm$. The data from the C(^{14}B,^{9}Li+n) reaction is shown in Fig. 1b, where the peak around 500 keV corresponds very probably to the population of the *p*-wave resonance [4]. Data for ^{9}He was limited by statistics - removal of second proton corresponds to an order of magnitude decrease in the cross section. A distribution that

---

[1] Present affiliations: Triumf, Vancouver, Canada; Saint Mary's University, Halifax, Canada.

CP1165, *Nuclear Structure and Dynamics '09*
edited by M. Milin, T. Nikšić, D. Vretenar, and S. Szilner
© 2009 American Institute of Physics 978-0-7354-0702-2/09/$25.00

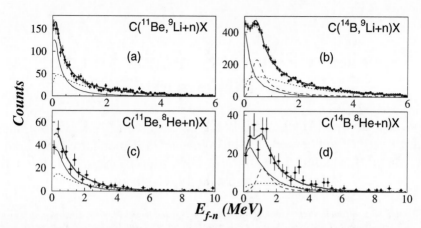

FIGURE 1. ^{8}He-n and ^{9}Li-n IM spectra after one-proton knockout from ^{11}Be (left panels) and breakup from ^{14}B (right panels). The thick solid lines correspond to the best adjustment to the data. The thin solid lines correspond to an s-wave virtual state and the dashed lines to a p-wave resonance. The dotted lines represent the uncorrelated distribution obtained by event-mixing.

consists entirely of event mixed events fails to describe the distribution at low relative energy for the reactions with both beams. The results obtained with the ^{11}Be beam are shown in Fig. 1c. The best description was obtained using an s-wave virtual state with a scattering length $a_s \geq -3 fm$. For the data acquired from the breakup of ^{14}B (Fig. 1d), the IM spectrum exhibits also a resonance around $E_{f-n} = 1.2$ MeV, which most probably corresponds to an excited $1/2^-$ state in ^{9}He [5]. These results suggest that the level inversion also occurs in ^{9}He, but with a much weaker core-n interaction than for ^{10}Li.

In summary, we have confirmed the $\nu s_{1/2}$ character of the ^{10}Li ground state and found good evidence for the existence of similar low-lying s-wave strength in ^{9}He. In addition, the results obtained for ^{10}Li with the ^{11}Be beam provide support for the simple selection rule arguments - that is, only final-state strength of the same character as the projectile neutron configuration was observed for removal of a deeply bound proton from a system with a very weakly bound valence neutron.

Finally, the author would like to thank his colleagues in the E483 collaboration.

REFERENCES

1. B. Jonson, *Phys. Rep.* **389**, 1 (2004).
2. T. Aumann, et al., *Phys. Rev. Lett.* **84**, 35 (2000), and references therein.
3. L. Chen, et al., *Phys. Lett.* **505B**, 21 (2001).
4. H. Jeppesen, et al., *Phys. Lett.* **642B**, 449 (2006); H. Simon, et al., *Nucl. Phys.* **A 791**, 267 (2007).
5. K. Seth, et al., *Phys. Rev. Lett.* **58**, 1930 (1987); H. Bohlen, et al., *Z. Phys.* **A 330**, 227 (1988); W. von Oertzen, et al., *Nucl. Phys.* **A 588**, 129c (1995).

Superheavy Element Synthesis And Nuclear Structure

D.Ackermann[a], S. Antalic[b], M. Block[a], S. Hofmann[a,c,1], H.-G. Burkhard[a],
S.Heinz[a], F.P. Heßberger[a], J. Khuyagbaatar[a], I. Kojouharov[a], M. Leino[d],
R. Mann[a], J. Maurer[a], K. Nishio[e], A.G. Popeko[f], S. Šaro[b], M. Venhart[b],
A.V. Yeremin[f] and J. Uusitalo[d]

[a]GSI Helmholtzzentrum für Schwerionenforschung GmbH, Planckstr.1 , D-64921 Darmstadt, Germany
[b]Department of Nuclear Physics, Comenius UniversitySK-84248 Bratislava, Slovakia
[c] Institut für Physik, Johann Wolfgang Goethe-Universität, D-60438 Frankfurt, Germany
[d]Department of Physics, University of JyväskyläFIN-40351 Jyväskylä, Finland
[e]Japan Atomic Energy Agency (JAEA),Tokai, Ibaraki 319-1195, Japan
[f]Flerov Laboratory of Nuclear Reactions, JINR Ru-141 980 Dubna, Russia

Abstract. After the successful progress in experiments to synthesize superheavy elements (SHE) throughout the last decades, advanced nuclear structure studies in that region have become feasible in recent years thanks to improved accelerator, separation and detection technology. The means are evaporation residue(ER)-α-α and ER-α-γ coincidence techniques complemented by conversion electron (CE) studies, applied after a separator. Recent examples of interesting physics to be discovered in this region of the chart of nuclides are the studies of K-isomers observed in ^{252,254}No and in ^{270}Ds.

Keywords: Superheavy Element, fusion, fusion/fission, transactinides, nuclear structure, K-isomers.
PACS: 21.10.-k, 21.10.Hw, 23.20.Lv, 23.60.+e, 25.60.Pj, 25.70.-z, 25.70.Gh, 25.70.Jj.

INTRODUCTION

The most recent activities at GSI concerning the search for SHE were the successful production of ^{283}112 in the reaction ^{48}Ca + ^{238}U confirming an earlier result from FLNR, and the attempt to synthesize an isotope with Z=120 in the reaction ^{64}Ni + ^{238}U. The superheavy elements, however, are a nuclear structure phenomenon. They owe their existence to shell effects. In recent years the development of advanced experimental set-ups allowed for detailed nuclear structure studies of nuclei with Z≥100. Reviews of those recent achievements are given in [1,2].

DECAY CHAINS, *K*-ISOMERS, MASSES AND MORE

The combination of the UNILAC accelerator at GSI with the velocity filter SHIP and its decay spectroscopy setup is among the best suited set-ups for these studies

---

[1] Joseph Buchmann –Professor Laureatus

CP1165, *Nuclear Structure and Dynamics '09*
edited by M. Milin, T. Nikšić, D. Vretenar, and S. Szilner
© 2009 American Institute of Physics 978-0-7354-0702-2/09/$25.00

worldwide. Xu et al. predict high *K*-isomers to be a general feature for prolate deformed heavy nuclei [3]. Experimentally about 14 cases have been identified in the region of $Z \geq 96$ as shown in Fig. 1. *K*-isomers have been found, apart from Sg and Hs, for all even-*Z* elements with $Z=100$ to 110. We could recently establish and/or confirm such states for ^{252,254}No [4,5]. The heaviest nucleus where such a state was found is ^{270}Ds [6]. *K*-isomers were found also in even-odd and odd-even isotopes in this region like e.g. ^{251}No [7], ^{253}No [8], ^{255}No [9], and ^{255}Lr [10,11].

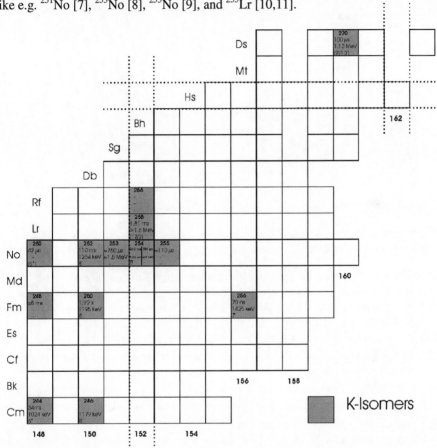

FIGURE 1. Excerpt of the chart of nuclides indicating the K-isomers observed for heavy nuclei in the region Z≥96 Half-life, decay energy, spin and parity values are given for K-Isomers only.

For ^{254}No two isomeric states had been observed at JYFL [12] and ANL [13] with excitation energies and half-lives of 1293-1297 keV and ≈ 2.5 MeV, and 275 ± 7 ms and ≈ 180 μs, respectively. In both publications four-quasiparticle configurations are proposed for this state with $K^{\pi} = 16^+$ [12] and 14^+ [13]. In a recent measurement at SHIP we could establish the band structure above the long lived isomer as well as the link to the g.s. rotational band by means of delayed γ-γ coincidences between the 606 keV line, the intra band transitions and decays of the 275 ms isomer, and the observation of the two new transitions with 778 keV and 856 keV [14]. We measured a half-life of 198±13 µs for the second isomeric state. The resulting decay scheme is shown in Fig. 2a.

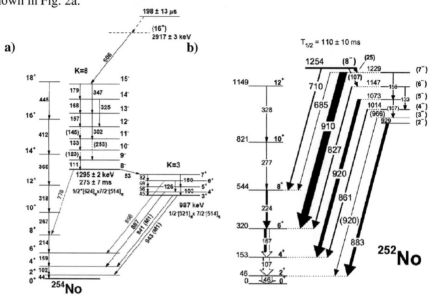

FIGURE 2. a) Decay scheme of ^{254}No with the placement and spin assignment of the two K-isomers [14]. **b)** Decay scheme of ^{252}No with the placement and spin assignment of the K-isomer [4].

For ^{252}No we observed for the first time a *K*-isomer with a half-life of 110±10 ms at an excitation energy of 1254 keV [4]. Beyond the spin and parity assignment of this isomer we could establish also a detailed decay scheme including a side band below the isomeric state and its connection to the g.s. rotational band (Fig. 2b).

For the reaction ^{64}Ni + ^{207}Pb we observed a total of 8 decay chains which we all attributed to the production of ^{270}Ds followed by the sequential emission of two α-particles leading two the daughter ^{266}Hs and the granddaughter ^{262}Sg which eventually decayed by spontaneous fission [6]. We observed for the ^{270}Ds α decay two groups with half-lives of $6.0^{+8.2}_{-2.2}$ ms and 100^{+140}_{-40} µs, respectively. We assigned our decay data to various decay paths as illustrated in Fig. 3. Note: our interpretation is based on theoretical expectations [3,15,16]. However, more detailed experimental information is needed to test this scenario. For ^{262}Sg we observed only spontaneous fission decay. Its α-sf branching ratio is expected to be around 15% [6]. The observation of the α-branch would connect this decay chain down to ^{254}No for which we obtained recently a precise mass value at SHIPTRAP [17]. So we would establish, together with the $Q_α$ values from the decay chain, an experimental mass excess value for the even-even ^{270}Ds and produce a valuable anchor point for theory in the SHE region.

OUTLOOK: TOWARDS THE ISLAND OF STABILITY

Low cross sections, the advances in nuclear structure investigations, reaction mechanism studies, chemistry and SHE synthesis experiments with a steadily increasing demand for higher beam intensities, more sensitive and more sophisticated detection set-ups and new methods determine the road map for future SHE investigations. Intensity increase is one of the major issues in this context. A dedicated continuous wave (CW) accelerator which would increase the beam by intensity an order of magnitude is a fundamental development needed for further progress in SHE research. Mass determination by an adequate spectrometer, complementary to precise trap mass measurements, would be extremely helpful for a final confirmation of the unconnected chains obtained in the ^{48}Ca induced reactions on actinide targets.

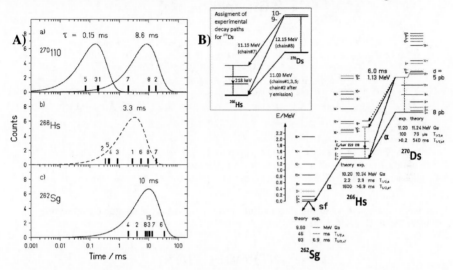

Figure 3: A) Time distribution of the subsequent α-decays (a, b) and fission events (c) measured in the reaction ^{64}Ni + ^{207}Pb [6]). **B)** Proposed decay and level scheme for ^{270}Ds and its decay products on the basis of our observations [6] and calculations [16], taken from [15]. The insert illustrates the decay paths for ^{270}Ds by the chain numbers from [6].

REFERENCES

1. R.-D. Herzberg and P.T. Greenlees, *Prog. Part. Nuc. Phys.* **61**, 674 (2008).
2. M. Leino and F.P. Heßberger, *Ann. Rev .Nucl .Part .Sci.* **54**, 175 (2004).
3. F.R. Xu, et al., *Phys. Rev. Lett.* **92**, 252501 (2004).
4. B. Sulignano, et al., *Eur. Phys. J.* **A 33**, 327 (2007).
5. B. Sulignano, Ph.D. thesis, Johannes Gutenberg-University Mainz (2007).
6. S. Hofmann, et al., *Eur. Phys. J.* A **10**, 5 (2001).
7. F.P. Heßberger, et al., *Eur. Phys. J.* A **30**, 561 (2006).
8. F.P. Heßberger, et al., *Eur. Phys. J.* A **29**, 165 (2006).
9. F.P. Heßberger, *Phys. At. Nuc.* **8**, 1445 (2007)
10. K. Hauschild, et al., *Phys. Rev.* C **78**, 021302 (2008).
11. S. Antalic, et al., *Eur. Phys. J.* A **38**, 219 (2008).
12. R.-D. Herzberg, et al., *Nature* **442**, 896 (2006).
13. S.K. Tandel, et al., *Phys. Rev. Lett.* **97**, 082502 (2006).
14. F.P. Heßberger, et al., submitted to *Eur. Phys. J. A*.
15. S. Hofmann, *Jour. Nucl. and Radiochem. Sc.* **4**, R1 (2003).
16. S. Cwiok, W. Nazarewicz and P.H. Heenen, *Phys. Rev. Lett.* **83**, 1108 (1999).
17. M. Block, et al., to be published.; M. Block, et al., *Eur. Phys. J.* **A 25**, 49 (2005).

Heavy-Ion Fusion Mechanism and Predictions of Super-Heavy Elements Production

Yasuhisa Abe[a] , Caiwan Shen[b] , David Boilley[c] , Bertrand G. Giraud[d] and Grigory Kosenko[e]

[a]RCNP, Osaka University, Ibaraki (Osaka), 567-0047, Japan.
[b]School of Science, Huzhou Teachers College, Huzhou (Zhejiang), 313000, China,
[c]GANIL,CEA/DSM-CNRS/IN2P3, BP 55027, F-14076, France,
and Univ. Caen, BP 5186, F-14032 Caen, France,
[d]IPT, CEA/DSM, CEA-Saclay, Gif-sur-Yvette, F-91191, France,
[e]Department of Physics, Omsk University, Omsk, RU-644077, Russia

Abstract. Fusion process is shown to firstly form largely deformed mono-nucleus and then to undergo diffusion in two-dimensions with the radial and mass-asymmetry degrees of freedom. Examples of prediction of residue cross sections are given for the elements with Z=117 and 118.

Keywords: **Heavy-ion Fusion; Fusion hindrance; Super-heavy elements; Cross section.**
PACS: 25.70, Jj, 25.70. Lm, 27.90. +b

INTRODUCTION

Theoretical prediction of optimum incident system, optimum incident energy, and absolute value of maximum cross section for Super-Heavy Elements (SHE) production is a long-standing challenging problem in nuclear physics. It becomes more and more important when we go to heavier elements, because residue cross section becomes smaller and smaller, down to the order of pb to fb. Why such extremely small cross sections? One immediately thinks of fragility of SHE, that is, the fact that there is no barrier against fission within the Liquid Drop Model (LDM), and only so-called shell correction energy in the ground state sustains nucleus of SHE against fission decay, as is understood by the fissility being close to 1. The feature is correctly taken into account through Ignatyuk prescription of the level density parameter [1,2]. As expected, the survival probability for SHE is very small, even if compound nucleus is formed. The small cross section, however, is not only due to the survival probability, but also due to small fusion probability, which is expected from so-called fusion hindrance [3,4]. Existence of the hindrance has been known experimentally since many years ago, but its mechanism was not understood yet.

Recently the mechanism has been clarified [5-7]. The theory is based on the observation that di-nucleus configuration formed by the incident projectile and target is located outside the fission saddle or the ridge-line, because the configuration has a very large deformation as a compound nucleus, while the saddle point configuration is

CP1165, *Nuclear Structure and Dynamics '09*
edited by M. Milin, T. Nikšić, D. Vretenar, and S. Szilner
© 2009 American Institute of Physics 978-0-7354-0702-2/09/$25.00

close to the spherical shape in heavy nuclei with fissility close to 1. One more essential point in the theory is an assumption that dissipation is very strong, strong enough for the relative kinetic energy to dissipate already at the contact distance of the projectile and the target, which is confirmed with the Surface Friction Model (SFM) [8,9]. Thus, we employ Smoluchowski or over-damped Langevin equation for the multi-dimensional fusion dynamics for overcoming of the saddle point or the ridge-line to the spherical compound nucleus.

DYNAMICS FROM DI-NUCLEUS TO MONO-NUCLEUS[10, 11]

For the description of nuclear shapes of the composite system formed by the incident channel, we employ Two-Center Parameterization (TCP), which encompasses di-nucleus as well as mono-nucleus configurations [12-14].

There are three essential parameters: distance between two centers of the oscillator potentials R, mass-asymmetry α, and neck correction ε. The neck correction ε is defined as a ratio between the smoothed peak height at the connection point of the right and left oscillator potentials and the height of the potential without correction, i.e., that of the potential spike at the connection point. So, ε can vary between 1 and 0. The value 1 corresponds to the smoothed peak with the same height as the original spike, while the value 0 does to no peak, i.e., to a single wide flat potential. In other words, the former describes the touching configuration of the incident channel, while the latter does mono-nucleus with very large deformation. In the TCP, the three degrees of freedom are almost independent, especially in di-nucleus configurations, though they are not normal modes. Thus, it is meaningful to analyze each degree of freedom separately. Of course, there are couplings between them, and friction tensor also induces couplings, for which we use so-called One-Body Dissipation model (OBD) [15]. The coupling effects will be discussed elsewhere, starting with multi-dimensional Smoluchowski equation [16].

Firstly, we take up the case with mass-symmetric entrance channel, which makes the problem simpler with only two degrees of freedom left. The radial motion is already solved and analyzed in detail [17,18]. Fusion probability, i.e., a probability for passing over the saddle point into the spherical shape is shown to be given by the fluctuation of Langevin trajectories, or by a tail of diffusion, and to be well approximated by an error function. In cases with strong dissipation such as OBD, the function can be approximated by an Arrhenius function, $\exp(-V_{sad}/T)$, where V_{sad} denotes the saddle point height measured from the energy of the touching configuration and T the temperature of the composite system. This clearly explains the feature of the fusion hindrance, i.e., an extremely small and slow increase of the fusion probability known experimentally [3,4]. Here, it is worth to notice that V_{sad} =0.0 defines a border between the normal and the hindered fusions. Interestingly, the border line obtained is found to be consistent with the measured data on $^{100}Mo+^{100}Mo$ (normal) and $^{110}Pd+^{110}Pd$ (hindered) systems [19]. Next, the time evolution of the radial fusion was analyzed, which shows that the fusion process is undertaken in time scale around several in unit of $\hbar/MeV$ [18].

During the diffusion process in the radial motion, how does the neck degree of freedom evolve? In order to answer that question, we solve the neck motion with the

129

starting point ε equal to 1.0. The radial variable is fixed at the contact distance of two nuclei for the moment. The potential for ε is calculated with LDM, which is almost linear, and the friction is calculated with OBD. Corresponding 1-dimensional Smoluchowski equation was already solved by him [20].

We apply it to our present case and find that the neck degree of freedom quickly reaches to the equilibrium distribution in the space [0.0, 1.0], in one order of magnitude shorter than the time scale of the radial diffusion [10]. The average value of the neck variable is about 0.1, near the bottom of the potential. This indicates that the fusion process proceeds to firstly filling-out of the neck cleft or crevice of the di-nucleus toward the formation of the mono-nucleus with the very large deformation, and then to diffusion in the radial degree of freedom.

The mass-asymmetry degree of freedom is also analyzed in the same way for ^{48}Ca induced reactions, and turns out that its time scale is close to that of the radial one, not to the neck one [11].

PRELIMINARY RESULTS OF RESIDUE CROSS SECTIONS

In realistic calculations of fusion probability, we numerically solve two-dimensional Langevin equation for radial and mass-asymmetry degrees of freedom [21] with neck parameter being fixed at the average value of 0.1. We also need to know the probability for overcoming of the usual Coulomb barrier in the entrance channel. There are two ways to employ: the empirical formula for capture cross sections [22] and Langevin calculations of trajectories with SFM [8,9]. As for the survival probability, we use the theory of statistical decay for particle emission and fission decay [23,24].

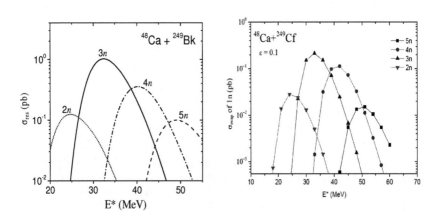

FIGURE 1. Predictions of xn residue cross sections for Z=117 and 118 elements. No arbitrary parameter is introduced except the reduction factor for the shell correction energy in the calculations of the survival probabilities

Preliminary studies are made on ^{48}Ca + Bk isotopes [25] as well as ^{48}Ca + ^{249}Cf systems, where mass table predicted by P. Möller et al. is used [26]. In Fig. 1, the results for the systems for Z=117 and 118 are shown, where the empirical formula is used for the capture probability with certain modification of the parameters suitable for ^{48}Ca induced reactions, and a part of the code HIVAP for the statistical decay.

Since there are still ambiguities in fusion probabilities, we premise that absolute values of residues cannot be reproduced. We adjust it expediently by reducing the shell correction energy, i.e., fission barrier height and thus the survival probability. The calibration of the factor is made with the Dubna data on ^{48}Ca + ^{248}Cm system [27]. The factor turns out to be 0.45, which we keep in use for the other systems [25]. In order to eliminate the factor, more detailed investigations are necessary on fusion dynamics in very heavy systems, which are under way [16].

ACKNOWLEDGMENTS

The present work is supported by JSPS grant No. 18540268 and National Science Foundation of China grant No. 10675046.

REFERENCES

1. A.V. Ignatyuk, et al., *Sov. J. Nucl. Phys.* **21**, 255 (1975).
2. Y. Abe and B. Bouriquet, *Acta Physica Polonica* **B34**, 1927-1945 (2003).
3. W. Westmeier, et al., *Phys. Lett.* **B117**, 163-166 (1982).
4. W. Reisdorf, *J. Phys. G; Nucl. Part.* **20**, 1297 (1994).
5. Y. Abe, *Eur. Phys. J.* **A13**, 143 (2002).
6. Y. Abe, et al., *Prog. Theor. Phys. Suppl.* **No. 146**, 104- 109 (2002)..
7. Y. Abe, et al., *Acta Physica Polonica* **B34**, 2091-2105 (2003).
8. D.H.E. Gross and H. Kalinowski, *Phys. Reports* **45**, 175 (1978).
9. G. Kosenko, C.W. Shen, and Y. Abe, *J. Nucl. Radiochem. Sci.* **3,** 19-22 (2002).
10. Y. Abe, et al., *Intern. J. Mod. Phys.* **E17**, 2214-2220 (2008).
11. Y. Abe, et al., to appear in *Intern. J. Mod. Phys.*
12. J. Maruhn and W. Greiner, *Z. Phys.* **251**, 431 (1972).
13. S. Suekane, et al., *JAERI-memo* **5918** (1974).
14. A. Iwamoto, et al., *Prog. Theor. Phys.* **55**, 115 (1976).
15. J. Blocki, et al., *Ann. Phys.* **113**, 330 (1978).
16. D. Boilley, et al., publication under preparation.
17 Y. Abe, et al., *Phys. Rev.* **E61**, 1125-1133 (2000).
18. D. Boilley, Y. Abe and J.D. Bao, *Eur. Phys. J.* **A18**, 627-631 (2003).
19. C. Shen, et al., to appear in *Science in China Series G.*
20 M. V. Smoluchowski, *Physik Zeit.* **17,** 585 (1916).
21. C. Shen, G. Kosenko and Y. Abe, *Phys. Rev.* **C66**, 061602 (2002),
    B. Bouriquet, G. Kosenko, and Y. Abe, *Eur. Phys. J.* **A22**, 9-12 (2004).
22. K. Siwek-Wilczynska, E. Siemaszko and J. Wilczynski, *Acta Physica Polonica* **B33**, 451 (2002).
23. HIVAP code, W. Reisdorf.
24. B. Bouriquet, D. Boilley and Y. Abe, *Comp. Phys. Comm.* **159**, 1-18 (2004).[KEWPIE I]
    A. Marchix, PhD thesis, Univ. Caen, 2007. [KEWPIE II]
25. C. Shen, et al., *Intern. J. Mod. Phys.* **E17, suppl.** 66-79 (2008).
26. P. Möller, et al., *Atom. Data Nucl. Data Tables* **59**, 185 (1995).
27. Yu. Ts. Oganessian, et al., *Phys. Rev.* **C74** 044602 (2006), and references therein.

Possibility of Production of New Superheavy Nuclei in Complete Fusion Reactions

G.G. Adamian[*,†], N.V. Antonenko[*], V.V. Sargsyan[*,**] and W.Scheid[‡]

[*]Joint Institute for Nuclear Research, 141980 Dubna, Russia
[†]Institute of Nuclear Physics, 702132 Tashkent, Uzbekistan
[**]Yerevan State University, 0025 Yerevan, Armenia
[‡]Institut für Theoretische Physik der Justus–Liebig–Universität, D–35392 Giessen, Germany

Abstract. For superheavy elements (SHE) with $Z=112$-116 and 118, the obtained dependence of the survival probability on Z indicates the next doubly magic nucleus beyond ^{208}Pb at $Z \geq 120$.

Keywords: Complete fusion; Superheavy nucleus; Magic numbers
PACS: 25.70.Jj, 24.10.-i, 24.60.-k

The experimental evaporation residue cross sections σ_{xn} in the ^{48}Ca+Actinide complete fusion reactions do not depend strongly on the atomic number Z of SHE and are on the picobarn level. As known, the cross section of compound nucleus formation strongly decreases with increasing $Z_1 \times Z_2$. Since the absolute value of evaporation residue cross section is ruled by the product of complete fusion cross section and survival probability, the loss in the formation probability of compound nucleus in actinide-based reactions can be compensated by the gain in the survival probability of SHE. Then, one can reveal the behavior of the survival probability with increasing Z of SHE by using the experimental evaporation residue cross sections and calculated fusion cross sections.

The cross section of the production of SHE as the evaporation residues in the xn-evaporation channel is written as a sum over all partial waves J

$$\sigma_{xn}(E_{\text{c.m.}}) = \sum_J \sigma_{fus}(E_{\text{c.m.}},J)W_{xn}(E_{\text{c.m.}},J), \qquad (1)$$

$$\sigma_{fus}(E_{\text{c.m.}},J) = \int_0^{\pi/2} \int_0^{\pi/2} d\cos\Theta_1 d\cos\Theta_2 \, \sigma_c(E_{\text{c.m.}},J,\Theta_i)P_{CN}(E_{\text{c.m.}},J,\Theta_i).$$

Here, the averaging over the orientations of statically deformed interacting nuclei (Θ_i (i=1,2) are the orientation angles with respect to the collision axis) is taken into consideration. The value of $\sigma_c(E_{\text{c.m.}},J,\Theta_i) = \frac{\pi\hbar^2}{2\mu E_{\text{c.m.}}}(2J+1)T(E_{\text{c.m.}},J,\Theta_i)$ defines the transition of the colliding nuclei over the Coulomb barrier with the probability T and the formation of dinuclear system (DNS). T is calculated with the Hill-Wheeler formula.

The DNS model [1] is successful in describing the complete fusion reactions especially related to the production of heavy and superheavy nuclei. In the DNS model the compound nucleus is reached by a series of transfers of nucleons from the light nucleus to the heavy one. The dynamics of the DNS is considered as a combined diffusion in the degrees of freedom of the mass asymmetry $\eta = (A_1 - A_2)/(A_1 + A_2)$ (A_1 and A_2 are the mass numbers of the DNS nuclei) and of the internuclear distance R. The diffusion in R

CP1165, *Nuclear Structure and Dynamics '09*
edited by M. Milin, T. Nikšić, D. Vretenar, and S. Szilner
© 2009 American Institute of Physics 978-0-7354-0702-2/09/$25.00

occurs towards the values larger than the sum of the radii of the DNS nuclei and finally leads to the quasifission (decay of the DNS). After the capture stage, the probability of complete fusion

$$P_{CN} = \lambda_\eta^{Kr}/(\lambda_\eta^{Kr} + \lambda_{\eta_{sym}}^{Kr} + \lambda_R^{Kr})$$

depends on the competition between the complete fusion in η, diffusion in η to more symmetric DNS and quasifission. This competition can strongly reduce the value of $\sigma_{fus}(E_{c.m.}, J)$ and, correspondingly, the value of $\sigma_{xn}(E_{c.m.})$.

The survival probability $W_{xn}(E_{c.m.}, J)$ estimates the competition between fission and neutron evaporation in the excited compound nucleus. In expression (1) the contributing angular momentum range is limited by W_{xn}. In the case of highly fissile SHE, W_{xn} is a narrow function of J different from zero in the vicinity of $J=0$ for all bombarding energies. The angular momentum dependence can be separated as

$$W_{xn}(E_{c.m.}, J) \approx W_{xn}(E_{c.m.}, J=0) \exp[-\frac{J^2}{J_m^2(x)}], \tag{2}$$

where $J_m(x) = 10(1 + \frac{1}{\sqrt{2}} + ... + \frac{1}{\sqrt{x}})^{-1}$.

Using Eqs. (1) and (2), and replacing the sum over J by the integral, we obtain the following factorization:

$$\sigma_{xn}(E_{c.m.}) = \sigma_{fus}^{eff}(E_{c.m.})W_{xn}(E_{c.m.}, J=0), \tag{3}$$

$$\sigma_{fus}^{eff}(E_{c.m.}) = \frac{\pi\hbar^2}{\mu E_{c.m.}} \int_0^\infty \int_0^{\pi/2} \int_0^{\pi/2} dJ d\cos\Theta_1 d\cos\Theta_2 \, J e^{-\frac{J^2}{J_m^2(x)}} T P_{CN},$$

where σ_{fus}^{eff} is the effective fusion cross section because it contains the angular momentum dependence of survival probability. Using Eq. (3), one can extract the value of survival probability at the zero angular momentum from experimental cross section $\sigma_{xn}^{exp}(E_{c.m.})$ as

$$W_{xn}(E_{c.m.}, J=0) = \sigma_{xn}^{exp}(E_{c.m.})/\sigma_{fus}^{eff}(E_{c.m.}). \tag{4}$$

The fusion probability and, correspondingly, the effective fusion cross section $\sigma_{fus}^{eff}(E_{c.m.})$ decreases by about 2 orders of magnitude with increasing the charge number of compound nucleus from $Z=112$ to $Z=118$ (Fig. 1). The fusion hindrance is due to a strong competition between complete fusion and quasifission in the DNS. The contribution of quasifission to the reaction cross section strongly increases with Z due to the increasing Coulomb repulsion in the DNS. As seen in Fig. 1, the cross section of compound nucleus formation in the ^{48}Ca-induced reactions with actinide targets is substantially higher than the evaporation residue cross section reducing by the survival probability.

In Fig. 2 the extracted values of W_{3n} and W_{4n} with Eq. (4) deviates from the expected magic proton number $Z=114$. This indicates an increase of the stability of SHE beyond $Z=114$. The experimental error-bars result the error-bars in the deduced W_{xn}. Since the fission barrier is determined by the shell correction energy, the absolute value of the

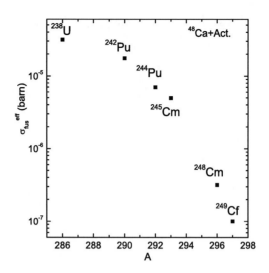

FIGURE 1. Effective fusion cross section as a function of mass number of the compound nucleus in ^{48}Ca-induced complete fusion reactions at bombarding energies supplying the maximal yield of evaporation residues in $3n$-channel. The actinide targets are indicated.

shell correction energy is expected to be increased with Z. The shell correction energy strongly depends on that how the neutron and proton numbers of the compound nucleus are closed to the magic proton and neutron numbers. The found experimental trend of the Q_α-values in α-decay chains also indicates the monotonic increase of the amplitude of the ground state shell correction energy with charge number in the region $Z=112$-118 [3]. One can expect the increasing stability of nuclei approaching the closed neutron $N=184$ shell. However, in Fig. 2 $W_{3n}\binom{296}{180}116) < W_{3n}\binom{297}{179}118)$. This probably indicates that $Z=114$ is not a proper proton magic number and the next doubly magic nucleus beyond ^{208}Pb is the nucleus with $Z \geq 120$. The shell closure at $Z \geq 120$ may influence stronger on the stability of the SHE than the sub-shell closure at $Z=114$. Note that the experimental uncertainties seem to be too small to overcome the trends presented in Fig. 2.

In conclusion, the found enhancement of W_{xn} with increasing charge number of the SHE from $Z=114$ to 118 indicates that the ground state shell corrections growth with Z. Thus, the present experimental a magic proton shell is at $Z \geq 120$. If the survival probability of compound nucleus with $Z \geq 120$ may be much higher than the one of compound nucleus with $Z=114$, there is some hope to synthesize new SHE with $Z \geq 120$ by using the present experimental set up and the actinide-based reactions with neutron-rich stable projectiles heavier than ^{48}Ca.

This work was supported in part by DFG (Bonn), and RFBR (Moscow). The IN2P3-JINR and Polish-JINR Cooperation programs are gratefully acknowledged.

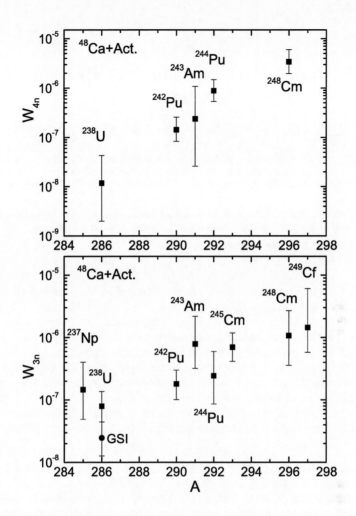

FIGURE 2. The survival probabilities of SHE in 3n-and 4n-channels, extracted with Eq. (4) and experimental σ_{xn}^{exp} from Ref.[2], as functions of mass number of the compound nucleus. For the reaction ^{48}Ca+^{238}U, the experimental σ_{3n}^{exp} from GSI [3] is used as well.

REFERENCES

1. V.V. Volkov, *Part. Nucl.* **35**, 797 (2004).
2. Yu.Ts. Oganessian, *J. Phys. G* **34**, R165 (2007).
3. S. Hofmann, *Lect. Notes Phys.* **764**, 203 (2009).

Isotopic Dependence of Isomeric States in Heavy Nuclei

G.G. Adamian[*,†], N.V. Antonenko[*] and W. Scheid[**]

[*]Joint Institute for Nuclear Research, 141980 Dubna, Russia
[†]Institute of Nuclear Physics, 702132 Tashkent, Uzbekistan
[**]Institut für Theoretische Physik der Justus-Liebig-Universität, Giessen, Germany

Abstract. The isotopic trends of isomer states in heavy nuclei are treated. Two-quasiparticle states of Fm and No isotopes as well as one-quasipartical states of Es and Md are discussed.

Keywords: isomer states, heavy nuclei
PACS: 21.10.-k, 21.10.Hw, 27.90.+b

High-spin K-isomer states, which are usually assumed as two quasiparticle high-spin states, were observed in heavy nuclei ^{250,256}Fm, ^{252,254}No, ^{266}Hs, and ^{270}Ds [1]. The one-quasiparticle isomer states are also known among odd heaviest nuclei. In order to calculate the energies of isomer states, the two-center shell model [2] is used for finding the single-particle levels at the ground state of nucleus. The shape parameterization used in this model effectively includes all even multipolarities. The dependence of the parameters of **ls** and l^2 terms on A and $N - Z$ are modified for the correct description of the ground state spins of known odd actinides.

The contribution of an odd nucleon, occupying a single-particle state $|\mu >$ with energy e_μ, to energy of a nucleus is described by the one-quasiparticle energy $\sqrt{(e_\mu - e_F)^2 + \Delta^2}$. Here, the Fermi energy e_F and the pairing-energy gap parameter Δ are calculated with the BCS approximation. The values of Δ obtained in our calculations differ from those in Refs. [3, 4] within 0.05–0.1 MeV.

The microscopical corrections, quadrupole parameters of deformation calculated with the two-center shell model are close to those obtained with the microscopic-macroscopic approaches in Refs. [3, 4].

In order to demonstrate the quality of our calculations of the excited energies of two-quasiparticle states, in Fig. 1 the calculated energies of the states with $K^\pi = 8_\nu^-$ (the index ν (π) denotes neutron (proton) pair) are compared with the experimental values [5] in the isotones from ^{176}Yb till ^{184}Pt. The maximal disagreement does not exceed 200 keV that is quite satisfactory. Therefore, without the claim of precise quantity description one can qualitatively describe the isotonic and isotopic trends in energy of two-quasiparticle states with the two-center shell model.

The calculated energies of two-quasiparticle states with $K \geq 4$ in several even isotopes of Fm are shown in Fig. 2. In the recent experiment [6] the state $8_\nu^- (9/2^- [734] \otimes 7/2^+ [624])$ was observed in ^{250}Fm at 1.199 MeV that is closed to our result. In ^{248,250}Fm the relatively low-lying isomer states with $K = 7_\nu^-$ and 8_ν^- are expected. In ^{248}Fm the relatively low-lying isomer states with $K^\pi = 6_\nu^+$ and 7_ν^- are

CP1165, *Nuclear Structure and Dynamics '09*
edited by M. Milin, T. Nikšić, D. Vretenar, and S. Szilner
© 2009 American Institute of Physics 978-0-7354-0702-2/09/$25.00

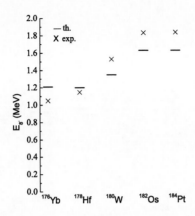

FIGURE 1. The calculated (th) and experimental (exp) [5] energies of two-quasiparticle states with $K^\pi = 8_\nu^-$ are compared for the isotones from ^{176}Yb till ^{184}Pt.

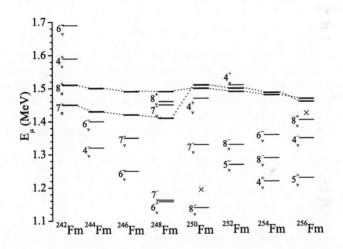

FIGURE 2. The calculated energies of two-quasiparticle states in the indicated even isotopes of Fm. The states created by the break of proton and neutron pairs are indicated by thick and thin lines, respectively.

expected. In ^{242,244}Fm the isomer states with $K \geq 6$ are above 1.38 MeV that is larger than the energies of the revealed K-isomers in ^{252,254}No [1, 7]. In order to observe these isomers, one should produce the neutron-deficient Fm isotopes with the statistics larger than those for the nuclei ^{252,254}No that is, of course, time consuming.

The calculated two-quasiparticle states with $K \geq 4$ in several even isotopes of No are presented in Fig. 3. While in ^{250,252}No the states related to the break of neutron pair are

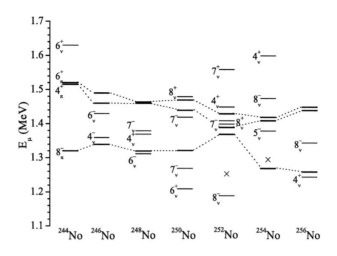

FIGURE 3. The same as in Fig. 2, but for the indicated even isotopes of No. The crosses indicate the experimental results [1].

well lower in energy than the states related to the break of proton pair, in ^{244,246,254}No the lowest two-quasiparticle states are related to the break of proton pair. This is in a good agreement with available experimental data [1]. Because of the subshell closure at $N = 152$, the energy of the lowest two-quasineutron isomer in ^{254}No is larger than in ^{252}No. In ^{250}No the two-quasineutron state 6^+ was attributed with the experimentally observed isomer state [8]. ^{256}No seems to be other good candidate to study the low-lying isomer states with $K = 8_\pi^-$.

In the isotones ^{252}Fm and ^{254}No with $N = 152$ the low-lying two-quasineutron states are similar (Figs. 2 and 3). If the isomers 5_ν^- and 8_ν^- would be revealed in one of these nuclei, they should be also in other. In ^{252}Fm ($4n$ channel of the ^{18}O$+^{238}$U reaction) the γ-transitions from these isomer states are expected. In the isotones ^{250}Fm and ^{252}No with $N = 150$ the low-lying 8_ν^- states exist.

In Fig. 4 the values of Q_α for the α-decays from the ground-states of Md isotopes are compared with the available experimental data [9]. Within the accuracy of our calculation the calculated results are in a good agreement with the experiment. The values of Q_α corresponding to α-decays into the rotational states $7/2^-$ are presented in Fig. 4 as well.

As the isotonic dependence in even-Z nuclei, the isotopic dependence of the energy of one-quasiparticle state with certain K^π is rather smooth. For several isotopes of Es, the energies of rotational states $9/2^+$, $7/2^-$, and $9/2^-$ were calculated with the formalism of Ref. [10]. The isotopic dependence of the energy of rotational state $7/2^-$ is steeper than the one of the energy of one-quasiproton state $7/2^-[514]$. The ground-state of odd isotopes of Md is $7/2^-[514]$ that is in a good agreement with the recent experiment [9].

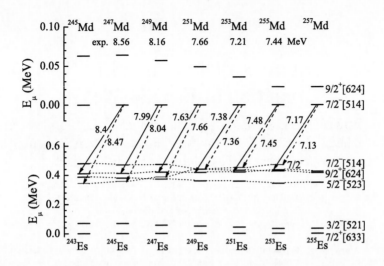

FIGURE 4. The proposed α-decays from the ground-states of indicated isotopes of Md. The α-decays to the rotational $7/2^-$ states are shown by dashed arrows. The calculated values of Q_α, which are near the corresponding arrows, can be compared with the presented experimental data [9].

For the isotopes of Es, we obtained $7/2^+[633]$ as the ground-state. In the isotopes of Es the M2 gamma-transitions between the states $3/2^-[521]$ and $7/2^+[633]$ would occur with $T_\gamma \approx 0.4$ ms, i.e. $3/2^-$ states can be treated as isomers.

Concluding, the used modified two-center shell model is suitable to describe structure properties of heaviest nuclei and to predict the isotopic trends of K-isomer states. Note that the calculated values of Q_α and, correspondingly, the estimated values of α-decay half-lives seem to be in a satisfactory agreement with the experimental data.

We thank F. P. Hessberger and J. Khuyagbaatar for fruitful discussions. This work was supported in part by DFG and RFBR. The IN2P3–JINR and Polish–JINR Cooperation Programmes are gratefully acknowledged.

REFERENCES

1.   R. -D. Herzberg, P. T. Greenlees, *Prog. Part. Nucl. Phys.* **61**, 674 (2008).
2.   J. Maruhn, and W. Greiner, *Z. Physik* **251**, 431 (1972).
3.   A. Parkhomenko, and A. Sobiczewski, *Acta Phys. Pol.* **36**, 3115 (2005); **365**, 2447 (2004).
4.   P. Möller, J. R. Nix, W. D. Myers, and W. J. Swiatecki, *At. Data Nucl. Data Tables* **59**, 185 (1995).
5.   http://www.nndc.bnl.gov/ensdf/
6.   P. T. Greenlees *et al.*, *Phys. Rev. C* **78**, 021303 (2008).
7.   B. Sulignano *et al.*, *Eur. Phys. J. A* **33**, 327 (2007).
8.   D. Peterson *et al.*, *Phys. Rev. C* **74**, 014316 (2006).
9.   F. P. Hessberger *et al.*, *Eur. Phys. J. A* **26**, 233 (2005).
10. T. M. Shneidman, G. G. Adamian, N. V. Antonenko, and R. V. Jolos, *Phys. Rev. C* **74**, 034316 (2006).

Fragment mass distribution in ^{238}U(d, pf) reaction at E$_d$=124 MeV

A. Bogachev^a, E. Kozulin^a, E. Chernyshova^a, D. Gorelov^a, G. Knyazheva^a, L. Krupa^a, S. Smirnov^a, J.Äystö^b, V.A.Rubchenya^b, W.H.Trzaska^b, L. Calabreta^c and E.Vardaci^d

^{a}FLNR, Joint Institute for Nuclear Research, 141980 Dubna, Russia
^{b}Department of Physics, University of Jyväskylä, Finland
^{c}Laboratori Nazionali del Sud, Catania, Italy
^{d}Dipartamento di Fisica and INFN , Napoli, Italy

Abstract. Mass-energy distributions of fission fragments in the deuteron induced fission of ^{238}U at deuteron energy 124 MeV have been measured. For better understanding of the reaction mechanism, the inclusive proton spectra and proton spectra in coincidence with fission fragments were measured in the experiment.

Keywords: ^{238}U(d,f), ^{238}U(d,pf), inclusive proton spectra
PACS: 25.85.-w

Interest to the study of the deuteron induced reactions at intermediate energies (E$_d$ > 30 MeV) is rising presently. More than fifty years ago Serber [1] considered the deuteron-breakup process as a source of high neutrons. The (d, pf) reaction is possible way for investigation of the neutron-induced fission in wide region of fast neutron energies [2].

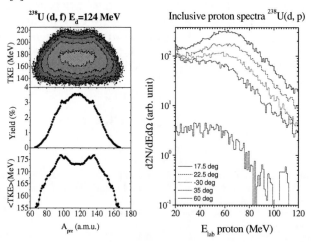

FIGURE 1. Left panel from top to bottom: two-dimensional Mass – Total Kinetic Energy distribution of fission fragments formed in deuteron induced fission of ^{238}U, and its mass distribution and average total kinetic energy distribution. Right panel: inclusive proton spectra at different angles respect to the beam axis.

CP1165, *Nuclear Structure and Dynamics '09*
edited by M. Milin, T. Nikšić, D. Vretenar, and S. Szilner
© 2009 American Institute of Physics 978-0-7354-0702-2/09/$25.00

The mass and energy of fission fragments were measured by CORSET spectrometer [3], which consists of two identical arms included compact start detectors and position sensitive stop detector based on micro-channel plates. The arms of the spectrometer were symmetrically installed with respect to the beam axis at angles 88°-88°. Light Charged Particles (LCP), mostly deuterons and protons, were detected by telescopes from 8π LP apparatus. Each LCP telescope consists of 300 μm Si detector backed by a 15 mm CsI(Tl) crystal and has an active area of 25 cm^2. All telescopes give 1% coverage for LCP detection.

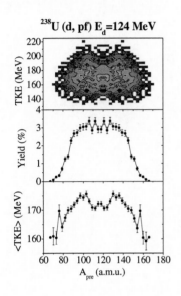

^{238}U (d, pf) E_d=124 MeV

In the figure 1 (left panel) the mass-energy distributions of fission fragments formed in deuteron induced fission of ^{238}U are shown. The local maxima are visible at mass 133 and 143 a.m.u. that correspond to nuclear shells N=82 and N=88. The inclusive proton spectra measured in this experiment at different angles are presented in figure 1 (right panel). At forward angles proton spectra have peaks near the proton energy 60 MeV which disappear at backward angles. These spectra are similar to that measured by Ridicas et al. [4] in the same reaction at deuteron energy 100 MeV.

The mass-energy distributions of fission fragments accompanied by high energy protons are shown in figure 2. In the mass region A=78-80 a.m.u. the enhanced yield of the fission fragments is observed. This could be explained by the influence of nuclear shells N=50 and Z=28.

FIGURE 2. From top to bottom: two-dimensional Mass – Total Kinetic Energy distribution of fission fragments formed in deuteron induced fission of ^{238}U in coincidence with registered high energy protons, and its mass distribution and average total kinetic energy distribution.

ACKNOWLEDGMENTS

The work was supported by the Russian Foundation for Basic Research (Grant Number 07-02-00439a and 07-02-00943a).

REFERENCES

1. R. Serber, *Phys. Rev.* **72**, 1008 (1947).
2. V.A. Rubchenya, et al., in Proceedings of the second international conference "fission and properties of neutron-rich nuclei", St. Andrew, Scotland, 1999, Ed. By J.H. Hamilton, W.R. Philips, and H.K. Carter (World Sci., Singapure, 2000), p. 484.
3. E.M. Kozulin, et al., *Instr. Exp. Tech.* **51**, 44 (2008) .
4 . D.Ridicas, et al., *Phys. Rev.* C **63**, 014610 (2000).

NUCLEAR COLLECTIVE MOTION

Relativistic Point Coupling Model for Vibrational Excitations in the Continuum

P. Ring*, J. Daoutidis*, E. Litvinova†, T. Nikšić**, N. Paar** and D. Vretenar**

*Physics Department Technical University Munich, 85748 Garching, Germany
†Gesellschaft für Schwerionenforschung mbH, 64291 Darmstadt, Germany
**Physics Department, Faculty of Science, University of Zagreb, Croatia

Abstract. An implementation of the relativistic random phase approximation with the proper treatment of the continuum has been developed for the relativistic point coupling model and applied to investigate collective excitations in spherical nuclei. The results are compared with the spectral implementation of the same model. In heavy nuclei, where the escape width is negligible, we find an excellent agreement between both methods in the region of giant resonance and some discrepancies in the region of low-lying pygmy resonance. The differences are more pronounced in light nuclei due to the larger values of the escape widths.

Keywords: Continuum RPA, point-coupling models
PACS: 21.30.Fe, 21.60.Jz, 21.65.+f, 21.10.-k

INTRODUCTION

The entire framework of the relativistic density functional theory (RDFT) has attracted considerable interest in the past decade. Because these models are based on the Lorentz covariance, the spin and spatial degrees of freedom are treated in the consistent manner, thus providing a relatively simple, but still very successful description of nuclei all over the periodic table [1, 2]. Moreover, the same functionals can be used to study both statical as well as dynamical properties of finite nuclei, such as collective vibrations. Self-consistent relativistic RPA (RRPA) calculations have a long history. The initial studies based on the meson-exchange model with non-linear meson couplings [3], carried out in the spectral representation including only particle-hole (*ph*) pairs with particles above the Fermi energy and holes in the Fermi sea, could not reproduce empirical data. Soon it has been realized that a fully self-consistent treatment necessitates the inclusion of antiparticle-hole (*ah*) pairs with antiparticles in the Dirac sea and holes in the Fermi sea [4], thus increasing the numerical effort considerably. Therefore, apart from a few investigations using continuum RPA [5] most of the applications of the RRPA model have been carried out in the spectral representation [6]

In order to reduce the numerical burden in beyond mean-field calculations, several attempts have been made to develop relativistic point coupling (PC) models with forces of zero range [7], in analogy to non-relativistic Skyrme-functionals, but only recently parameter sets have been found, which are comparable in quality to the modern meson-exchange models [8, 9, 10]. This model has so far been successfully applied to configuration mixing calculations in the framework of the generator coordinate method [11],

CP1165, *Nuclear Structure and Dynamics '09*
edited by M. Milin, T. Nikšić, D. Vretenar, and S. Szilner
© 2009 American Institute of Physics 978-0-7354-0702-2/09/$25.00

and we plan to implement it also in the particle vibrational coupling calculations [12].

This talk is devoted to recently developed implementation of the RRPA with an exact treatment of the coupling to the continuum. The point-coupling effective interaction PC-F1 [8] is employed, since previous studies using the spectral implementation of the RRPA suggest that it provides an accurate description of the collective excitations in finite nuclei [13].

THEORETICAL FRAMEWORK

The starting point is an implementation of the relativistic point-coupling model, developed in Ref. [8]. The model is based on an energy functional $E[\hat{\rho}]$, where $\hat{\rho}$ denotes the relativistic single-particle density matrix. In order to derive the time-dependent relativistic mean-field (TDRMF) equations of motion for the single-particle density $\rho(t)$, we use the time-dependent variational principle and calculate the response to an external field $\hat{F}$ oscillating with the frequency ω. The cross section of this process is proportional to the strength function

$$S(\omega) = -\frac{1}{\pi} Im \int d^3 r d^3 r' \, F^*(\mathbf{r}) R(\mathbf{r}, \mathbf{r}'; \omega) F(\mathbf{r}'), \qquad (1)$$

where $F(\mathbf{r})$ is the operator inducing the reaction and $R(\mathbf{r}, \mathbf{r}'; \omega)$ is the response function. The TDRMF-equations lead in the small amplitude limit to the linearized Bethe Salpeter equation for the response function [14]

$$R(\omega) = R^0(\omega) + R^0(\omega) V^{ph} R(\omega). \qquad (2)$$

The relativistic residual interaction V^{ph} is calculated as the second derivative of the energy density functional with respect to the density matrix

$$V^{ph} = \frac{\delta^2 E[\hat{\rho}]}{\delta \hat{\rho} \delta \hat{\rho}}. \qquad (3)$$

Starting from the energy density functional PC-F1 and neglecting for the moment the Coulomb force, the following form of the zero-range residual interaction is obtained

$$V^{ph}(1,2) = \sum_c \Gamma_c^{(1)} \, \delta(\mathbf{r}_1 - \mathbf{r}_2) v_c(\mathbf{r}_1) \Gamma_c^{\dagger(2)}, \qquad (4)$$

where Γ_c is a combination of Dirac- and isospin matrices characterizing the various channels of the relativistic zero-range interaction (scalar-isoscalar, vector-isoscalar and vector-isovector in this particular implementation). Using angular momentum coupling techniques and combining all the dependence on spin, isospin and angular variables to the local single particle operators

$$Q_c^{(1)}(r) = \frac{\delta(r - r_1)}{r r_1} \left[\Gamma_c^{(1)} Y_L(\Omega_1) \right]_J, \qquad (5)$$

146

we can express the interaction (4) as a sum (or integral) of separable terms of the following form

$$V^{ph}(1,2) = \sum_c \int_0^\infty dr \, Q_c^{(1)}(r) \, v_c(r) \, Q_c^{\dagger(2)}(r). \tag{6}$$

Thus, for the point-coupling implementation of the RMF models, the residual interaction V^{ph} in Eq. (2) can be handled relatively easily. In general, $v_c(r)$ contains the density-dependent coupling constants and/or various rearrangement terms, depending on the form of the particular effective interaction used. Also, it contains the Laplacian originating from the derivative terms of the point-coupling model. By discretizing the radial coordinate, the Laplacian is expressed by a matrix and the integral equation is transformed to a matrix equation. In a similar manner the direct part of the Coulomb interaction is fully taken into account.

The essential part of the model is the calculation of the response function R^0 that describes the response of a non-interacting system. Formally it can be expressed by the matrix elements of the operators Q_c in the following way

$$R_{cc'}^0(r,r';\omega) = \sum_{ph} \frac{\langle h|Q_c^+(r)|p\rangle\langle p|Q_{c'}(r')|h\rangle}{\omega - \varepsilon_p + \varepsilon_h + i\eta} - \frac{\langle p|Q_c^+(r)|h\rangle\langle h|Q_{c'}(r')|p\rangle}{\omega + \varepsilon_p - \varepsilon_h + i\eta}, \tag{7}$$

where h stands for occupied (hole) and p for all unoccupied (particle) states of the static solution of the RMF equations. ε_p and ε_h are the corresponding single particle energies, i.e., the eigenvalues of the Dirac equation. It is important to notice that the index p runs over the discrete states above the Fermi energy and below the continuum limit, the discrete states in the negative energy Dirac sea and both the positive as well as the negative energy continuum. This corresponds to the *no-sea* approximation used in the relativistic density functional theory, which implies that the levels in the Dirac sea are unoccupied. There are essentially two ways to calculate the free response function $R_{cc'}^0(r,r';\omega)$:

a) In the so-called "spectral representation" of the random phase approximation the continuum is discretized (DRPA). The empty levels in the continuum are obtained either by solving the ground state Dirac equation within a box of finite radius R, or by expanding the Dirac spinors in a finite oscillator basis [15]. It is evident that in this method one has to include a tremendous number of *ph*-states in the sums of Eq. (7). For vanishing η the response function $R^0(\omega)$ is real and vanishes at the real ω-axis except for the large number of poles located at the *ph*-energies. In numerical applications this problem is solved usually by using a finite imaginary part $\eta = \Gamma/2$. The poles are smeared by a Lorentzian of width Γ and the decay-width, which anyhow cannot be taken into account on the RPA level, is treated in a phenomenological way. This approach has been applied in most of the present relativistic RPA and QRPA calculations. However, it does not include the proper treatment of the continuum.

b) The proper treatment of the continuum has first been introduced by Shlomo and Bertsch [16]. The sum over p can be safely extended to run over the full space, since terms of the form $\sum_{hh'}$ vanish due to the cancellation of forward and backward going parts. Using the completeness relations one can express the response function

$$R^0_{cc'}(r,r';\omega) = \sum_h \{ \langle h(r)|Q^+_c G(r,r';\omega+\varepsilon_h)Q_{c'}|h(r')\rangle$$

$$+ \langle h(r')|Q_{c'}G(r',r;-\omega+\varepsilon_h)Q^+_c|h(r)\rangle \} . \qquad (8)$$

in terms of the single particle Green's function $G(E) = 1/(E - \hat{h})$ of the Dirac Hamiltonian $\hat{h}$. The sum over many discretized particle states is avoided by this method, and one is left with the sum over the relative few bound states with the radial Dirac spinors $|h(r)\rangle$. The Dirac sea is treated automatically in full agreement with the *no-sea* approximation. In order to include the effects of the continuum, one has to apply the regular boundary conditions at the origin and the out-going wave boundary conditions for large r. For positive energies the outgoing wave boundary conditions induce automatically an imaginary part for $R^0(\omega)$ and the continuum is treated exactly without discretization.

APPLICATIONS

Isovector Giant Dipole Resonances

The isovector giant dipole resonance (IVGDR) is the first mode of collective excitations to be observed experimentally. It has been a subject of numerous investigations, both theoretical and experimental. An external electromagnetic field causes protons and neutrons to oscillate in opposite phases, leading to a pronounced peak in the photoabsorption cross section. In Fig. 1 we show the distribution of the isovector dipole strength in the doubly magic nucleus ^{132}Sn. Results obtained with CRPA (red curve) are compared with the solutions from the spectral representation (blue lines). As far as the resonance position and the overall strength distribution are concerned, the agreement between two methods is excellent. Moreover, the energy weighted sum rule obtained with CRPA

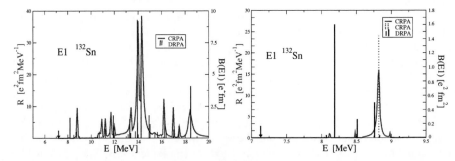

FIGURE 1. (Color online) Left panel: the isovector dipole strength distribution in ^{132}Sn calculated with the parameter set PC-F1. The red curve corresponds to the strength distribution (units on the l.h.s.) obtained by a non-spectral representation without smearing ($\Gamma = 0$), the blue lines give the discrete B(E1)-values (units on the r.h.s.) obtained by the spectral representation with the same force. In order to distinguish the continuum (red curve) and the discrete (blue lines) calculations we have used here a small smearing parameter $\Gamma = 10$ keV in the continuum calculation. The black arrow indicates the theoretical neutron emission threshold. Right panel: the E1 pygmy resonance (PDR).

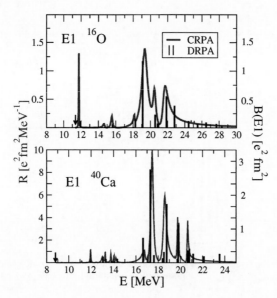

FIGURE 2. (Color online) The isovector dipole strength distribution in the nuclei ^{16}O and ^{40}Ca. Details are the same as in Fig. 1. The theoretical neutron separation energies, indicated by black arrow are $E_{th} = 11.33$ MeV for ^{16}O and $E_{th} = 8.91$ MeV for ^{40}Ca.

$m_1(E1) = 563.60$ [MeV·fm^2], is in very good agreement with the spectral representation calculations $m_1(E1) = 591.02$ [MeV·fm^4], exhausting 122,9% of the Thomas-Reiche-Kuhn sum rule [14].

For nuclei with large neutron excess one can observe low-lying dipole strength close to the neutron emission threshold. This mode of excitation is usually called pygmy dipole resonance (PDR) and can be interpreted as a vibrational mode with dipole character where the neutron skin oscillates against the proton-neutron core with T=0. In the right panel of Fig. 1 we display the isovector dipole strength distribution in the low-energy region for ^{132}Sn isotope. We notice discrepancies in the fine details of the strength distribution between the CRPA and spectral representation approach. This is a consequence of the proximity of the neutron emission threshold, leading to the larger continuum effects.

In Fig. 2 we show the electric dipole strength distribution of two lighter nuclei ^{16}O and ^{40}Ca. The strength obtained in CRPA calculations (red curves) is compared with the B(E1)-values resulting from discrete DRPA calculations (blue lines). The position of the corresponding peaks and poles with large strength are again in rather good agreement. Also, we notice how the escape widths increase as the mass of the nucleus decreases.

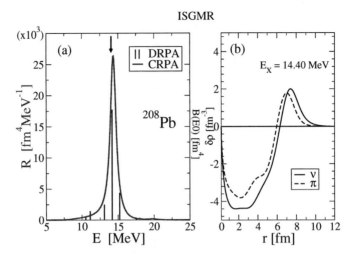

FIGURE 3. (Color online) (a) The isoscalar monopole spectrum in ^{208}Pb, calculated with the parameter set PC-F1. The red curve corresponds to the strength distribution (units on the l.h.s.) obtained by a non-spectral representation without smearing ($\Gamma = 0$), the blue lines give the discrete B(E0)-values (units on the r.h.s.) obtained by the spectral representation with the same force. The black arrow indicates the experimental centroid energy of the resonance. (b) the neutron and proton transition densities at the peak with the energy $E = 14.40$ MeV.

Isoscalar Giant Monopole Resonances

The isoscalar monopole mode of excitation is induced by the following external field operator

$$F_{L=0}^{T=0} = \sum_i^A r_i^2. \tag{9}$$

In Fig. 3 we show the isoscalar monopole strength distribution for the nucleus ^{208}Pb calculated with CRPA (full red line) and compare it with the discrete B(E0) values (blue) obtained by the spectral representation of the response function for the same parameter set PC-F1. No additional smearing has been used, which means that the observed width of the CRPA strength corresponds entirely to the escape width. Due to the relatively high Coulomb and centrifugal barriers in heavy nuclei, the escape width is very small in the Pb region. The spectral representation approach, of course, provides no width at all. Otherwise, both methods give very similar results. In particular, the energy weighted sum rule obtained in CRPA $m_1(E0) = 5.448 \cdot 10^5$ [MeV·fm^4] is in almost perfect agreement with both the spectral representation result $m_1(E0) = 5.446 \cdot 10^5$ [MeV·fm^4], as well as the classical value $m_1(E0) = 4A\hbar/2m\langle r^2 \rangle = 5.453 \cdot 10^5$ [MeV·fm^4]. This shows that the results obtained in the literature by relativistic RPA calculations using the spectral method are very reliable for such heavy nuclei [6].

In the right panel of Fig. 3, we display the neutron and proton transition densities for the excitation corresponding to the isoscalar giant monopole resonance. They emphasize the isoscalar character of the collective breathing mode, with proton and neutron densities oscillating in phase over the entire volume of the nucleus.

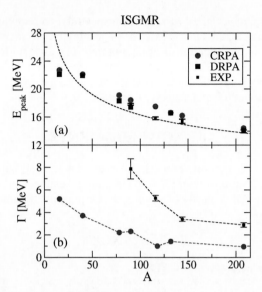

FIGURE 4. (Color online) (a) The ISGMR centroid energies as a function of the mass number, (b) The total experimental and calculated escape widths of the ISGMR as a function of the mass number. Details are given in the text.

Fig. 4 summarizes the results for the isoscalar monopole strength distributions as a function of the mass number A. In the upper panel, we plot the centroid energies for both CRPA (red dots) and spectral representation RPA (blue dots), together with the experimental centroid energies. We also show the phenomenological A-dependence $\bar{E}_{1^-} \approx 31.2A^{-1/3} + 20.6A^{-1/6}$ by the dashed line. It becomes clear that CRPA can successfully reproduce collective excitations over the known range of nuclei.

In the lower panel of Fig. 4 we plot the calculated escape widths $\Gamma^{\uparrow}$ of the E0 resonances in comparison with the total experimental widths. The red values correspond to the full width half maximum (FWHM) of the peak, obtained with CRPA , while the experimental values are indicated in black. The evident disagreement is not surprising, because only $1p1h$ configurations are taken into account in the continuum RPA, whereas the largest part of the damping correspond to the spreading width that originates from more complex configurations, i.e. $2p2h$ and higher order excitations.

CONCLUSIONS

Starting from a point coupling Lagrangian, we have used the non-spectral relativistic RPA approach to examine the corresponding excitation spectra and we have compared the results with spectral calculations based on the same Lagrangian. This method has several advantages. The coupling to the continuum is treated consistently using the relativistic single particle Green's function at the appropriate energy. In this way, complicated sums over unoccupied states are avoided. This is particularly important for relativistic applications since the Dirac sea is now automatically treated properly.

Comparing calculations with spectral and non-spectral representations of the response function for the same Lagrangian we find, that in general the spectra are well reproduced within the spectral approximation, if an appropriate phenomenological smearing parameter is used and if a sufficiently large number of ph-configurations is taken into account. However, we find differences in the vicinity of the neutron threshold, where the coupling to the continuum is not properly reproduced in the spectral method.

ACKNOWLEDGMENTS

This research has been supported the Gesellschaft für Schwerionenforschung (GSI), Darmstadt, the Bundesministerium für Bildung und Forschung, Germany under project 06 MT 246 and by the DFG cluster of excellence "Origin and Structure of the Universe".

REFERENCES

1. P. Ring, *Prog. Part. Nucl. Phys.* **37**, 193 (1996).
2. D. Vretenar, A. V. Afanasjev, G. A. Lalazissis, and P. Ring, *Phys. Rep.* **409**, 101 (2005).
3. Z.-Y. Ma, N. Van Giai, H. Toki, and M. L'Huillier, *Phys. Rev.* **C55**, 2385 (1997).
4. P. Ring, Z.-Y. Ma, N. Van Giai, D. Vretenar, A. Wandelt, and L.-G. Cao, *Nucl. Phys.* **A694**, 249 (2001).
5. J. Piekarewicz, *Phys. Rev.* **C62**, 051304(R) (2000).
6. D. Vretenar, A. Wandelt, and P. Ring, *Phys. Lett.* **B487**, 334 (2000); Z.-Y. Ma, N. Van Giai, A. Wandelt, D. Vretenar, and P. Ring, *Nucl. Phys.* **A686**, 173 (2001); D. Vretenar, N. Paar, P. Ring, and G. A. Lalazissis, *Nucl. Phys.* **A692**, 496 (2001); D. Vretenar, N. Paar, P. Ring, and T. Nikšić, *Phys. Rev.* **C65**, 021304 (2002); Z.-Y. Ma, A. Wandelt, N. Van Giai, D. Vretenar, P. Ring, and L.-G. Cao, *Nucl. Phys.* **A703**, 222 (2002); T. Nikšić, D. Vretenar, and P. Ring, *Phys. Rev.* **C66**, 064302 (2002); N. Paar, P. Ring, T. Nikšić, and D. Vretenar, *Phys. Rev.* **C67**, 034312 (2003); N. Paar, D. Vretenar, and P. Ring, *Phys. Rev. Lett.* **94**, 182501 (2005); N. Paar, D. Vretenar, T. Nikšić, and P. Ring, *Phys. Rev.* **C74**, 037303 (2006); A. Ansari, and P. Ring, *Phys. Rev.* **C74**, 054313 (2006); D. Peña Arteaga, and P. Ring, *Phys. Rev.* **C77**, 034317 (2008); N. Paar, D. Vretenar, E. Khan, and G. Coló, *Rep. Prog. Phys.* **70**, 691 (2007).
7. P. Manakos, and T. Mannel, *Z. Phys.* **A334**, 481 (1989).
8. T. Bürvenich, D. G. Madland, J. A. Maruhn, and P.-G. Reinhard, *Phys. Rev.* **C65**, 044308 (2002).
9. T. Nikšić, D. Vretenar, G. A. Lalazissis, and P. Ring, *Phys. Rev.* **C77**, 034302 (2008).
10. T. Nikšić, D. Vretenar, and P. Ring, *Phys. Rev.* **C78**, 034318 (2008).
11. T. Nikšić, D. Vretenar, and P. Ring, *Phys. Rev.* **C73**, 034308 (2006); T. Nikšić, D. Vretenar, and P. Ring, *Phys. Rev.* **C74**, 064309 (2006); T. Nikšić, D. Vretenar, G. A. Lalazissis, and P. Ring, *Phys. Rev. Lett.* **99**, 092502 (2007).
12. E. Litvinova, and P. Ring, *Phys. Rev.* **C73**, 044328 (2006); E. Litvinova, P. Ring, and D. Vretenar, *Phys. Lett.* **B647**, 111 (2007); E. Litvinova, P. Ring, and V. I. Tselyaev, *Phys. Rev.* **C75**, 064308 (2007); E. Litvinova, P. Ring, and V. I. Tselyaev, *Phys. Rev.* **C78**, 014312 (2008).
13. T. Nikšić, D. Vretenar, and P. Ring, *Phys. Rev.* **C72**, 014312 (2005).
14. P. Ring, and P. Schuck, *The Nuclear Many-Body Problem*, Springer-Verlag, Berlin, 1980.
15. Y. K. Gambhir, P. Ring, and A. Thimet, *Ann. Phys. (N.Y.)* **198**, 132 (1990).
16. S. Shlomo, and G. F. Bertsch, *Nucl. Phys.* **A243**, 507 (1975).

The Giant Monopole Resonance in Unstable Nuclei and the Role of Superfluidity

E. Khan

Institut de Physique Nucléaire, Université Paris-Sud,IN2P3-CNRS, F-91406 Orsay Cedex, France

Abstract. Superfluidity is found to favour the compressibitiy of nuclei using a full self-consistent treatment. Pairing correlations may explain why doubly magic nuclei such as ^{208}Pb are stiffer compared to open-shell nuclei. Pairing gap dependence of the nuclear matter incompressibility should also be investigated. Both the GMR and the GQR have been measured in an unstable nuclei for the first time.

Keywords: nuclear incompressibility, superfluidity, giant monopole resonance
PACS: 21.10.Re,24.30.Cz,21.60.Jz

INTRODUCTION

Superfluidity initially referred to a system with a dramatic drop of its viscosity [1]: it could be suspected that a super-fluid would be easier to compress than a normal fluid. This question has been investigated in Fermionic atoms traps [2], but this signal remains to be confirmed. Theoretically both microscopic and hydrodynamical investigations show no variation of the compression mode between the normal and the superfluid phases [3], but the analysis is complicated by the temperature change between the two phases. In nuclear physics, the study of the role of superfluidity in the compressibility can also be performed: the isoscalar Giant Monopole Resonance (GMR) is a compression mode, allowing to probe for related superfluid effects. Ideal tools are especially isotopic chains, where pairing effects are evolving from normal (doubly-magic) nuclei to superfluid (open-shell) ones [4].

INCOMPRESSIBILITY AND SUPERFLUIDITY

A previous study on the role of superfluidity on nuclear incompressibility has been performed, finding a negligible effect [5], but the theoretical approach was not self-consistent. Indeed self-consistency is crucial since pairing effects are expected to be small in the GMR: this high energy mode is mainly built from particle-hole configurations located far from the Fermi level, where pairing do not play a major role. However giant resonances are known to be very collective [6] and pairing can still have a sizable effect on the GMR properties: around 10 % on the centroid, which is the level of accuracy of present analysis on the extraction of K_∞ [7, 8]. This requires the advent of accurate microscopic models in the pairing channel, such as fully self-consistent Quasiparticle Random Phase Approximation (QRPA) [9, 10], achieved only recently.

CP1165, *Nuclear Structure and Dynamics '09*
edited by M. Milin, T. Nikšić, D. Vretenar, and S. Szilner
© 2009 American Institute of Physics 978-0-7354-0702-2/09/$25.00

When considering the available GMR data from which the K_∞ value has been extracted, ^{208}Pb is stiffer than the Sn, Zr and Sm nuclei: K_∞ is about 20 MeV larger, both in non-relativistic and in relativistic approaches [7, 8]. The question may not be "why Tin are so soft ?"[11] but rather "why ^{208}Pb is so stiff ?". Recently the GMR was measured on the stable Tin isotopic chain (from ^{112}Sn to ^{124}Sn) [12]. Once again it has been noticed that it is not possible to describe the GMR both in Sn and in Pb with the same functional, Tin being softer than Pb [11, 9]. In the non relativistic case, fully self-consistent QRPA calculations on Sn isotopes lead to $K_\infty \simeq 215$ MeV. The relativistic DDME2 paremeterisation using QRPA describes well the GMR in the Sn isotopes [13], but predict a low value of the GMR in ^{208}Pb compared to the experiment, as can be seen on Fig. 8 of Ref. [8]. On the contrary, a recent relativistic functional describes well the ^{208}Pb GMR, but systematically overestimates the Sn GMR values of about 1 MeV [14]. Finally, attempts have been performed in order to describe Sn GMR data with relativistic functionals having a lower incompressibility and hence different density dependence and K_{sym} value [15]. Once again the ^{208}Pb and Tin GMR cannot be described at the same time: this puzzling situation is due to the higher value of K_∞ extracted from ^{208}Pb, compared to Tin, Sm and Zr nuclei.

In Ref. [9] it has been found that including pairing effects in the description of the GMR allows to explain part of the Sn softness : pairing may decrease the predicted centroid of the GMR of few hundreds of keV, located at ~ 16 MeV. The consequences of superfluidity on nuclear incompressibility may solve the above mentioned puzzle: pairing is vanishing in the doubly magic ^{208}Pb nucleus, unlikely the other nuclei. It is therefore necessary to use a fully microscopic method including an accurate pairing approach. In order to predict the GMR in a microscopic way we use the constrained HF method, extended to the full Bogoliubov pairing treatment (CHFB). The CHF(B) method has the advantage to very precisely predict the centroid of the GMR using the m_{-1} sumrule [16, 17]. The whole residual interaction (including spin-orbit and Coulomb terms) is taken into account and this method is by construction the best to predict the GMR centroid [17]. Introducing the monopole operator as a constraint, the m_{-1} value is obtained from the derivative of the mean value of this operator. The m_1 sumrule is extracted from the usual double commutator, using the Thouless theorem [18]. Finally the GMR centroid is given by $E_{GMR}=\sqrt{m_1/m_{-1}}$. All details on the CHF method can be found in [16, 7]. The extension of the CH method to the CHFB case has been recently demonstrated in [19]. Skyrme functionals and a zero-range surface pairing interaction are used in the present work. The magnitude of the pairing interaction is adjusted so to describe the trend of the neutron pairing gap in Tin isotopes. This interaction is known to describe a large variety of pairing effects in nuclei [20].

Fig. 1 displays the GMR energy obtained from the Sn measurements, compared to CHFB predictions using the SkM* interaction [21]. The striking feature is the increase of the GMR centroid, located at the doubly magic ^{132}Sn nucleus, using the CHFB predictions. This indicates that pairing effects should be considered to describe the behaviour of nuclear incompressibility, and that vanishing of pairing make the nuclei stiffer to compress, confirming our previous statement on the stiffness of ^{208}Pb. It would be very interesting to measure the GMR in the ^{132}Sn unstable nucleus. The importance of pairing effect can be understood in a simple way: since the nuclear incompressibility is defined as the second derivative of the energy functional at saturation density [22], there is no obvious reason why the pairing terms of the functional would play no role in the nuclear incompressibility. It is not possible to describe the GMR centroid of both ^{208}Pb and other nuclei with the same functional, as stated above. The puzzle of the stiffness of ^{208}Pb may come from its doubly magic behaviour.

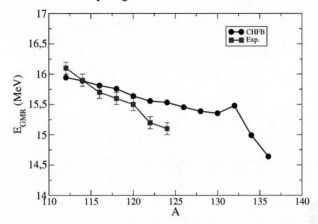

FIGURE 1. (Color online) Excitation energies of the GMR in $^{112-136}$Sn isotopes stable $^{112-124}$Sn isotopes with constrained HFB methods and the SkM* interaction, compared to the data of Ref. [12].

To further investigate the role of pairing on nuclear incompressibility, Fig. 2 displays the GMR energy with respect to the average pairing gap calculated using the HFB approach, from ^{112}Sn to ^{132}Sn. A clear correlation is observed: the more superfluid the nuclei, the lower the incompressibility. Hence it may be easier to compress superfluid nuclei. This may be the first evidence of the role of superfluidity on the compressibility of a Fermionic system. The decrease of incompressibility in superfluid nuclei raises the question of a similar effect in infinite nuclear matter: for now, incompressibility is given independently from the pairing part of the functional. However, considering present results, equations of state used for neutron star and supernovae predictions should take into account pairing to provide their incompressibility value. The comparison with GMR data shows, as mentioned above, that the functional as a whole (including pairing effects) is probed. The question of the behaviour of K_∞ with respect to the pairing gap is raised: it seems clear from nuclear data that nuclei incompressibility decreases with increasing pairing gap. This should be investigated in nuclear matter.

MEASUREMENT OF THE GMR AND GQR IN UNSTABLE NUCLEI

Experimentally, the measurement of the GMR on an isotopic chain facilitates the study of superfluidity on the GMR properties [12], and the possibility to measure the GMR in unstable nuclei emphasizes this feature: with only 15 hours of effective data taking and an average beam intensity of 5.10^4 pps, isoscalar GMR and GQR resonances were

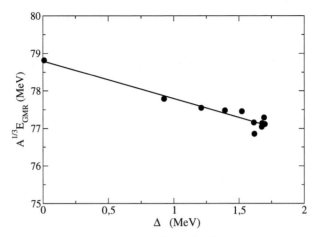

FIGURE 2. Nuclear incompressibilities $E_{GMR}.A^{1/3}$ in $^{112-132}$Sn isotopes calculated with the CHFB method and the SkM* interaction, as a function of the pairing gap Δ predicted by the HFB calculation.

measured in the ^{56}Ni unstable nucleus [23]. It was the first measurement of these resonances in a short-lived nucleus. This makes the method relevant with respect to predicted production rates of a large domain of exotic nuclei at current and future facilities: measurements on heavier nuclei such as ^{132}Sn would complement the systematics already existing for stable tin isotopes. Measurements in neutron-rich Nickel isotopes up to ^{68}Ni is scheduled at GANIL in order to probe the evolution of giant resonances along an isotopic chain.

CONCLUSION

Superfluidity favours the compressibility of nuclei, using a fully microscopic CHFB approach on the Tin isotopic chain. This may be the first evidence of a sizable effect of superfluidity on the compressibility of a Fermionic system. Doubly magic nuclei exhibit a specific increase of the GMR energy, due to the collapse of pairing. ^{208}Pb is therefore the "anomalous" data compared to the others. This difficulty to describe with a single functional both doubly magic and other nuclei has already been observed on the masses, namely the so-called "mutually enhancement magicity" (MEM), described in [24, 25]: functionals designed to describe masses of open-shell nuclei cannot predict the masses of doubly magic nuclei such as ^{132}Sn and ^{208}Pb, which are systematically more bound that predicted. In order to consider MEM, it may be necessary to take into account quadrupole correlation effects due to the flatness of the potentials for open-shell nuclei [26]. It would be useful to find a way to predict the GMR beyond QRPA by taking into account such quadrupole correlations. This may solve the current puzzle of the stiffness of ^{208}Pb. Experimentally, measurements of the GMR in unstable nuclei should be performed in doubly magic ^{132}Sn, as well as extending the measurement on the Sn and Pb isotopic chains. The pairing gap dependence on the nuclear matter incompressibility should also be investigated, since it is shown that incompressibility decreases with increasing pairing gap in nuclei.

ACKNOWLEDGMENTS

The author thanks G. Colò, M. Grasso, J. Margueron, Nguyen Van Giai, P. Ring, H. Sagawa and M. Urban for fruitful discussions about the results of this work.

REFERENCES

1. A.J. Leggett, *Rev. Mod. Phys.* **71** (1999) S318.
2. J. Kinast, A. Turlapov and J.E. Thomas, *Phys. Rev.* **A70** (2004) 051401(R).
3. M. Grasso, E. Khan, M. Urban, *Phys. Rev.* **A72** (2005) 043617.
4. A. Bohr and B. Mottelson, *Nuclear Structure*, Benjamin Inc. (1969), New York.
5. O. Civitarese et al., *Phys. Rev.* **C43** (1991) 2622.
6. M. Harakeh, A. Van der Woude, *Giant Resonances*, Oxford University Press (2001).
7. G. Colò and Nguyen Van Giai, *Nucl. Phys.* **A731** (2004) 15.
8. D. Vretenar, T. Niksic and P. Ring, *Phys. Rev.* **C68** (2003) 024310.
9. J. Li, G. Colò and J. Meng, *Phys. Rev.* **C78** (2008) 064304.
10. N. Paar et al., *Phys. Rev.* **C67** (2003) 034312.
11. J. Piekarewicz, *Phys. Rev.* **C76** (2007) 031301(R)
12. T. Li *et al.*, *Phys. Rev. Lett.* **99** (2007) 162503.
13. G.A. Lalazissis, *Proc. of Nuclear Structure and Dynamics*, Dubrovnik (2009).
14. T. Niksic, D. Vretenar and P. Ring, *Phys. Rev.* **C78** (2008) 034318.
15. J. Piekarewicz and M. Centelles, *Phys. Rev.* **C79** (2009) 054311.
16. O. Bohigas, A.M. Lane and J. Martorell, *Phys. Rep.* **5** (1979) 267.
17. G. Colò et al., *Phys. Rev.* **C70** (2004) 024307.
18. D.J. Thouless, *Nucl. Phys.* **22** (1961) 78.
19. L. Capelli, G. Colò and J. Li, *Phys. Rev.* **C79** (2009) 054329.
20. M. Bender, P.-H. Heenen and P.-G. Reinhard, *Rev. Mod. Phys.* **75**, 121 (2003).
21. J. Bartel et al., *Nucl. Phys.* **A386** (1982) 79.
22. J.P. Blaizot, *Phys. Rep.* **64** (1980) 171.
23. C. Monrozeau et al., *Phys. Rev. Lett.* **100** (2008) 042501.
24. N. Zeldes, T.S. Dumitrescu and H.S. Köhler, *Nucl. Phys.* **A399** (1983) 11.
25. D. Lunney, J.M. Pearson and C. Thibault, *Rev. Mod. Phys.* **75**, 1021 (2003).
26. M. Bender, G.F. Bertsch and P.-H. Heenen, *Phys. Rev. Lett.* **94** (2005) 102503.

Nuclear Pairing and Pairing Vibrations in Stable and Neutron-Rich Nuclei

M. Grasso, E. Khan and J. Margueron

Institut de Physique Nucléaire, Université Paris-Sud, IN2P3-CNRS, F-91406 Orsay Cedex, France

Abstract. Pairing interactions with various density dependencies (surface/volume mixing) are constrained with the two-neutron separation energy in the Tin isotopic chain. The response associated with pairing vibrations in very neutron-rich nuclei is sensitive to the density dependence of the pairing interaction. Using the same pairing interaction in nuclear matter and in Tin nuclei, the range of densities where the LDA is valid in the pairing channel is also studied.

Keywords: Pairing functional, pair transfer excitation modes
PACS: 21.60.Jz,21.65.Cd,25.40.Hs,25.60.Je

PAIRING GAP IN NUCLEAR MATTER AND IN NUCLEI

Studies on pairing effects in both nuclear matter and finite nuclei have recently known intensified interest [1]. Within the Skyrme mean field approach, a Skyrme interaction is employed in the particle-hole channel and a zero-range density-dependent interaction is currently used in the pairing channel [2]. An interesting observable that could help in constraining the pairing functional is pairing vibrations, measured through two-particle transfer [3, 4]. The first microscopic calculations for pairing vibrations have been performed recently [5, 6, 7]. Nuclear matter could also help in constraining the pairing functional. This requires however to bridge nuclei and nuclear matter through LDA in the pairing channel [8, 6, 9, 10, 11]. In nuclear matter the medium polarization increases the pairing gap at low densities in symmetric matter, whereas it reduces the gap in neutron matter, indicating an isospin dependence of the pairing functional [12]. The application to finite nuclei of extended pairing density functional have shown the relevance of the LDA in the pairing channel [13].

We consider here surface and various mixed paring interactions. We choose ^{124}Sn and ^{136}Sn nuclei: these are spherical nuclei where pairing vibrations are likely to occur [4]. One is stable and the second has a large neutron excess. The microscopic calculations for the ground state are based on the Hartree-Fock-Bogoliubov (HFB) model with the Skyrme interaction SLy4. The adopted pairing interaction is the usual zero-range density-dependent interaction

$$V_{pair} = V_0 \left[1 - \eta \left(\frac{\rho(r)}{\rho_0} \right)^{\alpha} \right] \delta \left(\mathbf{r_1} - \mathbf{r_2} \right), \qquad (1)$$

where η provides the surface/volume character of the interaction. We set $\alpha = 1$ and $\rho_0 = 0.16$ fm^{-3}. The numerical cutoff is given by $E_{max} = 60$ MeV (in quasiparticle

CP1165, *Nuclear Structure and Dynamics '09*
edited by M. Milin, T. Nikšić, D. Vretenar, and S. Szilner
© 2009 American Institute of Physics 978-0-7354-0702-2/09/$25.00

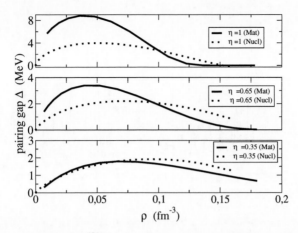

FIGURE 1. Pairing gap for ^{124}Sn and in nuclear matter for different pairing interactions

energies) and $j_{max} = 15/2$. For each value of η, V_0 is chosen to fit the known experimental two-neutron separation energies for Sn isotopes.

The relation between the pairing gap in uniform matter at a given density and the pairing field at a given radius in nuclei has been explored in Ref. [13]. It has been found that in the case of mixed interactions, the LDA is in good agreement with the full microscopic HFB calculation. Fig. 1 displays the gap calculated with the HFB approach as a function of the density in ^{124}Sn. Nuclear matter gap is surimposed, where the neutron-proton asymmetry are taken as the one of ^{124}Sn. One can see that the LDA is not valid for a pure surface pairing, but becomes valid in the $\eta = 0.35$ mixed case, for densities lower than 0.1 fm^{-3}. Therefore, in this case, the low density part of the nuclear matter gap could be constrained by nuclei calculations of the gap.

PAIRING VIBRATIONS WITHIN QRPA

As stated above, it may be useful to consider pairing vibrations as additional observables in order to constrain the pairing interaction. It should be noted that a related study will also be performed in [14]. We refer to [3, 4] for details on pairing vibrations. Basically, these modes corresponds to the (collective) filling of subshells, in transition from an A to A+2 nuclei.

The QRPA equations are solved here in coordinate space, using the Green's function formalism and calculating the response function in the pp channel.

Fig. 2 (left panel) shows the QRPA response for ^{124}Sn, with a pure surface and the two mixed interactions. As expected the residual interaction plays a similar role in all the cases, gathering strength and shifting it to lower energy. In the case of ^{124}Sn, a peak around 9 MeV is the strongest for the surface pairing interaction, to be compared with the one around 10 MeV for the other interactions. Hence it is expected that the pairing vibration transition strength should be larger in the case of a pure surface force. However it is known that it is difficult to describe accurately the magnitude of these transitions, especially for absolute cross section calculations [15]. It is necessary to rely

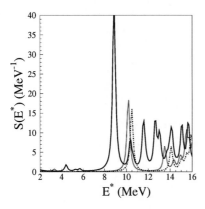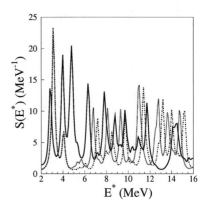

FIGURE 2. QRPA response function for ^{124}Sn (left panel) and ^{136}Sn (right panel) in the two neutrons 0^+ addition mode. The pure surface mode is in solid line, the $\eta=0.65$ mode is in dotted line, and the $\eta=0.35$ mode in dashed-dotted lines.

on the angular distribution [4]. The pairing transition density allows to calculate the form factor in the zero-range DWBA approximation. We have checked that in the pairing transition densities for ^{124}Sn the differences (for different pairing interactions) are not very large and may be eventually overruled by the experimental uncertainties.

For the ^{136}Sn neutron-rich nucleus, the low-energy spectrum displayed in the right panel of Fig. 2 is dramatically changed from using surface to other interactions, on a several MeV area. A three peaks structure appears in the surface case, compared to the 2 peak structure of the other cases. The integrated strength is also larger in the surface

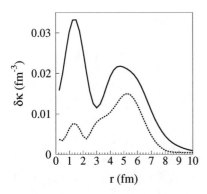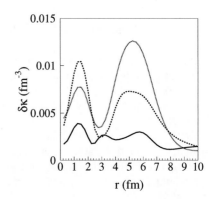

FIGURE 3. Neutron transition density in the two neutrons addition mode for ^{136}Sn for the first two peaks of the strength, in the case of the mixed $\eta=0.65$ interaction (left panel) and for the first 3 peaks of the strength, in the case of the pure surface interaction (right panel).

case.

Fig. 3 illustrates the corresponding transition densities. They exhibit very different shapes. For instance in the case of the most intense peak, the central part is dominant in the transition density for the mixed case, whereas the surface part of the transition density dominates in the pure surface interaction.

CONCLUSIONS

The impact of various pairing interactions on pairing vibrations predictions has been analysed using a HFB+QRPA approach. They should provide a good sensitivity from a pure surface interaction compared to mixed interactions, especially in the case of very neutron-rich nuclei such as ^{136}Sn. Moreover nuclear matter gap calculations show that the LDA in the pairing channel is valid in the surface of the nuclei for the $\eta = 0.35$ mixed surface/volume interaction, but not rigorously valid in other cases.

REFERENCES

1. D. J. Dean and M. Hjorth-Jensen, *Rev. Mod. Phys.* **75**, 607 (2003).
2. E. Garrido, P. Sarriguren, E. Moya de Guerra, P. Schuck, *Phys. Rev* **C60** (1999) 064312.
3. R. A. Broglia, O. Hansen and C. Riedel, *Advances in Nuclear Physics, NY Plenum* **6**(1973) 287.
4. W. von Oertzen, A. Vitturi, *Rep. Prog. Phys.* **64** (2001) 1247.
5. E. Khan, N. Sandulescu, Nguyen Van Giai, M. Grasso, *Phys. Rev.* **C69** 014314 (2004).
6. M. Matsuo, K. Mizuyama, Y. Serizawa, *Phys. Rev.* **C71** 064326 (2005).
7. B. Avez, C. Simenel, Ph. Chomaz, *Phys. Rev.* **C78** 044318 (2008).
8. N. Pillet, N. Sandulescu, P. Schuck, *Phys. Rev.* **C76** 024310 (2007).
9. K. Hagino and H. Sagawa, Phys. Rev. **C76**, 047302 (2007).
10. M. Matsuo, *Phys. Rev.* **C73** 044309 (2006).
11. K. Hagino, H. Sagawa, J. Carbonell, and P. Schuck *Phys. Rev. Lett.* **99**, 022506 (2007).
12. J. Margueron, H. Sagawa, K. Hagino, *Phys. Rev.* **C76** 064316 (2007).
13. J. Margueron, H. Sagawa, K. Hagino, *Phys. Rev.* **C77** 054309 (2008).
14. M. Matsuo, *Proc. of COMEX 3 Conference* (2009) .
15. M. Igarashi, K. Kubo and K. Yagi, *Phys. Rep.* **199**(1991) 1.

Monopole Modes of Excitation in Deformed Neutron-rich Mg Isotopes

Kenichi Yoshida

RIKEN Nishina Center for Accelerator-Based Science, Wako, Saitama 351-0198, Japan

Abstract. The giant monopole resonance (GMR) and the low-frequency mode of monopole excitation in neutron-rich magnesium isotopes close to the drip line are investigated by means of the deformed Hartree-Fock-Bogoliubov and quasiparticle random-phase approximations. It is found that the GMR has a two-peak structure due to the deformation. The lower-energy resonance is generated associated with the coupling to the $K^\pi = 0^+$ component of the giant quadrupole resonance. Besides the GMR, we obtain the soft $K^\pi = 0^+$ mode below the neutron emission threshold energy.

Keywords: nuclear density functional theory, collective modes of excitation, giant resonances, $^{34, 36, 38, 40}$Mg
PACS: 21.60.Jz, 21.10.Re, 27.30.+t, 27.40.+z

INTRODUCTION

Physics of nuclei located far from the stability line has been one of the active fields in nuclear physics. Collective motion in unstable nuclei especially has raised a considerable interest both experimentally and theoretically. This is because low-frequency modes of excitation are quite sensitive to the shell structure near the Fermi level and the detail of surface structure, and we can expect unique excitation modes to emerge associated with the new spatial structures such as neutron skins and the novel shell structures that generate new regions of deformation.

In order to investigate the possibility of emergence of collective excitation modes unique in deformed neutron-rich nuclei, we perform the quasiparticle random-phase approximation (RPA) calculation on top of the deformed Hartree-Fock-Bogoliubov mean field.

MONOPOLE EXCITATIONS

We show in the upper panel of Fig. 1 the isoscalar transition strengths for the monopole excitation. We employed the SkM* energy-density functional for the p-h channel and the mixed-type pairing energy-density functional for the p-p channel in the present calculation. Details of the calculation scheme are described in Ref. [1].

The characteristic features in the response functions are a two-peak structure of the giant monopole resonance (GMR) and a prominent peak below the threshold. The

CP1165, *Nuclear Structure and Dynamics '09*
edited by M. Milin, T. Nikšić, D. Vretenar, and S. Szilner
© 2009 American Institute of Physics 978-0-7354-0702-2/09/$25.00

peak energy for the GMR is given approximately by $80A^{-1/3}$ MeV [2]. The higher-energy peaks reproduce the systematic ($\simeq 25$ MeV).

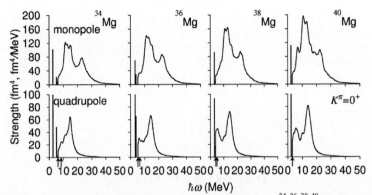

FIGURE 1. Isoscalar monopole and quadrupole transition strengths in $^{34,\,36,\,38,\,40}$Mg. The continuous line is obtained by smearing the transition strengths using a Lorentzian function with a width of 2 MeV. The arrows indicate the neutron emission threshold energies.

In order to understand the origin of the lower-energy peak at around 15 MeV, we show the $K^{\pi} = 0^+$ component of the transition strengths for the isoscalar quadrupole excitation in the lower panel of Fig. 1. At the same energy region where the lower-energy GMR appears, we can see a resonance structure. The peak energy for the GQR is given by $60 - 65A^{-1/3}$ MeV [2]. The resonance structure appearing at around 15 MeV in the lower panel of Fig. 1 has slightly smaller energy compared to the systematic. It should be noted here that the neutron-rich Mg isotopes under investigation are prolately deformed, so that the GQR splits into three resonances of the $K^{\pi} = 0^+, 1^+$ and 2^+ components, and the resonance peak of the $K^{\pi} = 0^+$ component lowers in energy.

Therefore, we can consider the lower-energy resonance of the GMR is generated associated with the coupling to the $K^{\pi} = 0^+$ component of the GQR, which is unique in deformed systems.

TABLE 1. Ratios of the neutron and proton matrix elements M_{ν}/M_{π} divided by N/Z for the lowest $K^{\pi} = 0^+$ mode.

^{34}Mg	^{36}Mg	^{38}Mg	^{40}Mg
1.57	1.58	1.82	1.91

In Ref. [3], we have studied the low-frequency $K^{\pi} = 0^+$ quadrupole vibrational mode in deformed neutron-rich nuclei. The low-lying collective modes appearing below the threshold in the present calculation have the same structure as those in Ref. [3]. In a deformation region, where the up- and down-sloping orbitals exist around the Fermi level, the collectivity emerges due to the coupling between the β vibration and pairing vibration. This condition is realized for the system with neutron numbers $N = 22$ and 28. The interesting feature of the low-frequency $K^{\pi} = 0^+$ mode here is the enhanced transition strength of neutrons as summarized in Table 1. As the neutron dip line is approached, the neutron excitation increases.

SUMMARY

We have investigated the monopole excitation modes in deformed neutron-rich Mg isotopes close to the drip line by means of the deformed quasiparticle RPA on top of the coordinate-space HFB. Because of the deformation, excitation modes with the angular momenta $l = 0$ and 2 can mix in the $K^\pi = 0^+$ channel, and accordingly we obtained a two-peak structure of the GMR. Besides the GMR, we found the emergence of the soft $K^\pi = 0^+$ mode below the threshold associated with the coupling between the quadrupole vibration and pairing vibration of neutrons.

ACKNOWLEDGMENTS

The author is supported by the Special Postdoctoral Researcher Program of RIKEN. The numerical calculations were performed on the NEC SX-8 supercomputer at the Yukawa Institute for Theoretical Physics, Kyoto University and the NEC SX-8R supercomputer at the Research Center for Nuclear Physics, Osaka University.

REFERENCES

1. K. Yoshida and N. Van Giai, *Phys. Rev. C* **78**, 064316 (2008).
2. P. Ring and P. Schuck, "The Nuclear Many-Body Problem", Springer, 1980.
3. K. Yoshida and M. Yamagami, *Phys. Rev. C* **77**, 044312 (2008).

QRPA Calculations for Spherical and Deformed Nuclei With the Gogny Force

S. Péru

CEA, DAM, DIF, F-91297 Arpajon, France

Abstract. Fully consistent axially-symmetric-deformed Quasi-particle Random Phase Approximation (QRPA) calculations have been performed with the D1S Gogny force. Dipole responses have been calculated in Ne isotopes to study the existence of soft dipole modes in exotic nuclei. A comparison between QRPA and generator coordinate method with Gaussian overlap approximation results is done for low lying 2^+ states in N=16 isotones and Ni isotopes.

Keywords: Mean field, Giant resonances, Pygmy resonances
PACS: 21.60.Jz, 24.30.Cz, 27.30.+t

INTRODUCTION

The study of giant resonance states provides key information on nuclear finite-system properties. The impact of the low-energy components of the isovector dipole resonance on nucleosynthesis has been shown [1]. Then a fully consistent description of resonances in both spherical and axially-deformed nuclei has been undertaken using the Quasi-particle Random Phase Approximation (QRPA) based on Hartree-Fock-Bogolyubov states (HFB) calculated with the Gogny D1S effective force [2]. An outline of the method is first presented. Then, the impact of deformed shell effects on the nuclear responses of the light Si and Mg isotopes, dipole response in Ne isotopes are discussed. Finally, a comparison is made between results obtained with the QRPA approach and the generator coordinate method using Gaussian overlap approximation.

THE METHOD

RPA calculations with the D1S Gogny force performed in spherical doubly magic nuclei [3] have been extended to the QRPA treatment of axially deformed nuclei. In both paired and unpaired regimes the QRPA matrix is expressed in terms of the two quasi-particle states obtained via HFB calculations at the minimum of the axially deformed potential energy surfaces [1]. The same force is used in the HFB calculations and QRPA residual interaction, which ensures a full consistency of our method [2].

---

[1] axially symmetric HFB+D1S results for the whole nuclear chart can been found on the web site http://www-phynu.cea.fr/science_en_ligne/carte_potentiel_ microscopiques/carte_potentiel_nucleaire.htm.

[2] More details concerning formalism, consistency and spurious states can been found in a previous publication [4].

CP1165, *Nuclear Structure and Dynamics '09*
edited by M. Milin, T. Nikšić, D. Vretenar, and S. Szilner
© 2009 American Institute of Physics 978-0-7354-0702-2/09/$25.00

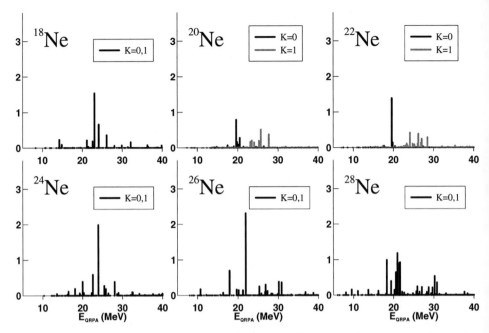

FIGURE 1. B(E1) transition probabilities in $e^2 fm^4$ unit as functions of QRPA energy values in Ne isotopes. For deformed isotopes the K=0 and K= ±1 components are drawn in black and gray respectively.

RESULTS

In previous calculations [4], results on giant resonances have been obtained in $^{22-28}$Mg and $^{26-30}$Si isotopes. In this study the role of the ground state intrinsic deformation on strength distributions has been shown. In deformed nuclei, theoretical isovector dipole responses have been found split into two major components corresponding to vibrations along the major and the minor axes. The strength was predicted to be almost equally distributed between these two energy components. K = 0 components were found in the low energy part of the spectra in the prolate nuclei, and in the high energy part in the oblate ones. Isoscalar monopole resonances also display a splitting in deformed nuclei. Isoscalar quadrupole and octupole resonances are found to be well fragmented in particular in well-deformed nuclei. The ordering of the K components as a function of energy is found to be related to the sign of the quadrupole moment. The comparison of the QRPA results in ^{24}Mg and in ^{28}Si with experimental data has shown that a fully microscopic QRPA approach of deformed nuclei including D1S Gogny force is able to provide a wealth of structure information.

As first extension of the previous work to phenomena such as pygmy states, Fig. 1 shows the response of the isovector dipole mode for Ne isotopes. For deformed stable $^{20-22}$Ne nuclei the strength is found to be split into two K components, as expected. For exotic ^{18}Ne and $^{24-28}$Ne isotopes, soft dipole states are predicted. For the proton rich ^{18}Ne isotope a first soft state is found at 14.2 MeV with a transition probability to the ground state B(E1)=0.237 $e^2 fm^2$ and a second one at 14.83 MeV with

TABLE 1. Energies and transition probabilities of the first 2^+ obtained with different method and compare with experimental data

	$E(2^+)$ 5DCH	$E(2^+)$ QRPA	$E(2^+)$ Exp.	B(E2) 5DCH	B(E2) QRPA	B(E2) Exp.
^{30}Si	2.11	2.22	2.23	220	243	215
^{28}Mg	1.50	1.80	1.47	202	216	350
^{26}Ne	1.19	2.20	2.02	228	99	228

a smaller B(E1)=0.095 $e^2 fm^2$ value. A small strength is obtained at 13.56 MeV with B(E1)= 0.045 $e^2 fm^2$ in ^{24}Ne. For ^{26}Ne one single state is found at 10.69 MeV with B(E1)=0.173 $e^2 fm^2$ three times smaller than the experimental value [5]. In ^{28}Ne four soft states are obtained at 8.13, 9.53, 11.07, and 13.99 MeV with B(E1)= 0.050, 0.111, 0.085 and 0.106 $e^2 fm^2$ values, respectively. The calculated neutron and proton transition densities for these states are in agreement with the usual picture of a neutron or proton skin vibrating against a nuclear core. A large isovector contribution coming from the core is also predicted. Let us note that the largest 2qp contribution to the pygmy state in ^{26}Ne is identified as $(2p\frac{3}{2}, 2s\frac{1}{2})$ which is in agreement with [6].

In a previuos study [7] using the five dimensional collective Hamiltonian (5DCH) formalism[3] with the same D1S Gogny force some predictions have already benn obtained for ^{26}Ne, ^{28}Mg and ^{30}Si. In Table 1 we compare QRPA and 5DCH predictions on the first 2^+ excited state for these N=16 isotones with experimental data.

These values show that the two approaches are able to provide structure information in harmonic nuclei with quite the same accuracy. Nevertheless the 5DCH approach is not able to describe nuclei which are too rigid against the deformation and, the QRPA approach is not able to take into account secondary minima as illustrated in Fig. 2. In this figure, QRPA[4] and 5DCH calculated energies of the first 2^+ excited states in Ni isotopes are compared with experimental values. It appears that 5DCH are in good agreement with the data except for ^{56}Ni and ^{68}Ni isotopes. For these isotopes there is a mutual enhancement of magicity coming from the Z=28 magic shell added to the N=28 magic shell and N=40 sub-shell, respectively. For these two "magic" isotopes QRPA results are in agreement with the data. QRPA predictions for ^{66}Ni, ^{70}Ni, ^{72}Ni overestimate the experimental data but are coherent with 5DCH predictions and allow to reproduce the experimental behavior of the first 2^+ energy between ^{66}Ni and ^{72}Ni.

CONCLUSION

To summarize, a fully consistent microscopic axially-symmetric-deformed QRPA approach has been developed and applied to light even-even s-d shell nuclei, namely Ne Mg and Si isotopes. In deformed nuclei, theoretical isovector dipole responses are

---

[3] the generator coordinate method with Gaussian overlap approximation in triaxial plane with three rotational degree of freedom

[4] QRPA results are not reported for $^{52-54}$Ni, $^{58-44}$Ni and $^{74-76}$Ni, since the method is not appropriate for these nuclei.

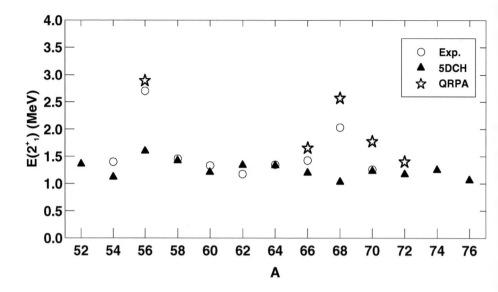

FIGURE 2. First 2^+ states energies for Ni isotopes : comparison between the two different approaches and experimental data.

found split into two major components, as expected. For exotic Ne isotopes soft dipole strengths are predicted by QRPA predictions. These "pygmy "states have been interpreted as a oscillation of an Isovector core against a neutron skin. Comparisons between QRPA results and 5DCH ones for N=16 isotones show that there is a good coherence between these approaches when the nuclei are good harmonic oscillators. On the contrary, results obtained for ^{56}Ni and ^{68}Ni show the importance of the mutual enhancement of magicity.

REFERENCES

1. S. Goriely, E. Khan anf M. Samyn, *Nucl. Phys.* **A739**, 331 (2004).
2. J. Dechargé and D. Gogny, *Phys. Rev.* **C21**, 1568 (1980); J.F. Berger, M. Girod, and D. Gogny, *Comp. Phys. Comm.* **63**, 365 (1991).
3. S. Péru, J.-F. Berger, and P.-F. Bortignon, *Eur. Phys.* **A26**, 25 (2005).
4. S. Péru, and H. Goutte , *Phys. Rev.* **C77** , 044313 (2008).
5. J. Gibelin et al., *Phys. Rev. Lett.* **101**, 212503 (2008).
6. Li-Gang Cao and Zhong-Yu Ma, Phys. Rev. C **71**, 034305 (2005).
7. A. Obertelli, S. Péru, J.-P. Delaroche, A. Gillibert, M. Girod, and H. Goutte, *Phys. Rev.* **C71**, 024304 (2005).

Evolution of the Low-lying E1 Strength With Deformation

D. Peña Arteaga*, E. Khan* and P. Ring†

*Institut de Physique Nucléaire, Université Paris-Sud, IN2P3-CNRS, F-91406 Orsay Cedex, France
†Physikdepartment, Technische Universität München, D-85748, Garching, Germany

Abstract. The evolution of low-lying E1 strength is systematically analyzed in very neutron-rich Sn nuclei, beyond ^{132}Sn until ^{166}Sn, within the Relativistic Quasiparticle Random Phase Approximation. The great neutron excess favors the appearance of a deformed ground state for $^{142-162}$Sn. The evolution of the low-lying strength in deformed nuclei is determined by the interplay of two factors, isospin asymmetry and deformation: while greater neutron excess increases the total low-lying strength, deformation hinders and spreads it. Very neutron rich deformed nuclei may not be as good candidates as spherical nuclei like ^{132}Sn for the experimental study of low-lying E1 strength.

Keywords: mean field, RMF, relativistic, QRPA, deformed nuclei
PACS: 21.10.-k, 21.30.Fe, 21.60.Jz, 24.30.Cz, 25.20.Dc, 27.30.+t

INTRODUCTION

Collective modes of excitation are an universal feature of nuclei [1]. Involving a large majority of nucleons, these modes can provide crucial insight into exotic nuclear structure. For instance, the pygmy mode is known to be a consequence of a vibration involving the neutron skin [2] against a core composed of neutrons and protons. Originally the increase of the low-lying strength was thought to be an exclusive phenomenon of heavy nuclei with large isospin asymmetry. But very similar excitation patterns have also been observed in light nuclei [3, 4], and in medium-heavy stable nuclei [5, 6]. It has been also thoroughly investigated within a variety of theoretical tools [7, 2, 8], including the RHB+RQRPA in spherical symmetry [9], with different degrees of success in comparison with experimental data.

The present work is focused on the influence of deformation of the low-lying E1 response in neutron rich nuclei. In particular, in very neutron-rich Tin isotopes, from ^{132}Sn (where current experimental data stands) up to ^{166}Sn. Several experimental and theoretical studies available on the less neutron-rich part of the isotopic chain: in general, it has been found that the low-lying E1 strength increases with neutron excess [2], and that the collectivity of the excitations is rather high. Tin nuclei are spherical because of the Z=50 shell closure. However it is believed that for strong neutron excess (beyond A ∼ 140), the large diffuseness of the density induces a weakening of the spin-orbit potential [10, 11], and therefore the disappearance of the shell closure, allowing for deformed nuclei. This means that in order to describe low energy excitations beyond ^{132}Sn, a model including both pairing and deformation is required.

CP1165, *Nuclear Structure and Dynamics '09*
edited by M. Milin, T. Nikšić, D. Vretenar, and S. Szilner
© 2009 American Institute of Physics 978-0-7354-0702-2/09/$25.00

The recently developed [12] relativistic deformed QRPA approach (RQRPAZ), provides a viable microscopical framework where deformation and pairing correlations are included in a fully self-consistent fashion.

DEFORMED RHB AND QRPA FORMALISM

A detailed discussion of deformed Relativistic Hartree Bogoliubov (RHB) and Relativistic Quasiparticle Random Phase Approximation for axially deformed systems can be found in references [13] and [12], respectively.

The starting point is an effective Lagrangian density including: the free nucleons (neutrons and protons), the free meson fields and the electromagnetic field, and the nucleon-meson interactions. The fields used are the σ, ω, and ρ mesons, that correspond to scalar isoscalar, vector isoscalar and vector isovector interactions, respectively. Also, the model employed in this investigation includes non-linear self-interaction terms in the σ-meson [14] that provide proven quantitative predicting power.

A Hamiltonian density can be derived from the Lagrangian density, leading the to an energy functional $E[\hat{\rho}, \phi, \kappa]$ (for details see Ref. [15]). The results presented in this investigation where obtained with a simple monopole-monopole interaction in the pairing channel. To study vibrational excitations, one introduces small harmonic oscillations around the ground state generalized density and expands the equation of motion [16] up to linear order, gaining the QRPA approximation, which provides the response of the system to external fields, and allows the study of collective excitation phenomena.

THE VERY NEUTRON-RICH SN ISOTOPES

The parameter set used in the Lagrangian is NL3 [18] with a monopole-monopole force in the pairing channel. The coupling constants of the pairing interaction were adjusted to reproduce the gap values predicted by the DS1 Gogny force [17]. Due to the weakening of the Z=50 shell closure, and coincidentally with the fill up of the $1h_{9/2}$ level, a minimum with axial deformation appears for ^{142}Sn. Deformation remains moderate ($\beta < 0.2$) until ^{150}Sn, where the $2f_{7/2}$ fills. From then on, deformation varies smoothly until ^{164}Sn, where tin nuclei become spherical again due to the proximity of the N=126 shell closure.

The analysis of the evolution of the Giant Dipole Resonance along an isotopic chain confirms that the Relativistic Quasiparticle Approximation reproduces basic features predicted by macroscopic hydrodynamical models, namely the reduction of the centroid position and the splitting of the response in two modes due to deformation. However the hydrodynamical models and the RQRPA GDR position predictions are at variance in the case of very neutron-rich nuclei. Furthermore, it has been confirmed that the GDR splitting depends linearly on the deformation.

Regarding the low lying E1 response, to assert if, and to what extent, it is in general affected by deformation, it is interesting to concentrate on the total low lying strength and discard details. The left panel of Fig. 1 shows the summed response energy versus the mass number for the whole isotopic chain. Circles and squares refer to spherical

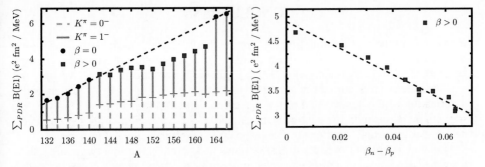

FIGURE 1. (Color online) Left: total pygmy strength versus the mass number. Blue circles and red squares denote spherical and deformed nuclei, respectively. The dashed orange vertical lines indicate the share of the total strength provided by the $K^\pi = 0^-$ mode, while the solid light blue ones are the same but for the $K^\pi = 1^-$ mode. Right: Total pygmy strength dependence on the difference of deformations for protons and neutrons $\beta_n - \beta_p$ for the deformed nuclei in the $^{132-166}$Sn chain.

and deformed nuclei, respectively. The dashed line, which is a least squares fit to the data points for spherical nuclei, clearly indicates that, all things being equal, for spherical nuclei the low-lying strength increases almost linearly with neutron number. It is important to note, however, that studies within the RQRPA [2] in the spherical $^{100-132}$Sn nuclei, show that this trend is reversed near a shell closure, in this particular isotopic chain approaching the neutron number N=82. Between ^{126}Sn and ^{132}Sn one finds a decrease of the PDR strength (see [2] and references therein).

However, the left panel of Fig. 1 also shows that the linear link between the addition of neutrons and an increase in total low-lying strength is no longer kept for deformed nuclei, where the growth is less pronounced. Furthermore, for nuclei where deformation most dramatically increases, from ^{148}Sn to ^{150}Sn, and to the most deformed ^{152}Sn, the summed low-lying strength even decreases with the addition of two neutrons.

Analysis of the transition densities in the intrinsic and laboratory systems of reference confirms its pygmy character, which is also present in spherical nuclei: in the nuclear interior both neutrons and protons oscillate in-phase, out-of-phase with the skin where only neutrons contribute.

In the right panel of Fig. 1 is plotted the total pygmy strength dependence on the difference of deformations for the ground state neutron and proton densities $\beta_n - \beta_p$. It shows that both are linearly linked. This result is equivalent to the situation found in spherical nuclei, where the neutron skin thickness $r_n - r_p$ determines the total low-lying pygmy strength. However, in the deformed case, in addition to the difference in neutron-proton densities radii, the difference of quadrupole deformations $\beta_n - \beta_p$ comes into play to determine the total low-lying strength. For prolate nuclei, this produces a reduction in strength in the $K^\pi = 1^-$ mode. For oblate nuclei with different proton and neutron deformations it is therefore plausible to expect a similar reduction in overall low-lying strength, caused in this case by a reduced neutron skin along the symmetry axis and thus a reduced $K^\pi = 0^-$ strength. In summary, the reduced skin thickness along a perpendicular of the symmetry axis, caused by the different deformations of the neutron and proton densities, might explain the reduction in strength of the $K^\pi = 1^-$ pygmy resonance in the deformed Tin isotopes under study.

CONCLUSIONS

It has been found that deformation hinders the dipole strength in the low-lying region on Tin isotopes. This effect has been linked to the suppression of vibrations along a perpendicular of the symmetry axis ($K^\pi = 1^-$ mode) for prolate deformed systems, and explained by the reduction of neutron skin in this direction caused by the difference in deformation of the neutron and proton densities. On the other hand, the low-lying E1 strength increases with the neutron number, and thus the interplay of these two effects determines the actual low-lying dipole response in deformed nuclei.

The analysis of the excitation peaks shows that, even if the low-lying strength is quenched and spread in deformed nuclei, it nevertheless shows pygmy character, with a neutron skin oscillating against neutron-proton core. The number of contributing qp-pairs is comparable to that found in the GDR region, and in agreement with other RQRPA studies in spherical nuclei. It is therefore concluded that for deformed nuclei with extreme isospin asymmetry the pygmy mode subsists, but is more spread than in spherical nuclei. Hence, prominent pygmy modes may be a specific characteristic of spherical neutron rich nuclei which are not too far from the valley of stability.

ACKNOWLEDGMENTS

The authors wish to thank D. Vretenar for fruitful discussions. This proceeding has also been supported by the Bundesministerium für Bildung und Forschung, Germany under project 06 MT 246 and by the DFG cluster of excellence "Origin and Structure of the Universe" (www.universe-cluster.de).

REFERENCES

1. M. N. Harakeh and A. van der Woude, *Giant Resonances* (Oxford University Press, Oxford, 2001).
2. N. Paar, D. Vretenar, E. Khan, and G. Coló, *Rep. Prog. Phys.* **70**, 691 (2007).
3. T. Aumann et al., *Nucl. Phys.* **A649**, 297c (1999).
4. E. Tryggestad et al., *Nucl. Phys.* **A687**, 231c (2001).
5. D. Savran, M. Babilon, A. M. van den Berg, M. N. Harakeh, J. Hasper, A. Matic, H. J. Wörtche, and A. Zilges, *Phys. Rev. Lett.* **97**, 172502 (2006).
6. D. Savran, M. Fritzsche, J. Hasper, K. Lindenberg, S. Muller, V. Y. Ponomarev, K. Sonnabend, and A. Zilges, *Phys. Rev. Lett.* **100**, 232501 (2008).
7. P.-G. Reinhard, *Nucl. Phys.* **A649**, 105c (1999).
8. M. Tohyama and A. S. Umar, *Phys. Lett.* **B516**, 415 (2001).
9. N. Paar, P. Ring, T. Nikšić, and D. Vretenar, *Phys. Rev.* **C67**, 034312 (2003).
10. J. Dobaczewski, I. Hamamoto, W. Nazarewicz, and J.A. Sheikh, *Phys. Rev. Lett.* **72**, 981 (1994).
11. G. A. Lalazissis, D. Vretenar, W. Pöschl, and P. Ring, *Phys. Lett.* **B418**, 7 (1998).
12. D. Pena Arteaga and P. Ring, *Phys. Rev.* **C77**, 034317 (2008).
13. Y. K. Gambhir, P. Ring, and A. Thimet, *Ann. Phys. (N.Y.)* **198**, 132 (1990).
14. J. Boguta and A. R. Bodmer, *Nucl. Phys.* **A292**, 413 (1977).
15. D. Vretenar, A. V. Afanasjev, G. A. Lalazissis, and P. Ring, *Phys. Rep.* **409**, 101 (2005).
16. P. Ring and P. Schuck, *The Nuclear Many-Body Problem* (Springer, Heidelberg, 1980).
17. S. Hilaire, J.-F. Berger, M. Girod, W. Satula, and P. Schuck, *Phys. Rev. Lett.* **B531**, 61 (2002).
18. G. A. Lalazissis, J. König, and P. Ring, *Phys. Rev.* **C55**, 540 (1997).

Finite Amplitude Method and Systematic Studies of Photoresponse in Deformed Nuclei

Takashi Nakatsukasa*,†, Tsunenori Inakura†,* and Kazuhiro Yabana†,*

*RIKEN Nishina Center, Wako-shi 351-0198, Japan
†Center for Computational Sciences, University of Tsukuba, Tsukuba 305-8571, Japan

Abstract. Finite amplitude method, which we have recently proposed, is applied to calculations of photoabsorption cross sections in light-mass to medium-mass nuclei. The method provides an alternative and feasible approach to fully self-consistent calculations of the random-phase approximation. We adopt a Skyrme functional and solve the linear-response equations directly in the three-dimensional coordinate space representation. As an example of our recent results, systematic behavior of the peak energies of the giant dipole resonance are shown for even-even nuclei.

Keywords: Random-phase approximation, Giant resonances
PACS: 21.60.Jz, 24.30.Cz, 25.20.Dc

INTRODUCTION

The random-phase approximation (RPA), as a small amplitude limit of the time-dependent Hartree-Fock (TDHF) or time-dependent density-functional theory (TDDFT), is known to well describe giant resonances as well as the low-lying excitations. However, since the calculation of the residual interaction are rather tedious for the realistic interaction, it has been common to ignore some terms in practice and to sacrifice the full self-consistency. Recently several groups have reported fully self-consistent RPA calculations [1, 2, 3, 4, 5]. However, they are achieved only for spherical nuclei.

We are performing systematic and fully self-consistent RPA calculations of photoabsorption cross sections for wide mass region ($A \leq 100$), for both spherical and deformed nuclei. To facilitate an implementation of the full self-consistency, we employ the finite amplitude method (FAM)[6]. The FAM allows us to calculate the matrix elements of the residual field, $\delta h = \delta h / \delta \rho \cdot \delta \rho$, using the finite difference. This does not require excessive programming but can be done by employing the program of the static Hartree-Fock calculation. In this report, we will show systematic behaviors of peak energies of giant dipole resonances (GDR) calculated with the Skyrme functional of SkM*.

FINITE AMPLITUDE METHOD

In this section, we recapitulate the FAM. For more details, readers are referred to the reference [6].

The linear-response RPA equation to a weak external field with a fixed frequency, $V_{ext}(\omega)$, can be expressed in terms of the forward and backward amplitudes, $|X_i(\omega)\rangle$

CP1165, *Nuclear Structure and Dynamics '09*
edited by M. Milin, T. Nikšić, D. Vretenar, and S. Szilner
© 2009 American Institute of Physics 978-0-7354-0702-2/09/$25.00

and $\langle Y_i(\omega)|$.

$$\omega |X_i(\omega)\rangle = (h_0 - \varepsilon_i) |X_i(\omega)\rangle + \hat{P} \{V_{\text{ext}}(\omega) + \delta h(\omega)\} |\phi_i\rangle, \tag{1}$$

$$-\omega \langle Y_i(\omega)| = \langle Y_i(\omega)| (h_0 - \varepsilon_i) + \langle \phi_i| \{V_{\text{ext}}(\omega) + \delta h(\omega)\} \hat{P}. \tag{2}$$

where the subscript i indicates the occupied orbitals ($i = 1, 2, \cdots, A$) and the operator $\hat{P}$ denotes the projector onto the particles space, $\hat{P} = 1 - \sum_i |\phi_i\rangle\langle\phi_i|$. Usually, the residual interaction $\delta h(\omega)$ is expanded to the first order with respect to $|X_i(\omega)\rangle$ and $|Y_i(\omega)\rangle$. This leads to the well-known matrix form of the linear-response equation and calculation of these matrix elements is most time-consuming in practice. Instead, we utilize the fact that the linearization is numerically realized for $\delta h(\omega) = h[\rho_0 + \delta\rho(\omega)] - h_0$ within the linear approximation. In order to perform this numerical differentiation in the program, we use a small trick in the calculation of single-particle Hamiltonian $h[\rho]$.

First, we should notice that the $\delta h(\omega)$ depends only on the forward "ket" amplitudes $|X_i(\omega)\rangle$ and backward "bra" ones $\langle Y_i(\omega)|$. In other words, it is independent of bras $\langle X_i(\omega)|$ and kets $|Y_i(\omega)\rangle$. This is related to the fact that the transition density $\delta\rho(\omega)$ depends only on $|X_i(\omega)\rangle$ and $\langle Y_i(\omega)|$.

$$\delta\rho(\omega) = \sum_i \{|X_i(\omega)\rangle\langle\phi_i| + |\phi_i\rangle\langle Y_i(\omega)|\}. \tag{3}$$

Then, we can calculate the residual fields in a following manner [6]:

$$\delta h(\omega) = \frac{1}{\eta} (h[\tilde{\rho}_\eta] - h_0), \tag{4}$$

where η is a small real parameter to realize the linear approximation. $\tilde{\rho}_\eta$ are defined by

$$\tilde{\rho}_\eta \equiv \sum_i \{(|\phi_i\rangle + \eta |X_i(\omega)\rangle)(\langle\phi_i| + \eta \langle Y_i(\omega)|)\}. \tag{5}$$

Once $|X_i(\omega)\rangle$ and $\langle Y_i(\omega)|$ are given, the calculation of $h[\tilde{\rho}_\eta]$ is an easy task. This does not require complicated programming, but only needs a small modification in the calculation of $h[\rho]$. Of course, eventually, we need to solve Eqs. (1) and (2) to determine the forward and backward amplitudes. We use an iterative algorithm to solve this problem. Namely, we start from initial amplitudes $|X_i^{(0)}\rangle$ and $\langle Y_i^{(0)}|$, then update them in every iteration, $(|X_i^{(n)}\rangle, \langle Y_i^{(n)}|) \to (|X_i^{(n+1)}\rangle, \langle Y_i^{(n+1)}|)$, until the convergence. In each step, we calculate $\delta h(\omega)$ using the FAM as Eq. (4).

Numerical results

Numerical details

We adopt the three-dimensional grid representation for the spherical model space with radius of 15 fm. All the wave functions are represented by these grid points:

$$\{\phi_i(\vec{r}_k, \sigma), X_i(\omega; \vec{r}_k, \sigma), Y_i(\omega; \vec{r}_k, \sigma)\}_{k=1,\cdots,N_{\text{grid}}}^{i=1,\cdots,A;\ \sigma=\text{up,down}}. \tag{6}$$

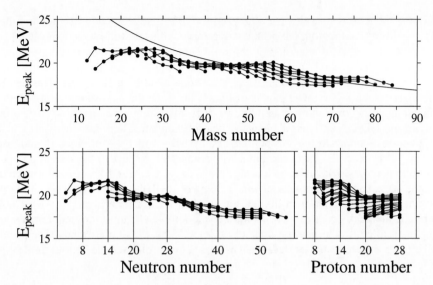

FIGURE 1. Calculated GDR peak energies as a function of mass number (top), neutron number (bottom left), and proton number (bottom right). The isotopic chains are connected by lines for the top and bottom left panels, while the isotonic chains are shown in the bottom right.

Since the results are not sensitive to mesh spacing in a region outside of the interacting region, the adaptive grid representation is used to reduce the number of grid points N_{grid} [7]. In the present calculations, $N_{grid} = 11,777$.

We use complex frequency, $\omega = E + i\Gamma/2$, with $\Gamma = 1$ MeV, which introduces an artificial damping of the width of 1 MeV. This smoothing is necessary in two reasons: To obtain smooth strength functions, and to speed up the convergence for the iterative procedure. For the iterative algorithm, we employ the generalized conjugate residual method with a restart in every twenty iterations. Typical number of iteration is order of hundred to thousand. It turns out that larger number of iteration is necessary for larger frequency ω.

GDR peak energies

We have carried out the systematic calculation of electric dipole response up to Nickel isotopes. We show evolution of the GDR peak positions in Fig. 1. The GDR peak position is estimated by the average energy

$$E_{peak} = \frac{m_1}{m_0}, \quad m_k = \int_0^{\omega_{max}} d\omega\, \omega^k S(\omega; D_{E1}) \tag{7}$$

where $S(\omega; D_{E1}) = \sum_n |\langle n|D_{E1}|0\rangle|^2 \delta(\omega - E_n)$. The maximum frequency is $\omega_{max} \approx 40$ MeV. The $E1$ operator is defined with the recoil charges for protons, Ne/A and for neutrons, $-Ze/A$. In deformed nuclei, since the peak energy depends on direction of the $E1$ operator (x, y, and z), their averaged value is adopted in Fig. 1. The upper panel shows the GDR peak energies from Oxygen to Nickel, as a function of mass number. In the medium-mass region, the peak energies approximately follow the empirical low, $21A^{-1/3} + 31A^{-1/6}$ MeV, denoted by the solid curve. However, in each isotopic chain, the peak energies in stable nuclei are the highest, while they are decreasing as leaving from the stability line. In addition, we can see some kind of shell effects. This can be more clearly seen in the lower panels of Fig. 1, in which the peak energies are plotted as functions of neutron and proton numbers. It is interesting to see a strong correlation between the GDR peak energies and the neutron number. There are cusps at $N = 14$ and 28 corresponding to the subshell closure of $1d_{5/2}$ and $1f_{7/2}$ orbitals. In addition, the peak energies at the cusps are almost invariant with respect to the proton number. While the neutron Fermi level is located in the $1d_{5/2}$, $1f_{7/2}$ and $1g_{9/2}$ orbitals, which have the largest angular momentum j in each major shell, the peak energies are roughly constant against the variation of the neutron number. Then, after filling up those highest-j orbitals, the peak energies start decreasing monotonically, until the major shells are completely filled. In the neutron-deficient side, we may see similar trends. The proton shell effects seem to be not as significant as those of neutrons (see the bottom right panel of Fig. 1).

CONCLUSION

We have applied the finite amplitude method (FAM) to realistic Skyrme functional of SkM* to study nuclear photoresponse. The method is so feasible that we may perform fully self-consistent RPA calculations in the three-dimensional grid representation. The systematics of peak energies of the giant dipole resonance are presented for Oxygen to Nickel isotopes. The calculation indicates a strong correlation between neutron shell effects and GDR peak energies.

ACKNOWLEDGMENTS

The work is supported by the PACS-CS project and the Joint Research Program (07b-7, 08a-8, 08b-24, 09a-25) at Center for Computational Sciences, University of Tsukuba. The work is also supported by the Large Scale Simulation Program (07-20 (FY2007), 08-14 (FY2008), 09-16 (FY2009)) of KEK, and by the Grand-in-Aid for Scientific Research (B) No. 21340073.

REFERENCES

1. J. Terasaki, J. Engel, M. Bender, J. Dobaczewski, W. Nazarewicz, and M. Stoitsov, *Phys. Rev.* **C71**, 034310 (2005).
2. J. Terasaki, and J. Engel, *Phys. Rev.* **C74**, 044301 (2006).
3. S. Fracasso, and G. Colò, *Phys. Rev.* **C72**, 064310 (2005).
4. T. Sil, S. Shlomo, B. K. Agrawal, and P.-G. Reinhard, *Phys. Rev.* **C73**, 034316 (2006).
5. N. Paar, P. Ring, T. Nikšić, and D. Vretenar, *Phys. Rev.* **C67**, 034312 (2003).
6. T. Nakatsukasa, T. Inakura, and K. Yabana, *Phys. Rev.* **C76**, 024318 (2007).
7. T. Nakatsukasa, and K. Yabana, *Phys. Rev.* **C71**, 024301 (2005).

A Variational Approach to Mass Fluctuations

P. D. Stevenson and J. M. A. Broomfield

Department of Physics, University of Surrey, Guildford, GU2 7XH

Abstract. The variational method of Balian and Vénéroni is applied to dynamic simulations of giant resonances using the effective Skyrme interaction. Fluctuations in the mass following the resonance decay are compared in the time-dependent Hartree-Fock and Balian-Vénéroni approaches. The Balian-Vénéroni results are consistly higher than the Time-Dependent Hartree-Fock results.

Keywords: Giant Resonances, TDHF, Balian-Vénéroni
PACS: 21.10.-k, 21.30.Fe, 21.60.Jz, 27.90.+b

INTRODUCTION

Time-dependent Hartree-Fock (TDHF) is a venerable method for the calculation of dynamics in many fermion systems [1]. It has found much application in nuclear physics, for example in the calculation of giant resonances [2, 3], or heavy-ion collisions [4, 5]. TDHF is the time-dependent extension of the Hartree-Fock mean-field approach in which the motion of each nucleon is determined by its interaction with the other nucleons via the mean field and its wavefunction represented by a single determinant at all times. The TDHF equations can be derived from a least-action variational principle [6] which leads to the 'best' determinant wavefunction and its time-development cosistent with the least-action principle. A system following such a path is optimised to the calculation of one-body observables, such as the mean particle number, but is not a priori expected to give physical results for two- or higher-body observables.

THE BALIAN-VÉNÉRNONI VARIATIONAL PRINCIPLE

Instead of using a least-action approach, other variational principles can lead to mean-field description of collective motion, which are optimised to observables of ones choice. In particular, Balian and Vénéroni have proposed methods for the evaluation of correlations between and fluctuations of one-body observables, while retaining the simplicity of a mean-field picture [7, 8, 9] and implemented the technique in a nuclear physics context [10]. Their approach overcomes, at least as far as it is possible in the space of Slater Determinants, the underestimantion of widths of mass distributions of final fragments in TDHF [11]. The source of this underestimation can be understood from the semiclassical nature of TDHF. Although the trajectories of the individual nucleons are quantum in nature, the collective path of the nucleus as a whole is classical in TDHF. If one, therefore, set a nuclear process in motion in a TDHF simulation, the collective path taken by the system will be fixed for all time as the path of least action. In reality, one expects that the nucleus will explore other nearby paths and take them with some

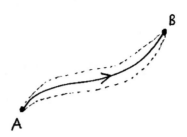

FIGURE 1. Path in collective space. A is it initial condition B is the final state. The solid line represents the fixed TDHF path. The dashed lines represent a sample of alternate paths that TDHF does not explore.

probability as demanded by quantum mecahincs. The outcome in nature will be more spread out amongst the final states than TDHF predicts. A schematic picture is given in Figure 1 in which the TDHF collective path between initial and final states A and B is shown as the solid line. Alternative paths are given by dashed lines, but these are not explored by a TDHF time-evolution.

The Balian-Vénéroni method is one approach amongst several that takes into account deviations from the classical path. Part of its utility is that it can make use of existing TDHF codes with minimal changes. Its realisation in such codes involves preparing a nucleus in an initial state at t_0 (i.e. at A in the figure) with initial conditions chosen according to the collective motion of interest, evolving to a later time t_1 (B in the figure), applying an infinitesimal operator to the wavefunctions at t_1 and evolving back in time along a different path. Finally, back at t_0 the overlap between the initial wavefunction and the forward and backward wavefunction. In particular, the fluctuation in a one-body observable $\hat{Q}$ at t_1 is given by [10]

$$\Delta \hat{Q}^2\big|_{t_1} = \frac{1}{2} \lim_{\varepsilon \to 0} \mathrm{Tr} \left[\frac{\rho(t_0) - \sigma(t_0, \varepsilon)}{\varepsilon} \right]^2 \tag{1}$$

where $\sigma(t, \varepsilon)$ is given by the final condition

$$\sigma(t_1, \varepsilon) = e^{i\varepsilon \hat{Q}} \rho(t_1) e^{-i\varepsilon \hat{Q}} \tag{2}$$

The Balian-Vénéroni method, in which small amplitude response to deviations from the classical path, can be related to RPA, and to stoachastic approaches [13].

IMPLEMENTATION AND RESULTS

The Balian-Vénéroni variational technique has been implemented for the caluclations of mass fluctuations following the decay of giant resonances and following heavy-ion collisions in a full three-dimensional mean-field code using the Skyrme interaction [12] with all the time-odd terms that arise from the Skyrme mean field (though the recent variants including the tensor interaction are not considered).

For either a giant resonance or a collision, the basic mode of running is to generate a static ground state for the system in question, exciting it either in a multipole giant

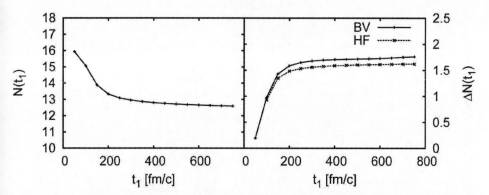

FIGURE 2. Dependence on t_1 of the number of emitted particles (left panel) and the fluctuation in the number (right panel) from the Balian-Vénéroni (BV) and Time-Dependent Hartree-Fock (HF) approaches.

resonance, or into translation motion for a collision, then running forward and backwards in time with the Balian-Vénéroni transformation (2) performed at t_1.

A sample calculation is presented in Figure 2, which shows the number of particles (left) and fluctuations (right) for a isovector giant dipole resonance in ^{16}O. The nucleus was excited at time $t = 0$ with an instantaneous boost which set the protons and neutrons moving out of phase. From the left plot one sees the rapid dripping of particles from the nuclues, ending with a total loss of between three and four nucleons by $t_1 = 750$ fm/c. This shows that the results have stabalised reasonably well by the end of the run. The extrapolation error related to sending the paramter ε in (1) to zero has been calculated, but is smaller than the marks in the figure. Many other numerical tests have been performed to ensure safe convergence and independence of model parameters [14].

The BV technique clearly results in a higher fluctuation in particle number than TDHF alone. In the above example, at $t_1 = 750$ fm/c, there is an increase of 8.6% in the fluctuation. This is a rather modest increase, though it depends on the size of the excitation given (50 fm^{-1} in this case) and the system in question. In other cases, the fluctuation can be larger [15, 16, 17] yet they only go some way to meeting the experimental values [14, 18]. Further systematic work to explore these results is underway.

ACKNOWLEDGMENTS

The authors acknowledge support from the UK Science and Technology Facilities Council.

REFERENCES

1. P. A. M. Dirac, *Proc. Cam. Phil. Soc.* **26**, 376 (1930).
2. P. D. Stevenson, M. R. Strayer, J. Rikovska Stone and W. G. Newton, *Int. J. Mod. Phys.* **E13**, 181 (2004).
3. T. Nakatsukasa and K. Yabana, *Eur. Phys. J.* **A25**, s1.527 (2005).
4. K. T. R. Davies and M. R. Strater, *Phys. Rev. Lett.* **44**, 23 (1980).
5. J. A. Maruhn, P.-G. Reinhard, P. D. Stevenson and M. R. Strayer, *Phys. Rev.* **C74**, 027601 (2006).
6. A. K. Kerman and S. E. Koonin, *Ann. Phys.* (NY) **100**, 332 (1976).
7. Roger Balian and Marcel Vénéroni, *Ann. Phys.* (NY) **187**, 29 (1988).
8. Roger Balian and Marcel Vénéroni, *Ann. Phys.* (NY) **216**, 351 (1992).
9. Roger Balian and Marcel Vénéroni, *Ann. Phys.* (NY) **281**, 65 (2000).
10. Roger Balian, Paul Bonche, Hubert Flocard and Marcel Vénéroni, *Nucl. Phys.* **A428**, 79c (1984).
11. C. H. Dasso, T. Dœssing and H. C. Pauli, *Z. Phys.* **A289**, 395 (1979).
12. Michael Bender, Paul-Henri Heenen and Paul-Gerhard Reinhard, *Rev. Mod. Phys.* **75**, 121 (2003).
13. Sakir Ayik, Kouhei Washiyama and Denis Lacroix, *Phys. Rev.* **C79**, 054606 (2009).
14. J. M. A. Broomfield, *Ph. D. thesis*, University of Surrey (2009).
15. P. D. Stevenson and J. M. A. Broomfield, *Proceedings of XXXII Symposium on Nuclear Physics*, Cocoyoc, Mexico, arXiv 0903.0130 (2009).
16. J. M. A. Broomfield and P. D. Stevenson, *AIP Conf. Proc.* **1098**, 133 (2009).
17. J. M. A. Broomfield and P. D. Stevenson, *J. Phys.* **G35**, 095102. (2008).
18. J. M. A. Broomfield and P. D. Stevenson, *to be published.*

Pygmy Dipole Strength in Exotic Nuclei and the Equation of State

A. Klimkiewicz[*,†], N. Paar[**], P. Adrich[*,†], M. Fallot[*], T. le Bleis[‡,§],
D. Rossi[¶], K. Boretzky[*], T. Aumann[*,‖], H. Alvarez-Pol[††], F. Aksouh[*],
J. Benlliure[††], T. Berg[¶], M. Boehmer[‡‡], E. Casarejos[††], M. Chartier[§§],
A. Chatillon[*], D. Cortina-Gil[††], U. Datta Pramanik[*], Th.W. Elze[§],
H. Emling[*], O. Ershova[*], B. Fernando-Dominguez[§§], H. Geissel[*],
M. Gorska[*], M. Heil[*], M. Hellström[*], H. Johansson[*,¶¶], K.L. Jones[*],
A. Junghans[***], O. Kiselev[¶], J.V. Kratz[¶], R. Kulessa[†], N. Kurz[*],
M. Labiche[†††], R. Lemmon[‡‡‡], Y. Litvinov[*], K. Mahata[*], P. Maierbeck[‡‡],
T. Nilsson[¶¶], C. Nociforo[¶], R. Palit[§], S. Paschalis[§§], R. Plag[*,§],
R. Reifarth[*,§], H. Simon[*], K. Sümmerer[*], G. Surówka[†], D. Vretenar[**],
A. Wagner[***], W. Waluś[†], H. Weick[*] and M. Winkler[*]

[*]*GSI Helmholtzzentrum für Schwerionenforschung GmbH, Darmstadt, Germany*
[†]*Uniwersytet Jagielloński, Kraków, Poland*
[**]*University of Zagreb, Croatia*
[‡]*University of Strassbourg, France*
[§]*Johann Wolfgang Goethe - Universität, Frankfurt am Main, Germany*
[¶]*Johannes Gutenberg - Universität, Mainz, Germany*
[‖]*ExtreMe Matter Institute EMMI, GSI, Darmstadt, Germany*
[††]*Universidade de Santiago de Compostela, Spain*
[‡‡]*Technische Universität München, Germany*
[§§]*University of Liverpool, UK*
[¶¶]*Chalmers University of Technology, Sweden*
[***]*FZD, Rossendorf, Germany*
[†††]*University of Paisley, UK*
[‡‡‡]*CCLRC Daresbury Laboratory, UK*

Abstract. A concentration of dipole strength at energies below the giant dipole resonance was observed in neutron-rich nuclei around ^{132}Sn in an experiment using the FRS-LAND setup. This so-called "pygmy" dipole strength can be related to the parameters of the symmetry energy and to the neutron skin thickness on the grounds of a relativistic quasiparticle random-phase approximation. Using this ansatz and the experimental findings for ^{130}Sn and ^{132}Sn, we derive a value of the symmetry energy pressure of $\overline{p}_0 = 2.2 \pm 0.5$ MeV/fm^3. Neutron skin thicknesses of R_n-$R_p = 0.23 \pm 0.03$ fm and 0.24 ± 0.03 fm for ^{130}Sn and ^{132}Sn, respectively, have been determined. Preliminary results on ^{68}Ni from a similar experiment using an improved setup indicate an enhanced cross section at low energies, while the results for ^{58}Ni are in accordance with results from photoabsorption measurements.

Keywords: ^{130,132}Sn, ^{58,68}Ni, Pygmy Dipole Resonance, Giant Dipole Resonance, Electromagnetic excitation, Neutron Skin, Symmetry Energy
PACS: 21.65.Cd, 21.65.Ef, 24.30.Cz, 25.60.-t, 25.70.De, 27.60.+j, 27.50.+e

CP1165, *Nuclear Structure and Dynamics '09*
edited by M. Milin, T. Nikšić, D. Vretenar, and S. Szilner
© 2009 American Institute of Physics 978-0-7354-0702-2/09/$25.00

INTRODUCTION

A new collective dipole mode in neutron-rich nuclei was predicted in the early 1990s [1, 2], usually referred to as pygmy dipole resonance (PDR). Recent random-phase approximation (RPA) calculations for medium-heavy and heavy neutron-rich nuclei [3, 4, 5, 6] support the picture, that the concentration of strength below the GDR region close to the particle threshold originates from the formation of a neutron skin. The resonance is viewed as a collective dipole oscillation of these neutrons from outer orbitals versus the isospin-saturated core of the nucleus. The formation of a skin is driven by the density dependence of the nuclear symmetry energy [7, 8]. Thus, experimental findings on the PDR strength can be used to derive valuable information about the parameters of the symmetry energy and thus on the evolution of the neutron-skin thickness. The experimental data for the PDR in ^{130}Sn and ^{132}Sn [9] are used to determine the parameters of the symmetry energy and the neutron skin thicknesses based on the relativistic Hartree- Bogoliubov (RHB) model plus relativistic quasiparticle random-phase approximation (RQRPA) [10]. Preliminary data from a similar experiment on the Nickel isotopes $^{57-72}$Ni are presented comparing the stable nucleus ^{58}Ni to the neutron-rich case ^{68}Ni.

ELECTROMAGNETIC RESPONSE OF TIN ISOTOPES

The dipole response has been studied for the tin isotopic chain $^{129-132}$Sn and for ^{133,134}Sb using Coulomb dissociation of exotic secondary beams at kinetic energies of about 500 A MeV and the LAND complete- kinematics reaction setup at GSI. The measured four-momenta of all decay products, i.e. heavy fragment, neutron(s), and γ-rays, allow to determine the excitation energies E* of the projectiles applying the invariant-mass method. Besides the dominant giant dipole resonance (GDR) strength at energies around 15 MeV, the excitation-energy spectra for all Sn and Sb isotopes investigated exhibit additional strength at low energies close to the particle threshold, exhausting several percent of the energy-weighted (TRK) sum rule [9, 10]. For the even nuclei ^{130}Sn and ^{132}Sn, the observed low-lying strength exhausts 7(3)% and 4(3)% of the TRK sum rule, respectively. The low-lying strength found for all studied exotic Sn- and Sb-isotopes is compared in the left panel of fig. 1 to results derived from ($\gamma, \gamma\prime$) experiments as a function of the asymmetry of the nuclei.

In order to explore the correlation between the PDR transition strength and the density dependence of the nuclear symmetry energy, a systematic theoretical investigation was performed. A series of fully self-consistent RHB+RQRPA calculations of ground-state properties and dipole-strength distributions has been performed [5]. The density dependence of the symmetry energy has been varied by changing the parametrization of the effective density-dependent meson-exchange (DD-ME) interactions used as a basis for theoretical modeling [17, 10]. An almost linear dependence between the relative PDR- strength $\Sigma B_{PDR}(E1)/\Sigma B_{GDR}(E1)$ and the symmetry energy parameter a$_4$ (symmetry energy per nucleon in pure neutron matter), the symmetry energy pressure p$_0$, and the neutron skin thickness is obtained, exemplarily shown in the right panel of fig. 1 for ^{132}Sn.

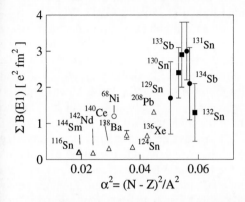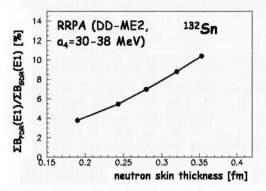

FIGURE 1. Left: Integrated pygmy dipole strength from the LAND-FRS experiment for unstable ^{129,130,131,132}Sn and ^{133,134}Sb nuclei (solid symbols) in comparison with data (open symbols) available for stable nuclei [11, 12, 13, 14, 15] and with data derived from a (γ, γ') experiment on ^{68}Ni based upon a branching ratio as explained in [16]. Right: The RHB+RQRPA relation between the relative dipole strength of the PDR mode over the GDR mode and neutron skin thickness, calculated for the set of effective DD-ME interactions with a_4= 30-38 MeV.

By application of the above mentioned framework, an average value for the symmetry energy pressure of $\overline{p}_0 = 2.2 \pm 0.5$ MeV/fm^3 was derived, using the experimental results for the relative PDR-strength for ^{130}Sn and ^{132}Sn. Due to the strong correlation between p_0 and a_4 observed in the RHB+RQRPA calculations, a_4 is determined accordingly to $\overline{a}_4 = 32.0 \pm 1.8$ MeV. In the next step, by employing these averaged findings, we extract from the calculations the neutron skin sizes. Values of R_n-$R_p = 0.23 \pm 0.03$ fm and of 0.24 ± 0.03 fm follow for ^{130}Sn and ^{132}Sn, respectively [10]. We note that the determination of the neutron skin thickness following this approach yields relatively high accuracies. However, we emphasize, that further experimental and theoretical investigations need to confirm the reliability of our approach and to estimate systematic uncertainties.

DIPOLE STRENGTH IN NI ISOTOPES

In a next experiment, the question of the dipole response in the continuum was addressed for the Nickel isotopic chain. A similar, improved setup and detection technique and relativistic secondary beams of $^{57-72}$Ni with energies of approximately 500 AMeV were used. The data are currently under analysis and the results presented in the following are preliminary. As a benchmark, we measured the stable ^{58}Ni beam in order to compare to the data derived in real photon absorption [18]. Applying the virtual photon description allows a direct comparison of these data to our new results. Since the photoabsorption data contain in the one neutron channel implicitly the neutron-proton channel (n,p), we analyze both decay channels within our experiment. However, the fraction of the (n,p) channel is of the order of 10% only, when integrating up to 30 MeV. The energy differential cross sections of the two datasets are in good agreement, and the integral values agree within the error bars. The analysis of the energy-differential cross sections for the neutron-rich Ni isotopes is ongoing. The preliminary results on the integral cross sections for ^{68}Ni give a first hint on additional strength below the GDR. While the two-neutron cross section is in good accordance with a Coulomb excitation calculation using the GDR with standard parameters derived from systematics over the mass region, we see a substantial excess of cross section for the measured one-neutron cross section. The results on ^{68}Ni are of particular interest, since in a complementary $(\gamma, \gamma\prime)$ experiment using the RISING setup at GSI evidence for sizeable strength at 11 MeV was reported [16]. A detailed study of the different decay channels can deliver a deeper insight into the decay characteristics and thus the nature of the PDR mode.

The question of the nature of the PDR also motivated an experiment investigating the dipole response of "proton-rich" nuclei. In these nuclei, the evolution of a skin is much less favoured due to the presence of the Coulomb barrier. Nevertheless, modern calculations have predicted the existence of PDR modes for light and medium-heavy nuclei, e.g. in ^{32}Ar [19, 20]. Using the LAND setup after an upgrade, now including also detectors for projectile-like protons, an experiment was performed recently investigating the dipole response in the neutron-deficient nuclei ^{32,34}Ar.

REFERENCES

1. Y. Suzuki, K. Ikeda, and H. Sato, *Prog. Theor. Phys.* **83**, 180 (1990).
2. P. V. Isacker, M. A. Nagarajan, and D. D. Warner, *Phys. Rev.* **C45**, R13 (1992).
3. D. Vretenar, N. Paar, P. Ring, and G. A. Lalazissis, *Nucl. Phys.* **A692**, 496 (2001).
4. N. Tsoneva, H. Lenske, and C. Stoyanov, *Phys. Lett.* **B586**, 213 (2004).
5. N. Paar, P. Ring, T. Nikšić, and D. Vretenar, *Phys. Rev.* **C67**, 34312 (2003).
6. N. Paar, D. Vretenar, E. Khan, and G. Colò, Rep. Prog. Phys. **70**, 691 (2007).
7. B.A. Brown, *Phys. Rev. Lett.* **85**, 5298 (2000).
8. R.J. Furnstahl, *Nucl. Phys.* **A706**, 85 (2002).
9. P. Adrich et al., *Phys. Rev. Lett.* **95**, 132501 (2005).
10. A. Klimkiewicz et al., *Phys. Rev.* **C76**, 051603 (R) (2007).
11. A. Zilges et al., *Phys. Lett.* **B542**, 43 (2002).
12. S. Volz et al., *Nucl. Phys.* **A779**, 1 (2006).
13. N. Ryezayeva et al., *Phys. Rev. Lett.* **89**, 272502 (2002).
14. K. Govaert et al., *Phys. Rev.* **C57**, 2229 (1998).
15. D. Savran et al., *Phys. Rev. Lett.* **100**, 232501 (2008).
16. O. Wieland et al., *Phys. Rev. Lett.* **102**, 092502 (2009).
17. D. Vretenar, T. Nikšić, and P. Ring , *Phys. Rev.* **C68**, 024310 (2003).
18. S.C. Fultz et al., *Phys. Rev.* **C10**, 608 (1974).
19. N. Paar, D. Vretenar and P. Ring, *Phys. Rev. Lett.* **94**, 182501 (2005).
20. C. Barbieri et al., *Phys. Rev.* **C77**, 024304 (2008).

Pygmy Resonances Within a Microscopic Multiphonon Approach

Edoardo G. Lanza

Istituto Nazionale di Fisica Nucleare - Sezione di Catania, Via S. Sofia 64, I-95123 Catania, Italy

Abstract. We perform HF+RPA calculations for nuclei with neutron excess in order to study the nature of the low lying dipole states. A detailed analysis shows that they are composed by many particle hole configurations in a cooperative, rather than collective, way. Nevertheless they represent a different mode as is evidenced by their transition densities. Relativistic Coulomb calculations within a microscopic multiphonon approach show that the contributions to the inelastic cross section of low lying multiphonon states is of the order of 10 %.

Keywords: pygmy resonance, giant resonance, collective states, multiphonon states
PACS: 21.60De, 21.60Jz, 24.30Cz, 25.70De, 24.30.Gd, 25.60.-t

INTRODUCTION

Among the different effects of the neutron skin on the collective properties of neutron rich nuclei the so called pygmy dipole resonances (PDR) have been the ones that have aroused more interest [1]. Early calculations [2] of strength distribution in neutron rich nuclei have shown that as far as the neutrons number increases the strength is moving towards the low energy region. Whether or not such strength at low energy corresponds to a collective mode is still under discussion.

From the experimental side, evidence for Coulomb excitation of PDR in heavy ion collisions at high energies on ^{132}Sn has been reported [3]. The motivation of the present study is due to our findings on multiphonon investigation at low energy excitation [4]. Indeed, we have found that a few low lying multiphonon states are excited in the collision process with a quite strong probability. These states lay in the region of the pygmy resonances and the question is whether they may contribute to the observed peak. Results of such calculations as well as many more details can be found in ref. [5].

NATURE OF DIPOLE LOW LYING STATES

We have performed HF plus RPA calculations for several closed shell Sn isotopes and with two Skyrme interaction: SGII [6] and SLY4 [7]. The RPA dipole strength distributions for some Sn isotopes and for the two Skyrme force considered show that, moving towards larger neutron excess, some dipole strength develops at low energies [5]. A way to see whether these low lying states are collective is to look at the number of particle-hole configurations that participate to their construction. But, most important, one should look at some other quantity that could describe the coherent aspect which is fundamental in the description of a collective state. Such quantity could be the reduced

CP1165, *Nuclear Structure and Dynamics '09*
edited by M. Milin, T. Nikšić, D. Vretenar, and S. Szilner
© 2009 American Institute of Physics 978-0-7354-0702-2/09/$25.00

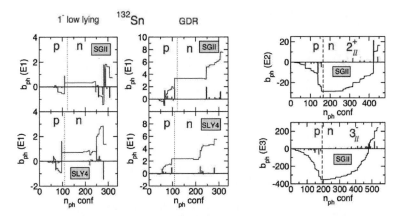

FIGURE 1. (Color online) Partial contributions b_{ph} of the reduced transition probability vs. the order number of the p-h configurations used in the RPA calculations for the two Skyrme interactions. The dotted lines divide the protons from the neutron configurations. The order goes from the most to the less bound ones. The bars corresponds to the individual b_{ph} contributions while the continuous thin line is the cumulative sum of the contributions. The figures on the left, middle and right columns correspond to two low lying dipole states, the GDR states and to the low lying 2^+ and 3^- states, respectively.

transition probability from the ground state to the excited state ν which can be written as

$$B(E\lambda) = |\sum_{ph} b_{ph}(E\lambda)|^2 = |\sum_{ph}(X_{ph}^\nu - Y_{ph}^\nu)T_{ph}^\lambda|^2 \qquad (1)$$

where T_{ph}^λ are the 2^λ multipole transition amplitudes associated with the elementary p-h configurations and the X and Y are the RPA amplitudes.

In fig. 1 we show the partial contributions b_{ph} versus the order number of the p-h configurations used in the RPA calculations for the two Skyrme interactions. The figures on the left column are for two low lying dipole states while the ones on the middle column are the GDR states obtained with the two Skyrme interaction as indicated in the figure. The states of the third column correspond to the low lying 2^+ and 3^-. The behaviour of the GDR as well as that of the 2^+ and 3^- contrasts with the one of the low lying dipole states: for these states there are several p-h configurations participating to the formation of the $B(E1)$ but some of them have opposite sign giving rise to a final value which is small tough, as we will see below, strong enough to get excited by the Coulomb interaction. From our novel analysis, it emerges that although the low-lying states cannot be considered as collective as the GDR states they cannot be described as single p-h configuration.

That the nature of the two low lying dipole states discussed above is qualitatively different and that they correspond to a new mode is illustrated in fig. 2 where the proton, neutron, isoscalar and isovector transition densities for three states, calculated with the SGII interaction, are reported. The behaviour for the low-lying state is similar to what has been found by several authors [2, 8, 9, 10]. In the surface region, the proton and neutron densities are not out of phase and the isoscalar transition density dominates over the isovector one; in the external region only the neutrons give a contribution to both

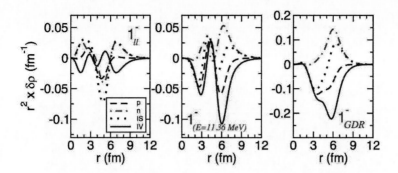

FIGURE 2. (Color online) Transition densities for the low-lying dipole state (upper), the dipole state at E=11.36 MeV (middle) and for the GDR state (lower) for the ^{132}Sn isotope calculated with the SGII interaction. We show the proton, neutron, isoscalar and isovector parts (as indicated in the legend).

isoscalar and isovector transition densities which have the same magnitude. For the GDR case the proton and neutron transition densities oscillate out of phase; the isovector part is much larger than the isoscalar one for all distances beyond the radius of the nucleus. For the state at E=11.36 MeV the transition density pattern resemble very much the one of the GDR: protons and neutrons start to oscillate out of phase. So, although the strengths of the two low lying states are very similar and their energies are close, they are two different nuclear modes as it is manifested in their different transition densities behaviour.

COULOMB EXCITATION CROSS SECTION

Starting from a microscopic approach based on RPA, mixing of two-phonon states among themselves and with one-phonon states is considered within a boson expansion approach with Pauli corrections. By diagonalizing a quartic microscopic Hamiltonian in the space of one-, two- and three-phonon states we generate mixed eigenstates with an anharmonic spectrum. Non linear terms are also taken into account in the external mean field of one of the nuclei which is responsible of the excitation of the other partner of the reaction [4].

The inelastic cross section are calculated by solving semiclassical coupled channel equations, the channels being superpositions of one-, two- and three-phonon states. The calculations have been done for two different Skyrme interactions (SGII and SLY4) and for the reactions ^{A}Sn + ^{280}Pb at 500 MeV/A [5]. We found an increase of the cross section in the low lying PDR energy region which is mainly due to the excitation of several states whose population is strongly suppressed by selection rules when anharmonicities and non-linearities are neglected. The increase varies from 3% up to 21% depending on the isotope considered and the Skyrme force used.

In table 1 we report the cross sections of the dipole low lying states as well as the contributions of the double phonon states lying in the energy regions of the PDR. Their contribution, for both interaction considered, is of the order of 10%. We note that they are not excited in the harmonic and linear approximation.

TABLE 1. Contributions to the total inelastic cross sections of the states listed in the first column, for the system ^{132}Sn + ^{208}Pb at 500 MeV/A (for both SGII and SLY4 interactions). Values for the harmonic and linear case (H-L) should be compared with the ones obtained in the anharmonic and non linear calculations (A-NL).

	SGII			SLY4		
	E (MeV)	H-L	A-NL	E (MeV)	H-L	A-NL
1^- low lying	8.7	28.4	30.1	9.5	29.8	31.3
$(2^+ \otimes 3^-)_{1^-}$	10.4	-	0.4	11.8	-	3.0
$(2^+ \otimes 2^+)_{2^+}$	9.7	-	0.9	10.6	-	0.3
$(3^- \otimes 3^-)_{3^-}$	11.3	-	3.1	13.1	-	1.1

SUMMARY

Within a HF + RPA calculations we select dipole low lying states which are a new excitation mode. This is clearly shown in their transition densities. The properties of these new states have been analyzed in detail in order to establish whether they are collective. Our investigation shows that they are due to a co-operative, although not collective, effect of several particle-hole excitations.

Relativistic Coulomb calculations show that, taking into account the presence of multiphonon states in the low lying dipole region, the excitation cross section may increase by a 10%.

ACKNOWLEDGMENTS

This work has been done in collaboration with M. A. Andrés, F. Catara, Ph. Chomaz and D. Gambacurta.

REFERENCES

1. N. Paar, D. Vretenar, E. Khan and G. Colò, *Rep. Prog. Phys.* **70**, 691 (2007) and referencs therein.
2. F. Catara, E. G. Lanza, M. A. Nagarajan and A. Vitturi *Nucl. Phys.* **A614**, 86 (1997); *Nucl. Phys.* **A624**, 449 (1997).
3. P. Adrich et al. (LAND-FRS Collaboration), *Phys. Rev. Lett.* **95**, 132501 (2005); A. Klimkiewicz et al. (LAND-FRS Collaboration), *Nucl. Phys.* **A788**, 145 (2007). A. Klimkiewicz et al. (LAND Collaboration) *Phys. Rev.* **A76**, 051603(R) (2007).
4. E. G. Lanza, M. V. Andrés, F. Catara, Ph. Chomaz and C. Volpe, *Nucl. Phys.* **A613**, 445 (1997); *Nucl. Phys.* **A636**, 452 (1998); *Nucl. Phys.* **A654**, 792c (1999); M.V. Andrés, F. Catara, E.G. Lanza, Ph. Chomaz, M. Fallot and J. A. Scarpaci, *Phys. Rev.* **A65**, 014608 (2001); E. G. Lanza, M.V. Andrés, F. Catara, Ph. Chomaz, M. Fallot and J. A. Scarpaci, *Phys. Rev.* **A74**, 064614 (2006).
5. E. G. Lanza, F. Catara, D. Gambacurta, M. V. Andrés and Ph. Chomaz, *Phys. Rev.* **A79**, 054615 (2009).
6. N. Van Giai and H. Sagawa, *Phys. Lett.* **B106**, 379 (1981); N. Van Giai, N. Sagawa, *Nucl. Phys.* **A371**, 1 (1981).
7. E. Chabanat, P. Bonche, P. Haensel, J. Meyer and R. Schaeffer, *Nucl. Phys.* **A635**, 231 (1998).
8. D. Vretenar, N. Paar, P. Ring, G.A. Lalazissis, *Nucl. Phys.* **A692**, 496 (2001).
9. D. Sarchi, P.F. Bortignon, G. Coló, *Phys. Lett.* **B601**, 27 (2007).
10. N. Tsoneva and H. Lenske, *Phys. Rev.* **C77** 024321 (2008).

Relativistic Random Phase Approximation At Finite Temperature

Y. F. Niu*,†, N. Paar†, D. Vretenar† and J. Meng*

*State Key Laboratory for Nuclear Physics and Technology, School of Physics, Peking University, Beijing 100871, China
†Physics Department, Faculty of Science, University of Zagreb, Croatia

Abstract. The fully self-consistent finite temperature relativistic random phase approximation (FTRRPA) has been established in the single-nucleon basis of the temperature dependent Dirac-Hartree model (FTDH) based on effective Lagrangian with density dependent meson-nucleon couplings. Illustrative calculations in the FTRRPA framework show the evolution of multipole responses of ^{132}Sn with temperature. With increased temperature, in both monopole and dipole strength distributions additional transitions appear in the low energy region due to the new opened particle-particle and hole-hole transition channels.

Keywords: random phase approximation, hot nuclei, pygmy resonances, giant resonances
PACS: 21.30.Fe, 21.60.Jz, 24.30.Cz

One of the major topics in studies of hot nuclei is to explore how the nuclear response is affected by the temperature, mainly the energy and the width of giant resonances. In recent years, the low-energy pygmy dipole resonance (PDR), corresponding to a neutron skin oscillation mode, has attracted considerable interest in experimental and theoretical studies[1]. Since the PDR could have a pronounced effect on neutron capture rates in the r-process nucleosynthesis, it is important to develop the microscopic model to study the low-energy modes in hot nuclei.

The aim of this work is to extend the Relativistic RPA (RRPA) based on effective interactions with a phenomenological density dependence to finite temperature. The studies of evolution of monopole and dipole responses of ^{132}Sn with temperature are performed. We focus on the analysis of the structure of the low energy modes.

The present calculation uses the finite temperature relativistic random phase approximation(FTRRPA) based on the temperature dependent Dirac-Hartree model (FTDH)[2]. The temperature is included self-consistently both in the ground state and RRPA. The FTRRPA equation is derived from the linear response theory. The occupation factors of Fermi-Dirac distribution for each single particle state appear in the RRPA matrix elements, so the configuration space is enlarged by particle-particle and hole-hole pairs.

As an illustrative example, the isoscalar monopole and isovector dipole responses of ^{132}Sn at temperatures 0, 1, 2 MeV are displayed in Fig. 1. In both cases, the giant resonances are only slightly modified at finite temperature. However, more pronounced effect can be observed in the low energy region. For monopole excitations, there are no transition peaks in the low energy region at $T = 0$ MeV. However, with increased temperature, new low-lying states appear. At $T = 2$ MeV, there are two main peaks at $E = 5.45$ and $E = 7.02$ MeV. They originate from the single particle transitions of neutrons from thermally unblocked orbitals, i.e., from $3p_{3/2}$ to $4p_{3/2}$ and from $2f_{7/2}$

CP1165, *Nuclear Structure and Dynamics '09*
edited by M. Milin, T. Nikšić, D. Vretenar, and S. Szilner
© 2009 American Institute of Physics 978-0-7354-0702-2/09/$25.00

to $3f_{7/2}$, correspondingly. In the case of dipole excitations, a pygmy resonance peak is obtained at $E = 7.75$ MeV at $T = 0$ MeV[1], while at $T = 2$ MeV the energy of this state is shifted to $E = 7.77$ MeV. Its configurations are only slightly modified. The contribution of the main neutron transition $3s_{1/2} \rightarrow 3p_{3/2}$ changes from 51.85% to 50.73% of the RRPA amplitudes. In addition, new low-energy transition strength appears at $E \approx 2 - 7$ MeV. These are particle-particle transitions involving neutron single particle states above the Fermi level which become thermally populated.

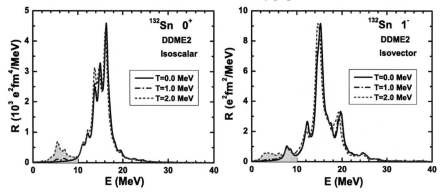

FIGURE 1. (Color online) Isoscalar monopole (left panel) and isovector dipole (right panel) responses of ^{132}Sn, calculated in the FTRRPA framework with DDME2 parameterization at temperatures 0, 1, 2 MeV.

In conclusion, we have developed the fully self-consistent FTDH+FTRRPA model based on effective Lagrangian with density dependent meson-nucleon couplings. The monopole and dipole responses are calculated at different temperatures for the representative nucleus ^{132}Sn. It is found that with increased temperature additional transitions appear in the low energy region due to new opened particle-particle and hole-hole transition channels. In the case of isoscalar monopole transition strength, low-energy states appear at finite temperature. In dipole excitation spectrum, the pygmy resonance retains its fundamental structure at $T = 2$ MeV, but it also becomes more distributed toward lower energies.

ACKNOWLEDGMENTS

This work was supported by the Unity through Knowledge Fund (UKF Grant No. 17/08) and MZOS - project 1191005-1010. Y. F. Niu acknowledges support from the National Foundation for Science, Higher Education and Technological Development of the Republic of Croatia. The work of J. M and D. V. was supported in part by the Chinese-Croatian project "Nuclear structure far from stability".

REFERENCES

1. N. Paar, D. Vretenar, E. Khan and G. Colò, *Rep. Prog. phys.* **70**, 691 (2007).
2. Y. F. Niu, N. Paar, D. Vretenar and J. Meng, submitted to *Phys. Lett.* **B**.

SYMMETRY DICTATED APPROACHES

Symmetry and Phase Transitions in Nuclei

Francesco Iachello

Center for Theoretical Physics, Sloane Physics Laboratory,
Yale University, New Haven, CT 06520-8120, USA

Abstract. The role of symmetry in (shape) quantum phase transitions (QPT) in nuclei is briefly reviewed. Finite size scaling is discussed. The concepts of excited state quantum phase transitions (ESQPT) and of critical matter are introduced. The *ab initio* theory of QPT is mentioned.

Keywords: algebraic models, phase transitions
PACS: 21.60.Fw;21.60.Ev;21.10.Re

INTRODUCTION

Nuclei are a fertile ground for the study of phase transitions. At nuclear matter density and zero temperature they are Fermi liquids. At approximately 20 MeV they undergo a phase transition from liquid to gas and at higher temperature (approximately 140 MeV) they are expected to undergo a transition from hadronic gas to quark-gluon plasma. At zero temperature, it has been suggested that as the density increases they undergo a phase transition to a solid, perhaps a color superconductor.

Finite nuclei are actually liquid drops with radius $R(\theta, \phi)$ and undergo phase transitions between different shapes similar in nature to the phase transitions between different crystal modifications. The phases have different symmetry. In the Landau example, the crystal $BaTiO_3$ undergoes a phase transition from cubic (symmetry O_h) to tetragonal (symmetry D_{4h}). In this example, the role of symmetry in phase transitions is very clear as it corresponds to the breaking of rotational invariance onto discrete crystal groups which determine the phases of the system. Shape phase transitions in nuclei are one of the best studied examples of phase transitions in physics, in particular, since nuclei are finite systems, the study of finite size scaling has played a very important role.

QUANTUM PHASE TRANSITIONS (QPT)

Quantum phase transitions are phase transitions that occur as a function of coupling constants, $\xi_1, \xi_2, ...$, called *control parameters*, that appear in the quantum Hamiltonian, H, that describes the system

$$H = \varepsilon (H_1 + \xi_1 H_2 + \xi_2 H_3 + ...). \tag{1}$$

Associated with phase transitions there are *order parameters*, the expectation values of suitable chosen operators that describe the state of the system $\langle O \rangle$. QPT's are also called *ground state* phase transitions [1] and/or *zero temperature* phase transitions.

CP1165, *Nuclear Structure and Dynamics '09*
edited by M. Milin, T. Nikšić, D. Vretenar, and S. Szilner

Theory of (shape) phase transitions in nuclei: the Interacting Boson Model

In this model [2], an even-even nucleus is described in terms of $N = n_s + n_d$ correlated pairs of nucleons with $J = 0, 2$ (s,d) treated as bosons. For the study of phase transitions it is sufficient to consider the simple Hamiltonian

$$H = \varepsilon_0 \left[(1 - \xi)\hat{n}_d - \frac{\xi}{4N} \hat{Q}^\chi \cdot \hat{Q}^\chi \right] \tag{2}$$

with $\hat{n}_d = (d^\dagger \cdot d)$ and $\hat{Q}^\chi = (d^\dagger \times \tilde{s} + s^\dagger \times \tilde{d})^{(2)} + \chi (d^\dagger \times \tilde{d})^{(2)}$. This model has algebraic structure U(6), *three phases*, with dynamic symmetries U(5), SU(3) and SO(6), corresponding to the breaking of U(6) into its subalgebras

$$
\begin{array}{cccll}
& U(5) & (I) & \xi = 0, \chi = anything & \text{Spherical} \\
\nearrow & & & & \\
U(6) \longrightarrow & SU(3) & (II) & \xi = 1, \chi = -\frac{\sqrt{7}}{2} & \text{Deformed axially} \qquad (3) \\
\searrow & & & & \\
& SO(6) & (III) & \xi = 1, \chi = 0 & \text{Deformed } \gamma\text{-unstable}
\end{array}
$$

and *two control parameters*, ξ and χ. (In general, the number of control parameters is the number of phases minus one.) A convenient *quantal order parameter* is

$$v_1 = \frac{\langle 0_1^+ | \hat{n}_d | 0_1^+ \rangle}{N}. \tag{4}$$

(For U(n) models, the number of independent order parameters is $n - 1$. If rotational invariance is imposed, the actual number is less than $n - 1$, for this problem two. The second quantal order parameter, v_2, will not be discussed here.)

Phase transitions and their order

A geometric description of the IBM can be obtained by the method of *number projected coherent states* [3], [4],

$$|N; \alpha_\mu\rangle \equiv \left(s^\dagger + \sum_\mu \alpha_\mu d_\mu^\dagger \right)^N |0\rangle. \tag{5}$$

Instead of using the variables α_μ, one can use the three Euler angles, $(\theta_1, \theta_2, \theta_3)$, and the two Bohr variables β, γ. The three phases correspond to shapes: spherical, U(5), deformed with axial symmetry, SU(3), and deformed γ-unstable, SO(6). The ground sate energy functional is

$$E(N; \beta, \gamma) = \frac{\langle N; \beta, \gamma | H | N; \beta, \gamma \rangle}{\langle N; \beta, \gamma | N; \beta, \gamma \rangle} \tag{6}$$

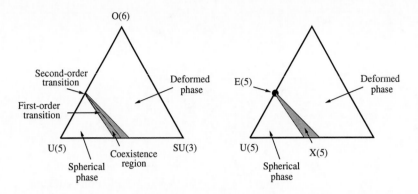

FIGURE 1. Shape phase diagram of IBM-1. Left: dynamical symmetries and phase transitions. Right: critical matter.

also called the Landau potential, $V(\beta, \gamma)$. Minimization of V with respect to β, γ provides the equilibrium values, β_e, γ_e (the *classical order parameters*). Evaluation of $V_{\min}$ and its derivatives with respect to the control parameters provides the order of the phase transition. This leads to the Ehrenfest classification in terms of discontinuties of the Landau potential as a function of the control parameters: discontinuity in $V_{\min}$ (0th order), in $\frac{\partial V_{\min}}{\partial \xi}$ (1st order), in $\frac{\partial^2 V_{\min}}{\partial \xi^2}$ (2nd order), ..., no discontinuity (crossover). In this way one can determine the shape phase diagram of nuclei in the Interacting Boson Model shown in Fig.1 [5].

Shape phase transitions in nuclei

Shape phase transitions have been observed unambiguously in two regions of the chart of nuclides, Region I: Nd-Sm-Gd-Dy and Region II: Sr-Zr. The control parameter, ξ, is proportional to the number of valence nucleons, i.e. $2N$, where N is the boson number. Some of the signatures of phase transitions that can be measured are

$$
\begin{array}{lc}
\text{Ground state energy} & E_0 \\
\text{Separation energy} & S_{2n}(N) = E_0(N+1) - E_0(N) \propto \frac{\partial E_0}{\partial \xi} \\
\text{Double separation energy} & S'_{2n}(N) = S_{2n}(N+1) - S_{2n}(N) \propto \frac{\partial^2 E_0}{\partial \xi^2} \\
\text{Radius} & \langle r^2 \rangle_0 = r_c^2 + DN + F \langle \hat{n}_d \rangle_{0_1^+} \\
\text{Isotope shift} & \delta \langle r^2 \rangle^{(N)} = D + F \left[\langle \hat{n}_d \rangle_{0_1^+}^{(N+1)} - \langle \hat{n}_d \rangle_{0_1^+}^{(N)} \right] \\
\text{B(E2)} & B(E2; 2_1^+ \to 0_1^+) \propto \langle \hat{n}_d \rangle_{0_1^+}^2
\end{array}
\tag{7}
$$

In Fig.2 (left panel) the 2n separation energies in the Sm isotopes are shown, indicating clearly the occurrence of a 1st order transition (discontinuity in $\frac{\partial E_0}{\partial \xi}$). In Fig.2 (right panel) the B(E2) values in Sm and Gd are shown (proportional to the square of the order parameter). There is a sharp increase in the order parameter between neutron number 88

195

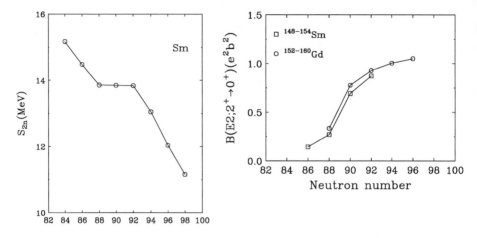

FIGURE 2. Left: experimental two-neutron separation energies in the Sm isotopes. Right: Experimental B(E2) values in the Sm and Gd isotopes.

and 90, indicating a phase transition, although it is not possible to distinguish between a first and a second order transition due to finite size effects.

Finite size scaling

The behavior of the order parameters with the number of particles, N, can be investigated numerically [6]. Qualitatively, in finite systems, the discontinuities are smoothed out: 0th order→1st order; 1st order→2nd order; 2nd order→crossover. Quantitatively, for second order transitions all physical quantities scale as a power law [7]

$$\Phi(\xi = \xi_c) \propto N^{-A_\Phi}. \tag{8}$$

The U(5)-SO(6) transition has been studied by the method of Continuous Unitary Transfomation (CUT) and compared with numerical studies [8]. It has been found that the scaling exponent for $\langle \hat{n}_d \rangle$ is 1/3 and thus for the order parameter $v_1 = \frac{\langle \hat{n}_d \rangle}{N}$ is -2/3.

EXCITED STATE QUANTUM PHASE TRANSITIONS (ESQPT)

Recently, the concept of QPT has been extended to excited state quantum phase transitions (ESQPT) [9]. These are phase transitions that occur as a function of excitation energy (also called *temperature dependent* phase transitions). Shape ESQPT are difficult to observe in nuclei because they rely on the measurement of highly excited states. (They have been recently observed in molecules [10].) A property of ESQPT that can be tested in nuclei is the collapse of all excitation energies to zero at the critical point. In finite systems, the energies do not go to zero and a gap, Δ, develops. Evidence for the gap in nuclei was found long ago [11].

CRITICAL MATTER

Matter at the critical value of a phase transition, $\xi = \xi_c$, is called *critical matter*. This is expected to be rather complex, especially at the critical point of a first order transition where two phases co-exist. Critical matter in the Interacting Boson Model is shown on the right hand side of Fig.1. Approximate *analytic expressions* for the energy spectra of nuclei at the critical point of a second order transition, called E(5), and along the line of first order transitions, called X(5), have been recently suggested [12], [13]. The energy levels are given in terms of zeros of Bessel functions. Evidence for critical nuclei is shown in [14], [15]. Other critical nuclei clearly identified are ^{150}Nd and ^{154}Gd [16].

MICROSCOPIC DESCRIPTIONS

Several attempts at a microscopic description of nuclear phase transitions were made in the 1980's. In view of the renewed interest in quantum phase transitions, new attempts are being made by making use of density functional theory, DFT [17], [18], [19], [20]. Both the U(5)-SO(6) transition and the U(5)-SU(3) transition have been investigated. In all these approaches, potential energy surfaces have been constructed. In the earlier calculations only the β dependence was studied, while in more recent calculations both the β and the γ dependence are included. All calculations appear to describe well the second order transition U(5)-SO(6) and produce potentials in agreement with those suggested by phenomenological IBM potentials along the U(5)-SO(6) line and the corresponding critical matter point E(5). Density functional theory appears however to describe the first order transition U(5)-SU(3) as a second order transition (no coexistence), although a flat bottom potential appears at the critical value X(5). This is in line with the IBM potentials which also show a 1st order transition with a small barrier between the coexisting minima and with the remarks in the preceeding section, where finite size scaling is discussed.

Several atempts have also been made to construct spectra from potentials along the transition lines making use of different approximate methods: Generator coordinate method, GCM, Collective Model, CM, and Interacting Boson Model, IBM. All calculation appear to reproduce spectra in qualitative agreement with phase transitional behavior, but, in some cases, in quantitative disagreement, especially for the gap Δ, which is over-predicted. As far as the signatures of phase transitions listed in (9) are concerned, calculations produce order parameters in agreement with experiment and with phenomenogical IBM calculations, but separation energies which do not show the sharp first order behavior observed in Fig.2 (left panel). Also some scale parameters are needed to go from the microscopic descriptions to the data and to the IBM phenomenological calculations. In any event, even with these drawbacks, considerable progress has been made towards an *ab initio* description of QPT in nuclei.

CONCLUSIONS

Shape phase transitions in nuclei have become a paradigm for QPT in Physics. Novel results for bosonic systems include:

1. The theory of shape QPT and their scaling behavior.

2. The theory of ESQPT and their scaling behavior.

3. *Experimental* evidence for phase transitional behavior in the order parameter, the gap, and the ground state energy.

4. *Experimental* evidence for critical matter.

5. *Ab initio* (DFT) theory of QPT.

The study of QPT has also become one of the most active areas of investigation in physics at the present time. Among the systems studied and not discussed here there are:

1. QPT in two-fluid systems, for nuclei proton-neutron systems with $U_\pi(6) \otimes U_\nu(6)$ symmetry [21], [22].

2. QPT in Bose-Fermi systems [23], [24].

3. QPT in configuration mixed models [25].

4. QPT in angle variables [26].

In conclusion, *symmetry* determines the *phases* of the system, and hence the phase transitions that may occur between them.

REFERENCES

1. R. Gilmore, *J. Math. Phys.* **20**, 891 (1979).
2. For a review, see, F. Iachello and A. Arima, *The Interacting Boson Model*, Cambridge University Press, 1987.
3. J.N. Ginocchio and M.W. Kirson, *Phys. Rev. Lett.* **44**, 1744 (1980).
4. A.E.L. Dieperink, O. Scholten and F. Iachello, *Phys. Rev. Lett.* **44**, 1747 (1980).
5. D.H. Feng, R. Gilmore and S.R. Deans, *Phys. Rev.* **C23**, 1254 (1981).
6. F. Iachello and N.V. Zamfir, *Phys. Rev. Lett.* **92**, 212501 (2004).
7. M.E. Fisher and M.N. Barber, *Phys. Rev. Lett.* **28**, 1516 (1972).
8. S. Dusuel, J. Vidal, J.M. Arias, J. Dukelsky, and J.E. Garcia-Ramos, *Phys. Rev.* **C62**, 011301 (R) (2005).
9. M. Caprio, P. Cejnar and F. Iachello, *Ann. Phys. (N.Y.)* **323**, 1106 (2008).
10. F. Perez-Bernal and F. Iachello, *Phys. Rev.* **A77**, 032115 (2008).
11. O. Scholten, F. Iachello and A. Arima, *Ann. Phys. (N.Y.)* **115**, 325 (1978).
12. F. Iachello, *Phys. Rev. Lett.* **85**, 3580 (2000).
13. F. Iachello, *Phys. Rev. Lett.* **87**, 052502 (2001).
14. R.F. Casten and N.V. Zamfir, *Phys. Rev. Lett.* **87**, 052503 (2001).
15. R.F. Casten and N.V. Zamfir, *Phys. Rev. Lett.* **85**, 3584 (2000).
16. R. Kruecken et al., *Phys. Rev. Lett.* **88**, 232501 (2002).
17. L.M. Robledo, R. Rodriguez-Guzman, and P. Sarriguren, *Phys. Rev.* **C78**, 034314 (2008).
18. T. Niksic, D. Vretenar, G.A. Lalazissis and P. Ring, *Phys. Rev. Lett.* **99**, 092502 (2007).
19. T.R. Rodriguez and J.L. Egido, *Phys. Lett.* **B663**, 49 (2008).
20. K. Nomura, N. Shimizu, and T. Otsuka, *Phys. Rev. Lett.* **101** 142501 (2008).
21. M.A. Caprio and F. Iachello, *Phys. Rev. Lett.* **93**, 242502 (2004).
22. J.M. Arias, J. Dukelsky, J.E. Garcia-Ramos, *Phys. Rev. Lett.* **93**, 212501 (2004).
23. J. Jolie, S. Heinze, P. van Isacker, and R.F. Casten, *Phys. Rev.* **C70**, 011305(R) (2004).
24. C.E. Alonso, J.M. Arias, L. Fortunato, and A. Vitturi, *Phys. Rev.* **C72**, 061302 (2005).
25. A. Frank, P. van Isacker and F. Iachello, *Phys. Rev.* **C73**, 061302 (2006).
26. F. Iachello, *Phys. Rev. Lett.* **91**, 132502 (2003).

Partial Dynamical Symmetry and Anharmonicity in Gamma-Soft Nuclei

A. Leviatan

Racah Institute of Physics, The Hebrew University, Jerusalem 91904, Israel

Abstract. Partial dynamical symmetry is shown to be relevant for describing the anharmonicity of excited bands in ^{196}Pt while retaining solvability and good $SO(6)$ symmetry for the ground band.

Keywords: Partial dynamical symmetry, anharmonicity, cubic terms, interacting boson model
PACS: 21.60.Fw, 21.10.Re, 21.60.Ev, 27.80.+w

Gamma-soft nuclei can be described in the interacting boson model (IBM) in its SO(6) dynamical symmetry (DS) limit [1]. The latter limit corresponds to the chain of nested algebras

$$U(6) \supset SO(6) \supset SO(5) \supset SO(3) \supset SO(2)$$
$$\downarrow \qquad \downarrow \qquad \downarrow \qquad \downarrow \qquad \downarrow \qquad (1)$$
$$[N] \qquad \langle \Sigma \rangle \qquad (\tau) \quad v_\Delta \quad L \qquad M$$

where, below each algebra, its associated labels of irreducible representations (irreps) are given and v_Δ is a multiplicity label. The eigenstates $|[N]\langle\Sigma\rangle(\tau)v_\Delta LM\rangle$ are obtained with a Hamiltonian with SO(6) DS which can be transcribed in the form

$$\hat{H}_{DS} = A\hat{P}_+\hat{P}_- + B\hat{C}_{SO(5)} + C\hat{C}_{SO(3)}. \qquad (2)$$

Here $\hat{C}_G$ denotes the quadratic Casimir operator of G, $\hat{P}_+ \equiv \frac{1}{2}(s^\dagger s^\dagger - d^\dagger \cdot d^\dagger)$, $4\hat{P}_+\hat{P}_- = \hat{N}(\hat{N}+4) - \hat{C}_{SO(6)}$ and $\hat{P}_- = \hat{P}_+^\dagger$. The monopole (s) and quadrupole (d) bosons represent valence nucleon pairs whose total number, $\hat{N} = \hat{n}_s + \hat{n}_d$, is conserved. The SO(6)-DS Hamiltonian, $\hat{H}_{DS}$, is completely solvable with eigenenergies

$$E_{DS} = \frac{1}{4}A(N-\Sigma)(N+\Sigma+4) + B\tau(\tau+3) + CL(L+1). \qquad (3)$$

The spectrum resembles that of a γ-unstable deformed rotor, where states are arranged in bands with SO(6) quantum number $\Sigma = N - 2v$, $(v = 0,1,2,\ldots)$. The in-band rotational structure is governed by the SO(5) and SO(3) terms in $\hat{H}_{DS}$ (2), with characteristic $\tau(\tau+3)$ and $L(L+1)$ splitting. A comparison with the experimental spectrum and E2 rates of ^{196}Pt [2] is shown in Fig. 1 and Table 1. It displays a good description for properties of states in the ground band $(\Sigma = N)$. This observation was the basis of the claim [3] that the SO(6)-DS is manifested empirically in ^{196}Pt. However, the resulting fit to energies of excited bands is quite poor. The 0_1^+, 0_3^+, and 0_4^+ levels of ^{196}Pt at excitation energies 0, 1403, 1823 keV, respectively, are identified as the bandhead states of the ground $(v = 0)$, first- $(v = 1)$ and second- $(v = 2)$ excited vibrational bands [3]. Their empirical anharmonicity, defined by the ratio $R = E(v=2)/E(v=1) - 2$, is found

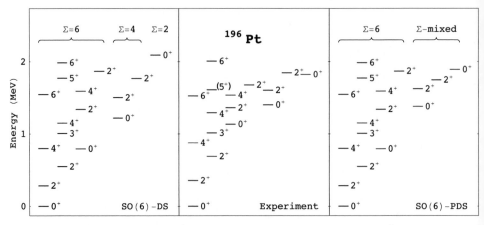

FIGURE 1. Observed spectrum of ^{196}Pt [2] compared with the calculated spectra of $\hat{H}_{DS}$ (2), with SO(6) dynamical symmetry (DS), and of $\hat{H}_{PDS}$ (6) with partial dynamical symmetry (PDS). The parameters in $\hat{H}_{DS}$ ($\hat{H}_{PDS}$) are $A = 174.2\,(122.9)$, $B = 44.0\,(44.0)$, $C = 17.9\,(17.9)$, and $\eta = 0\,(34.9)$ keV. The boson number is $N = 6$ and Σ is an SO(6) label. From [15].

to be $R = -0.70$. In the SO(6)-DS limit these bandhead states have $\tau = L = 0$ and $\Sigma = N, N-2, N-4$, respectively. The anharmonicity $R = -2/(N+1)$, as calculated from Eq. (3), is fixed by N. For $N = 6$, which is the appropriate boson number for ^{196}Pt, the SO(6)-DS value is $R = -0.29$, which is in marked disagreement with the empirical value. A detailed study of double-phonon excitations within the IBM, has concluded that large anharmonicities can be incorporated only by the inclusion of at least cubic terms in the Hamiltonian [4]. In the IBM there are 17 possible three-body interactions. One is thus confronted with the need to select suitable higher-order terms that can break the DS in excited bands but preserve it in the ground band. These are precisely the defining properties of a partial dynamical symmetry (PDS). The essential idea is to relax the stringent conditions of *complete* solvability, so that only part of the eigenspectrum retains all the DS quantum numbers. Various types of PDS are known to

TABLE 1. Observed [2] and calculated B(E2) values (in e^2b^2) for ^{196}Pt. For both the exact (DS) and partial (PDS) SO(6) dynamical symmetry calculations, the E2 operator is $e_b[(s^\dagger \times \tilde{d} + d^\dagger \times \tilde{s})^{(2)} + \chi(d^\dagger \times \tilde{d})^{(2)}]$ with $e_b = 0.151\ eb$ and $\chi = 0.29$. From [15].

Transition	Experiment	DS	PDS	Transition	Experiment	DS	PDS
$2_1^+ \rightarrow 0_1^+$	0.274 (1)	0.274	0.274	$2_3^+ \rightarrow 0_2^+$	0.034 (34)	0.119	0.119
$2_2^+ \rightarrow 2_1^+$	0.368 (9)	0.358	0.358	$2_3^+ \rightarrow 4_1^+$	0.0009 (8)	0.0004	0.0004
$2_2^+ \rightarrow 0_1^+$	3.10^{-8}(3)	0.0018	0.0018	$2_3^+ \rightarrow 2_2^+$	0.0018 (16)	0.0013	0.0013
$4_1^+ \rightarrow 2_1^+$	0.405 (6)	0.358	0.358	$2_3^+ \rightarrow 0_1^+$	0.00002 (2)	0	0
$0_2^+ \rightarrow 2_2^+$	0.121 (67)	0.365	0.365	$6_2^+ \rightarrow 6_1^+$	0.108 (34)	0.103	0.103
$0_2^+ \rightarrow 2_1^+$	0.019 (10)	0.003	0.003	$6_2^+ \rightarrow 4_2^+$	0.331 (88)	0.221	0.221
$4_2^+ \rightarrow 4_1^+$	0.115 (40)	0.174	0.174	$6_2^+ \rightarrow 4_1^+$	0.0032 (9)	0.0008	0.0008
$4_2^+ \rightarrow 2_2^+$	0.196 (42)	0.191	0.191	$0_3^+ \rightarrow 2_2^+$	< 0.0028	0.0037	0.0028
$4_2^+ \rightarrow 2_1^+$	0.004 (1)	0.001	0.001	$0_3^+ \rightarrow 2_1^+$	< 0.034	0	0
$6_1^+ \rightarrow 4_1^+$	0.493 (32)	0.365	0.365				

TABLE 2. $SO(6)$ decomposition of eigenstates of $\hat{H}_{\text{PDS}}$ (6), corresponding to bandhead states in ^{196}Pt.

Bandhead	$\Sigma = 6$	$\Sigma = 4$	$\Sigma = 2$	$\Sigma = 0$
$0^+(v=0)$	100 %			
$0^+(v=1)$		76.5 %	16.1 %	7.4 %
$0^+(v=2)$		19.6 %	18.4 %	62.0 %

be relevant to nuclear spectroscopy [5-11], to systems with mixed chaotic and regular dynamics [12, 13] and to quantum phase transitions [14]. In the present contribution we demonstrate the relevance of PDS to the anharmonicity of excited bands in ^{196}Pt [15].

Hamiltonians with $SO(6)$ PDS preserve the analyticity of only a *subset* of the states (1). The construction of interactions with this property requires n-boson creation and annihilation operators, $\hat{B}^\dagger_{[n]\langle\sigma\rangle(\tau)\ell m}$ and $\tilde{B}_{[n^5]\langle\sigma\rangle(\tau)\ell m}$, with definite tensor character in the basis (1). Of particular interest are n-boson annihilation operators which satisfy

$$\tilde{B}_{[n^5]\langle\sigma\rangle(\tau)\ell m}|[N]\langle N\rangle(\tau)v_\Delta LM\rangle = 0, \tag{4}$$

for all possible values of τ, L contained in the $SO(6)$ irrep $\langle N\rangle$. The annihilation condition (4) is satisfied for tensor operators with $\sigma < n$. This is so because the action of $\tilde{B}_{[n^5]\langle\sigma\rangle(\tau)\ell m}$ leads to an $(N-n)$-boson state that contains the $SO(6)$ irreps $\langle\Sigma\rangle = \langle N - n - 2i\rangle$, $i = 0, 1, \ldots$, which cannot be coupled with $\langle\sigma\rangle$ to yield $\langle\Sigma\rangle = \langle N\rangle$, since $\sigma < n$. Number-conserving normal-ordered interactions that are constructed out of such tensors (and their Hermitian conjugates) thus have $|[N]\langle N\rangle(\tau)v_\Delta LM\rangle$ as eigenstates with zero eigenvalue.

A systematic enumeration of all interactions with this property is a simple matter of $SO(6)$ coupling. For example, $SO(6)$ tensors, $\hat{B}^\dagger_{[n]\langle\sigma\rangle(\tau)\ell m}$, with $\sigma < n = 2$ or $\sigma < n = 3$ are found to be

$$\hat{B}^\dagger_{[2]\langle0\rangle(0)00} \propto \hat{P}_+, \quad \hat{B}^\dagger_{[3]\langle1\rangle(1)2m} \propto \hat{P}_+ d^\dagger_m, \quad \hat{B}^\dagger_{[3]\langle1\rangle(0)00} \propto \hat{P}_+ s^\dagger. \tag{5}$$

The two-boson $SO(6)$ tensor gives rise to a two-body $SO(6)$-invariant interaction, $\hat{P}_+\hat{P}_-$, which is simply the completely solvable $SO(6)$ term in $\hat{H}_{\text{DS}}$, Eq. (2). From the three-boson $SO(6)$ tensors one can construct three-body interactions with an $SO(6)$ PDS, namely, $\hat{P}_+\hat{n}_s\hat{P}_-$ and $\hat{P}_+\hat{n}_d\hat{P}_-$. Since the combination $\hat{P}_+(\hat{n}_s + \hat{n}_d)\hat{P}_- = (\hat{N}-2)\hat{P}_+\hat{P}_-$ is completely solvable in $SO(6)$, there is only one genuine partially solvable three-body interaction which can be chosen as $\hat{P}_+\hat{n}_s\hat{P}_-$, with tensorial components $\sigma = 0, 2$.

On the basis of the preceding discussion we propose to use the following Hamiltonian with $SO(6)$-PDS

$$\hat{H}_{\text{PDS}} = \hat{H}_{\text{DS}} + \eta\hat{P}_+\hat{n}_s\hat{P}_-, \tag{6}$$

where the terms are defined in Eqs. (2) and (5). The spectrum of $\hat{H}_{\text{PDS}}$ is shown in Fig. 1. The states belonging to the $\Sigma = N = 6$ multiplet remain solvable with energies given by the same DS expression, Eq. (3). As shown in Table 2, states with $\Sigma < 6$ are generally admixed but agree better with the data than in the DS calculation. Thus, although the ground band is pure, the excited bands exhibit strong $SO(6)$ breaking. The calculated

SO(6)-PDS anharmonicity for these bands is $R = -0.63$, much closer to the empirical value, $R = -0.70$. We emphasize that not only the energies but also the wave functions of the $\Sigma = N$ states remain unchanged when the Hamiltonian is generalized from DS to PDS. Consequently, the E2 rates for transitions among this class of states are the same in the DS and PDS calculations. This is evident in Table 1 where most of the E2 data concern transitions between $\Sigma = N = 6$ states. Only transitions involving states from excited bands (e.g., the 0_3^+ state in Table 1) can distinguish between DS and PDS.

A similar procedure can be implemented on a general dynamical symmetry chain

$$
\begin{array}{ccccc}
G_{\text{dyn}} & \supset & G & \supset \cdots \supset & G_{\text{sym}} \\
\downarrow & & \downarrow & & \downarrow \\
[h_N] & & \langle \Sigma \rangle & & \Lambda
\end{array}
\tag{7}
$$

where G_{dyn} and G_{sym} are, respectively, the dynamical and symmetry algebras of the system. For N identical particles the irrep $[h_N]$ is either symmetric $[N]$ (bosons) or antisymmetric $[1^N]$ (fermions). Hamiltonians which preserve the solvability of states with $\langle \Sigma \rangle = \langle \Sigma_0 \rangle$, involve n-particle annihilation tensor operators satisfying

$$
\hat{T}_{[h_n]\langle\sigma\rangle\lambda} \, |[h_N]\langle\Sigma_0\rangle\Lambda\rangle = 0,
\tag{8}
$$

for all possible values of Λ contained in the given G-irrep $\langle\Sigma_0\rangle$. The solution of condition (8) amounts to carrying out a G Kronecker product $\langle\sigma\rangle \times \langle\Sigma_0\rangle$. This establishes a generic and systematic procedure for identifying and selecting interactions, of a given order, with PDS. The resulting Hamiltonians break the DS but retain selected subsets of solvable eigenstates with good symmetry. As demonstrated in the present contribution, the advantage of using higher-order interactions with PDS is that they can be introduced without destroying results previously obtained with a DS for a segment of the spectrum.

This contribution is based on work done in collaboration with J.E. García-Ramos (Huelva) and P. Van Isacker (GANIL) and is supported by grants from the ISF and BSF.

REFERENCES

1. F. Iachello and A. Arima, *The Interacting Boson Model*, Cambridge Univ. Press, Cambridge, 1987.
2. H. Xiaolong, *Nuclear Data Sheets* **108**, 1093 (2007).
3. J. A. Cizewski *et al.*, *Phys. Rev. Lett.* **40**, 167 (1978); *Nucl. Phys.* **A323**, 349 (1979).
4. J. E. García-Ramos, J. M. Arias, and P. Van Isacker, *Phys. Rev.* **C62**, 064309 (2000).
5. A. Leviatan, *Phys. Rev. Lett.* **77**, 818 (1996).
6. P. Van Isacker, *Phys. Rev. Lett.* **83**, 4269 (1999).
7. A. Leviatan and J. N. Ginocchio, *Phys. Rev.* **C61**, 024305 (2000).
8. A. Leviatan and P. Van Isacker, *Phys. Rev. Lett.* **89**, 222501 (2002).
9. J. Escher and A. Leviatan, *Phys. Rev. Lett.* **84**, 1866 (2000); *Phys. Rev.* **C65**, 054309 (2002).
10. D. J. Rowe and G. Rosensteel, *Phys. Rev. Lett.* **87**, 172501 (2001); *Phys. Rev.* **C67**, 014303 (2003).
11. P. Van Isacker and S. Heinze, *Phys. Rev. Lett.* **100**, 052501 (2008).
12. N. Whelan, Y. Alhassid, and A. Leviatan, *Phys. Rev. Lett.* **71**, 2208 (1993).
13. A. Leviatan and N. D. Whelan, *Phys. Rev. Lett.* **77**, 5202 (1996).
14. A. Leviatan, *Phys. Rev. Lett.* **98**, 242502 (2007).
15. J. E. García-Ramos, A. Leviatan and P. Van Isacker, *Phys. Rev. Lett.* **102**, 112502 (2009).

Nuclear Shape Transitions From a Microscopic Approach

P. Sarriguren*, R.R. Rodríguez-Guzmán† and L.M. Robledo**

*Instituto de Estructura de la Materia, CSIC, Serrano 123, E-28006 Madrid, Spain
†Department of Physics, University of Jyväskylä, P.O. Box 35, FI-40014, Jyväskylä, Finland
**Departamento de Física Teórica C-XI, Universidad Autónoma de Madrid, 28049-Madrid, Spain

Abstract. Self-consistent Hartree-Fock-Bogoliubov calculations based on Gogny and Skyrme density-dependent interactions are performed to study the evolution of ground-state nuclear shapes with the number of nucleons in various isotopic chains. Signatures of critical point symmetries are analyzed by studying potential energy surfaces including triaxial degrees of freedom.

Keywords: Self-consistent mean field approach. Skyrme and Gogny forces. Nuclear deformation. Potential energy surfaces.
PACS: 21.60.Jz, 21.60.Fw

INTRODUCTION

The ground states of atomic nuclei are characterized by different equilibrium configurations which correspond to different geometrical shapes. The study of these equilibrium shapes, as well as the transition regions between them, has been the subject of a large number of theoretical and experimental studies [1].

Within the framework of algebraic models, the different nuclear shapes are associated with dynamic symmetries that provide a useful tool to describe properties of nuclei since they lead to exactly solvable problems and produce the results for observables in explicit analytic form. In nuclear physics, according to the Interacting Boson Model (IBM) [2], the dynamic symmetries are given by U(5), associated to spherical symmetry, SU(3), associated with axially deformed symmetry, and SO(6), describing γ-unstable shapes. The phase shape transitions correspond to the breaking of these dynamic symmetries and they occur as the number of nucleons change in the nucleus.

Iachello introduced some years ago, the E(5) and X(5) critical point symmetries, which provide parameter free (up to scale factors) predictions of excitation spectra and strengths for nuclei at the critical point of a phase shape transition [3]. These symmetries were obtained within the framework of the collective Bohr Hamiltonian under some simplifying approximations. In particular, the potential in the β degree of freedom was approximated by a simple square well potential, which is decoupled from the potential in the γ-variable. In the case of E(5), which corresponds to the transition from spherical vibrational U(5) to deformed γ-unstable O(6), the potential is constant in the γ-direction. In the case of X(5), related to the transition from U(5) to axially symmetric prolate SU(3), a harmonic oscillator potential is used in the γ-direction. Empirical evidence of these transitional symmetries at the critical points has been observed in several isotopes of Ba, Pd, and Xe for E(5), and in ^{152}Sm, ^{150}Nd, ^{154}Gd, and ^{156}Dy for X(5).

CP1165, *Nuclear Structure and Dynamics '09*
edited by M. Milin, T. Nikšić, D. Vretenar, and S. Szilner
© 2009 American Institute of Physics 978-0-7354-0702-2/09/$25.00

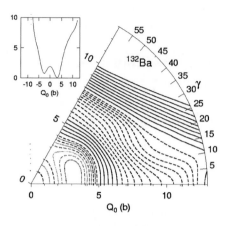

FIGURE 1. (Color online) Contour plot of the PES for ^{132}Ba with the Gogny D1S force.

In this work we investigate shape transitions within non-relativistic microscopic models, based on effective interactions between nucleons that provide a unified description of nuclear properties along the nuclear chart. In particular, we study whether the assumptions made on the β-γ potentials to construct the point symmetries are justified microscopically. Thus, we study various isotopic chains in which the occurrence of shape transitions has been predicted and we show results for the potential energy surfaces (PES) corresponding to constrained mean field calculations [4, 5, 6]. The theoretical framework is based on both the self-consistent Hartree-Fock-Bogoliubov approximation with the finite-range and density-dependent Gogny interaction (D1S) [7], and the self-consistent Hartree-Fock + BCS with short-range density-dependent Skyrme interactions (SLy4) [8] and a zero-range density-dependent interaction in the pairing channel. In the latter case we use the code EV8 [9] that solves the HF equations in a coordinate space mesh. The role of triaxiality is also considered and discussed in those nuclei where this degree of freedom could be relevant.

RESULTS

In Figure 1 we can see the contour plot of the PES for ^{132}Ba, obtained from Gogny D1S, as a function of the quadrupole moment Q_0 (b) and γ (deg) angle. This nucleus has been proposed as a candidate for the E(5) symmetry. Along with the triaxial plot, there is a cut corresponding to the axially symmetric shape ($\gamma = 0°$ and $\gamma = 60°$). We can see that the potential is rather flat in β (Q_0) around sphericity and that it is practically constant in the γ-direction at the β deformation corresponding to the minimum of the energy. It is also interesting to notice that the oblate axial minimum becomes a saddle point when triaxiality is considered. All of this makes this nucleus a reasonably good candidate to show E(5)-like behavior. Similar conclusions are obtained for the other candidates ^{108,110}Pd, $^{128-132}$Xe, and ^{130,134}Ba [4, 6].

In Figure 2 we have chosen Nd isotopes, and in particular ^{150}Nd, to study the X(5)-like behavior. Constrained HF+BCS calculations for Nd isotopes with SLy4 and zero-range

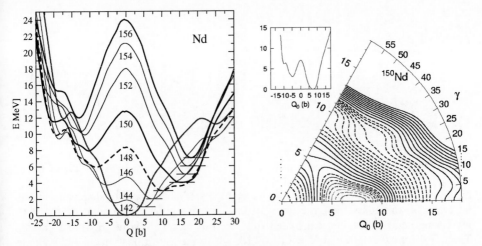

FIGURE 2. (Color online) Left: axial PES for Nd isotopes with SLy4. Right: contour plot of the PES for ^{150}Nd with the Gogny D1S force (notice that Q is defined as $2Q_0$).

pairing force are shown in the left panel of Figure 2. The axial plots show a clear shape transition from spherical ^{142}Nd (N=82) to well developed prolate shapes at $^{154-156}$Nd. The isotope ^{150}Nd (N=90) shows a transitional behavior with a shallow minimum on the prolate side and an additional minimum on the oblate side with an energy barrier rather high between these minima. On the right panel, the triaxial contour plot for the X(5) candidate ^{150}Nd shows that for Q_0 values in the vicinity of the prolate minimum, the energy presents a parabolic behavior as a function of γ. This is particularly true for low γ angles but tends to become flat as γ approaches $60°$. The oblate axial minimum becomes also a saddle point when looking at the $Q_0 - \gamma$ plane. Thus, the assumption of an infinite square well made on the β-potential to arrive to the X(5) symmetry, is not realized within non-relativistic microscopic calculations that produce systematically potential energy barriers at $\beta = 0$ excluding the spherical configuration from the coexisting shapes. Similar results are obtained for the other X(5) candidates ^{152}Sm, ^{154}Gd, and ^{156}Dy [4, 6]. The results also agree with those obtained from relativistic calculations [10]. We

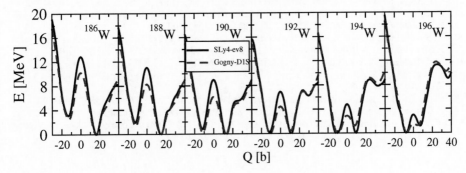

FIGURE 3. (Color online) Axial PES for $Z = 74$ isotopes from mean field calculations with various forces.

205

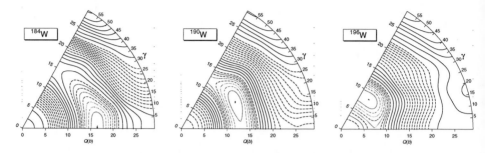

FIGURE 4. (Color online) Contour plots of the PES for W isotopes with the Skyrme SLy4 force.

next consider the chain of W isotopes as examples of oblate-prolate shape transitions occurring in this mass region. Figure 3 shows the evolution of the axial shapes with increasing neutron number, calculated from various forces, Skyrme SLy4 and Gogny. The transition takes place at $N = 116 - 118$, where the transitional nuclei show oblate and prolate minima coexisting at close energies. In Figure 4 we can see the contour plots of the PES with SLy4 in three isotopes where the transition is manifest. We can see in ^{184}W a real prolate minimum and an oblate saddle point. ^{190}W shows a triaxial ground state with the axial prolate and oblate minima transformed into saddle points. There is a very soft behavior of the PES along the γ degree of freedom, developing a very shallow triaxial minimum. Finally, in ^{196}W the oblate minimum has been developed, while the prolate one is now a saddle point. Similar results are found in Yb, Hf, Os, and Pt isotopes [5]. In general, for all of these isotopes, the transition from prolate to oblate shapes takes place at $N = 116 - 118$, where the energies of oblate and prolate shapes are nearly degenerate. The triaxial analysis of these nuclei shows that the axial oblate and prolate minima, which are separated by high potential barriers in the β degree of freedom, are linked very softly in the γ degree of freedom, making these minima saddle points in the extended $\beta - \gamma$ plane.

REFERENCES

1. J.L. Wood, K. Heyde, W. Nazarewicz, M. Huyse, and P. Van Duppen, *Phys. Rep.* **215**, 101 (1992).
2. F. Iachello and A. Arima, *The Interacting Boson Model* (Cambridge University Press, England, 1987).
3. F. Iachello, *Phys. Rev. Lett.* **85**, 3580 (2000); **87**, 052502 (2001).
4. R. Rodríguez-Guzmán and P. Sarriguren, *Phys. Rev.* **C76**, 064303 (2007).
5. P. Sarriguren, R. Rodríguez-Guzmán, and L.M. Robledo, *Phys. Rev.* **C77**, 064322 (2008).
6. L.M. Robledo, R. Rodríguez-Guzmán, and P. Sarriguren, *Phys. Rev.* **C78**, 034314 (2008).
7. J. Dehargé and D. Gogny, *Phys. Rev.* **C21**, 1568 (1980); J.F. Berger, M. Girod and D. Gogny, *Nucl. Phys.* **A428**, 23c (1984).
8. E. Chabanat, P. Bonche, P. Haensel, J. Meyer, and R. Schaeffer, *Nucl. Phys.* **A635**, 231 (1998).
9. P. Bonche, H. Flocard, and P.-H. Heenen, *Comput. Phys. Comm.* **171**, 49 (2005).
10. R. Fossion, D. Bonatsos, and G.A. Lalazissis, *Phys.Rev.* **C73**, 044310 (2006); T. Niksic, D. Vretenar, G.A. Lalazissis, and P. Ring, *Phys. Rev. Lett.* **99**, 092502 (2007).

Decoherence as a Signature of an Excited State Quantum Phase Transition in Two Level Boson Systems

J. E. García-Ramos*, J. M. Arias†, J. Dukelsky**, P. Pérez-Fernández† and A. Relaño**

*Departamento de Física Aplicada, Universidad de Huelva, 21071 Huelva, Spain
†Departamento de Física Atómica, Molecular y Nuclear, Facultad de Física, Universidad de Sevilla, Apartado 1065, 41080 Sevilla, Spain
**Instituto de Estructura de la Materia, CSIC, Serrano 123, E-28006 Madrid, Spain

Abstract. We analyze the decoherence induced on a single qubit by the interaction with a two-level boson system with critical internal dynamics. We explore how the decoherence process is affected by the presence of quantum phase transitions in the environment. We conclude that the dynamics of the qubit changes dramatically when the environment passes through a continuous excited state quantum phase transition. If the system-environment coupling energy equals the energy at which the environment has a critical behavior, the decoherence induced on the qubit is maximal and the fidelity tends to zero with finite size scaling obeying a power-law.

Keywords: Quantum decoherence, quatum phase transition, excited state quatum phase transition
PACS: 03.65.Yz, 05.70.Fh, 64.70.Tg

Real quantum systems always interact with the environment. This interaction leads to decoherence, the process by which quantum information is degraded and purely quantum properties of a system are lost [1].

The connection between decoherence and environmental quantum phase transitions has been recently investigated in [2]. In this contribution and in reference [3] we analyze the relationship between decoherence and an environmental excited state quantum phase transition (ESQPT).

An ESQPT is analogous to a standard quantum phase transition (QPT), but taking place in some excited state of the system, which defines the critical energy E_c at which the transition takes place. We can distinguish between different kinds of ESQPT, either first order or continuous [4]. In this contribution we will concentrate in the latter case, which usually entails a singularity in the density of states (for an illustration see Fig. 1).

These kinds of phase transitions have been identified in the Lipkin model, in the interacting boson model (IBM), and in more general boson or fermion two-level pairing Hamiltonians (for a complete discussion, including a semiclassical analysis, see [5]).

Here, we consider an environment having both QPTs and ESQPTs coupled to a single qubit. The Hamiltonian of the environment, defined as a function of a control parameter α, presents a QPT at a critical value α_c. We define a coupling between the central qubit and the environment that entails an effective change in the control parameter, $\alpha \rightarrow \alpha'$, making the environment to cross the critical point if $\alpha' > \alpha_c$. Moreover, the coupling also implies an energy transfer to the enviroment $E \rightarrow E'$, and therefore it can also make

CP1165, *Nuclear Structure and Dynamics '09*
edited by M. Milin, T. Nikšić, D. Vretenar, and S. Szilner
© 2009 American Institute of Physics 978-0-7354-0702-2/09/$25.00

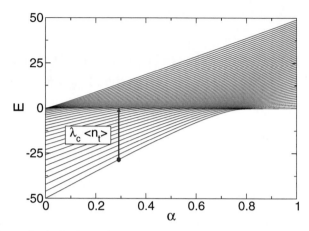

FIGURE 1. Energy levels for the environment Hamiltonian (3) with $L = 0$ and $N = 50$. The arrow shows the jump that the coupling with the central qubit produces in the environment.

the environment to reach the critical energy E_c of an ESQPT.

Following [2] we will consider our system composed by a spin $1/2$ particle coupled to a bosonic environment by the Hamiltonian H_{SE}:

$$H_{SE} = I_S \otimes H_E + |0\rangle \langle 0| \otimes H_{\lambda_0} + |1\rangle \langle 1| \otimes H_{\lambda_1}, \tag{1}$$

where $|0\rangle$ and $|1\rangle$ are the two components of the spin $1/2$ system, and λ_0, λ_1 the couplings of each component to the environment. The three terms H_E, H_{λ_0} and H_{λ_1} act on the Hilbert space of the environment.

With this kind of coupling, the environment evolves with an effective Hamiltonian depending on the state of the central spin $H_j = H_E + H_{\lambda_j}$, $j = 0, 1$. If the environment is initially in its ground state $|g_0\rangle$, the decoherence factor is determined, up to an irrelevant phase factor, by H_1, and its absolute value is equal to

$$|r(t)| = \left| \langle g_0| e^{-iH_1 t} |g_0\rangle \right|. \tag{2}$$

A value of $|r(t)|$ equal to zero implies that the qubit is no longer in a superposition of states $|0\rangle + |1\rangle$.

To be specific, let us consider a two level boson Hamiltonian, constructed out of scalar bosons, s, in the lowest level and bosons carrying an arbitrary angular momentum L in the upper level.

$$H_E = \alpha n_L - \frac{1-\alpha}{N} Q^\chi \cdot Q^\chi, \quad \text{with } Q_\mu^\chi = s^\dagger L + L^\dagger s + \chi [L^\dagger \times \tilde{L}]_\mu^{(L)}, \tag{3}$$

where n_L is the number of L bosons, N the total number of bosons and $\cdot$ stands for the scalar product.

This Hamiltonian has a second order QPT at $\alpha_c = 4/5$ for $\chi = 0$ [6], while experiences a first order phase transition for $\chi \neq 0$. We will focus in the case of $\chi = 0$. Using the coherent state formalism [6] it can be shown that for $\alpha > 4/5$ the environment

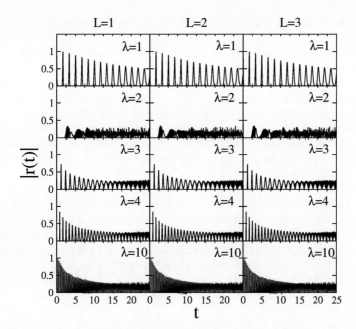

FIGURE 2. $|r(t)|$ for $\alpha = 0$, five different values of λ and $L = 1, 2, 3$. In all cases $N = 1000$.

is a condensate of s bosons corresponding to a symmetric phase. For $\alpha < 4/5$ the environment condensate mixes s and L bosons forming a non-symmetric phase.

Choosing $\lambda_0 = 0$ and $\lambda_1 = \lambda$ the coupling Hamiltonian reduces to a very simple form $H_{Coup} = \lambda n_L$, which results into the effective Hamiltonians for each component of the systems $H_0 = H_E$ and $H_1 = H_E(\alpha \rightarrow \alpha + \lambda)$. Therefore, the system-environment coupling parameter λ modifies the environment Hamiltonian. It is straightforward to show that H_1 goes through a second order QPT at $\lambda_* = 4 - 5\alpha$, for $\alpha < 4/5$, using [6]. Furthermore, a semiclassical calculation [5] shows that H_E also passes through an ESQPT at $E_c = 0$, if $\lambda < \lambda_*$. This phenomenon is illustrated in Fig. 1.

We start the evolution with the ground state of the environment $|g_0\rangle$. At $t = 0$ we switch on the interaction between the system and the environment, and let the system evolve under the complete Hamiltonian. By instantaneously switching on this interaction, the energy of the environment increases, and its state gets fragmented into a region with average energy equal to $E = \langle g_0 | H_1(\alpha) | g_0 \rangle$. Therefore, if $\langle g_0 | H_1(\alpha) | g_0 \rangle = 0$, the coupling with the central qubit induces the environment to *jump* into a region around the critical energy E_c. This is illustrated in Fig. 1. Starting from a state in the non-symmetric phase with $\alpha < \alpha_c$, the coupling with the qubit, $H_{\lambda_1} = \lambda n_L$ increases the energy of the environment up to the critical point E_c. Resorting to the coherent state approach [6], we can obtain a critical value of the coupling strength

$$\lambda_c(\alpha) = \frac{1}{2}(4 - 5\alpha), \quad \alpha < \frac{4}{5}. \tag{4}$$

In Fig. 2 we show the modulus of the decoherence factor $|r(t)|$ for $\alpha = 0$ and several values of λ and L, corresponding to the vibron ($L = 1$), the IBM ($L = 2$) and the octupole model ($L = 3$). First, we note that the presented behavior is independent on L. In four of the five cases of λ we can see a similar pattern, fast oscillations plus a smooth decaying envelope. The most striking feature of Fig. 2 is the panel corresponding to $\lambda = 2$, for which $|r(t)|$ quickly decays to zero and then randomly oscillates around a small value. We note that this particular case constitutes a singular point for both the shape of the envelope of $|r(t)|$ and the period of its oscillating part. Making use of Eq. (4) for $\alpha = 0$ we obtain precisely $\lambda_c = 2$, the value at which the coherence of the system is completely lost. Therefore, the existence of an ESQPT in the environment has a strong influence on the decoherence that it induces in the central system. We can summarize this result with the following conjecture:

If the system-environment coupling drives the environment to the critical energy E_c of a continuous ESQPT, the decoherence induced in the coupled qubit is maximal.

This conjecture has been checked for different values of $\alpha < \alpha_c = 4/5$ obtaining in all the cases that the rapid decay to zero of $|r(t)|$ always happens for $\lambda \approx \lambda_c$ (see Fig. 3 of reference [3]). It has been also checked how this magnitude behaves in the thermodinamical limit. The results displayed in Fig. 4 of reference [3] confirm that the presence of an ESQPT in the environment spectrum is clearly signaled by the qubit decoherence factor. Moreover, it can be defined an order parameter for the ESQPT related to $|r(t)|$.

To summarize, our main finding is that the decoherence is maximal when the system-environment coupling introduces in the environment the energy required to undergo a continuous ESQPT and that this results is independent on the value of L.

This work has been partially supported by the Spanish MEC (FEDER) under projects number FIS2006-12783-C03-01, FPA2007-63074 and FIS2008-04189, by Comunidad de Madrid and CSIC under project 200650M012, by Junta de Andalucía under projects FQM160, FQM318, P05-FQM437 and P07-FQM-02962 and by the Spanish Consolider-Ingenio 2010 Programme CPAN (CSD2007-00042). A.R. is supported by the Spanish program "Juan de la Cierva", and P. P-F., by a grant from the Spanish MEC.

REFERENCES

1. W. H. Zurek, *Rev. Mod. Phys.* **75**, 715 (2003); M. Nielsen and I. Chuang, *Quantum Computation and Quantum Information* (Cambridge University Press, Cambridge, UK, 2000).
2. H. T. Quan, Z. Song, X. F. Liu, P. Zanardi, and C. P. Sun, *Phys. Rev. Lett.* **96**, 140604 (2006); F. M. Cucchietti, S. Fernandez-Vidal, and J. P. Paz, *Phys. Rev.* **A75**, 032337 (2007); C. Cormick and J. P. Paz, *Phys Rev.* **A77**, 022317 (2008).
3. A. Relaño, J.M. Arias, J. Dukelsky, J.E. García-Ramos, and P. Pérez-Fernández, *Phys. Rev.* **A78**, 060102R (2008).
4. P. Cejnar, S. Heinze, and M. Macek, *Phys. Rev. Lett.* **99**, 100601 (2007).
5. M. A. Caprio, P. Cejnar, and F. Iachello, *Ann. Phys.* **323**, 1106 (2008).
6. J. Vidal, J. M. Arias, J. Dukelsky, J. E. García-Ramos, *Phys. Rev.* **C73**, 054305 (2006); J. M. Arias, J. Dukelsky, J. E. García-Ramos, and J. Vidal, *Phys. Rev.* **C75**, 014301 (2007).

Exact Analytic Study of Nuclear Shape Phase Transitions

G. Lévai

*Institute of Nuclear Research of the Hungarian Academy of Sciences (ATOMKI),
P. O. Box 51, H-4001 Debrecen, Hungary*

Abstract. The application of the sextic oscillator is proposed in the Bohr Hamiltonian to describe the phase transition between the spherical and γ-unstable shape phases. It is shown that exact results can be obtained for the energy eigenvalues and wave functions of the low-lying levels, as well as for electric quadrupole transition rates between them. The ^{134}Ba nucleus and the even Ru isotope chain are considered as examples. Possible generalizations of the model are also outlined.

Keywords: Nuclear shape phases, E(5) symmetry, sextic oscillator
PACS: 21.10.Re, 21.60.Ev, 03.65.Ge

INTRODUCTION

Certain regions of the nuclid chart are known to correspond to characteristic nuclear shapes, such as the spherical vibrator, the axially deformed rotor and the γ-unstable rotor. These systems are also known to be linked with the U(5), SU(3) and O(6) symmetries, respectively [1]. Within the geometric description these shapes correspond to the minima of the potential surface $V(\beta,\gamma)$ appearing in the Bohr Hamiltonian [2] that describes quadrupole collective excitations in terms of the intrinsic shape variables β, γ (in the full solution of this five-dimensional problem the Euler angles describing the spatial orientation of the nucleus also have to be considered, but these can be separated from the shape variables). The structure of the potential $V(\beta,\gamma)$ depends on various parameters characterizing the given nucleus, e.g. the proton and neutron number, etc. Changing these parameters gradually (e.g. by proceeding along an isotope chain) the potential shape and its minima change and may give rise to transition from one equilibrium shape (and symmetry) to another one. This procedure can be interpreted as a shape phase transition of some (typically first- or second-) order that goes through a critical point.

The first example for such a scenario has been proposed by Iachello [3], who considered the transition between the spherical (U(5)) and the γ-unstable (O(6)) shape phases by approximating the $V(\beta,\gamma)$ potential by the infinite square well in β. This flat potential mimics the situation in which a deformed minimum develops at $\beta > 0$ besides the spherical minimum at $\beta = 0$. This phase transition was found to be of second order and the E(5) symmetry was associated with it. This problem can be solved exactly in terms of Bessel functions, so parameter-free analytical results are available for the key spectroscopic properties (energy spectrum, electric quadrupole transition rates). Comparison of these quantities with the spectroscopic data of actual nuclei revealed that several nuclei (e.g. ^{134}Ba) are close to the E(5) critical point symmetry.

The transition between the spherical and axially deformed shape phases has also

CP1165, *Nuclear Structure and Dynamics '09*
edited by M. Milin, T. Nikšić, D. Vretenar, and S. Szilner
© 2009 American Institute of Physics 978-0-7354-0702-2/09/$25.00

been studied by Iachello [4]. In this case the γ shape variable also plays a role, and it was assumed that the $V(\beta,\gamma)$ potential can be separated to β- and a γ-dependent components. In order to solve the Bohr Hamiltonian further approximations have to be made, nevertheless, exact results can again be be obtained. In this case the phase transition was found to be of first order, and it was associated with the X(5) symmetry. Several nuclei have been proposed as candidates for X(5) symmetry. Further models making use of further approximations in the $V(\beta,\gamma)$ potential have been developed.

These results naturally emphasize the importance of exactly solvable Bohr Hamiltonians. Such solutions have been obtained for a number of potentials (see e.g. [5]). Some of these are the adaptations of the most well-known examples of solvable quantum mechanical potentials, e.g. the Kratzer and Davidson potentials as the generalizations to formally arbitrary angular momenta of the Coulomb and harmonic oscillator potentials. Having analytical solutions in hand allows not only the description of the collective excitations of certain nuclei near equilibrium shapes, but the analysis of transitions between shape phases also becomes possible if the potential is flexible enough.

Here we present the soultion of the Bohr Hamiltonian for the sextic oscillator in the β variable, which has rather flexible shape and thus can be applied to describe γ-independent potentials with one or two local minima [6]. This potential is ideal to describe phase transition between the spherical and the γ-unstable phases, but with certain modifications it might be considered also in more general $V(\beta,\gamma)$ potentials too.

THE SEXTIC OSCILLATOR AS A γ-INDEPENDENT POTENTIAL

The general Bohr Hamiltonian is written as [2]

$$H = -\frac{\hbar^2}{2B}\left(\frac{1}{\beta^4}\frac{\partial}{\partial\beta}\beta^4\frac{\partial}{\partial\beta} + \frac{1}{\beta^2\sin3\gamma}\frac{\partial}{\partial\gamma}\sin3\gamma\frac{\partial}{\partial\gamma} - \frac{1}{4\beta^2}\sum_k\frac{Q_k^2}{\sin^2(\gamma-\frac{2}{3}\pi k)}\right) + V(\beta,\gamma).$$
(1)

If the potential is assumed to be independent of the γ variable, i.e. $V(\beta,\gamma) = U(\beta)$, then the β-dependent part can be separated by the substitution $\Psi(\beta,\gamma,\theta_i) = \beta^{-2}\phi(\beta)\Phi(\gamma,\theta_i)$, leading to a form similar to the usual radial Schrödinger equation

$$-\frac{d^2\phi}{d\beta^2} + \left(\frac{(\tau+1)(\tau+2)}{\beta^2} + u(\beta)\right)\phi = \varepsilon\phi,$$
(2)

where $\varepsilon = \frac{2B}{\hbar^2}E$ and $u(\beta) = \frac{2B}{\hbar^2}U(\beta)$. In (2) τ originates from the angular equation and essentially plays the role of the angular momentum in five spatial dimensions.

In Ref. [6] the $u(\beta)$ potential was chosen ast the sextic oscillator

$$u(\beta) = (b^2 - 4ac)\beta^2 + 2ab\beta^4 + a^2\beta^6 + u_0,$$
(3)

which is a quasi-exactly solvable potential [7] meaning that only a subset of its lowest-energy solutions can be obtained exactly for certain combinations of the potential parameters. $u(\beta)$ in (3) depends on three parameters, a, b and c, but c is related to the τ parameter via $2c = (\tau + 2M + \frac{7}{2})$ in order to account for the "centrifugal" term. Here

TABLE 1. Spectroscopic results for the sextic oscillator, the infinite square well [3] and the β^4 potential [8] and their comparison with experiment.

	$\dfrac{E(4^+_{1,2})}{E(2^+_{1,1})}$	$\dfrac{E(0^+_{2,0})}{E(2^+_{1,1})}$	$\dfrac{E(6^+_{1,3})}{E(2^+_{1,1})}$	$\dfrac{B(E2;4^+_{1,2}\to2^+_{1,1})}{B(E2;2^+_{1,1}\to0^+_{1,0})}$	$\dfrac{B(E2;2^+_{2,0}\to2^+_{1,1})}{B(E2;2^+_{1,1}\to0^+_{1,0})}$	$\dfrac{B(E2;0^+_{1,3}\to2^+_{1,2})}{B(E2;2^+_{1,1}\to0^+_{1,0})}$
sextic osc.	2.39	3.68	3.70	1.70	1.03	2.12
E(5)	2.20	3.03	3.59	1.68	0.86	2.21
β^4	2.09	2.39	3.27	1.82	1.41	2.52
^{134}Ba (exp.)	2.31	3.57	3.65	1.56(18)	0.42(12)	

M is a non-negative integer number that sets the number $M+1$ of solutions that can be obtained exactly. In typical applications of the sextic oscillator in the Bohr Hamiltonian it is sufficient to consider $M=0$ and $M=1$: with this the most characteristic levels are included in the description. It has to be mentioned that since c depends on the $\tau+2M$ combination, $u(\beta)$ is slightly different for even and odd values of τ (c is $11/4$ and $13/4$ in the respective two cases). This minor difference shows up near the origin and can be kept small with the appropriate choice of the parameters. In practical applications we set the minima of the two potentials at the same value by using different values of u_0 for even and odd values of τ.

The solutions are written as

$$\phi_{n,\tau}(\beta) = N_{n,\tau}P_n(\beta^2)(\beta^2)^{\frac{c}{4}+1}\exp\left(-\frac{a}{4}\beta^4 - \frac{b}{2}\beta^2\right),\tag{4}$$

where $P_{n,\tau}(\beta^2)$ is a polynomial of the order $n=0$, 1 and the $N_{n,\tau}$ normalization constants can be expressed exactly in terms of parabolic cylinder functions. Note that normalizability requires $a\geq 0$ and that $a=0$ corresponds to the harmonic oscillator limit. The energy eigenvalues and the matrix elements necessary to calculate B(E2) values are also expressed exactly [6].

As can be seen from (3), the extrema of the potential depend on the sign of b and $b^2 - 4ac$. If $b^2 > 4ac$ and $b > 0$ hold then the potential has a minimum at $\beta = 0$ and it increases with β. When $b^2 < 4ac$ (irrespective of the sign of β) the minimum shifts to a finite value of β, while for $b^2 > 4ac$ and $b < 0$ (i.e. $b < -2\sqrt{ac}$) there will be two minima (one at $b=0$ and one at $\beta > 0$) separated by a local maximum.

In order to test the performance of the model the parameters a and b were set to values that result in an energy spectrum similar to that of the ^{134}Ba nucleus, the best candidate for the E(5) critical point symmetry. With the values $a=40000$ and $b=200$ the potential had a minimum with $\beta > 0$. The most important spectroscopic values are listed in Table 1. It is seen that the experimental data are reproduced rather well, but the results of the sextic oscillator are also close to the key values of the E(5) critical point symmetry.

As a further test of the sextic oscillator as a realistic model we fitted the low-lying energy spectrum of the even Ru isotopes with $A=98$ to 112 and analyzed the corresponding potentials. The values of a and b can be extracted exactly form the energy eigenvalue of the first 2^+ and the second 0^+ levels as $a = E_{1,2}(E_{2,0}-E_{1,2})/40$ and $b = E_{1,2}/2 - E_{2,0}/4$, where we used the notation $E_{n+1,\tau}$ [6]. The resulting parameters are displayed in Table 2. When (small) negative, i.e. unphysical value was obtained for a, we took the harmonic limit $a=0$. In these cases (A=98, 100 and 102) the energy

213

TABLE 2. Parameters a and b fitted to the ^{A}Ru spectra.

A	98	100	102	104	106	108	110	112
a	[0]	[0]	[0]	1496	4190	5154	14684	11563
b	347	318	283	216	143	114	-63	-36

spectrum was indeed rather close to a harmonic situation. The potential was rather flat for ^{104}Ru, which is in accordance with an expected phase transition and E(5) symmetry here [9]. The potential was still very flat for ^{106}Ru, but it had a finite minimum that became increasingly deeper and got further away from $\beta = 0$ as A increased to 112. These findings are in reasonable agreement with the expectations and confirm the applicability of the sextic potential in the Bohr Hamiltonian. It has to be mentioned though that in these calculations the same mass coefficient $\hbar^2/(2B)$ was used in (1), although this quantity is known to be scaled with the mass number as $\hbar^2/(2B) \sim A^{-5/3}$ [2]. Modifying the scale in this spirit would directly rescale the a and b parameters with $A^{10/3}$ and $A^{5/3}$, respectively.

Given the flexible shape of the sextic oscillator this potential might be useful in other regions too. Work is in progress to incorporate the dependence on the γ variable too in the formalism. For this techniques applied in other approaches might be useful. With the approximation applied in Ref. [4] to replace β^2 with its average value in the γ-dependent equation the situation close to the X(5) symmetry might be discussed. Another possibility is considering $V(\beta, \gamma) = V_1(\beta) + \mu/(\beta^2 \sin^2(3\gamma))$, i.e. a potential periodic in the γ variable [10]. Further approximations in the separation of the angular and shape variables as in Ref. [11] can also be considered.

ACKNOWLEDGMENTS

This work was supported by the OTKA grant No. T049646 (Hungary).

REFERENCES

1. F. Iachello and A. Arima, *The Interacting Boson Model*, Cambridge University Press, Cambridge (1987).
2. A. Bohr and B. Mottelson, *Nuclear structure Vol. II*, Benjamin, Reading, MA (1975).
3. F. Iachello, *Phys. Rev. Lett.* **85**, 3580 (2000).
4. F. Iachello, *Phys. Rev. Lett.* **87**, 052502 (2001).
5. L. Fortunato, *Eur. Phys. J.*, **A26**, 1 (2005).
6. G. Lévai and J. M. Arias, *Phys. Rev.* **C69**, 014304 (2004).
7. A. G. Ushveridze, *Quasi-exactly solvable models in quantum mechanics*, IOPP, Bristol (1994).
8. J. M. Arias, C. E. Alonso, A. Vitturi, J. E. Garcia-Ramos, J. Dukelsky and A. Frank, *Phys. Rev.* **C68**, 041302(R) (2004).
9. A. Frank, C. E. Alonso and J. M. Arias, *Phys. Rev.* **C65**, 014301 (2001).
10. S. De Baerdemacker, L. Fortunato, V. Hellemans and K. Heyde, *Nucl. Phys.* **A769**, 16 (2006).
11. L. Fortunato, S. De Baerdemacker, and K. Heyde, *Phys. Rev.* **C74**, 014310 (2006).

New Microscopic Derivation of the Interacting Boson Model and Its Applications to Exotic Nuclei

K. Nomura*, N. Shimizu* and T. Otsuka*,‡

*Department of Physics, University of Tokyo, Hongo, Bunkyo-ku, Tokyo 113-0033, Japan
†Center for Nuclear Study, University of Tokyo, Hongo, Bunkyo-ku, Tokyo 113-0033, Japan
**RIKEN Nishina center, Hirosawa, Wako-shi, 351-0198 Saitama, Japan
‡National Superconducting Cyclotron Laboratory, Michigan State University, East Lansing, MI

Abstract. We present a new scheme to determine the Hamiltonian of the Interacting Boson Model (IBM) microscopically, which starts from the mean-field model with Skyrme-type interactions. Surface deformations and the effects of the nuclear force are simulated in terms of bosons. By comparing Potential Energy Surface (PES) of the mean-field model with that of the IBM, the parameters of the IBM Hamiltonian can be obtained. By this method, quantum shape/phase transitions and critical-point symmetries can be reproduced. Systematic calculations on Xe and Ru isotopes shall be newly presented, which partly contain the predicted energy levels.

Keywords: Interacting Boson Model, Mean-field model, shape-phase transition
PACS: 21.10.Re, 21.60.Ev, 21.60.Fw, 21.60.Jz

INTRODUCTION

The Interacting Boson Model (IBM) [1] has been successful in phenomenological studies on surface deformations, among which the quadrupole collectivity is a prominent aspect, and their certain regularities. Besides, the IBM has its own microscopic foundation, starting from the nucleonic degrees of freedom. This has been done in terms of the nuclear shell model, and has been applied to mainly spherical shapes [2, 3, 4, 5]. For general cases, including strongly deformed configurations, however, the microscopic derivation of the IBM remains an open problem.

On the other hand, the mean-field (Skyrme) model has been successful in studying intrinsic-state properties [6], while it is not so easy to compute the levels and wave functions precisely, restoring the fundamental symmetries of the laboratory frame [7]. We have proposed a novel way to determine the parameters of the IBM Hamiltonian for general cases, starting from the mean-field theory [8], and now we shall apply it to several other situations of the quantum shape-phase transitions, *e.g.*, in Ru isotopes.

FRAMEWORK

Firstly, we carry out the constrained Skyrme Hartree-Fock plus BCS calculations in mesh representation [9] to obtain the fermionic PES's (denoted as HF PES), characterized by the quadrupole deformation parameters denoted as (β_f, γ_f). The constraint

CP1165, *Nuclear Structure and Dynamics '09*
edited by M. Milin, T. Nikšić, D. Vretenar, and S. Szilner
© 2009 American Institute of Physics 978-0-7354-0702-2/09/$25.00

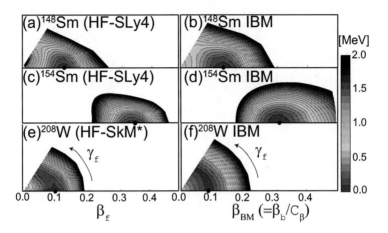

FIGURE 1. (Color online) Typical potential energy surfaces (PES's) of the HF (left) and the IBM (right) plotted up to 2 MeV from the minima. Contour spacing is 100 keV and minima can be identified by solid circles.

imposed here means the one with mass quadrupole moment.

In the present study, we take the consistent-Q formalism Hamiltonian [1] of the IBM-2 [2], $\hat{\mathcal{H}} = \varepsilon(\hat{n}_{d\pi} + \hat{n}_{dv}) + \kappa \hat{\mathcal{Q}}_\pi \cdot \hat{\mathcal{Q}}_v$. The bosonic PES (denoted as IBM PES) can be obtained in the coherent state formalism [10],

$$\langle \hat{\mathcal{H}} \rangle = \frac{\varepsilon(n_\pi + n_v)\beta_b^2}{1+\beta_b^2} + \frac{n_\pi n_v \kappa \beta_b^2}{(1+\beta_b^2)^2}\left[4 - 2\sqrt{\frac{2}{7}}(\chi_\pi + \chi_v)\beta_b \cos 3\gamma_b + \frac{2}{7}\chi_\pi \chi_v \beta_b^2\right], \quad (1)$$

where β_b and γ_b stand for the bosonic deformation variables being common to proton and neutron systems. We also assume $\beta_b = C_\beta \beta_f$ with C_β a coefficient and $\gamma_b = \gamma_f$. The HF PES is simulated by the IBM PES using five parameters, namely ε, κ, $\chi_{\pi,v}$ and C_β.

The IBM parameters are determined so that the IBM-PES reproduces the HF-PES up to 2 MeV from the energy minimum. For this comparison, an automated method can be applied using the wavelet transform technique, although we do not discuss in detail as it gives almost the same parameters as described below. The overall pattern of the HF PES reflects the effects of nuclear force and Pauli principle on corresponding deformed states. By reproducing the HF PES, the IBM one is expected to simulate such fundamental properties of fermions in a simple manner.

Examples of the PES's are shown in Fig. 1 for ^{148}Sm [near-spherical, or U(5) limit], ^{154}Sm [axial rotor, or SU(3) limit] and ^{208}W [γ-unstable]. Generally, the IBM PES reproduces the global pattern, including the position of the minimum, of the HF PES up to 2 MeV, but is somewhat flatter in β, which might be due to the limit of boson model space and cannot be improved any more.

Figure 2(a) shows an example of the derived IBM parameters for Sm isotopes as functions of the neutron number, N. While the parameters ε and κ change gradually, χ_v varies rather suddenly around $N=90$, reflecting the spherical to deformed transition. C_β changes smoothly with N, while χ_π is kept constant. These variations of the IBM

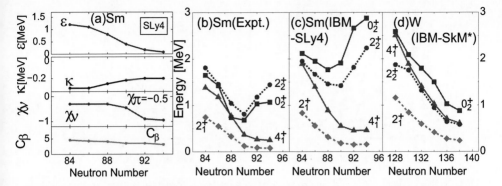

FIGURE 2. (Color online) (a)Evolution of parameters in Eq. (1) with the neutron number. $\chi_\pi = -0.5$ is kept constant. (b)Experimental and (c)calculated levels for Sm isotopes with the neutron number, studied with SLy4 force. (d) Calculated (IBM from SkM*) levels for W isotopes.

parameters produce levels consistent with experimental tendencies without adjustment to the data. Levels can be computed by diagonalizing the boson Hamiltonian [11].

NUMERICAL RESULTS

Figures 2(b) and 2(c) show the experimental and calculated low-lying levels of Sm isotopes, respectively. The spectra look like those of spherical vibrators around N=86. With the increase of N, levels come down consistently with the experiments particularly for yrast states. Around N=90, there seems to be the X(5) critical point [12], beyond which side-band levels go up in both experiment and calculation. Finally around N=92, the levels look like rotational band.

The present method can be applied to (experimentally) unknown territories. In Fig. 2(d), level evolution for W isotopes is shown. Around N=128, the levels look like those of spherical vibrators. As the magnitude of the deformation becomes larger with N, each level comes down, keeping the E(5)/O(6) pattern for N=130-138. Indeed, as reported in [8], level scheme for N=134 W nucleus is quite similar to ^{134}Ba, an example of the E(5) symmetry [13]. Such sustained E(5)/O(6) pattern seems to be a characteristic feature of exotic nuclei. A similar tendency can be found in exotic Os isotopes.

Systematic calculations have also been performed in a similar fashion for Xe and Ru isotopes for N=50-82 shell. Their calculated levels are compared with experiments in Fig. 3, where the Skyrme SkM* and SLy4 forces are used for Xe and Ru, respectively. The agreements of the calculated levels with experimental ones look basically fine, especially for yrast states. While the calculated results for both isotopic chains exhibit the transitions from spherical to weakly-deformed, or to γ-unstable states as the valence neutrons increase, result for Ru contains predicted levels for N >70 region, approaching the U(5)-like structure near the neutron N=82 shell closure.

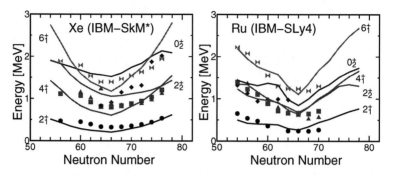

FIGURE 3. (Color online) Calculated spectra (solid lines) for Xe and Ru isotopes are compared with experimental data (dots) for *N*=54-78. Skyrme SkM* and SLy4 interactions are taken for Xe and Ru, respectively.

CONCLUSIONS

By using the recently proposed way to determine the IBM Hamiltonian, various situations of the shape-phase transitions can be described, including the systematics of levels for Ru and Xe isotopes. The mean-field model and the IBM may help each other, where the former has a role to provide us with the parameters and the latter calculates levels and wave functions precisely. What is more important is, unlike the existing IBM studies, the present method has quantitative predictive power on experimentally unknown nuclei, including those in unexplored territories of the nuclear chart.

ACKNOWLEDGMENTS

This work has been supported in part by Mitsubishi foundation, the JSPS core-to-core program (EFES) and JSPS Research Fellowship for Young Scientists.

REFERENCES

1. A. Arima and F Iachello, *Phys. Rev. Lett.* **35**, 1069 (1975) ; F. Iachello, and A. Arima, *The interacting boson model*, (Cambridge University Press, Cambridge, 1987).
2. T. Otsuka, A. Arima, F. Iachello, and I. Talmi, *Phys. Lett.* **76B**, 139 (1978) ; T. Otsuka, A. Arima, and F. Iachello, *Nucl. Phys.* **A309**, 1 (1978).
3. T. Mizusaki and T. Otsuka, *Prog. Theor. Phys.*, Suppl. **125**, 97 (1997).
4. M. Deleze, et al., *Nucl. Phys.* **A551**, 269 (1993).
5. K. Allaart, et al., *Nucl. Phys.* **A458**, 412 (1986).
6. T. H. R. Skyrme, *Nucl. Phys.* **9** 615 (1959) ; D. Vautherin and D. M. Brink, *Phys. Rev.* **C5** 626 (1972).
7. J. Dobaczewski, et al., *Phys. Rev.* **C76**, 054315 (2007).
8. K. Nomura, N. Shimizu, and T. Otsuka, *Phys. Rev. Lett.* **101**, 142501 (2008).
9. P. Bonche et al., *Comp. Phys. Comm.* **171**, 49 (2005).
10. A. E. L. Dieperink and O. Scholten, *Nucl. Phys.* **346**, 125 (1980) ; J. N. Ginocchio and M. Kirson, *Nucl. Phys.* **A350**, 31 (1980) ; A. Bohr and B. R. Mottelson, Phys. Scripta **22** 468 (1980).
11. T. Otsuka and N. Yoshida, *JAERI-M report* **85** (Japan At. Enr. Res. Inst., 1985).
12. F. Iachello, *Phys. Rev. Lett.* **87**, 052501 (2001).
13. F. Iachello, *Phys. Rev. Lett.* **85**, 3580 (2000).

Analysis of Nuclear Quantum Phase Transitions

Z. P. Li*, T. Nikšić†, D. Vretenar†, J. Meng*, G. A. Lalazissis** and P. Ring‡

*State Key Laboratory of Nuclear Physics and Technology, School of Physics, Peking University, Beijing 100871, China
†Physics Department, Faculty of Science, University of Zagreb, Croatia
**Department of Theoretical Physics, Aristotle University of Thessaloniki, GR-54124, Greece
‡Physik-Department der Technischen Universität München, D-85748 Garching, Germany

Abstract. A microscopic analysis, based on nuclear energy density functionals, is presented for shape phase transitions in Nd isotopes. Low-lying excitation spectra and transition probabilities are calculated starting from a five-dimensional Hamiltonian, with parameters determined by constrained relativistic mean-field calculations for triaxial shapes. The results reproduce available data, and show that there is an abrupt change of structure at $N = 90$, that corresponds to a first-order quantum phase transition between spherical and axially deformed shapes.

Keywords: Nuclear Quantum Phase Transitions, Nuclear Energy Density Functionals
PACS: 21.60.Jz, 21.60.Ev, 21.10.Re, 21.90.+f

Phase transitions in equilibrium shapes of atomic nuclei correspond to first- and second-order quantum phase transitions (QPT) between competing ground-state phases induced by variation of a non-thermal control parameter (number of nucleons) at zero temperature. Nuclear shape phase transitions have been the subject of numerous recent theoretical and experimental studies, and present a rapidly growing field of research [1]. Theoretical analyses have typically been based on phenomenological geometric models of nuclear shapes and potentials, or algebraic models of nuclear structure, but more recently several attempts have been made towards a fully microscopic description of shape QPT starting from nucleonic degrees of freedom [2, 3, 4].

In Refs. [3, 4] we have reported a microscopic study of nuclear QPT in the region $Z = 60, 62, 64$ with $N \approx 90$, based on constrained self-consistent relativistic mean-field (RMF) calculations of potential energy surfaces. While in Ref. [3] the generator coordinate method (GCM) was used to perform configuration mixing of angular-momentum and particle-number projected relativistic wave functions restricted to axial symmetry, in [4] collective excitation spectra and transition probabilities have been calculated starting from a five-dimensional Hamiltonian for quadrupole vibrational and rotational degrees of freedom, with parameters determined by constrained mean-field calculations for triaxial shapes, i.e. including both β and γ deformations [5]. The results reproduce available data, and show that there is an abrupt change of structure at $N = 90$ that can be approximately characterized by the X(5) analytic solution at the critical point of the first-order quantum phase transition between spherical and axially deformed shapes.

Fig. 1 (a) displays the self-consistent RMF plus BCS-pairing binding energy map of ^{150}Nd. The important feature of this energy surface is the extended flat prolate minimum in the interval $0.2 < \beta < 0.4$, which has been interpreted as a signature of possible phase transition. Starting from constrained self-consistent solutions, the parameters that deter-

CP1165, *Nuclear Structure and Dynamics '09*
edited by M. Milin, T. Nikšić, D. Vretenar, and S. Szilner
© 2009 American Institute of Physics 978-0-7354-0702-2/09/$25.00

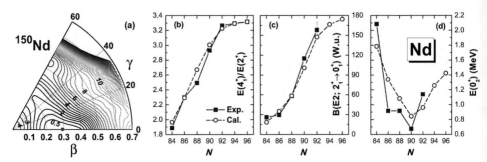

FIGURE 1. (Color online) (a) Self-consistent RMF+BCS binding energy map (in MeV) of ^{150}Nd in the $\beta - \gamma$ plane ($0° \leq \gamma \leq 60°$). All energies are normalized with respect to the absolute minimum. In (b), (c), and (d), evolution with neutron number in Nd isotopes of some characteristic collective observables: $R_{4/2} \equiv E(4_1^+)/E(2_1^+)$, B(E2; $2_1^+ \rightarrow 0_1^+$) (in Weisskopf units), and the first excited 0^+ state, respectively.

mine the collective Hamiltonian are calculated as functions of the deformations β and γ. The diagonalization of the resulting Hamiltonian yields the excitation energies and collective wave functions, that are used to calculate observables. In Fig. 1 (b), (c), and (d) we illustrate the evolution with neutron number in Nd isotopes of some characteristic collective observables: $R_{4/2} \equiv E(4_1^+)/E(2_1^+)$, B(E2; $2_1^+ \rightarrow 0_1^+$) (in Weisskopf units), and the first excited 0^+ state, respectively. Values calculated with the PC-F1 energy density functional are shown in comparison to data [6]. The calculation reproduces in detail the rapid increase of $R_{4/2}$ and B(E2; $2_1^+ \rightarrow 0_1^+$) from the spherical nucleus ^{144}Nd to the well-deformed rotors $^{152-156}$Nd. For the $N = 90$ isotope the calculated $R_{4/2} = 3$ and experimental $R_{4/2} = 2.93$ values are very close to the characteristic, parameter-free, $X(5)$ model prediction $R_{4/2} = 2.91$ [1]. On also notes that the empirical isotopic dependence of the first excited 0^+ state is reproduced and, in particular, the prediction that this state has the lowest excitation energy precisely at $N = 90$, in agreement with data.

ACKNOWLEDGMENTS

This work is partly supported by the Chinese-Croatian project "Nuclear structure far from stability", the NSFC under Grant No. 10775004, the Major State 973 Program 2007CB815000, and MZOS - project 1191005-1010.

REFERENCES

1. R. F. Casten, and E. A. McCutchan, *J. Phys. G: Nucl. Part. Phys.* **34**, R285–R320 (2007).
2. J. Meng, W. Zhang, S. G. Zhou, H. Toki, and L. S. Geng, *Eur. Phys. J.* **A25**, 23-27 (2005).
3. T. Nikšić, D. Vretenar, G. A. Lalazissis, and P. Ring, Phys. Rev. Lett. 99, 092502 (2007).
4. Z. P. Li, T. Nikšić, D. Vretenar, J. Meng, G. A. Lalazissis, and P. Ring, *Phys. Rev.* **C79**, 054301 (2009).
5. T. Nikšić, Z. P. Li, D. Vretenar, L. Próchniak, J. Meng, and P. Ring, *Phys. Rev.* **C79**, 034303 (2009).
6. NNDC National Nuclear Data Center, Brookhaven National Laboratory, http://www.nndc.bnl.gov/

Dynamic Chirality in Nuclei

D. Tonev*,†, G. de Angelis†, S. Brant**, P. Petkov* and A. Ventura‡

*Institute for Nuclear Research and Nuclear Energy, BAS, 1784 Sofia, Bulgaria
†INFN, Laboratori Nazionali di Legnaro, 35020 Legnaro, Italy
**Department of Physics, Faculty of Science, University of Zagreb, 10000 Zagreb, Croatia
‡ENEA, 40129 Bologna and INFN, Sezione di Bologna, Italy

Abstract. The possible chiral interpretation of twin bands in odd-odd nuclei was investigated in the Interacting Boson Fermion-Fermion Model. The analysis of the wave functions has shown that the possibility for angular momenta of the valence proton, neutron and core to find themselves in the favorable, almost orthogonal geometry is present, but not dominant. Such behaviour is found to be similar in nuclei where both the level energies and the electromagnetic decay properties display the chiral pattern, as well as in those where only the energies of the corresponding levels in the twin bands are close together. The difference in the structure of the two types of chiral candidates nuclei can be attributed to different β and γ fluctuations, induced by the exchange boson-fermion interaction of the Interacting Boson Fermion-Fermion Model. In both cases the chirality is weak and dynamic. The existence of doublets of bands in ^{134}Pr can be attributed to dynamic chirality dominated by shape fluctuations.

Keywords: Dynamic chirality, Interacting Boson Fermion-Fermion Model, Transition probabilities
PACS: 21.10.Re, 11.30.Rd, 21.60.Fw, 23.20.Lv

INTRODUCTION

In 1997 Frauendorf and Meng proposed chirality as a novel feature of rotating nuclei [1]. A spontaneous breaking of the chiral symmetry can take place for configurations where the angular momenta of the valence protons, valence neutrons and the core are mutually perpendicular. The non zero components of the total angular momentum on all the three axes can form either a left-handed or a right-handed set and therefore, the system manifests chirality [2]. Since the chiral symmetry is dichotomic, its spontaneous breaking by the axial angular momentum vector leads to doublets of closely lying rotational bands of the same parity [1, 2, 3]. Pairs of bands possibly due to the breaking of the chiral symmetry have been found in the A~105, A~130 and A~190 mass regions, the most typical examples being odd-odd nuclei in the A~130 mass region. There the yrast and side bands are built on the $\pi h_{11/2}$ particle-like $\otimes$ $\nu h_{11/2}$ hole-like configuration. In the present work we shall analyze this case, but the conclusions can be applied equally well to the other two mass regions.

Due to the underlying symmetry, the pair of chiral bands should exhibit systematic properties. The yrast and the side bands should be nearly degenerate. In the angular momentum region where chirality sets in the $B(E2)$ values of the electromagnetic transitions de-exciting analogue states of the chiral twin bands should be almost equal. Correspondingly the $B(M1)$ values should exhibit odd-even staggering, being for the $\pi h_{11/2} \nu h_{11/2}^{-1}$ configuration much bigger for transitions de-exciting states with odd spins

CP1165, Nuclear Structure and Dynamics '09
edited by M. Milin, T. Nikšić, D. Vretenar, and S. Szilner
© 2009 American Institute of Physics 978-0-7354-0702-2/09/$25.00

than for transitions de-exciting states with even spins. The $B(M1)$ values for $\Delta I = 1$ transitions connecting the side to the yrast band should have the odd-even staggering out of phase with respect to the $B(M1)$ staggering for transitions de-exciting states in the yrast and the side bands, i.e. for the $\pi h_{11/2}\, \nu h_{11/2}^{-1}$ configuration $B(M1)$ values for transitions de-exciting states with even spins have to be much bigger then for the ones de-exciting states with odd spins. The last condition means that the $B(M1)_{In}/B(M1)_{Out}$ staggering in the side band should be in phase with the $B(M1)$ staggering in the yrast band.

To investigate the setting up of chirality in a certain nucleus, it is crucial to determine the $B(E2)$ and $B(M1)$ values, and the works [4, 5] demonstrate that the $B(E2)$ and $B(M1)$ pattern is not a unique fingerprint of chirality. In ^{134}Pr [6, 7] the $B(M1)$ values in both partner bands behave similarly. In contrast, the intraband $B(E2)$ strengths within the two bands differ. The $B(M1)$ staggering in both bands and the $B(M1)_{In}/B(M1)_{Out}$ staggering are not observed. In the cases of ^{128}Cs [8] and ^{135}Nd [9] the electromagnetic decay properties display the expected chiral pattern. In the present article, for shortness, the structure in which a pair of twin bands is close in excitation energy, but the electromagnetic decay properties do not show the chiral pattern, will be denoted as *case A*. The structure where the pair of twin bands is close in excitation energy and the electromagnetic decay properties display the chiral pattern, will be denoted as *case B*. Odd-odd nuclei in the A~130 mass region can be classified as case A or case B nuclei. In all these nuclei the cores are γ-soft, their odd-proton odd-mass neighbours have also a similar structure and their odd-neutron odd-mass neighbours have a similar structure, too.

IBFFM CALCULATIONS AND DISCUSSION

In the case of ^{134}Pr the large fluctuations of the deformation parameters β and γ around the triaxial equilibrium shape enhance the content of achiral configurations in the wavefunctions [4, 6, 7]. In such a context the term dynamical chirality introduced in Ref. [7], refers to the possibility that the angular momenta of the proton, neutron and core in the odd-odd nucleus find themselves in the favorable geometry, as if they would in the equivalent triaxially deformed rotor. The present calculation will show that such possibility is present, but it is far from being dominant. The condition for the appearance of twin bands with wave functions realistic enough to reproduce the electromagnetic decay of the bands, is that the core is a γ-unstable rotor whose effective γ is in the range of triaxial values. The IBFFM calculation with $\Theta_3 = 0.022$ MeV, $\Gamma_0^\nu = 0.75$ MeV, $\Lambda_0^\nu = 1.6$ MeV, $A_0^\nu = 0.1$ MeV, $V_T = -22.0$ MeV and all other parameters as in Refs. [6, 7], gives the representative structure of the case A (left panels in Fig. 1). With the residual proton-neutron interaction the staggering of the signature $S(I)$ for yrast states with medium and high spin is in agreement with the experimental data in the A~130 mass region. It exhibits a weak staggering, being bigger for states with odd, than for those with even spins. The wave functions are not sizeably changed in respect to Refs. [6, 7]. Consequently, the $B(E2)$ and $B(M1)$ values (third row panels in Fig. 1), calculated with the same effective charges and gyromagnetic ratios as in Refs. [6, 7], are very close to the values obtained in those calculations.

The result of the IBFFM calculation predicts the structure typical for the case A: twin

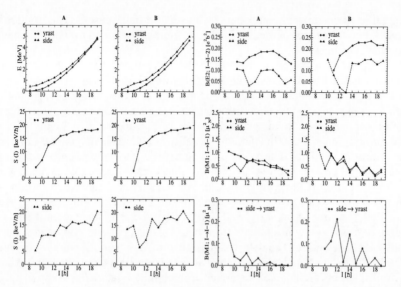

FIGURE 1. (Color online) Twin bands calculated in the IBFFM for the case A (first and third column panels) and the case B (second and fourth column panels). In the first two upper panels the excitation energies of levels in the yrast and side band are shown. In the first two middle panels, the signature $S(I)$ in the yrast band is displayed. In the first two panels on bottom the signature $S(I)$ in the side band is presented. $B(E2)$ and $B(M1)$ values calculated in the IBFFM for the case A (third column panels) and the case B (fourth column panels). In the upper panels $B(E2)$ values for transitions in the yrast and side band are shown. In the middle panels $B(M1)$ values for transitions in the yrast and side band are shown. In the panels on the bottom $B(M1)$ values for $\Delta I = 1$ transitions from side to yrast band are shown.

bands, correct signature, different $B(E2)$ values in the two bands, absence of $B(M1)$ staggering in both bands and a very weak $B(M1)$ staggering for $\Delta I = 1$ transitions from the side to the yrast band. This structure was attributed to a weak dynamic (fluctuation dominated) chirality in Ref. [7]. The same boson core allows to describe nuclei classified as case B provided one modifies the boson-fermion coupling parameters. The spin of the band head of the yrast band depends on the proton and neutron number of the nucleus through the occupation probabilities of fermion configurations. The total boson-fermion IBFFM interaction strength for each matrix element is a product of the interaction strength, the fermion configuration term and the BCS term that includes fermion occupation probabilities. Therefore, the spin value of the band head of the yrast band is a result of the interplay of many factors. The structure of intermediate and high spins in the yrast and side band is, on the other hand, mainly determined by the interaction strengths. Providing Γ_0, Λ_0 and A_0 for protons and neutrons from the calculations of neighbouring odd-mass nuclei, the structure of intermediate and high spin states in the yrast and side band can be investigated.

The β and γ distributions for cases A and B are sizeably different. The two bands have significantly more similar deformations in the case B than in the case A. The yrast bands in both cases are similar, with fluctuations on higher spins being somewhat smaller in the case B. In addition, in the case B components with γ closer to $\gamma = 30^o$ are more pronounced. The side band at medium and higher spins has far less shape fluctuations and a significant decrease of components with small β in the case B. The similarities, in the case B, between states of the two bands are more pronounced for odd spin members than for the even ones.

The distribution of the core structure in the wave functions reveals the dynamical mechanism. In the case B the weaker, but still strong, boson-fermion interactions (particularly the exchange interaction) for the neutron (hole-like) quasiparticle are not strong enough to admix big components from higher lying core bands. The ground and the $\gamma-$band components are dominant in the states of the side band. All chiral signatures are present, but large shape fluctuations sizeably reduce their magnitude in respect to the full chiral predictions. The $\gamma-$softness and shape fluctuations prevent the angular momenta of the proton, neutron and core to create chiral favorable geometry on the scale they would do if the core is triaxial. For nuclei where the boson-fermion interaction is stronger, other higher lying core structures admix into the states of the side (and partially into the yrast, too) band, increasing the shape fluctuations. This allows the admixture of more contributions from near axial shapes and consequently washes out the $B(M1)$ staggering. The only visible signature of dynamical chirality remains the vicinity of the excitation energies of the two bands. In both cases the left-handed and right-handed sectors are not well separated. The chirality materializes only in a dynamical way, as a slow anharmonic vibration. This finding can be extended equally well to the A∼105 and A∼190 mass regions.

The present IBFFM calculation shows that the exchange interaction, i.e. the antisymmetrization of odd fermions with the fermion structure of the bosons, has the dominant role in the formation of different types of chiral pattern. The coupling of the odd proton does not show any effect of such antisymmetrization because the occupation probability $v^2(\pi h_{11/2}) = 0.08$ is very low. For neutrons $v^2(\nu h_{11/2}) = 0.62$ is of the order where the full effect of the Pauli principle takes place. The physical foundation of chirality could therefore be beyond geometry and shape fluctuations, in the microscopic structure of twin bands.

ACKNOWLEDGMENTS

D.T. express his gratitude to Ivanka Necheva for her outstanding support. D.T. and P.P are indebted to the French Egide Organism and the French Ministère des Affaires Etrangères et Européennes under ECO-NET contract number 23331RH.

REFERENCES

1. S. Frauendorf and J. Meng, *Nucl. Phys.* **A617**, 131 (1997).
2. V.I. Dimitrov, S. Frauendorf, and F. Dönau, *Phys. Rev. Lett.* **84**, 5732 (2000).
3. K. Starosta et al., *Nucl. Phys.* **A682**, 375c (2001).
4. S. Brant, D. Tonev, G. de Angelis and A. Ventura, *Phys. Rev.* **C78**, 034301 (2008).
5. B. Qi, S.Q. Zhang, S.Y. Wang, J.M. Yao, and J. Meng, *Phys. Rev.* **C79**, 041302(R) (2009).
6. D. Tonev, et al., *Phys. Rev. Lett.* **96**, 052501 (2006).
7. D. Tonev, et al., *Phys. Rev.* **C76**, 044313 (2007).
8. E. Grodner, et al., *Phys. Rev. Lett.* **97**, 172501 (2006).
9. S. Mukhopadhyay, et al., *Phys. Rev. Lett.* **99**, 172501 (2007).

Isovector Quadrupole Excitations in the Valence Shell of Vibrational Nuclei

N. Pietralla*, T. Ahn†,*, M. Carpenter**, L. Coquard*, R.V.F. Janssens**, K. Heyde‡, K. Lister**, J. Leske*, G. Rainovski§ and S. Zhu**

*Institut für Kernphysik, Technische Universität Darmstadt, 64289 Darmstadt, Germany
†Wright Nuclear Structure Laboratory, Yale University, New Haven, CT 06250, USA
**Argonne National Laboratory, 9700 South Cass Avenue, Argonne, Illinois 60439, USA
‡Department of Subatomic and Radiation Physics, Proeftuinstraat,86 B-9000 Ghent, Belgium
§Faculty of Physics, St.Kliment Ohridski University of Sofia, 1164 Sofia, Bulgaria

Abstract. Coulomb excitation experiments in inverse kinematics of stable ions impinging on a carbon target at energies of about 85 % of the respective Coulomb barriers have been performed in the A∼130 region. Proton-neutron mixed-symmetry one-quadrupole phonon state $2^+_{1,ms}$ has been identified in several nuclides from the large $B(M1;2^+_{1,ms} \to 2^+_1)$ value as compared to typical $M1$ strengths between low-energy 2^+ states. As an example, we focuss on the evolution of the $2^+_{1,ms}$ state in the N=80 isotonic chain. The local proton-neutron quadrupole-quadrupole interaction strength is determined from these data within a schematic two-state mixing model.

Keywords: Coulomb excitation, IBM, Mixed Symmetry States, two-state mixing model
PACS: 20.21.Re

INTRODUCTION

Proton-neutron (pn) mixed-symmetry states (MSSs) are important sources of information on the effective proton-neutron interaction in collective nuclei. Their excitation energies are directly related to the proton-neutron interaction in the nuclear valence shell. This fact is obvious in the interacting boson model where the excitation energies of MSSs determine the strength of the Majorana interaction to which pn symmetric states at the yrast line are insensitive [1, 2]. Recent data on the N=80 isotones from our COULEX program at Argonne National Laboratory (ANL) quantify this statement [3] . Investigation of the proton-neutron interaction in the valence shell is an important subject of contemporary nuclear structure physics. Its evolution with neutron number represents a key-issue for future studies with intense beams of neutron-rich radioactive nuclides.

Vibrational nuclei exhibit a one-quadrupole phonon excitation as the lowest-lying state of mixed pn symmetry, i.e the $2^+_{1,ms}$ state. Its close relation to the 2^+_1 state is evident in the Q-phonon scheme [4], where the wave functions of the one-quadrupole phonon excitations are well approximated by the expressions

$$|2^+_1\rangle = [Q_\pi + Q_\nu] |0^+_1\rangle \tag{1}$$

$$|2^+_{1,ms}\rangle = N \left[\frac{Q_\pi}{N_\pi} - \frac{Q_\nu}{N_\nu} \right] |0^+_1\rangle \tag{2}$$

CP1165, *Nuclear Structure and Dynamics '09*
edited by M. Milin, T. Nikšić, D. Vretenar, and S. Szilner

where $Q_{\pi,\nu}$ $(N_{\pi,\nu})$ denote the proton and neutron quadrupole operators (boson numbers), $N=N_\pi + N_\nu$ and $|0_1^+\rangle$ is the ground state of a collective even-even nucleus. Despite its fundamental role in nuclear structure, the $2_{1,ms}^+$ state has only recently been studied systematically, e.g., [6–10]. The dominant fragments of the one-phonon $2_{1,ms}^+$ state are observed at about 2 MeV excitation energy. Due to their isovector character, MSSs decay rapidly by dipole transitions and are very short lived, typically a few tens of femtoseconds. Large $M1$ matrix elements of ≈ 1 μ_N are in fact the unique signatures for MSSs and, thus lifetime information is needed for making safe MS assignments. An overview of the mixed symmetry states in vibrational nuclei safely assigned from large absolute $M1$ strengths has been carried out in [5]. Coulomb excitation is an ideal method for this task. We have recently begun a research program on the $2_{1,ms}^+$ states at ANL with the nucleus ^{138}Ce as a case study. Crucial influence of sub-shell closures on mixed-symmetry structures was first observed [10, 11], which sensitively tests the effective proton-neutron interaction in microscopic valence shell models [12]. The one-phonon $2_{1,ms}^+$ state of ^{136}Ce has been identified from similar Coulomb excitation experiments at Gammasphere. Further experiments were performed on ^{124,126,128,130,132,134}Xe, ^{148,154}Sm, ^{94}Mo and ^{96}Ru. As an example, the results concerning the $N = 80$ isotonic chain will be presented below more in details.

RECENT EXPERIMENTAL RESULTS

The experiments have been performed at Argonne National Laboratory. The superconducting ATLAS accelerator provided the ion beams with energies corresponding to $\sim$ 85 % of the Coulomb barrier for a reaction on ^{12}C nuclei. The beam intensity amounted typically was $\sim$ 1pnA and was impinging on a stationnary ^{12}C target of thickness 1 mg/cm^2. Light target ions were chosen in order to favor the one-step Coulomb excitation process over multi-step processes. The γ-rays emitted by Coulomb-excited states of the beam nuclei were detected in the Gammasphere array which consisted of $\sim$ 100 high purity Compton supressed Germanium detectors arranged in 16 rings [13, 14]. An event was defined by a γ-ray of multiplicity 1 or higher. Two corrections had to be done in order to get the total single spectra displayed on the left side of Fig.1, namely the Doppler correction (recoiling velocity $\sim$ 6%) and the background subtraction (difference between the "in-beam" spectrum and the "off-beam" spectrum scaled to eliminate the 1461-keV ^{40}K line).

By calculating the Coulomb excitation cross sections for each excited state with the code CLX and fitting them to our experimental data (normalized to the 2_1^+ state), we deduced the $E2$ matrix elements corresponding to each transitions. The multipole mixing ratios of the $2^+ \rightarrow 2_1^+$ transitions were obtained from the γ-ray angular distribution if sufficient statistics have been obtained. The large B(M1) value, signature of the MSS, is then easily derived, as described in, e.g [3, 10, 15, 16].

In the stable $N = 80$ isotones, it is observed that the 2_1^+ state decreases in energy as one goes to higher proton boson number N_π while, in contrast, the $2_{1,ms}^+$ state increases in energy (see Fig. 1, right-hand side). Thus, the separation between the two levels becomes larger as a function of the product of valence particle pairs $N_\pi N_\nu$.

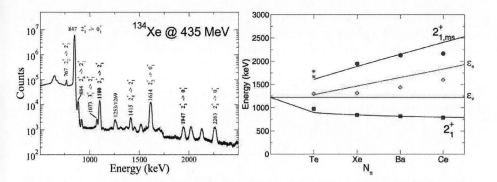

FIGURE 1. (Color online) Left: background subtracted and Doppler corrected singles γ-ray spectrum summed over all Ge detectors after the Coulomb excitation on ^{12}C targets for ^{134}Xe. On the right: fit of the energies of the $2^+_{1,\text{ms}}$ and 2^+_1 levels in the $N = 80$ isotones shown as filled circles and squares, respectively. The lines labeled ε_π and ε_ν represent the unperturbed energy of the proton and neutron state, respectively. The experimental energies of the 2^+_1 states in the corresponding $N = 82$ isotones are given as diamonds. Tentative 2^+ states in ^{132}Te are shown as asterisks.

According to the two-state mixing scheme outlined in Ref. [17], a fit was performed to the energy splitting of the observed 2^+_1 and $2^+_{1,\text{ms}}$ states. This scheme assumes that the observed 2^+_1 and $2^+_{1,\text{ms}}$ states arise from a mixing of unperturbed 2^+ proton (with energy ε_π) and neutron (with energy ε_ν) excitations. In this scheme [17], the interaction between the two states results from the proton-neutron quadrupole-quadrupole interaction and, thus, it can be parametrized as $V_{pn}(N_\pi, N_\nu) = \beta \sqrt{N_\pi N_\nu}$ in leading order.

The 2^+_1 states in the neighboring semi-magic $N = 82$ isotones increase in energy almost linearly with valence proton number, as can be seen in Fig. 1. Consequently, the energy of the proton state for $N = 80$ isotones was linearly parametrized by the expression $\varepsilon_\pi = a + b(N_\pi - 1)$ in a leading order approximation in N_π, where a was chosen to be equal to the energy of the 2^+_1 state in the neutron-closed shell nucleus ^{134}Te. Likewise, the value for ε_ν is taken to be constant for the $N = 80$ isotones, and is set equal to the energy of the 2^+_1 state of ^{130}Sn. The matrix to be diagonalized then becomes:

$$H = \begin{bmatrix} \varepsilon_\pi & V_{pn} \\ V_{pn} & \varepsilon_\nu \end{bmatrix} \Longrightarrow H_{Diag} = \begin{bmatrix} E(2^+_1) & 0 \\ 0 & E(2^+_{1,\text{ms}}) \end{bmatrix}$$

with eigenvalues:

$E(2^+_1) = \frac{\varepsilon_\pi + \varepsilon_\nu}{2} - \sqrt{\frac{(\varepsilon_\pi - \varepsilon_\nu)^2}{4} + \beta^2 N_\pi N_\nu}$ and $E(2^+_{1,\text{ms}}) = \frac{\varepsilon_\pi + \varepsilon_\nu}{2} + \sqrt{\frac{(\varepsilon_\pi - \varepsilon_\nu)^2}{4} + \beta^2 N_\pi N_\nu}$

as shown on the right of Fig.1 by thick curves. The parameters b and β are then fitted to the experimental data. The values for the parameters used were $a = \varepsilon_\pi(N_\pi = 1) = E_{2^+_1}(^{134}_{52}\text{Te}_{82}) = 1279$ keV, $\varepsilon_\nu(N_\nu = 1) = E_{2^+_1}(^{130}_{50}\text{Sn}_{80}) = 1221$ keV, $b = 0.23(4)$ MeV, and $\beta = 0.35(1)$ MeV.

A fit to the excitation energy of the 2^+_1 states of even-even Te, Xe, Ba, Ce nuclei for neutron numbers $60 \leq N \leq 80$ in Ref. [17], gave rise to a value of $\beta = 0.365$ MeV, a value quite close to the value derived here locally using nuclei with $N = 80$ only, considering now the excitation energies of both, the 2_1^+ and $2_{1,\text{ms}}^+$ states. Making use of the fitted value of $\beta = 0.35(1)$ MeV as derived in the present study and using the expression of β as derived from Eq. (3.4) of Ref. [17], a rather precise value of the local strength κ of the proton-neutron quadrupole-quadrupole residual interaction can be derived. Considering the limit $j \to \infty$ ($\kappa = -\frac{5}{12}\beta$), which is a good approximation for the large j values 7/2, 11/2 that are relevant in this mass region, a value of $\kappa = 0.15(1)$ MeV results, (see Fig. 2 and Table II in Ref. [18]).

CONCLUSION

The Coulomb excitation method in inverse kinematics is a powerful tool to study the lowest lying mixed-symmetry state $2_{1,\text{ms}}^+$. This technique, part of a research program at ANL, has been applied to explore the evolution of the MSSs in the A=130 mass region in order to better understand the pn valence shell interactions in vibrational nuclei. As an example, the strength of the quadrupole-quadrupole proton-neutron residual interaction in the $N = 80$ even-even stable isotones has been derived and results in a value of $\kappa = 0.15(1)$ MeV.

We would like to thank the staff at ANL for their support during the experiments and A. Poves, V. Werner, and F. Iachello for discussions. This work was partially supported by the U.S. Department of Energy, Office of Nuclear Physics, under Contract No. DE-AC02-06CH11357 and by the DFG under grant No. Pi 393/2-1 and under grant No. SFB 634 and by the Helmholtz International Center for FAIR. G. R acknowledges the support from the Bulgarian NSF under contract DO 02-219. K. H thanks the financial support from the "FWO Vlaanderen", the University of Ghent and IUAP (P5/07 and P6/23).

REFERENCES

1. F. Iachello and A. Arima, *The interacting boson model* (Cambridge University Press, 1987).
2. P. van Isacker, K. Heyde, J. Jolie, A. Sevrin, *Ann. Phys. (N.Y.)* **171**, 253 (1986).
3. T. Ahn et al., submitted to *Phys. Lett.* **B**.
4. T. Otsuka and K-H Kim, *Phys. Rev.* C **50**, 1768(R) (1994).
5. N. Pietralla, P. von Brentano, A.F Lisetskiy, *Prog. Part. Nucl. Phys.* **60**, 225 (2008).
6. N. Pietralla et al., *Phys. Rev. Lett.* **83**, 1303 (1999); *Phys. Rev. Lett.* **84**, 3775 (2000).
7. N. Pietralla et al., *Phys. Rev.* **C64**, 031301(R) (2001).
8. N. Pietralla and K. Starosta, *Nucl. Phys. News* **Vol. 13** No.2, 15 (2003).
9. N. Pietralla et al., Phys. Rev. **C68**, 031305(R) (2003).
10. G. Rainovski, N. Pietralla et al., Phys. Rev. Lett. **96**, 122501 (2006).
11. N. Pietralla et al., proceedings CGS12.
12. N. Lo Iudice et al.,*Phys. Rev.* **C77**, 044310 (2008).
13. I. Lee, *Nucl. Phys.* **A520**, 641 (1990).
14. P. Nolan, F. Beck, and D. Fossan, *Annu. Rev. Nucl. Part. Sci.* **45**, 561 (1994).
15. L. Coquard et al., *Proceedings Capture Gamma-Ray Spectroscopy 13*, 140 (2008).
16. L. Coquard et al., in preparation.
17. K. Heyde, J. Sau, *Phys. Rev.* **C33**, 1050 (1986).
18. R. Fossion et al., *Phys. Rev.* **C65** 044309 (2002).

Nuclear Spectra from Skyrmions

N.S. Manton

DAMTP, University of Cambridge, Wilberforce Road, Cambridge CB3 0WA, U.K.

Abstract. The structures of Skyrmions, especially for baryon numbers $4, 8$ and 12, are reviewed. The quantized Skyrmion states are compared with nuclear spectra.

Keywords: Skyrmions, Quantization, Nuclear Spectra
PACS: 12.39.Dc, 21.10.Hw, 21.60.Ev

Skyrmions and their quantization

At this conference there were several talks on the Skyrme force, a simple way to parametrize the interaction of a pair of nucleons inside a large nucleus. This talk was about Skyrme's other great idea – the Skyrmion [12]. Skyrme believed that the three-component pion field plays a dominant role in the nuclear medium, and found a way to replace the point-like nucleons by topological twists in the pion field. These are now called Skyrmions. The existence of Skyrmions is natural if the pion field is described by a nonlinear sigma model, that is, by four fields $\sigma, \pi_1, \pi_2, \pi_3$ subject to the constraint $\sigma^2 + \pi_1^2 + \pi_2^2 + \pi_3^2 = 1$, as the field then takes its values in a three-sphere, and field configurations in physical three-dimensional space (with boundary condition $\sigma = 1$) are characterized topologically by an integer B, which is identified as the baryon number. Skyrme proposed a relativistic Lagrangian and field equation which support a stable finite energy Skyrmion of baryon number $B = 1$. The Lagrangian has just three parameters: a mass scale, length scale, and dimensionless pion mass parameter. This particular Lagrangian must be regarded as a phenomenological model, but many variants have been considered, some of which include further meson fields. The Skyrmion solutions in these variant models are qualitatively very similar.

The Skyrmion is a localized, smooth classical solution, with a spherically symmetric energy density. The pion field varies from point to point, so the solution spontaneously breaks translation, rotation and isospin symmetry. These symmetries are restored by introducing collective coordinates. There is an effective, purely kinetic Lagrangian for a Skyrmion slowly moving and rotating in space and isospace. This can be quantized, rather as for a rigid body, and the resulting quantum states can be identified with the physical states of a nucleon, or delta resonance [1].

The precise quantization rules are subtly connected with the topology of the (infinite-dimensional) space of all Skyrme field configurations. Ultimately, because QCD has gauge group $SU(3)$, and baryons are made of three quarks, the quantized Skyrmion must have half-integer spin. The symmetry of the Skyrmion solution then correlates spin half with isospin half, thereby reproducing the required states of a nucleon.

Skyrme's intention was to understand nuclei in terms of Skyrmions, and for this one

CP1165, *Nuclear Structure and Dynamics '09*
edited by M. Milin, T. Nikšić, D. Vretenar, and S. Szilner
© 2009 American Institute of Physics 978-0-7354-0702-2/09/$25.00

FIGURE 1. Surfaces of constant baryon density for the $B = 4$, 8 and 12 Skyrmions.

needs to find and understand multi-Skyrmion solutions with arbitrarily large baryon numbers. The field equation is fully nonlinear, so this is analytically and numerically challenging. Over the last twenty years there has been striking progress, first for the Skyrmions of baryon numbers $B = 2, 3$ and 4 [6], then for Skyrmions in the massless pion limit with B up to 22 and beyond [4], and more recently for Skyrmions with a realistic positive pion mass. A surprise is that the pion mass has a big effect for $B = 8$ and up, and for B a multiple of 4 the solutions look like clusters of $B = 4$ Skyrmions [2].

These Skyrmion solutions are static, and of minimal potential energy for each baryon number. Some dynamical aspects have been investigated, including Skyrmion vibrational modes, the forces between well-separated Skyrmions, and some examples of classical Skyrmion collisions. But these studies have not led to a systematic quantized theory.

Instead, the programme of quantizing Skyrmions as rigid bodies in space and isospace has been pushed forward [3], developing the work pioneered by Braaten and Carson [5] and Kopeliovich [10] for $B = 2$, and Carson [7] for $B = 3$. The most interesting Skyrmions recently studied are those with $B = 8$ and $B = 12$, which have the structures, respectively, of two touching $B = 4$ Skyrmions, and an equilateral triangle of three $B = 4$ Skyrmions close together (see FIG. 1). The $B = 4$ Skyrmion itself has a cubic shape, and is little deformed by its neighbours. The classical mass and the inertia tensors of these solutions have been calculated numerically. The $B = 12$ solution, for example, behaves as an oblate rotor in space and as a slightly oblate rotor in isospace, very weakly coupled together.

The quantum states are classified as for a rigid body, with a total spin/parity J^P and isospin I, and the energy levels are calculated using methods well-known for rigid bodies. The topological constraints mean that only a selection of spin/isospin combinations occur [11]. For $B = 4$, and isospin 0, there is a 0^+ state representing the ground state of ^{4}He, and a highly excited 4^+ state. For $B = 8$ there is a rotational band of ^{8}Be states with $J^P = 0^+, 2^+, 4^+$, and for $B = 12$ a sequence of ^{12}C states with $J^P = 0^+, 2^+, 3^-, 4^-, 4^+, 5^-, 6^+, \ldots$. All this is unsurprising, given the similarity of the Skyrmions to the structures predicted by an alpha-particle model. However, more states occur, with isospins 1, 2 and beyond. This is the novelty of the Skyrmion approach, where isospin is treated as a collective excitation.

For $B = 4$ the lowest-energy isospin 1 state has $J^P = 2^-$. For $B = 8$ one finds low-lying $J^P = 0^-, 2^+, 2^-, 3^+, \ldots$ states with isospin 1, and for $B = 12$ there are states with

$J^P = 1^+(\text{twice}), 2^-, 2^+, ...$ This agrees well with the observed spectra of ^{4}H, ^{8}Li and ^{12}B, and their isobars, with the exception of the negative parity states of ^{8}Li. So here the Skyrme model makes a prediction that there are additional, low-lying 0^- and 2^- states of ^{8}Li, so far unseen. The isospin 2 states agree fairly well with the rather limited available data for ^{8}He, ^{12}Be and their isobars. For more details of all these states and their energies, see ref. [3].

Satisfactory spectra have also been found for the quantized $B = 6$ and $B = 10$ Skyrmions, though the energy spacings and orderings do not all come out right. $B = 5$ and $B = 7$ cause more difficulty, as Skyrmion analogues of an alpha-particle plus nucleon cluster structure, and alpha-particle plus triton, have not yet been pinned down.

Further states

One cannot match all the experimentally observed, excited states of nuclei with those obtained by quantizing the most stable Skyrmions as rigid rotors. For example, only the ground 0^+ state of ^{12}C can be obtained this way. To model higher-energy 0^+ states, like the 7.65 MeV Hoyle state, other Skyrmion configurations are required. There is a $B = 12$ Skyrmion solution, which appears to be locally stable, made from three $B = 4$ Skyrmions in a linear chain. It has a large moment of inertia for spatial rotations about two axes. When quantized, it has a 0^+ ground state, and a sequence of excited spin and isospin states. S.W. Wood is currently investigating how good a fit these give not only to the 2^+ and 4^+ states of ^{12}C that appear related to the Hoyle state, discussed by Freer at this conference [8], but also to the high-lying molecular band of ^{12}Be [9]. The classical solution can be interpreted as a dimer of $B = 6$ Skyrmions, so a quantized state with isospin 2 readily has an interpretation as a molecule of two ^{6}He nuclei.

Skyrmions and other models of nuclei

Various features of the Skyrmion picture of nuclei contrast with more conventional models:-

1.) Although a single $B = 1$ Skyrmion is spherical, all Skyrmions of higher baryon number have quite complicated intrinsic structure. The $B = 4$ Skyrmion is a hollow cube, for example. This contrasts with traditional views of nuclear structure, where deformations of an intrinsic spherical shape are possible, but the interior density remains fairly uniform. Of course, the simple quadrupole and octupole models of surface-deformed nuclei are phenomenologically useful, but they can only be approximations. The Skyrme model, through its nonlinear field equation allows one to explore mathematically what a more precise geometrical description could be.

In recent years, an ever larger number of nuclei are being regarded as having non-spherical structure, and there is also some evidence for dips in baryon density in the nuclear core. The various cluster models usually assume that the cluster components are spherical (at least the cartoon Ikeda diagram shows this) but probably these components are deformed in some way too. For example, if ^{16}O is a tetrahedral cluster of four alpha-particles, then the individual alpha-particles are probably not exactly spherical. The $B = 16$ Skyrmion suggests a more precise shape [2].

2.) The Skyrme model gives classical values to the pion fields, and isospin quantum numbers are not obtained until after quantization. This puts spin and isospin on a similar footing, and is only consistent with a picture of nuclei built of isospin-half nucleons if there is a significant component of nuclear wavefunctions in which nucleons are replaced by deltas (and even higher isospin states). Such a mixed wavefunction could approach the coherent intrinsic state where a classical direction in isospace is defined at each spatial point.

3.) A multi-Skyrmion solution can be obtained by starting with individual, separated $B = 1$ Skyrmions, and letting them approach and coalesce (it is understood how they should be oriented so as to attract). In this process the Skyrmions merge somewhat and lose their identities. The effect is interesting, but not dramatic, because it is highly unfavoured energetically for Skyrmions to be literally on top of each other. The $B = 4$ Skyrmion illustrates this. Though cubic, by a small deformation it can be pulled apart into a tetrahedral arrangement of four $B = 1$ Skyrmions. This can be done in two ways, because the vertices of a cube form two tetrahedral groupings. The oscillation between one tetrahedron and the other, passing through the cubic Skyrmion, was investigated by Walhout [13]. This kind of motion is rather inconceivable in a point-nucleon model.

The merging of structures occurs in all kinds of topological soliton dynamics, and not just for Skyrmions – it is perhaps the key distinction between a field theoretic model and a point-particle model – but phenomenological evidence for it in nuclear physics is not easy to pin down. A possible place to look is in the anomalies in the space-star cross sections. It is rather natural, from a Skyrmion point of view, for a head-on collision of a deuteron and nucleon to end with three nucleons emerging along trajectories separated by 120° (in the CM frame). Similarly, a head-on collision of two deuterons could easily result in four nucleons emerging in one plane, along trajectories separated by 90°. Classical motions of Skyrmions exhibit these behaviours, but the initial conditions have to be carefully arranged, and it has not yet been possible to work out, either classically or quantum mechanically, the differential cross sections. Since the classical Skyrmion orientations have to be aligned favourably for these processes to occur, one may predict that space-star formation is sensitive to the polarization states of the incoming deuterons/nucleons. For two deuterons approaching along the z-axis, space-star formation in the xy-plane should be favoured if the deuterons both have $J_3 = 0$.

Skyrmions and quarks

These final remarks are more speculative and concern the distribution of quarks in nuclei. The Skyrme model is best justified as arising from QCD in the large N_c limit, but this limit is not physical, as $N_c = 3$, and a nucleus of baryon number B has $3B$ quarks. Does the Skyrme model hint at where they are most likely to be located? The $B = 1$ Skyrmion is not by itself informative, but the structure of the $B = 4$ Skyrmion (which we have seen plays an important role as a substructure of larger Skyrmions) is more so. The cubic shape has its baryon density concentrated along twelve edges, so we propose that each edge centre is the location of one quark. If the $B = 4$ Skyrmion is tetrahedrally deformed, then this leads to four $B = 1$ Skyrmions centred near four of the eight original vertices, and as each of these new vertices unambiguously brings with it three neighbouring edges, each $B = 1$ Skyrmion will acquire three quarks. More mathematically, the proposal is that in general, a quark (or antiquark if the baryon density is negative) is associated to each point in space where the field components take one of the three special values $\pi_1 + i\pi_2 = e^{\pm i\pi/3}$ or -1, with $\sigma = \pi_3 = 0$. Quark confinement is automatic, since each field value occurs with the same (signed) multiplicity in a Skyrmion, so it costs infinite energy to isolate a fraction of a Skyrmion. This proposal implies that for a nucleon, the quarks are most likely to be found arranged as an equilateral triangle. Since the $B = 2$ Skyrmion is toroidal (not illustrated), the proposal implies that, at least when the proton and neutron are close together, the quarks in a deuteron are equally spaced around a circle.

This proposal needs considerable refinement, as it should incorporate some constraints on the distribution of the individual up and down quarks, and of their spin directions. Some constraints arise if one insists that the three quarks closest to each vertex of the $B = 4$ Skyrmion combine to form a nucleon, not a delta. The proposal also needs to be expressed in a form that respects rotational and isospin symmetry. Following an isospin rotation of the $B = 4$ Skyrmion by $60°$ about the 3rd isospin axis, the field values above will occur at the six face centres of the cube, with multiplicity two. The alpha-particle wavefunction therefore interpolates betwen a 12-quark state and a 6-diquark state. The probability of finding diquarks is however suppressed, because of the vanishing baryon density at the face centres.

This proposal may help lattice QCD theorists construct approximate nuclear states.

ACKNOWLEDGMENTS

I thank Richard Battye, Paul Sutcliffe and Stephen Wood for their contributions to the work presented here, and also Tony Kennedy and Nigel Buttimore for discussions.

REFERENCES

1. G. S. Adkins, C. R. Nappi, and E. Witten, *Nucl. Phys.* **B 228**, 552 (1983).
2. R. A. Battye, N. S. Manton, and P. M. Sutcliffe, *Proc. R. Soc. A* **463**, 261 (2007).
3. R. A. Battye, N. S. Manton, P. M. Sutcliffe, and S. W. Wood, arXiv:0905.0099 (2009).
4. R. A. Battye, and P. M. Sutcliffe, *Rev. Math. Phys.* **14**, 29 (2002).
5. E. Braaten, and L. Carson, *Phys. Rev.* D **38**, 3525 (1988).
6. E. Braaten, S. Townsend, and L. Carson, *Phys. Lett.* **235B**, 147 (1990).
7. L. Carson, *Phys. Rev. Lett.* **66**, 1406 (1991).
8. M. Freer, these proceedings.
9. M. Freer et al., *Phys. Rev. Lett.* **82**, 1383 (1999).
10. V. B. Kopeliovich, *Yad. Fiz.* **47**, 1495 (1988).
11. S. Krusch, *Annals Phys.* **304**, 103 (2003).
12. T. H. R. Skyrme, *Proc. R. Soc. Lond. A* **260**, 127 (1961).
13. T. S. Walhout, *Nucl. Phys.* **A 547**, 423 (1992).

Study of Even-Even/Odd-Even/Odd-Odd Nuclei in Zn-Ga-Ge Region in the Proton-Neutron IBM/IBFM/IBFFM

N. Yoshida*, S. Brant† and L. Zuffi**

*Faculty of Informatics, Kansai University, Takatsuki 569-1095, Japan
†Department of Physics, Faculty of Science, University of Zagreb, 10000 Zagreb, Croatia
**Dipartimento di Fisica dell'Università di Milano
and Istituto Nazionale di Fisica Nucleare, Sezione di Milano, Via Celoria 16, Milano 20133, Italy

Abstract. We study the even-even, odd-even and odd-odd nuclei in the region including Zn-Ga-Ge in the proton-neutron IBM and the models derived from it: IBM2, IBFM2, IBFFM2. We describe ^{67}Ga, ^{67}Zn, and ^{68}Ga by coupling odd particles to a boson core ^{66}Zn. We also calculate the β^+-decay rates among ^{68}Ge, ^{68}Ga and ^{68}Zn.

Keywords: interacting boson model, interacting boson-fermion model, interacting boson-fermion-fermion model, O(6) symmetry, E(5) symmetry, odd-odd nuclei, beta decay
PACS: 21.60.Fw, 21.60.Ev, 23.40.-s, 27.50.+e

INTRODUCTION

The nuclei in the region with mass $60 \lesssim A \lesssim 70$ are known to show various symmetry characters ranging from the U(5) to the O(6) limits of the interacting boson model (IBM) [1, 2], including the critical symmetry E(5) [3]. We study these nuclei in the proton-neutron version of IBM models, namely, the proton-neutron interacting boson model (IBM2), interacting boson-fermion model (IBFM2), and interacting boson-fermion-fermion model (IBFFM2). The IBFFM2 is more suited for precise description than IBFFM1 because the connection to the shell model is stronger.

STRUCTURE OF ^{66}ZN, ^{67}GA, ^{67}ZN, AND ^{68}GA

The even-even Zn isotopes are known to be in the transitional region between the U(5) and the O(6) limits. It was recently pointed out that ^{64}Zn shows the properties of the E(5) critical symmetry [4], while the isotope ^{66}Zn was treated as O(6) type [5] and was discussed in connection with F-spin symmetry. Starting from the similar O(6)-type Hamiltonian, we have made some adjustments. The result is shown in Fig. 1. A reasonable agreement is seen. The F-spin mixed states are shown by dashed lines connected with the candidates of corresponding experimental levels. Electromagnetic transitions are in reasonable agreement.

CP1165, *Nuclear Structure and Dynamics '09*
edited by M. Milin, T. Nikšić, D. Vretenar, and S. Szilner

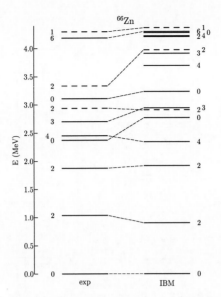

FIGURE 1. Energy levels in ^{66}Zn. The experimental levels are compared with IBM2 calculation. The levels shown by dashed lines can be interpreted as F-spin mixed symmetry states.

For the boson core plus an odd proton ^{67}Ga, and the boson core plus an odd neutron ^{67}Zn nucleus, the IBFM2 Hamiltonian

$$H = H^{\mathrm{B}} + H^{\mathrm{F}} + V^{\mathrm{BF}} \tag{1}$$

consists of the IBM2 Hamiltonian H^{B}, the Hamiltonian of the odd fermion H^{F}, and the boson-fermion interaction V^{BF}. The boson-fermion interaction includes the quadrupole-quadrupole interaction and the exchange interaction. We use same boson Hamiltonian as that for the ^{66}Zn core. The $p_{3/2}$, $p_{1/2}$, and $f_{5/2}$ orbitals are included for both the odd proton and the odd neutron. The boson-fermion interactions have been varied. The results are shown in Fig. 2. Good agreement between the experimental and the calculated energy levels is obtained.

The odd-odd nucleus ^{68}Ga is treated as a system consisting of the boson core ^{66}Zn plus an odd proton and an odd neutron. Although this nucleus has been studied in the IBFFM1 [6] with identical bosons [7, 8], the IBFFM2 description [9, 10] will be important in explaining higher levels and β-decay where difference between proton and neutron bosons must be properly taken into account. The Hamiltonian:

$$H = H^{\mathrm{B}} + H_{\pi}^{\mathrm{F}} + H_{\nu}^{\mathrm{F}} + V_{\pi}^{\mathrm{BF}} + V_{\nu}^{\mathrm{BF}} + V_{\mathrm{RES}} \tag{2}$$

consists of the boson Hamiltonian, the fermion Hamiltonians, and the interactions between the odd fermions and bosons that are the same as those in the odd-even neighbors. The residual interaction V_{RES} between the odd proton and the odd neutron consists of the delta interaction, the spin-spin interaction, the spin-spin-delta interaction, and the tensor interaction. They are determined from the experimental energy levels and electromagnetic properties. Fig. 3 compares the energy levels.

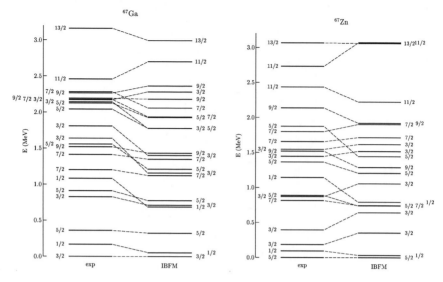

FIGURE 2.   Energy levels in ^{67}Ga and ^{57}Zn.

BETA-DECAY

We describe the β-decay (β^+ and electron capture) in the chain: ^{68}Ge $\rightarrow$ ^{68}Ga $\rightarrow$ ^{68}Zn. The calculation is performed using the β-decay operators that Dellagiacoma and Iachello derived for odd-even nuclei [11]. In Table 1, we compare the resulted log-ft values. In spite of the fact that the β-decay operators have no free parameters, the agreement between the experiment and the calculation is very good for the transitions among the ground states. In the transitions to higher states in ^{68}Zn, there are some deviations. More detailed calculations are in progress.

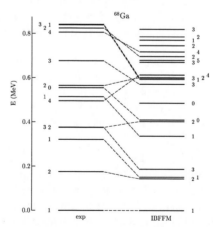

FIGURE 3.   Energy levels in ^{68}Ga.

236

TABLE 1. The $\log_{10}$-ft values in the transitions to ^{68}Ga and to ^{68}Zn.

nuclei	final state	$\log_{10}$-ft (exp.)	$\log_{10}$-ft (cal.)
^{68}Ge $\rightarrow$ ^{68}Ga	1_1^+	5.00 (6)	4.69
^{68}Ga $\rightarrow$ ^{68}Zn	0_1^+	5.2	5.36
	0_2^+	6.2	6.99
	2_1^+	5.5	4.94
	2_2^+	5.9	4.84
	2_3^+	5.7	4.76
	2_4^+	5.1	5.72

SUMMARY

We have studied even-even, odd-even and odd-odd nuclei in the Zn region in IBM2, IBFM2, and IBFFM2. The β-decay in Ge-Ga-Zn is described , and reasonable agreement in log-ft values is obtained for $1_1^+ \rightarrow 0_1^+$ and $1_1^+ \rightarrow 0_2^+$ in ^{68}Ge$\rightarrow$^{68}Zn. For other transitions, wave functions need improvement. Detailed analysis is being carried out with the E(5) symmetry and F-spin mixed-symmetry states taken into account.

REFERENCES

1. F. Iachello and A. Arima, *The interacting boson model*, Cambridge Univ. Press, Cambridge, 1987.
2. F. Iachello and P. Van Isacker, *The interacting boson-fermion model*, Cambridge Univ. Press, Cambridge, 1991.
3. F. Iachello, *Phys. Rev. Lett.* **85**, 3580 (2000).
4. C. Mihai, N. V. Zamfir, D. Bucurescu, G. Căta-Danil, I. Căta-Danil, D. G. Ghiţă, M. Ivaşcu, T. Sava, L. Stroe, and G. Suliman, *Phys. Rev.* **C75**, 044302 (2007).
5. A. Gade, H. Klein, N. Pietralla, and P. von Brentano, *Phys. Rev.* **C65**, 054311 (2002).
6. J. Timár, T. X. Quang, Zs. Dombrádi, T. Fényes, A. Krasznahorkay, S. Brant, V. Paar, and Lj. Šimičić, *Nucl. Phys.* **A552**, 170 (1993).
7. V. Paar, "New coupling limits, dynamical symmetries and microscopic operators of IBM/TQM" in *Capture Gamma-Ray Spectroscopy and Related Topics-1984*, edited by S. Raman, AIP Conference Proceedings 125, American Institute of Physics, New York, 1985, pp. 70–88; S. Brant, V. Paar, and D. Vretenar, *Z. Phys.* **A319**, 355 (1984); V. Paar, D. K. Sunko, and D. Vretenar, *Z. Phys.* **A327**, 291 (1987).
8. S. Brant, and V. Paar, *Z. Phys.* **A329**, 151 (1988).
9. N. Yoshida, H. Sagawa, and T. Otsuka, *Nucl. Phys.* **A567**, 17 (1994).
10. S. Brant, N. Yoshida, and L. Zuffi, *Phys. Rev.* **C74**, 024303 (2006).
11. F. Dellagiacoma and F. Iachello, *Phys. Lett.* **B218**, 399 (1989).

Collective Structures Within the Cartan-Weyl Based Geometrical Model

S. De Baerdemacker*,†, K. Heyde* and V. Hellemans**

*Department of Subatomic and Radiation Physics, Ghent University, Proeftuinstraat 86,
B-9000 Ghent, Belgium
† University of Toronto, Department of Physics, 60 St. George St., Toronto, ON, M5S 1A7, Canada.
**University of Notre Dame, Department of Physics, Notre Dame, IN, 46556-5670, USA

Abstract. The algebraic derivation of the matrix elements of the quadrupole collective variables within the canonical basis of $SU(1,1) \times SO(5)$ is applied to a simple fermionic system with $j = 1/2$ to illustrate the method.

Keywords: Lie algebras, Collective models, matrix elements
PACS: 02.20Qs, 21.50Ev, 21.60Cs

For studying the structure of atomic nuclei, the theoretical description of physical observables involves the matrix elements of the corresponding operator within a suitable basis. For the quadrupole collective model [1], this basis can have an algebraic origin and many studies were devoted to determine the matrix elements of the collective quadrupole variables $\alpha_{2\mu}$ within the $SU(1,1) \times SO(5)$ basis, such as e.g. the recently developed Cartan-Weyl based Collective model [2, 3, 4] (we refer to these articles for an overview of the literature). This method is purely algebraic because the explicit construction of the basis functions in α can be bypassed. Every matrix element is treated as a variable within a set of algebraic equations, which are derived from the tensorial properties of the operators in the canonical basis. Nevertheless, the complexity of the system of equations will rise with the order of the Lie algebra and the dimension of its representations, and it is unclear at present whether the method is viable for a general Lie-algebra. Therefore, it is instructive to apply these ideas to a conceptual and algebraically less complex system, however retaining sufficient relevant degrees of freedom. A good candidate for this purpose is a fermionic spin $j = 1/2$ system as it constitutes an antisymmetric representation of an SO(5) algebra.

It is readily verified that a fermionic $j = 1/2$ system can give rise to 4 possible scenarios, i.e. an empty level ($n = 0$), 1 particle ($n = 1$) with possible spin projections $m = \pm 1/2$, or a completely occupied level ($n = 2$). These 4 situations can be algebraically characterised through the following $SU(2)_J \times SU(2)_S$ generators

$$J_0 = \tfrac{1}{2}(\hat{n}_{\frac{1}{2}} - \hat{n}_{-\frac{1}{2}}), \qquad J_+ = a^{\dagger}_{\frac{1}{2}} a_{-\frac{1}{2}}, \qquad J_- = a^{\dagger}_{-\frac{1}{2}} a_{\frac{1}{2}},$$
$$S_0 = \tfrac{1}{2}(\hat{n}_{\frac{1}{2}} + \hat{n}_{-\frac{1}{2}} - 1), \qquad S_+ = a^{\dagger}_{\frac{1}{2}} a^{\dagger}_{-\frac{1}{2}}, \qquad S_- = a_{-\frac{1}{2}} a_{\frac{1}{2}}, \tag{1}$$

with a_m, $a^{\dagger}_m$ the standard annihilation/creation operators and $\hat{n}_m$ the particle-number operator with spin projection m. From these definitions, we notice that the genera-

CP1165, *Nuclear Structure and Dynamics '09*
edited by M. Milin, T. Nikšić, D. Vretenar, and S. Szilner
© 2009 American Institute of Physics 978-0-7354-0702-2/09/$25.00

tors $\{J_0, J_\pm\}$ constitute the spin algebra $SU(2)_J$ whereas the generators $\{S_0, S_\pm\}$ span the quasi-spin algebra $SU(2)_S$, known from the algebraic treatment of the pairing interaction. [5]. The Cartan algebra $\{J_0, S_0\}$, complemented with the Casimir operators $\{\mathscr{C}_2[SU(2)_J], \mathscr{C}_2[SU(2)_S]\}$ of both $SU(2)$ algebras, defines a basis $|J, M; S, M_S\rangle$, such that the 4 situations can be represented by

$$
\begin{aligned}
&|n=0, m=0\rangle = |0,0; \tfrac{1}{2}, -\tfrac{1}{2}\rangle, \quad |n=1, m=\tfrac{1}{2}\rangle = |\tfrac{1}{2}, \tfrac{1}{2}; 0, 0\rangle, \\
&|n=2, m=0\rangle = |0,0; \tfrac{1}{2}, \tfrac{1}{2}\rangle, \quad\;\; |n=1, m=-\tfrac{1}{2}\rangle = |\tfrac{1}{2}, -\tfrac{1}{2}; 0, 0\rangle.
\end{aligned}
\tag{2}
$$

For the description of e.g. 2-particle transfer, we can rely on standard angular momentum theory [6] to calculate the matrix element of $S_\pm$ in this basis. The situation is different for 1-particle transfer as the creation/annihilation operators are not contained in the algebra. For this purpose, we define the following operators $T_{\mu\nu}$

$$
T_{\frac{1}{2}\frac{1}{2}} = \tfrac{1}{2}a^\dagger_{\frac{1}{2}}, \quad T_{\frac{1}{2}, -\frac{1}{2}} = \tfrac{1}{2}a_{-\frac{1}{2}}, \quad T_{-\frac{1}{2}, \frac{1}{2}} = \tfrac{1}{2}a^\dagger_{-\frac{1}{2}}, \quad T_{-\frac{1}{2}, -\frac{1}{2}} = -\tfrac{1}{2}a_{\frac{1}{2}},
\tag{3}
$$

extending the algebra $SU(2)_J \times SU(2)_S$ towards $SO(5)$ under the standard commutation relations. As a result, the 4 basis states can be identified as a $(0,1)$ antisymmetric representation of $SO(5)$ [7]. It is known that the short root generators $T_{\mu\nu}$ have good bitensorial character under $SU(2)_J \times SU(2)_S$ [2, 8], so we can exploit this when calculating the 1-particle transfer matrix elements. Starting from the expectation value of the anticommutation relation

$$
1 = a^\dagger_{\frac{1}{2}} a_{\frac{1}{2}} + a_{\frac{1}{2}} a^\dagger_{\frac{1}{2}} = -4\left(T_{\frac{1}{2}\frac{1}{2}} T_{-\frac{1}{2}-\frac{1}{2}} + T_{-\frac{1}{2}-\frac{1}{2}} T_{\frac{1}{2}\frac{1}{2}} \right),
\tag{4}
$$

in the state $|0,0; \tfrac{1}{2}, -\tfrac{1}{2}\rangle$, we insert the complete set of 4 basis states in between the $T_{\mu\nu}$ operators. The bitensorial properties then dictate that all but one term in the sum vanish. Hence, we derive (with $T^\dagger_{\mu\nu} = (-)^{\mu+\nu} T_{-\mu-\nu}$)

$$
1 = 4|\langle \tfrac{1}{2}, \tfrac{1}{2}; 0, 0|T_{\frac{1}{2}\frac{1}{2}}|0,0; \tfrac{1}{2}, -\tfrac{1}{2}\rangle|^2,
\tag{5}
$$

and the matrix element is obtained as the solution of the trivial algebraic equation (5). It is clear that this procedure can be generalised for any other matrix element.

Financial support comes from the 'FWO-Vlaanderen', Ghent University and the Interuniversity Attraction Pole (IUAP) under projects P5/07 and P6/23. S.D.B. acknowledges a travel grant from the "FWO Vlaanderen" for a stay at the University of Toronto.

REFERENCES

1. A. Bohr, and B. Mottelson, *Nuclear Structure, Vol.2*, World Scientific, Singapore, 1998.
2. S. De Baerdemacker, K. Heyde, and V. Hellemans, *J. Phys. A: Math. Theor.* **40**, 2733 (2007).
3. S. De Baerdemacker, K. Heyde, and V. Hellemans, *J. Phys. A: Math. Theor.* **41**, 304039 (2008).
4. S. De Baerdemacker, K. Heyde, and V. Hellemans, *Phys. Rev.* **C79**, 034305 (2009).
5. I. Talmi, *Simple models of complex nuclei*, Harwood academic publishers, Chur, 1993.
6. M. E. Rose, *Elementary theory of angular momentum*, John Wiley and Sons, Inc., New York, 1957.
7. F. Iachello, *Lie algebras and applications, Lecture notes in physics*, Springer Verlag, Berlin, 2006.
8. T. M. Corrigan, F. J. Margetan, and S. A. Williams, *Phys. Rev.* **C14**, 2279 (1976).

NUCLEAR ENERGY DENSITY FUNCTIONALS

Non-empirical Nuclear Energy Functionals, Pairing Gaps and Odd-Even Mass Differences

T. Duguet[*,†] and T. Lesinski[**,‡]

[*]CEA, Centre de Saclay, IRFU/Service de Physique Nucléaire, F-91191 Gif-sur-Yvette, France
[†]National Superconducting Cyclotron Laboratory and Department of Physics and Astronomy,
Michigan State University, East Lansing, MI 48824, USA
[**]Department of Physics and Astronomy, University of Tennessee, Knoxville, TN 37996, USA
[‡]Physics Division, Oak Ridge National Laboratory, Oak Ridge, TN 37831, USA

Abstract. First, we briefly outline some aspects of the starting project to design non-empirical energy functionals based on low-momentum vacuum interactions and many-body perturbation theory. Second, we present results obtained within an approximation of such a scheme where the pairing part of the energy density functional is constructed at first order in the nuclear plus Coulomb two-body interaction. We discuss in detail the physics of the odd-even mass staggering and the necessity to compute actual odd-even mass differences to analyze it meaningfully.

Keywords: Non-empirical energy density functional, finite nuclei, odd-even mass staggering
PACS: 21.60.Jz, 21.10.Dr

INTRODUCTION AND ELEMENTS OF FORMALISM

Like-particle pairing is an essential ingredient of nuclear-structure models, in particular regarding the description of exotic nuclei [1]. Also, superfluidity plays a key role in neutron stars, e.g. it impacts post-glitch timing observations [2] or their cooling history [3].

Within a single-reference (SR) implementation of the energy density functional (EDF) formalism [4], pairing is incorporated through the breaking of the $U(1)$ symmetry associated with particle-number conservation. As a result, the binding energy $\mathscr{E}_{SR}$ of the many-body system is postulated to be a functional of both the one-body density matrix $\rho_{ji} \equiv \langle \Phi | c_i^\dagger c_j | \Phi \rangle$ and the pairing tensor $\kappa_{ji} \equiv \langle \Phi | c_i c_j | \Phi \rangle$, the dependence on the latter being allowed by the use of an auxiliary product state of reference $|\Phi\rangle$ that mixes particle numbers (of given parity) [5].

Modern empirical parameterizations of existing EDFs, e.g. Skyrme or Gogny, provide a fair description of bulk and certain spectroscopic properties of known nuclei [4]. On the other hand, they lack predictive power away from known data and a true spectroscopic quality, in particular regarding the part that drives superfluidity. As a result, several groups currently work on empirically improving the analytical form and the fitting of functionals, e.g. see Refs. [6, 7] for recent attempts to pin down the isovector content of purely local pairing functionals.

Along with improving the phenomenology at play, the quest for predictive EDFs starts to benefit from a complementary approach [8] that does not primarily rely on fitting known data but that roots the analytical form of the functional and the value of its couplings into underlying low-momentum two- and three-nucleon (NN and NNN) interac-

CP1165, *Nuclear Structure and Dynamics '09*
edited by M. Milin, T. Nikšić, D. Vretenar, and S. Szilner

tions [9, 10] through the application of many-body perturbation theory[1] (MBPT) [14]. The overall goal of such a project is (i) to bridge with *ab-initio* many-body techniques applicable to light nuclei, (ii) calculate properties of heavy/complex nuclei from basic vacuum interactions and (iii) perform controlled calculations with theoretical error bars. First results following such a route are currently being reported [15]. It is an objective of the present contribution to expose results of such an effort to build the pairing part of the EDF non-empirically [16, 17, 18, 19, 20].

We propose to write the energy functional at a given order in (Goldstone) MBPT under a generic form that is convenient to bridge with existing phenomenological EDFs

$$\mathscr{E}_{SR}[\{\rho_{ij}\},\{\kappa_{ij}\},\{\kappa_{ij}^*\};\{E_k\}] \equiv \sum_{ij} t_{ij}\rho_{ji} \tag{1}$$

$$+\frac{1}{2}\sum_{ijkl} \bar{v}_{ijkl}^{\rho\rho} \rho_{ki}\rho_{lj} + \frac{1}{4}\sum_{ijkl} \bar{v}_{ijkl}^{\kappa\kappa} \kappa_{ij}^* \kappa_{kl}$$

$$+\frac{1}{6}\sum_{ijklmn} \bar{v}_{ijklmn}^{\rho\rho\rho} \rho_{li}\rho_{mj}\rho_{nk} + \frac{1}{4}\sum_{ijklmn} \bar{v}_{ijklmn}^{\rho\kappa\kappa} \rho_{li} \kappa_{jk}^* \kappa_{mn}$$

$$+\frac{1}{24}\sum_{ijklmnop} \bar{v}_{ijklmnop}^{\rho\rho\rho\rho} \rho_{mi}\rho_{nj}\rho_{ok}\rho_{pl} + \cdots,$$

where all dependencies on ρ and $\kappa^*\kappa$ have been made explicit. The *effective vertices* $\bar{v}_{ijkl}^{\rho\rho}$, $\bar{v}_{ijkl}^{\kappa\kappa}$... thus introduced are expressed in terms of the vacuum two-, three-,... body interactions and on quasi-particle energies E_k that are to be determined self-consistently through a chosen procedure. More precisely, a term of given power in ρ and/or $\kappa^*\kappa$ in Eq. 1 receives contributions from different perturbative orders and/or many-body forces. To exemplify this, we can write the vertices arising at second order in the NN interaction $\bar{v}^{NN}$, in a perturbation theory that does not account for pairing explicitly[2]

$$\bar{v}_{ijklijkl}^{\rho\rho\rho\rho} \equiv 6\frac{|\bar{v}_{ijkl}^{NN}|^2}{\varepsilon_i + \varepsilon_j - \varepsilon_k - \varepsilon_l} \;\; ; \;\; \bar{v}_{ijkijk}^{\rho\rho\rho} \equiv \frac{1}{2}\sum_l \bar{v}_{ijklijkl}^{\rho\rho\rho\rho} \;\; ; \;\; \bar{v}_{ijij}^{\rho\rho} \equiv \bar{v}_{ijij}^{NN} + \frac{1}{6}\sum_k \bar{v}_{ijkijk}^{\rho\rho\rho} \;, \tag{2}$$

where ε_i denotes single-particle energies to be determined self-consistently. The EDF form of Eq. 1 may naively suggests that it results from the average value, in the unperturbed vacuum, of an (hypothetical) effective Hamilton operator containing two-body (second line), three-body (third line),... pieces. However, Eq. 2, that provides microscopic expressions for the matrix elements of $\bar{v}_{ijkl}^{\rho\rho}$, $\bar{v}_{ijkl}^{\kappa\kappa}$, $\bar{v}_{ijklmn}^{\rho\rho\rho}$..., demonstrates that re-extracting an (effective) Hamilton operator from the energy density has no foundation[3] and can at best be the result of approximations.

---

[1] Infinite resummation of certain categories of diagrams and/or a redefinition of the unperturbed vacuum $|\Phi\rangle$ are always possible. Switching from conventional hard-core potentials to low-momentum interactions is essential to make a perturbative approach viable, e.g. second-order calculations performed in terms of low-momentum interactions provide satisfactory results for bulk correlations [11, 12, 13].

[2] The single-particle basis solution of Eq. 3 also diagonalizes the density matrix ρ of $|\Phi\rangle$ in this case.

[3] Note for instance that symmetry properties of $\bar{v}_{ijkl}^{\rho\rho}$, $\bar{v}_{ijklmn}^{\rho\rho\rho}$ and $\bar{v}_{ijklijkl}^{\rho\rho\rho\rho}$ under the exchange of fermionic indices are *not* as expected from two-, three- and four-body operators.

Forms as given by Eq. 1 are known as orbital-dependent energy functionals [21] in electronic systems density functional theory (DFT), with the important subtlety that DFT implies that quasi-particle wave functions (U_k, V_k) and quasi-particle energies E_k are generated through the variationally optimum local one-body potential, i.e. the optimal effective potential (OEP) [22]. We do not insist on that here to rely on a framework that embraces empirical Gogny functionals whose associated one-body fields are non-local [4]. Of course, none of the existing empirical functionals do depend on quasi-particle energies E_k. It remains to be seen in the future whether such an extension is necessary and tractable.

An alternative to OEP that is closer to what is currently done with empirical EDFs consists of determining quasi-particle wave-functions and energies through the minimization of $\mathscr{E}_{SR}$ with respect[5] to independent matrix elements of ρ and κ, under the constraint to have given neutron and proton numbers in average. This leads to solving Hartree-Fock-Bogoliubov-like (HFB) [5] equations

$$\begin{pmatrix} h - \lambda & \Delta \\ -\Delta^* & -h^* + \lambda \end{pmatrix} \begin{pmatrix} U_k \\ V_k \end{pmatrix} = E_k \begin{pmatrix} U_k \\ V_k \end{pmatrix} . \tag{3}$$

The one-body field h that drives the correlated single-particle motion and the shell structure, as well as the field Δ that drives superfluidity, are defined as

$$h_{ij} \equiv \frac{\delta \mathscr{E}_{SR}}{\delta \rho_{ji}} \equiv t_{ij} + \Sigma_{ij} \equiv t_{ij} + \sum_{kl} \bar{v}^{ph}_{ikjl} \, \rho_{lk} \quad ; \quad \Delta_{ij} \equiv \frac{\delta \mathscr{E}_{SR}}{\delta \kappa^*_{ij}} \equiv \frac{1}{2} \sum_{kl} \bar{v}^{pp}_{ijkl} \, \kappa_{kl} , \tag{4}$$

through which two effective vertices $\bar{v}^{ph}$ and $\bar{v}^{pp}$ are introduced that can be expressed in terms of $\bar{v}^{\rho\rho}_{ijkl}$, $\bar{v}^{\kappa\kappa}_{ijkl} \ldots$, i.e. they themselves possess a diagrammatic expansion in terms of $\bar{v}^{NN}$ and $\bar{v}^{NNN}$.

Our immediate focus is on the pairing part of the EDF. Beyond enhancing its predictive power, our aim is to understand better the microscopic processes that build superfluidity in finite nuclei. Typical questions relate to (i) the contribution from the direct NN and NNN interactions, its breaking down in partial waves (essentially $^1S_0$, $^3P_1$, $^1D_2$ in decreasing order of expected importance), as well as (ii) the role of higher-order effects associated with the coupling to (collective) fluctuations.

To answer the first of these two questions, our current target is to perform reliable finite-nuclei calculations at first order in low-momentum NN and NNN interactions generated through renormalization group techniques [9, 10]. The upper row of Tab. 1 shows the corresponding diagrammatic for the one-body fields, omitting for simplicity contributions from the NNN interaction. As an intermediate step, we present here approximate results such that $\bar{v}^{ph}$ and the part of the EDF that depends only on ρ is empirically provided by the SLy4 Skyrme parametrization [24], while pairing vertices $\bar{v}^{\kappa\kappa} = \bar{v}^{pp}$ are

---

[4] Using a non-local pairing field, as in the present work, renormalizes from the outset the ultraviolet divergence that arises when using a (quasi-)local pairing field.

[5] Not only the present scheme does not insist on obtaining a local potential but also partial derivatives with respect to quasi-particle energies are omitted. The so-called Krieger-Li-Iafrate approximation to the OPE [23] also omits such functional derivatives.

TABLE 1. Perturbative expansion scheme to first (up) and second (down) order. The dashed line denotes the free-space NN interaction. Diagrams with more than one anomalous propagator are not shown.

$\Sigma^{(1)} =$	$\Delta^{(1)} =$
$\Sigma^{(2)} =$	$\Delta^{(2)} =$

computed at first order in the Coulomb plus nuclear[6] NN interaction [16, 17, 18, 19]. Such an EDF contains neither energy dependencies nor ρ-κ cross terms. Only the dominant $^1S_0$ partial-wave of the NN is included in $\bar{v}^{\kappa\kappa}$ whereas the effect of $^3P_1$ and $^1D_2$ is discussed in Ref. [26]. Also, the first-order contribution of the NNN interaction to $\bar{v}^{\rho\kappa\kappa}$ and $\bar{v}^{\rho\rho}$ will be reported on in Ref. [20]. Note that for such a calculation to be a decent approximation of the targeted first-order one, at least as for extracting pairing gaps, it is crucial that the empirical Skyrme parametrization that drives the underlying shell structure is characterized by an isoscalar effective k-mass $m_0^* \approx 0.7\, m$ at saturation density [19]. Of course, we eventually aim at calculating h at lowest-order in *both* the NN and the NNN low-momentum interactions, possibly making use of the density matrix expansion [27, 28]. Eventually, higher-order contributions are left out for future works[7].

EXPERIMENTAL VERSUS THEORETICAL PAIRING GAPS

We limit ourselves to discussing the odd-even mass staggering (OEMS) whereas other observables are reported on in Ref. [20]. The OEMS is dominated by the deficit of binding energy of the unpaired nucleon in odd nuclei, i.e. the "pairing gap". In the SR-EDF formalism, such a staggering relates to the description of odd nuclei through the excitation of a quasi-particle on top of an even-number parity vacuum. Experimentally, the OEMS is extracted through n-points difference-mass formulae $\Delta_q^{(n)}(N/Z)$ [29]. However, because of the technical difficulty to compute odd nuclei, data are often compared to purely theoretical pairing gaps extracted from the calculation of just one even-even nucleus. One such theoretical pairing gap is the *Lowest Canonical State* gap Δ_{LCS}, defined as the diagonal matrix element of the pairing field Δ in the *canonical* single-particle state whose associated quasi-particle energy E_k is the lowest [17, 18].

The difficulty with such comparisons is not only that (i) finite-difference mass formulae are contaminated by contributions other than the targeted "pairing gap" [30, 31, 32, 33, 34] but also that (ii) the "pairing gap" that makes the actual OEMS is itself an average of Δ_{LCS} extracted from the even-even and (blocked) odd-even nuclei involved in the finite-difference mass formula [33, 34]. As a result, comparisons based on theoretical gaps extracted from one even-even nucleus can only be of semi-quantitative character, which is often fine as empirical pairing functionals are not yet targeting a nucleus by

---

[6] We use the low-momentum NN interaction V_{lowk} [9] built from the Argonne v_{18} NN potential [25] at a renormalization cut-off $\Lambda = 2.5\,\mathrm{fm}^{-1}$.

[7] See the lower row of Tab. 1 for the second-order contributions to the one-body fields.

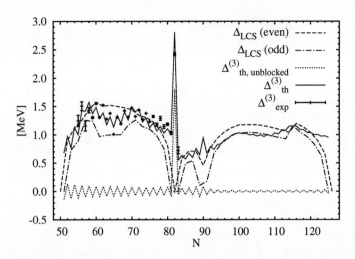

FIGURE 1. (Color online) Experimental three-point mass differences (crosses) for neutrons along the tin isotopic chain versus several theoretical measures of the OEMS: Δ^n_{LCS}(even) (dashed-line), Δ^n_{LCS}(odd) (dashed-dotted line) and $\Delta^{(3)}_n(N)$ (full line). Theoretical three-point mass differences are also shown for odd-even nuclei computed using an even-number-parity vacuum as a reference state, i.e. a HFB state without *any* quasi-particle blocking as if odd-even nuclei had the same structure as even-even ones (dotted line) [33, 34].

nucleus agreement with experiment. However, aiming at such a level of agreement in the (distant?) future and at doing so in a non-empirical fashion requires the comparison of apples with apples, i.e. to compare theoretical and experimental odd-even mass differences. At the price of requiring a good understanding of the different contributions to the OEMS [33, 34], doing so allows more fruitful comparisons between theory and experiment, e.g. to analyze the interplay between pairing and the underlying shell structure. This is what we wish to briefly exemplify in the present contribution.

RESULTS

In Refs. [17, 18], experimental $\Delta^{(3)}_q$(odd) were compared to Δ^q_{LCS}(even). Main results were that neutron and proton pairing gaps computed from the Skyrme plus non-empirical pairing energy functional were close to data for a large set of semi-magic light-, medium- and heavy-mass nuclei. Implications of such results were also discussed. Here, we wish to analyze the qualitative modifications brought about by comparing directly theoretical and experimental three-point mass differences. To do so, we computed odd-even nuclei through the self-consistent blocking procedure performed within the filling approximation [35, 36]. Results for neutron gaps along the tin isotopic chain are reported in Fig 1.

To analyze meaningfully the OEMS [34], the ground state of odd nuclei is best understood as a quasi-particle excitation on top of an even-number parity vacuum that shares the structure of even-even nuclei but that has the odd number of particles on average. In this way, the quasi-particle excitation is performed at (almost) constant particle number [33]. The even-number parity vacuum provides the smooth part of the energy while the quasi-particle excitation, that is dominated by the static "pairing gap", generates the relative lack of binding of odd nuclei without which no interesting physics would be extracted from odd-even mass differences. The dotted line in Fig. 1 shows the contribution of the smooth part of the energy to $\Delta_n^{(3)}(N)$, i.e. when odd-even isotopes are described as if they had the structure of even-even ones. One sees that such a contribution, which reflects the curvature of the smooth part of the energy, oscillates *symmetrically around zero* and accounts exactly for the odd-even oscillation of $\Delta_n^{(3)}(N)$. This demonstrates that contributions other than the targeted "pairing gap" contaminate $\Delta_n^{(3)}(N)$ *in an opposite way* for odd and even N [34], which contradicts the usual belief [30] that $\Delta_n^{(3)}(\text{odd})$ is free from such contaminations[8].

In Fig. 1, the comparison between experimental data for odd N and $\Delta_{LCS}^q(\text{even})$ (dashed line) recalls the results of Refs. [17, 18] and sets the stage for what comes next. Those two curves are consistent with each other, with a slight overestimation (underestimation) of the data at mid-shell (just below and above the $N = 82$ shell closure). Such a situation is representative of the results obtained along other semi-magic isotopic and isotonic chains. Still, certain features that are visible in the data, i.e. (i) the lowering around $N = 65$, (ii) the flat trend as one approaches the $N = 82$ shell closure and (iii) the finite jump from $N = 81$ to $N = 83$, are not reproduced by the bell-shaped curve provided by $\Delta_{LCS}^q(\text{even})$. At best, one can talk of an overall semi-quantitative agreement and wonder whether the remaining discrepancies are due to limitations of (a) the pairing part of the EDF, (b) the use of $\Delta_{LCS}^q(\text{even})$ as a measure of the OEMS and/or (c) SR calculations that miss dynamical correlations associated with particle number restoration and collective pairing vibrations that are of importance in the weak pairing regime, e.g. near shell closures.

The full-fledged comparison of experimental (stars) and theoretical (full line) three-point mass differences is also provided in Fig. 1. The most striking feature is the ability of the calculation to grasp quantitatively the three non-trivial features seen in the data and outlined in the previous paragraph. As a result, one goes from a semi-quantitative agreement with experiment across the major shell using $\Delta_{LCS}^q(\text{even})$ to the ability to compare on a nucleus by nucleus basis. In particular, there were hints that the lowering of the pairing gaps around $N = 65$ could be partly due to dynamical pairing fluctuations [37]. Here, such a feature is well reproduced at the SR-HFB level. It will be of interest to study whether using the non-empirical pairing functional computed from the finite-range and non-local V_{lowk} interaction is essential to obtain such a pattern or if it is entirely driven by the interplay with the underlying shell structure, independently of the detailed characteristics of the pairing functional employed.

A similar situation occurs regarding the behavior of the OEMS towards and across the $N = 82$ shell closure. One sees from Δ_{LCS}^q that static pairing correlations collapse in

---

[8] It was suggested in Ref. [34] to use $\Delta_n^{(3)}(\text{odd})$ as a measure of the sole "pairing gap" *because* the contribution from the time-odd reversal symmetry breaking, not discussed in the present paper, possibly cancels out the contribution from the smooth part of the energy in this case. This is however subject to revision due to the current lack of knowledge regarding time-odd terms in the nuclear EDF.

the immediate vicinity of $N = 82$, i.e. in ^{131,132,133}Sn, while the experimental $\Delta_n^{(3)}(N)$ sustains a non-zero value down to $N = 81$ and 83^9. Surprisingly enough, when going from Δ_{LCS}^q to theoretical three-point mass differences, the experimental trend is well captured down to $N = 81$ and across the $N = 82$ where the OEMS jumps by 350 keV, $\Delta_n^{(3)}(83)$ being the last piece of available data. Due to the collapse of Δ_{LCS}^q close to the shell closure, it is usually stated that $\Delta_n^{(3)}$ is dominated by other contributions than static pairing in this regime, i.e. by dynamical pairing fluctuations and contributions associated with the discreteness of the underlying shell structure. Regarding the former, we just saw that a SR calculation omitting entirely dynamical pairing fluctuations[10] can account for the data. Regarding the latter, it is to be noted that (i) in the (hypothetical) zero-pairing limit, and still assuming spherical symmetry, $\Delta_n^{(3)}(N)$ is zero from $N = 70$ to $N = 81$ as one fills the highly degenerate $h_{11/2}$ shell and that (ii) the regularly oscillating contribution of the smooth part of the energy seen in Fig. 1 demonstrates that the structure of odd nuclei is still best understood, down to ^{131}Sn and ^{133}Sn, as a quasi-particle excitation on top of a *statically paired* even-number parity vacuum. Eventually, the energy of the quasi-particle excitation that builds $\Delta_n^{(3)}(81,83)$ and leads to the unpaired *blocked* state is dominated by pairing correlations, i.e. it would be zero in the zero-pairing limit. Although dynamical pairing correlations are likely to renormalize the OEMS, the present results implies that odd-even mass differences might be less impacted by such correlations than other observables in the vicinity of shell closures.

CONCLUSIONS

We discuss pairing gaps obtained in tin isotopes using an energy density functional whose pairing part is constructed at first order in the nuclear plus Coulomb interaction. Only the (dominant) $^1S_0$ partial wave of the two-nucleon force is incorporated whereas the contributions from $^3P_1$ and $^1D_2$ [26], as well as from the three-nucleon interaction [20], will be reported on soon. Most importantly, we discuss in detail the physics of the odd-even mass staggering and the necessity to compute actual odd-even mass differences to analyze it meaningfully and compare with data on a nucleus-by-nucleus basis. In particular, an excellent description of the odd-even mass staggering is obtained in the vicinity of magic shell closures *prior* to incorporating dynamical pairing correlations associated with particle number restoration and pairing vibrations.

ACKNOWLEDGMENTS

We wish to thank K. Bennaceur, K. Hebeler, J. Meyer and A. Schwenk for our fruitful collaboration on designing non-empirical pairing energy density functionals. This

---

9 One must remove $\Delta_n^{(3)}(82)$ from the analysis as it measures the $N = 82$ shell gap rather than static pairing correlations. Contrarily, $\Delta_n^{(3)}(81,83)$ are *not* influenced by the $N = 82$ shell gap.

10 The Lipkin-Nogami procedure is not used in the present calculation.

work was supported by the U.S. Department of Energy under Contract Nos. DE-FG02-96ER40963, DE-FG02-07ER41529 (University of Tennessee) and DE-AC05-00OR22725 with UT-Battelle, LLC (Oak Ridge National Laboratory).

REFERENCES

1. J. Dobaczewski, W. Nazarewicz, *Prog. Theor. Phys. Suppl.* **146**, 70 (2003).
2. P. Avogadro, F. Barranco, R. A. Broglia, E. Vigezzi, *Phys. Rev.* **C75** (2007) 012805.
3. H. Heiselberg, M. Hjorth-Jensen, *Phys. Rep.* **328**, 237 (2000).
4. M. Bender, P.-H. Heenen, P.-G. Reinhard, *Rev. Mod. Phys.* **75**, 121 (2003).
5. P. Ring, P. Schuck, *The Nuclear Many-Body Problem*, Springer, Berlin, Heidelberg, (2000).
6. J. Margueron, H. Sagawa, K. Hagino , *Phys. Rev.* **C77**, 054309 (2008).
7. M. Yamagami, Y. R. Shimizu, T. Nakatsukasa, arXiv:0812.3197
8. T. Duguet, K. Bennaceur, T. Lesinski, J. Meyer, *Opportunities with Exotic Beams*, Proceedings of the 3rd ANL/MSU/JINA/INT RIA Workshop, edited by T. Duguet, H. Esbensen, K. M. Nollett, C.D. Roberts (World Scientific, 2007) p. 21; nucl-th/0606037.
9. S. K. Bogner, T. T. S. Kuo, A. Schwenk, *Phys. Rep.* **386**, 1 (2003).
10. R. Roth, S. Reinhardt, H. Hergert, *Phys. Rev.* **C77**, 064003 (2008).
11. S. K. Bogner, A. Schwenk, R. J. Furnstahl, A. Nogga, *Nucl. Phys.* **A763**, 59 (2005).
12. S. K. Bogner, R. J. Furnstahl, A. Nogga, A. Schwenk, arXiv:0903.3366.
13. R. Roth, P. Papakonstantinou, N. Paar, H. Hergert, T. Neff, H. Feldmeier, *Phys. Rev.* **C73**, 044312 (2006).
14. P. Nozières, *Theory of interacting Fermi systems*, Westview press, Advanced Book Classics, (1964).
15. J. E. Drut, R. J. Furnstahl, L. Platter, arXiv:0906.1463, and references therein.
16. T. Duguet, *Phys. Rev.* **C69**, 054317 (2004).
17. T. Duguet, T. Lesinski, *Eur. Phys. Jour. ST* **156**, 207 (2008).
18. T. Lesinski, T. Duguet, K. Bennaceur, J. Meyer, *Eur. Phys. J.* **A40** (2009).
19. K. Hebeler, T. Duguet, T. Lesinski, A. Schwenk, arXiv:0904.3152.
20. T. Lesinski, T. Duguet, K. Bennaceur, J. Meyer, in preparation.
21. E. Engel, *A Primer in Density Functional Theory*, edited by C. Fiolhais, F. Nogueira and M. Marques, (Springer, Berlin, 2003), p. 56.
22. J. D. Talman, W. F. Shadwick, *Phys. Rev.* **A14**, 36 (1976).
23. J. B. Krieger, Y. Li, G. J. Iafrate, *Phys. Lett.* **A146**, 256 (1990).
24. E. Chabanat, P. Bonche, P. Haensel, J. Meyer, R. Schaeffer, *Nucl. Phys.* **A635**, 231 (1998).
25. R. B. Wiringa, V. G. J. Stoks, R. Schiavilla, *Phys. Rev.* **C51**, 38 (1995).
26. S. Baroni, A. Schwenk, in preparation.
27. S. K. Bogner, R. J. Furnstahl, L. Platter, arXiv:0811.4198.
28. B. Gebremariam, S. K. Bogner, T. Duguet, in preparation.
29. A. Bohr and B. R. Mottelson, *Nuclear Structure* (Benjamin, New York, 1969), Vol. 1.
30. W. Satula, J. Dobaczewski, W. Nazarewicz, *Phys. Rev. Lett.* **81**, 3599 (1998).
31. K. Rutz, M. Bender, P.-G. Reinhard, J. A. Maruhn, *Phys. Lett.* **B468**, 1 (1999).
32. M. Bender, K. Rutz, P.-G. Reinhard, J. A. Maruhn, *Eur. Phys. J.* **A8**, 59 (2000).
33. T. Duguet, P. Bonche, P.-H. Heenen, J. Meyer, *Phys. Rev.* **C65**, 014310 (2001).
34. T. Duguet, P. Bonche, P.-H. Heenen, J. Meyer, *Phys. Rev.* **C65**, 014311 (2001).
35. S. Perez-Martin, L. M. Robledo, *Phys. Rev.* **C78**, 014304 (2008).
36. G. F. Bertsch, C. A. Bertulani, W. Nazarewicz, N. Schunck, M. V. Stoitsov, *Phys. Rev.* **C79**, 034306 (2009).
37. M. Anguiano, J. L. Egido, L. M. Robledo, *Phys. Lett.* **B545**, 62 (2002).

250

Energy Functional Based on Natural Orbitals and Occupancies for Static Properties of Nuclei.

Denis Lacroix

GANIL, CEA and IN2P3, Boîte Postale 5027, 14076 Caen Cedex, France

Abstract. The possibility to use functionals of occupation numbers and natural orbitals for interacting fermions is discussed as an alternative to multi-reference energy density functional method. An illustration based on the two-level Lipkin model is discussed.

Keywords: Energy Density Functional, Density Matrix Functional Theory
PACS: 21.60.Jz, 21.10.Dr

INTRODUCTION

The nuclear many-body problem of N interacting nucleons can be solved exactly only in very specific cases or for very small particle numbers. This is due to the large number of degrees of freedom involved in such a complex system. Let us for instance consider particles interacting through n-body Hamiltonian written as

$$H = \sum_{ij} t_{ij} a_i^+ a_j + \frac{1}{4} \sum_{ijkl} \tilde{v}_{ijkl} a_i^+ a_j^+ a_l a_k + \cdots \tag{1}$$

Then the exact ground state energy can be written as

$$E_{\text{Exact}}(\gamma^{(1)}, \gamma^{(2)}, ...) = \sum_{ij} t_{ij} \gamma_{ji}^{(1)} + \frac{1}{4} \sum_{ijkl} \tilde{v}_{ijkl} \gamma_{kl,ij}^{(2)} + \cdots, \tag{2}$$

where $\gamma_{ji}^{(1)} \equiv \langle a_i^+ a_j \rangle$, $\gamma_{kl,ij}^{(2)} \equiv \langle a_i^+ a_j^+ a_l a_k \rangle$, ... denote the one-, two-, ... body density matrices that contain all the information on the one-, two-...body degrees of freedom respectively. A natural way to reduce the complexity of this problem is to assume that at a given level, the $k-$body (and higher-order) density matrices becomes a functional of the lower-order ones. This is what is done for instance in the Hartree-Fock (HF) approximation where all k-body density matrices (with $k \geq 2$) become a functional of $\gamma^{(1)}$. Unfortunately, the HF theory applied to the nuclear many-body problem in terms of the vacuum Hamiltonian is a poor approximation and Many-Body theories beyond HF are necessary.

The introduction of Energy Density Functional (EDF) approaches in the 70's was a major breakthrough (see for instance [1] for a recent review). In its simplest form, the EDF formalism starts with an energy postulated as a functional of $\gamma^{(1)}$, the latter being built out of a Slater Determinant. Then the ground state energy is obtained by minimizing

CP1165, *Nuclear Structure and Dynamics '09*
edited by M. Milin, T. Nikšić, D. Vretenar, and S. Szilner
© 2009 American Institute of Physics 978-0-7354-0702-2/09/$25.00

the energy with respect to $\gamma^{(1)}$, i.e.

$$E_{\text{Exact}} \simeq \mathscr{E}_{\text{MF}}(\gamma^{(1)}) \tag{3}$$

Parameters are generally adjusted on specific experimental observations and therefore encompass directly many-body correlations. Current EDF uses a generalization of eq. (3) obtained by considering quasi-particle vacua as trial states. By making explicit use of symmetry breaking, such a functional called hereafter Single-Reference (SR-) EDF is able to account for static correlation associated with pairing and deformation. Actual SR-EDF takes the form [1]:

$$E_{\text{Exact}} \simeq \mathscr{E}_{\text{MF}}(\gamma^{(1)}) + \mathscr{E}_{\text{Cor}}(\kappa.\kappa^*) \tag{4}$$

where κ denotes the anomalous density. To restore symmetries and/or incorporate dynamical correlations, guided by the Generator Coordinate Method (GCM), a second level of EDF implementation, namely Multi-Reference (MR-) EDF is introduced. Recently, difficulties with the formulation and implementation of have been encountered in MR-EDF. A minimal solution has been proposed in ref. [2, 3, 4]. Besides these problems, the authors of ref. [2] have pointed out the absence of a rigorous theoretical framework for the MR EDF approach. At the heart of the problem is the possibility to break symmetries in functional theories and then restore them using configuration mixing. This issue needs to be thoroughly addressed in the future.

In this context, it is interesting to see if extensions of the functional used at the SR-EDF level can grasp part of the effects that for standard functionals require the MR level. It is worth realizing that, in the canonical basis for which $\gamma^{(1)} = \sum_i |\varphi_i\rangle n_i \langle \varphi_i|$, we have

$$\mathscr{E}_{\text{Cor}}(\kappa.\kappa^*) = \mathscr{E}_{\text{Cor}}[\{\varphi_i, n_i\}] = \frac{1}{4}\sum_{i,j} \bar{v}_{iijj}^{\kappa\kappa}\sqrt{n_i(1-n_i)}\sqrt{n_j(1-n_j)}, \tag{5}$$

and therefore, the energy can be regarded as a functional of natural orbitals φ_i and occupation numbers n_i. As a matter of fact, for electronic systems, Gilbert has generalized the Kohn-Sham theory and shown that the exact energy of a system can be obtained by minimizing such a functional [5] leading to the so-called Density Matrix Functional Theory (DMFT). The possibility to consider occupation numbers as building blocks of the nuclear energy functional has recently been discussed in ref. [6, 7]. Two levels of theory can be developed along the line of Gilbert's idea (i) either, functionals in the strict Gilbert framework can be designed. In that case, since the density identify with the exact density at the minimum, it should respect all symmetries of the bare Hamiltonian. (ii) or we exploit the concept of symmetry breaking. In the latter case, similarly to the SR-EDF, strictly speaking we cannot anymore rely on the theorem, but we may gain better physical insight with relatively simple functionals.

---

[1] Note that the denomination "mean-field" or the separation into a "mean-field" like and "correlation" like is completely arbitrary since, as we mention previously, the so-called "mean-field" part already contains correlation much beyond a pure Hartree-Fock approach.

APPLICATION TO THE LIPKIN MODEL AND DISCUSSION

The descriptive power of DMFT is illustrated here in the two-level Lipkin model [10]. In this model, the Hartree-Fock (HF) theory fails to reproduce the ground state energy whereas configuration mixing like Generator Coordinate Method (GCM) provides a suitable tool [8, 9]. Therefore, the two-level Lipkin model is perfectly suited both to illustrate that DMFT could be a valuable tool and to provide an example of a functional for system with a "shape" like phase-transition. In this model, one considers N particles distributed in two N-fold degenerated shells separated by an energy ε. The associated Hamiltonian is given by $H = \varepsilon J_0 - \frac{V}{2}(J_+J_+ + J_-J_-)$ where V denotes the interaction strength while J_0, $J_\pm$ are the quasi-spin operators defined as $J_0 = \frac{1}{2}\sum_{p=1}^{N}(c_{+,p}^\dagger c_{+,p} - c_{-,p}^\dagger c_{-,p})$, $J_+ = \sum_{p=1}^{N} c_{+,p}^\dagger c_{-,p}$ and $J_- = J_+^\dagger$. $c_{+,p}^\dagger$ and $c_{-,p}^\dagger$ are creation operators associated with the upper and lower levels respectively. Due to the specific form of the Lipkin Hamiltonian, $\gamma^{(1)}$ simply writes in the natural basis as $\gamma^{(1)} = \sum_{p=1}^{N}\left\{|\varphi_{0,p}\rangle n_0\langle\varphi_{0,p}| + |\varphi_{1,p}\rangle n_1\langle\varphi_{1,p}|\right\}$ with $n_1 = (1 - n_0)$. Introducing the angle α between the state $|-,p\rangle$ and $|\varphi_0,p\rangle$, leads to the following mean-field functional [11]

$$\mathscr{E}_{MF}(\{\varphi_{i,p}, n_i\}) = \mathscr{E}_{MF}(\alpha, n_0) = -\frac{\varepsilon}{2}N\left\{\cos(2\alpha)(2n_0 - 1) + \frac{\chi}{2}\sin^2(2\alpha)(2n_0 - 1)^2\right\}. \quad (6)$$

where $\chi = V(N-1)/\varepsilon$. This expression is easily obtained by generalizing the Hartree-Fock case (recovered here if $n_0 = 1$). The main challenge of the method is to obtain an accurate expression for $\mathscr{E}_{Cor}$. To get the functional, clearly identified cases from which properties of the functional could be inferred have been used [11], namely the $N = 2$ case and the large N limit. In the two-particles case, the correlation energy can be analytically obtained and reads

$$\mathscr{E}_{Cor}^{N=2}(\alpha, n_0) = -2V\left\{\sin^2(2\alpha)n_0(1 - n_0) + \left(\sin^4(\alpha) + \cos^4(\alpha)\right)\sqrt{n_0(1 - n_0)}\right\} \quad (7)$$

A simple extension of the $N = 2$ case for larger number of particles is to assume that each pair contributes independently from the others leading to $\mathscr{E}_{Cor}^N = [N(N-1)/2]\mathscr{E}_{Cor}^{N=2}$. However, such a simple assumption leads to a wrong scaling behavior in the large N limit. Indeed, in this case, $\mathscr{E}_{Cor}^N \propto N^2$ as N tends to infinity while a $N^{4/3}$ scaling is expected [12]. To obtain the correct limit, a semi-empirical factor $\eta(N)$ can be introduced such that

$$\mathscr{E}_{Cor}^{N\geq3}(\alpha, n_0) = \eta(N)\frac{N(N-1)}{2}\mathscr{E}_{Cor}^{N=2}(\alpha, n_0), \quad (8)$$

with $\eta(N) = cN^{-2/3}$. The value $c = 1.5$ has been retained using a fitting procedure. Examples of results obtained by minimizing the functional given by Eqs. (6) and (8) are shown in Fig. 1 for different particle numbers and interaction strengths. In all cases, a very good agreement, much better than the HF case is found.

The Lipkin example suggests that DMFT can be a valuable tool for describing ground state of a many-body system when symmetry breaking plays a significant role. The functional designed here is exact only in the $N = 2$. Note that the functional proposed

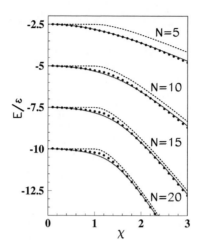

FIGURE 1. (Color online) Exact ground state energy (solid lines) displayed as a function of χ for $N = 5$ to 20 resp. from top to bottom. In each case, the corresponding HF (dashed line) and DMFT (filled circle) minimum energy are shown. The DMFT calculation is performed using the mean-field and correlation energy resp. given by Eq. (6) and Eq. (8) with $\eta(N) = 1.5\,N^{-2/3}$ (Adapted from [11]).

here breaks signature symmetry and therefore enters into the level (ii) of functional discussed in the introduction. The Lipkin model is however rather schematic and cannot be used as a guidance for realistic situations. The possibility to design a new accurate functional for nuclei remains a challenging problem.

ACKNOWLEDGMENTS

The author thanks M. Assié, B. Avez, M. Bender, T. Duguet, C. Simenel, O. Sorlin and P. Van Isacker for enlightening discussions at different stages of this work and T. Papenbrock for useful remarks on the scaling behavior in the Lipkin model.

REFERENCES

1. M. Bender, P.-H. Heenen, and P.-G. Reinhard, *Rev. Mod. Phys.* **75**, 121 (2003).
2. D. Lacroix, T. Duguet, and M. Bender, *Phys. Rev.* **C79**, 044318 (2009).
3. M. Bender, T. Duguet, and D. Lacroix, *Phys. Rev.* **C79**, 044319 (2009).
4. T. Duguet, M. Bender, K. Bennaceur, D. Lacroix, and T. Lesinski, *Phys. Rev.* **C79**, 044320 (2009).
5. T. L. Gilbert, *Phys. Rev.* **B 12**, 2111 (1975).
6. T. Papenbrock and A. Bhattacharyya, *Phys. Rev.* **C75**, 014304 (2007).
7. M. G. Bertolli and T. Papenbrock, *Phys. Rev.* **C78**, 064310 (2008).
8. P. Ring and P. Schuck, *The Nuclear Many-Body Problem* (Springer-Verlag, New-York, 1980).
9. A. P. Severyukhin, M. Bender, P.-H. Heenen *Phys. Rev.* **C74**, 024311 (2006).
10. H. J. Lipkin, and N. Meshkov, *Nucl. Phys.* **A62**, 188 (1965).
11. D. Lacroix, *Phys. Rev.* **C79**, 014301 (2009).
12. S. Dusuel and J. Vidal, it Phys. Rev. Lett. **93**, 237204 (2004).

Spatial Two-neutron Correlations Induced by Pairing in Finite Nuclei

N. Pillet[*], N. Sandulescu[†], P. Schuck[**] and J.-F. Berger[*]

[*]CEA, DAM, DIF, F-91297 Arpajon, France
[†]Institute of Physics and Nuclear Engineering, 76900 Bucharest, Romania
[**]Institut de Physique Nucléaire, CNRS, UMR8608, Orsay, F-91406, France
Université Paris-Sud, Orsay, F-91505, France

Abstract. We study spatial properties induced by pairing correlations, within the HFB approach employing the finite range D1S Gogny interaction. Local and non-local quantities, such as pairing tensor, are discussed as well as the coherence length. We show that a generic feature occurs at the surface of nuclei, namely Cooper pair of small size characterized by a coherence length $\xi \sim 2fm$.

Keywords: Pairing correlations. Spatial properties.
PACS: 21.30.Fe, 21.10.Re, 21.60.Jz, 24.30.Cz

HFB FORMALISM

The importance of elementary excitations using as building blocks pairs of nucleons for the description of nuclear excitations has been known for a long time in superfluid nuclei. During the sixties, it was already realized that pairing correlations are relevant in the description of two-particle transfer reactions. In such kind of reactions, as for example those induced by light ions, a di-nucleon system preferentially coupled to J=0 and confined in space is added/removed to/from the target nucleus. In such cases, it is then important to understand correlations in spin, isopsin, angular momentum and spatial properties, in particular at the surface of the nucleus where Pauli is weaker, that favors the appearance of correlated substructure of few nucleons.

A quantity that is usually studied when one deals with spatial properties of pairs is the coherence length, that is a measure of the spatial extent of correlated pairs. A simple estimate of this quantity is obtained using uncertainty relation. One finds that it is proportional to the ratio between the Fermi momentum and the energy gap resulting from the binding of the pairs:

$$\xi \sim \frac{\hbar k_F}{\pi \Delta} \tag{1}$$

When the proportionality coefficient is equal to one, this relation is known as the Pippard's formula. A simple numerical application with $\Delta = 1MeV$ and $k_F = 1.34fm^{-1}$ leads to a coherence length of the order or larger than 50 fm, that corresponds to very long range pairing correlations.

In order to investigate microscopically and in a systematic way spatial properties of pairing in finite nuclei, we have used the HFB approach [1], employing the finite range

CP1165, *Nuclear Structure and Dynamics '09*
edited by M. Milin, T. Nikšić, D. Vretenar, and S. Szilner
© 2009 American Institute of Physics 978-0-7354-0702-2/09/$25.00

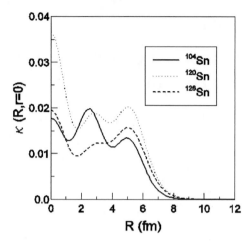

FIGURE 1. Local pairing tensor for ^{120}Sn.

D1S Gogny force [2]. The HFB equations are written as:

$$\begin{pmatrix} h(\rho) - \mu & \Delta(\kappa) \\ -\Delta(\kappa) & -h(\rho) + \mu \end{pmatrix} \begin{pmatrix} u_k \\ v_k \end{pmatrix} = E_k \begin{pmatrix} u_k \\ v_k \end{pmatrix} \qquad (2)$$

where ρ and κ are the usual one-body density and pairing tensor, respectively. We solve the HFB equations in a harmonic oscillator basis. Then, the spatial representation of ρ and κ is achieved by the spatial representation of harmonic oscillator wave functions. In order to describe spatial properties according to the center of mass $\vec{R}$ and the relative $\vec{r}$ coordinate of pairs, a Moshinsky transformation between harmonic oscillator wave functions is applied.

RESULTS

Considering the two-body density matrix in the HFB approximation, one sees that correlated pairs can be adequately studied through the Cooper pair probability $|\kappa|^2$.

In Figure 1 is displayed the local pairing tensor $\kappa(R, r = 0)$ in few Sn isotopes. On this pattern, one clearly identifies shell effects, that is, pairing features strongly depend on the nature of single particle orbitals, closest to the Fermi level. For example, in ^{120}Sn one easily recognizes in the center of the nucleus (small values of R) the contribution associated with the $3s_{1/2}$ orbitals.

The non-local part of the pairing tensor $\kappa(R, r)$, expressed in terms of the center of mass and relative coordinates, is a very interesting quantity as it gives a first estimation of the spatial extent of of pairs. In the left panel of Figure 2, the results are shown for Sn, Ni and Ca isotopes. In all the patterns, one notices a rather strong concentration of the Cooper pairs along the R-axis, with a distribution in R that is rather different from various isotopes due to shell effects. For example, one observes depression of pair

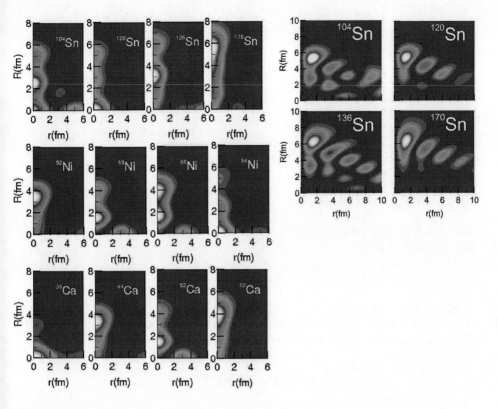

FIGURE 2. (Color online) Non-local pairing tensor for Sn, Ni and Ca isotopes (left panel). Probability distribution of pairing $P(R, r)$ for Sn isotopes (right panel).

probability in the center of the nucleus when there is no s-state in the major shell. This strong concentration of Cooper pairs is not a trivial effect. It comes from parity mixing of orbitals.

The probability distribution of pairing defined as:

$$P(R, r) = R^2 r^2 |\kappa(R, r)|^2 \tag{3}$$

is an important quantity as it enters in the calculation of the mean value of two-body operators, as pairing energy or spectroscopic factor used in the evaluation of cross sections for two-nucleon transfer reaction. The right panel of figure 2 displays P(R,r) for a few isotopes. The striking feature is that for all nuclei the same scenario emerges: the probability distribution of pairing is strongly concentrated in the surface due to the parity mixing and the finite size of nuclei introduced through the phase space factor $r^2 R^2$.

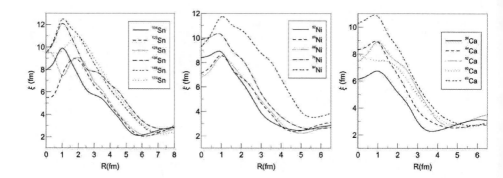

FIGURE 3. Coherence length for Sn (left), Ni (middle) and Ca (right) isotopes.

To calculate the coherence length, we have used the standard formula of the root mean squared radius.

$$\xi(R) = \sqrt{\frac{\int |\kappa(R,r)|^2 \ r^2 \ d^3r}{\int |\kappa(R,r)|^2 \ d^3r}} \qquad (4)$$

In Figure 3, one identifies a generic feature: well defined and pronounced minima for the coherence length ξ are obtained for R of the order of the surface radius of all nuclei. This minimum value of ξ has the surprisingly small value of $\sim 2fm$.

CONCLUSION

In this study, we have seen that pairing correlations depend, of course, strongly on the shell structure around the Fermi level. But a somewhat surprising generic effect occurs in the nuclear surface where shell effects play no role. Namely, nucleons of Cooper pairs are spatially strongly correlated in the surface of nuclei with a coherence length that is of the order of $\xi \sim 2fm$, two times smaller than the smallest value in infinite matter. Consequences on two-particle transfer form factor in superfluid nuclei are expected.

REFERENCES

1. N. Pillet, N. Sandulescu, P. Schuck, *Phys. Rev.* **C76**, 024310 (2007).
2. J. Dechargé and D. Gogny, *Phys. Rev.* **C21**, 1568 (1980); J.-F. Berger, M. Girod and D. Gogny, *Comp. Phys. Comm.* **63**, 365 (1991).

Phenomenological Relativistic Energy Density Functionals

G.A. Lalazissis*, T. Niksic†, S. Kartzikos*, N. Paar†, D. Vretenar† and P. Ring**

*Physics Department, Aristotle University of Thessaloniki, Greece
†Physics Department, University of Zagreb, Croatia
**Physics Department, TU Muenchen, Garching, Germany

Abstract. The framework of relativistic nuclear energy density functionals is applied to the description of a variety of nuclear structure phenomena, not only in spherical and deformed nuclei along the valley of β-stability, but also in exotic systems with extreme isospin values and close to the particle drip-lines. Dynamical aspects of exotic nuclear structure is explored using the fully consistent quasiparticle random-phase approximation based on the relativistic Hartree-Bogoliubov model. Recent applications of energy density functionals with explicit density dependence of the meson-nucleon couplings are presented.

Keywords: relativistic energy density functional, quasiparticle random phase approximation, superheavy nuclei
PACS: 21.30.Fe; 21.60.Jz

Various implementations of density functional theory have become a standard tool in nuclear structure calculations. The adopted functionals are universal in the sense that they can be used to describe both statical and dynamical properties of nuclei all over the periodic table. Relativistic mean field (RMF) model represents a particular implementation of a density functional theory based on a Lorentz covariance. Basically, the nucleus is described as a system of Dirac nucleons coupled to the exchange mesons and the electromagnetic field through an effective Lagrangian. The isoscalar scalar σ-meson, the isoscalar vector ω-meson, and the isovector vector ρ-meson build the minimal set of meson fields that together with the electromagnetic field is necessary for a quantitative description of bulk and single-particle nuclear properties. In order to describe the properties of finite nuclei on a quantitative level, it is necessary to introduce additional density-dependence in the model, either through the non-linear meson terms, or the density-dependent couplings. The functional form of the meson-nucleon vertices can be deduced from in-medium Dirac-Brueckner interactions, obtained from realistic free-space NN interactions, or a phenomenological approach can be adopted, with the density dependence for the σ, ω and ρ meson-nucleon couplings adjusted to properties of nuclear matter and a set of spherical nuclei. In addition, the model can be easily extended to include pairing correlations in a self-consistent manner, leading to the relativistic Hartree-Bogoliubov model (RHB). For a detailed description of this model we refer the reader to [1] and references cited therein.

Recently, we have introduced effective interaction with explicit density dependence of the σ, ω, and ρ meson-nucleon couplings (DD-ME2) [2], which provides a significant improvement in the description of the nuclear structure, especially in the isovector

CP1165, *Nuclear Structure and Dynamics '09*
edited by M. Milin, T. Nikšić, D. Vretenar, and S. Szilner
© 2009 American Institute of Physics 978-0-7354-0702-2/09/$25.00

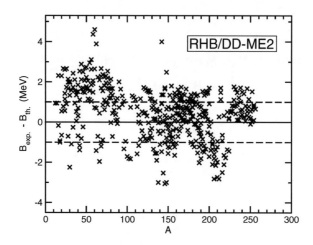

FIGURE 1. Absolute deviations of the binding energies calculated with the DD-ME2 interaction from the experimental values [3].

channel. In the following we present some of the applications of the new DD-ME2 effective interaction.

We have calculated ground states of 420 even-even nuclei within the framework of the RHB model. The DD-ME2 effective interaction has been used in the particle-hole (ph), and the D1S Gogny interaction [4] in the particle-particle (pp) channel. The binding energies are compared with experimental data [3] in Fig. 1. Except for a few Ni isotopes with $N \approx Z$ that are notoriously difficult to describe in a pure mean-field approach, and several transitional medium-heavy nuclei, the calculated binding energies are in very good agreement with the experimental data. Although this illustrative calculation cannot be compared with microscopic mass tables that include more than 9000 nuclei [5], we emphasize that the rms error including all the masses shown in Fig. 1 is about 970 keV. This is a significant improvement compared with the previous RMF calculations, where the rms error was about 2.5 MeV [6].

The fully self-consistent relativistic (quasiparticle) random phase approximation R(Q)RPA [7, 8] has been used to describe collective excitations in both doubly-closed and open-shell nuclei. The RQRPA is formulated in the canonical basis of the RHB model. It should be emphasized that both in the ph and pp channels, the same interactions are used in the RHB equations that determine the canonical quasiparticle basis, and in the matrix equations of the RQRPA. For ^{208}Pb the RRPA results for the monopole and isovector dipole response are displayed in the left panels of Fig. 2 The calculated peak energies of the isoscalar giant monopole resonance (ISGMR): 13.9 MeV, and isovector giant dipole resonance (IVGDR): 13.5 MeV should be compared with the experimental excitation energies: $E = 14.1 \pm 0.3$ MeV [9] for the monopole resonance, and $E = 13.3 \pm 0.1$ MeV [10] for the dipole resonance, respectively. In the right panels of Fig. 2 we compare the RQRPA results for the Sn isotopes with experimental data on IVGDR excitation energies [11] and ISGMR energies [12]. The arrows denote the location of the experimental peak energies. Obviously, the RHB+RQRPA calculation

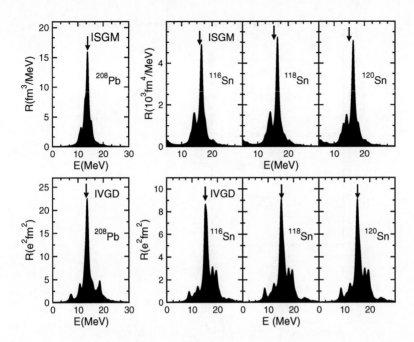

FIGURE 2. The isoscalar monopole (upper panels) , and the isovector dipole strength distributions in ^{208}Pb and in ^{116,118,120} Sn nuclei calculated with the effective interaction DD-ME2.

with the DD-ME2 interaction reproduces in detail the experimental excitation energies and the isotopic dependence of both the IVGDR and ISGMR.

Another important field of applications of self-consistent mean-field models includes the structure and decay properties of superheavy nuclei. Already from Fig. 1 we can conclude that the DD-ME2 effective interaction reproduces the masses of heavy and superheavy nuclei with high accuracy. Previous calculations have shown that the results for the Q_α values [2] and the fission barriers [13] are also very close to the empirical values. In Fig. 3 we display the fission barriers in superheavy nuclei with Z=112,114,116. Experimental data, which specify the lower limit of the heights of fission barriers, are taken from Ref. [14]. Since the assignment of the neutron number has some uncertainty, the same experimental barrier appears for the nuclei with the same proton number Z. The calculated barriers are very close to the empirical data. The results are compared with those obtained using an older NL3 effective interaction [15] which clearly underestimates the height of the fission barrier. The difference should be attributed to improved

TABLE 1. Comparison of the RHB results and data on excitation energies E (in MeV) of fission isomers.

Nucleus	E(NL3)	E(DD-ME2)	E(exp)
^{236}U	1.72	2.24	2.75
^{238}U	1.70	1.54	2.557
^{240}Pu	2.29	2.21	~ 2.8

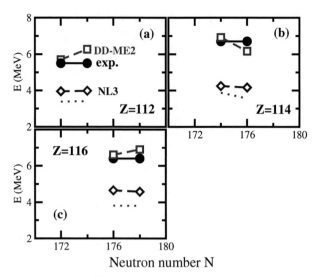

FIGURE 3. (Color online) The experimental and calculated heights of the inner fission barriers in superheavy nuclei.

isospin and surface properties of the DD-ME2 effective interaction.

Finally, in Tab. 1 we compare the energies of the fission isomers in several even-even nuclei, for which reliable data [16] are available. Although both effective interactions provide similar results, those obtained using the DD-ME2 force are somewhat closer to experiment. In conclusion, we have reported selected applications of the covariant nuclear density functional framework. By adjusting the universal functional to nuclear matter and ground-state properties of a small number of spherical nuclei, this approach is very successful in describing a variety of structure phenomena throughout the table of nuclides.

REFERENCES

1. D. Vretenar, A.V. Afanasjev, G.A. Lalazissis, P. Ring *Phys. Rep.*409 (2005) 101.
2. G. A. Lalazissis, T. Nikšić, D. Vretenar, and P. Ring, *Phys. Rev.* 71 (2005) 024312.
3. G. Audi, A. H. Wapstra, and C. Thibault, *Nucl. Phys.*A729 (2003) 337.
4. J. F. Berger, M. Girod, and D. Gogny, *Nucl. Phys.*A428 (1984) 23.
5. S. Goriely, M. Samyn, M. Bender, and J. M. Pearson, *Phys. Rev.*68 (2003) 054325.
6. G.A. Lalazissis, S. Raman and P. Ring, *Atom. Data Nucl. Data Tables* 71 (1999) 1.
7. T. Nikšić, D. Vretenar, and P. Ring, *Phys. Rev.* 66 (2002) 064302.
8. N. Paar, P. Ring, T. Nikšić and D. Vretenar, *Phys. Rev.* 67 (2003) 034312.
9. D.H. Youngblood, H.L. Clark, and Y.W. Lui, *Phys. Rev. Lett.* 82 (1999) 691.
10. J. Ritman et al., *Phys. Rev. Lett.* 70 (1993) 533.
11. B. L. Berman and S. C. Fultz, *Rev. Mod. Phys.* 47 (1975) 713.
12. T. Li et al., *Phys. Rev. Lett.* 99 (2007) 162503.
13. S. Karatzikos et al, (2009) submitted for publication
14. M. G. Itkis, Yu. Ts. Oganessian, V.I. Zagrebaev, *Phys. Rev.* C65 (2002) 044602.
15. G.A. Lalazissis, J. König, and P. Ring, *Phys. Rev.*55 (1997) 540.
16. B. Singh, R. Zywina, and R. B. Firestone, *Nucl. Data Sheets* 97 (2002) 241.

Coupled-Channel Green's Function Approach to Deformed Continuum Hartree-Fock-Bogoliubov Theory

Hiroshi Oba* and Masayuki Matsuo*,†

*Graduate School of Science and Technology, Niigata University, Niigata 950-2181, Japan
†Department of Physics, Faculty of Science, Niigata University, Niigata 950-2181, Japan

Abstract. We formulate the coordinate-space Hartree-Fock-Bogoliubov (HFB) theory which can describe pair correlated and deformed nuclei near the drip-lines. In order to ensure the correct asymptotic form of weakly bound and continuum quasiparticle states, we adopt a coupled-channel representation based on the partial wave expansion, and then we construct the exact HFB Green's function. Numerical examples calculated for ^{38}Mg are discussed.

Keywords: Hartree-Fock-Bogoliubov theory, pair correlation, deformed drip-line nuclei
PACS: 21.10.Gv,21.10.Pc,21.60.Jz,27.30.+t

INTRODUCTION

The RI-beam facilities in the new generation will enable us to explore nuclei near the neutron-drip line in the $10 \lesssim Z \lesssim 20$ and $N \gtrsim 20$ region. A new feature is that near-drip-line nuclei in this region are predicted to be systematically deformed [1]. In other words, these species may be called "weakly bound deformed nuclei", with which we can study interplays among the presence of weakly bound neutrons, the deformation, and other many-body correlations such as the pairing phenomenon. What we need then are appropriate theoretical methods which fit to the new circumstance.

If we consider the Hartree-Fock-Bogoliubov (HFB) theory, it must be formulated in such a way that the theory ensures the correct asymptotic wave functions of the quasi-particle states which are either bound or scattering under the influence of deformed mean-field potentials. As the first achievement in this direction, we refer to the work by Stoistov et al. [2] which employed the PTG basis set having analytic forms both for bound and scattering states. Alternatively the quasiparticle wave functions are expanded in the partial waves, and the differential equations of a coupled-channel form for the set of radial wave functions may be directly solved with the correct asymptotic boundary conditions. The latter was undertaken in the works by Hamamoto [3], which however lacked the scheme to determine the selfconsistent pair potential. This is the point that we overcome in the present work. To achieve this goal, we employ the Green's function formalism by Belyaev et al. [4] and we extend it to deformed cases. Details of the present work can be found in Ref. [5].

CP1165, *Nuclear Structure and Dynamics '09*
edited by M. Milin, T. Nikšić, D. Vretenar, and S. Szilner
© 2009 American Institute of Physics 978-0-7354-0702-2/09/$25.00

DEFORMED CONTINUUM HFB THEORY IN THE GREEN'S FUNCTION FORMALISM

The wave function $\phi(r\sigma,E)$ of the quasiparticle state at excitation energy E may be expanded with respect to the partial waves $L \equiv (jlm)$:

$$\phi(r\sigma,E) = \sum_L \phi_L(r,E)Y_L(\hat{r}\sigma), \quad \phi_L(r,E) = \begin{pmatrix} \phi_L^{(1)}(r,E) \\ \phi_L^{(2)}(r,E) \end{pmatrix}. \tag{1}$$

Here we need two components because of the presence of the pair correlation, and the radial wave functions $\phi_L(r,E)$ for different L's couple due to the deformation. Accordingly the HFB equation, which governs $\phi_L(r,E)$, is written as a coupled-channel differential equation. We solve the coupled-channel HFB equation with the boundary conditions, one at the origin and the other in the asymptotic region. For the latter we have solutions satisfying

$$\phi_L^{(out)}(r,E) \xrightarrow{r\to\infty} \begin{pmatrix} \frac{H_l^+(k_+,r)}{r}\delta_{LL'} \\ 0 \end{pmatrix}, \quad \begin{pmatrix} 0 \\ \frac{H_l^+(k_-,r)}{r}\delta_{LL'} \end{pmatrix}, \tag{2}$$

where $H^+(k,r)$ is the out-going Hankel/Coulomb functions, and the wave number is given by $k_\pm(E) = \sqrt{2m(\lambda \pm E)}/\hbar$ with an appropriate branch cut. Here λ is the Fermi energy for neutrons/protons. Let N be the truncated number of partial waves, then there are $2N$ independent solutions satisfying the above boundary conditions. We combine these solutions to form a single $2N \times 2N$ matrix function $\Phi^{(out)}(r,E)$. Similarly the solutions satisfying the boundary condition at the origin can be represented as a $2N \times 2N$ matrix wave function $\Phi^{(in)}(r,E)$.

We can construct the Green's function $G(E)$ for the HFB equation. In the coordinate-space representation it is written as

$$G(r\sigma,r'\sigma',E) = \sum_{LL'} Y_L(\hat{r}\sigma)g_{LL'}(r,r',E)Y_{L'}^*(\hat{r}'\sigma'), \tag{3}$$

where $g_{LL'}(r,r',E)$ is a 2×2 matrix corresponding to the two-component structure, but we can also define a $2N \times 2N$ matrix form $g(r,r',E)$ combining all $g_{LL'}(r,r',E)$. The exact Green's function is then given by

$$g(r,r',E) = \Phi^{(in)}(r,E)C^{(in)^T}(r',E)\theta(r'-r) + \Phi^{(out)}(r,E)C^{(out)^T}(r',E)\theta(r-r'), \tag{4}$$

where $C^{(out/in)}(r,E)$ is a subsidiary $2N \times 2N$ matrix function which should satisfy

$$\begin{pmatrix} \Phi^{(in)}(r,E) & -\Phi^{(out)}(r,E) \\ -\frac{d}{dr}\Phi^{(in)}(r,E) & \frac{d}{dr}\Phi^{(out)}(r,E) \end{pmatrix} \begin{pmatrix} C^{(in)^T}(r,E) \\ C^{(out)^T}(r,E) \end{pmatrix} = \frac{1}{r^2}\begin{pmatrix} 0 \\ 1 \end{pmatrix}. \tag{5}$$

Once the exact HFB Green's function is given, we can perform the summation over all the quasiparticle states to calculate the generalized density matrix $R(r\sigma,r'\sigma')$ by means

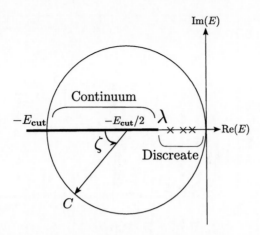

FIGURE 1. The contour path C in the complex plane of the quasiparticle energy E, which is used in Eq. (6).

of a contour integral in the complex E plane [4]:

$$R(r\sigma, r'\sigma') = \sum_{LL'} Y_L(\hat{r}\sigma)R_{LL'}(r, r')Y_{L'}^*(\hat{r}'\sigma'), \quad R_{LL'}(r, r') = \frac{1}{2\pi i}\int_C g_{LL'}(r, r', E)dE. \quad (6)$$

We choose the contour C as a circular path shown in Fig.1. This choice makes the integrand smooth, and we can perform the numerical integration efficiently using the higher-order Gauss-Legendre quadrature. It is noted that we sum up the contributions from continuum quasiparticle states, which are located along the real E axis both in $E < \lambda$ and $E < |\lambda|$, without introducing any discretization of the continuum quasiparticle spectrum.

NUMERICAL EXAMPLES

We have developed a new C++ code which employs the above formalism to obtain a converged HFB solution by a standard iterative procedure. Calculations have been carried out for ^{38}Mg, which is located close to the neutron drip-line. We use a deformed Woods-Saxon potential with a prolate deformation $\beta = 0.3$, and the density-dependent delta interaction (DDDI) to derive the pair potential. The parameters of the DDDI is fixed [6] to reproduce i) the scattering length $a = -18.5$fm of the bare nuclear force in the ^1S channel, ii) the BCS pairing gap in neutron matter at low densities, and iii) the experimental pairing gap in ^{120}Sn.

Fig. 2 shows the calculated pair number density $\tilde{n}(E)$ as a function of the quasiparticle energy E. It is the contribution of the quasiparticle state at energy E to the pair number $\tilde{N} = \int dr\tilde{\rho}(r) = \int_0^{E_{cut}} dE\tilde{n}(E)$, where $\tilde{\rho}(r)$ is the pair density. The peaks having narrow widths are resonances corresponding to the bound Nilsson orbits. The most noticeable feature seen in Fig. 2 is that there exist significant contributions from the "non-resonant" quasiparticle states, especially for the states with $\Omega = \frac{1}{2}$ and $\frac{3}{2}$. The contribution of the

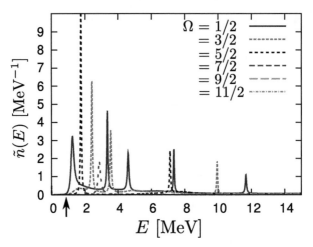

FIGURE 2. (Color online) The pair number density $\tilde{n}(E)$ as a function of the quasiparticle energy E, calculated for neutrons in ^{38}Mg.

non-resonant continuum quasiparticle states to the pair density (or the pair number) has been overlooked in the previous work [3].

The non-resonant continuum quasiparticle states play important roles as the Fermi energy λ approaches to zero (as the binding of the system becomes weaker). To observe this trend we have examined how the pairing properties varies by changing the depth of the Woods-Saxon potential with the other parameters kept constant. The average pairing gap increases by about $10 - 30\%$ (depending on its definition) when λ varies from $\lambda = -2.8$ to -0.47MeV, and we found that this is due to the increase of the contribution of the non-resonant quasiparticle states.

In the present model the energy of the neutron Nilsson orbit $[310]\frac{1}{2}$ is very close to the Fermi energy λ. The lowest energy peak in Fig. 2 corresponds to this orbit. It has been argued in Ref. [3] that the wave function of a weakly bound $\Omega = \frac{1}{2}$ state decouples from the pair field, and therefore the state-dependent pairing gap associated with this kind of orbit decreases as the binding energy becomes smaller. In the present calculation, however, we did not see the decrease of the state-dependent pairing gap of the weakly bound orbit $[310]\frac{1}{2}$. This is because the enhancement of the neutron pairing due to the weak binding effect compensates the decrease due to the decoupling effect.

REFERENCES

1. See for instance, J. Dobaczewski, M. V. Stoitsov, W. Nazarewicz, "Skyrme-HFB deformed nuclear mass table", *AIP Conference Proceedings* **726**, ed. R. Bijker, R. F. Casten, A. Frank, (American Institute of Physics, New York, 2004), p. 51.
2. M. Stoitsov, N. Michel and K. Matsuyanagi, *Phys. Rev.* **C77**, 054301 (2008).
3. I. Hamamoto, *Phys. Rev.* **C71**, 037302 (2005); *ibid.* **73**, 044317 (2006).
4. S. T. Belyaev, A. V. Smirnov, S. V. Tolokonnikov and S. A. Fayans, *Sov. J. Nucl. Phys.* **45**, 783 (1987).
5. H. Oba, and M.Matsuo, preprint arXiv:0905.3206[nucl-th].
6. M. Matsuo, Y. Serizawa and K. Mizuyama, *Nucl. Phys.* **A788**, 307c (2007).

Interplay of Tensor Correlations and Vibrational Coupling for Nuclear Single-Particle States

Gianluca Colò*, Hiroyuki Sagawa[†] and Pier Francesco Bortignon*

*Dipartimento di Fisica, Università degli Studi di Milano and INFN, Sezione di Milano, via Celoria 16, 20133 Milano, Italy
[†]Center for Mathematics and Physics, University of Aizu, Aizu-Wakamatsu, Fukushima 965-8560, Japan

Abstract. In this contribution we introduce, for the first time, a fully microscopic approach to particle-vibration coupling (PVC) based on the use of the Skyrme effective interactions. The capability of these forces to describe single-particle states in atomic nuclei, is a longstanding issue; it is certainly clear that the fragmentation of the single-particle strength lies beyond any mean field framework. After describing the formalism on which our microscopic approach is based, we discuss few preliminary results for ^{40}Ca and ^{208}Pb. Some perspectives are presented.

Keywords: Effective interactions; single-particle levels; particle-vibration coupling
PACS: 21.10.Pc, 21.30.Fe, 21.60.Jz

INTRODUCTION

The nuclear implementation of Density Functional Theory (DFT) is believed to be the most microscopic way one can use to study in a systematic way the ground states and the excitations of medium-heavy and heavy nuclei (up to the limiting case of infinite nuclear matter). This in keeping with the fact that the so-called *ab-initio* approaches cannot be used except for mass number A smaller than $\approx$ 10-15. In the nonrelativistic framework, DFT is mostly realized by starting from parametrizations of an effective interaction, and building from it the energy density as expectation value of the effective Hamiltonian over the most general Slater determinant (compatible with the symmetries of the system under study). In this respect, we can still talk of self-consistent mean-field (SCMF) when we start from a Skyrme effective force.

There are groups who are intensively working on the development of a very general form for energy density functionals, which may not be necessarily extracted from an effective Hamiltonian. The basic question which remains to be solved is to what extent this will allow improving on the existing functionals like those based on Skyrme or Gogny forces.

In particular, we focus in this contribution on single-particle states. Recently, many groups [1, 2, 3, 4, 5, 6, 7, 8] have devoted much attention to the role played by the tensor terms of the effective interactions like the Skyrme one, following the claim of the authors of Ref. [9], that the tensor force is crucial for the understanding of single-particle evolution in exotic nuclei.

Single-particle energies are not probably the best observable to consider within the realm of DFT. While the energies of the highest occupied, or lowest unoccupied, single-

CP1165, *Nuclear Structure and Dynamics '09*
edited by M. Milin, T. Nikšić, D. Vretenar, and S. Szilner
© 2009 American Institute of Physics 978-0-7354-0702-2/09/$25.00

particle states lies could in principle be predicted by DFT, due to the Koopman's theorem, probably other states and certainly the fragmentation of the single-particle strength, lie outside the predictive power of DFT calculations. This should cause no surprise, in view of analogous situations in condensed matter physics (cf., e.g., the gap in semiconductor or insulating bulk materials [10]).

To describe that fragmentation, calculations based on the idea of the coupling between the single-particle states and the collective nuclear vibrations (particle-vibration coupling, or PVC, model) have been available for several decades. We can refer to the book by A. Bohr and B.R. Mottelson [11] as well as to the extensive review [12]. Most of the studies based on PVC rely on phenomenological inputs. In almost all the cases, the basic idea is that the vibrations to be coupled to the single-particle states, are density fluctuations whose energies and form factors should be extracted from experiment. Discarding the problem of the link with a microscopic effective Hamiltonian, lies at the basis of the Bohr-Mottelson's model.

In this contribution, we pursue a different line. We would like to assess to what extent single-particle states can be reasonably described in a calculation based on an effective force, and how relevant the corrections due to PVC are, *once they are consistently calculated*. There exist few calculations of this type. V. Bernard and N. Van Giai [13] performed calculations based on the Skyrme force SIII but they made the approximation of dropping, in the evaluation of the PVC vertex, the velocity-dependent terms.

The purpose of our work is to present calculations of the particle-vibration coupling (PVC) which are based on fully self-consistent Random Phase Approximation (RPA) for the phonons, but are also free of all approximations like those made in Ref. [13]. To start this program, we present in this contribution results obtained by including the velocity-dependent terms in the PVC vertex.

FORMALISM

If the RPA amplitudes X and Y of the phonon Ln (we consider here only natural parity phonons with angular momentum L, labelled by n) are known, the PVC vertex, namely the reduced matrix element between the configuration made up with a state j and the phonon, and the state i, reads

$$\langle i||V_{\mathrm{ph}}||j, Ln\rangle = \sum_{\mathrm{ph}} \left(X_{\mathrm{ph}}^{Ln} + Y_{\mathrm{ph}}^{Ln} \right) \sqrt{2L+1} \, V_L(ihjp), \tag{1}$$

where V_L is the p-h coupled matrix element of the full particle-hole interaction V_{ph}.

In the present work, we use this form for the vertex. Aside from this, the shifts ΔE are calculated within second-order perturbation theory as in Refs. [12, 13]. We start from the Hartree-Fock solution associated with a given Skyrme interaction (in particular, we have chosen the parameter set SLy5 [14]), and we solve self-consistently the RPA equations. On top of the HF solution, we add perturbatively both the effect of the tensor force and that of PVC.

Our approach does not include any free parameter. Strictly speaking, the only free parameter is the choice of the model space. Phonons with multipolarity L from 0 to 4

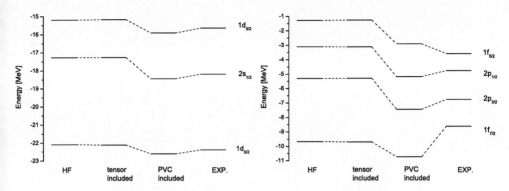

FIGURE 1. Results for ^{40}Ca. See the text for a discussion

(for Ca) and from 0 to 5 (for Pb), and natural parity, have been considered. Only those having energy smaller than 30 MeV, and fraction of total strength larger than 5% have been included in the calculations.

RESULTS

In Fig. 1, we display the results obtained for the nucleus ^{40}Ca. We consider, in this and in the following case, neutron states belonging to one major shell below and above the Fermi surface. As described above, on top of HF (first column of both panels), the tensor interaction with the same parameters of Refs. [3, 6] is added (second column) and then the particle-vibration coupling (third column). In the last column, the experimental results are shown. In this case, the tensor contribution is negligible, as it can be expected for a system in which both spin-orbit partners are filled. The shifts due to the particle-vibration coupling are not negligible, rather they are of the order of $\approx$ MeV. These shifts go in the direction of improving the agreement with experiment, except only for the $f_{7/2}$ state. The r.m.s. deviation between experimental and theoretical values improves from 1.40 MeV to 0.96 MeV if one includes the PVC contribution on top of HF plus tensor.

The results for ^{208}Pb have been obtained in the same way, and are displayed in Fig. 2 in full analogy with the previous figure. These results are more puzzling. Due to cancellations between diagrams (many angular momentum states are available and favour these cancellations), the energy shifts due to PVC are small (especially for the hole states). The r.m.s. deviation improves consequently only from 1.60 MeV to 1.25 MeV (as above, when including PVC on top of HF plus tensor).

In comparing with results obtained only with the velocity-independent part, we have found that the velocity-dependent contribution tend to cancel quite significantly the PVC vertex, and to reduce the obtained values of ΔE. This is the main outcome of the present investigation. For instance, if we consider the neutron hole state $f_{7/2}$ in ^{208}Pb, the value of ΔE=2.43 MeV obtained with only the (t_0, t_3) part, reduces to ΔE=0.63 MeV if one employs the full force. This latter and exact result is quite different from what one could obtain by approximating the (t_1, t_2) part taking the s.p. momentum fixed at k_F.

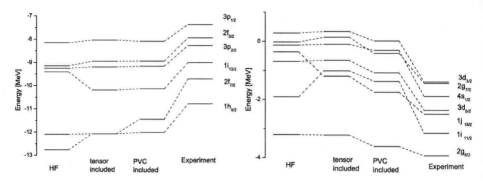

FIGURE 2. Results for ^{208}Pb. See the text for a discussion

PERSPECTIVES

The work presented here is a first step towards an exact calculation of the particle-vibration coupling (PVC) in nuclei, based on effective Hamiltonians including a zero-range interaction. Our main finding is that in case of the most used Skyrme forces, the velocity-dependent part tends to reduce the coupling strength of s.p. states with collective vibrations. We should now assess the role of the spin-orbit force, of the Coulomb force (in the case of protons), and of other modes than the density fluctuations. We already calculated spectroscopic factors (which are not discussed here for the sake of brevity). Going beyond second-order perturbation theory, and taking pairing into account, belong to a longer term perspective. The general aim is to have a unified understanding also of exotic nuclei.

REFERENCES

1. B. A. Brown, T. Duguet, T. Otsuka, D. Abe, and T. Suzuki, *Phys. Rev.* **C74**, 061303(R) (2006).
2. J. Dobaczewski, in *Proceedings of the Third ANL/MSU/JINA/INT RIA Workshop*, edited by T. Duguet, H. Esbensen, K. M. Nollett, and C. D. Roberts (World Scientific, Singapore, 2006).
3. G. Colò, H. Sagawa, S. Fracasso, and P. F. Bortignon, *Phys. Lett.* **B646**, 227 (2007).
4. D. M. Brink and F. Stancu, *Phys. Rev.* **C75**, 064311 (2007).
5. T. Lesinski, M. Bender, K. Bennaceur, T. Duguet, and J. Meyer, *Phys. Rev.* **C76**, 014312 (2007).
6. W. Zou, G. Colò, Z. Ma, H. Sagawa, and P. F. Bortignon, *Phys. Rev.* **C76**, 014314 (2007).
7. M. Grasso, Z. Ma, E. Khan, J. Margueron, and N. Van Giai, *Phys. Rev.* **C76**, 044319 (2007).
8. M. Zalewski, J. Dobaczewski, W. Satula, and T. R. Werner, *Phys. Rev.* **C77**, 024316 (2008).
9. T. Otsuka, T. Suzuki, R. Fujimoto, H. Grawe, and Y. Akaishi, *Phys. Rev. Lett.* **95**, 232502 (2005); T. Otsuka, T. Matsuo, abd D. Abe, *Phys. Rev. Lett.* **97**, 162501 (2006).
10. F. Aryasetiawan, O. Gunnarsson, *Rep. Progr. Phys.* **61**, 237 (1998).
11. A. Bohr, B. R. Mottelson, *Nuclear Structure, vol. II* (Benjamin, Reading Mass., 1975).
12. C. Mahaux, P. F. Bortignon, R. A. Broglia, C. H. Dasso, *Phys. Rep.* **120**, 1 (1985).
13. V. Bernard and N. Van Giai, *Nucl. Phys.* **A348**, 75 (1980).
14. E. Chabanat, P. Bonche, P. Haensel, J. Meyer, R. Schaeffer, *Nucl. Phys.* **A635**, 231 (1998).

Mean-Field and RPA Approaches to Stable and Unstable Nuclei with Semi-Realistic Interactions

H. Nakada

Department of Physics, Graduate School of Science, Chiba University,
Inage, Chiba 263-8522, Japan

Abstract. Semi-realistic NN interactions have been developed by modifying the M3Y interaction, which was derived from the G-matrix, so as to reproduce the saturation and the spin-orbit splittings in the mean-field regime. It is shown, in the mean-field and/or the random-phase approximations, that the nuclear matter properties and many properties of spherical nuclei are described to reasonable accuracy, by a semi-realistic interaction M3Y-P5$'$. Significance of the tensor force is confirmed for the N-dependence of proton single-particle levels in the Sn isotopes and for the $M1$ strength distribution in ^{208}Pb. Applying the semi-realistic interaction, we investigate magicity of $N = 32, 34$ and 40 in neutron-rich Ca nuclei.

Keywords: Semi-realistic interaction, Mean-field theory, Random-phase approximation, Tensor force, Unstable nuclei
PACS: 21.30.Fe, 21.60.Jz, 21.10.Dr, 21.10.Pc

INTRODUCTION

Whereas there has been notable progress in describing low energy phenomena of nuclei from the bare NN (and NNN) interactions, such approaches have not yet achieved full success in reproducing structure of medium- to heavy-mass nuclei [1, 2]. On the other hand, despite efforts to determine effective interactions (or energy density functionals) by fitting to the known data, it is questionable whether approaches in this line have predictive power for unknown nuclei. In today's nuclear structure physics, experimentally accessible region of nuclear chart is rapidly expanding, while predictablity is more and more needed in connection to astrophysics problems. Under these circumstances, we have proposed semi-realistic interactions [3, 4] and applied them to mean-field (MF) and RPA calculations, as an attempt for pursuing a practical and reliable theory.

We obtain semi-realistic interactions as follows. Starting from the M3Y interaction [5, 6] that was derived by fitting the Yukawa functions to Brueckner's G-matrix, we add a density-dependent contact term and modify some of the strength parameters, so as to reproduce the saturation and the spin-orbit splitting within the MF regime [3, 4]. The longest range part of the central force is kept identical to that of the one-pion exchange potential, $v_{\mathrm{OPEP}}^{(C)}$. The tensor force is not changed from that of the original M3Y interaction based on the Paris potential [6]. It is noted that a part of many-body force effects may be incorporated in the density-dependent term. The parameter-set M3Y-P5 was derived in Ref. [4]. In this paper we shall mainly show results obtained from the parameter-set M3Y-P5$'$, which is similar to M3Y-P5 but slightly modified in the pairing channel [7].

CP1165, *Nuclear Structure and Dynamics '09*
edited by M. Milin, T. Nikšić, D. Vretenar, and S. Szilner
© 2009 American Institute of Physics 978-0-7354-0702-2/09/$25.00

TABLE 1. Nuclear matter properties at the saturation point.

		SLy5	D1S	M3Y-P5$'$
k_{F0}	(fm^{-1})	1.334	1.342	1.340
$\mathcal{E}_0$	(MeV)	-15.98	-16.01	-16.14
$\mathcal{K}$	(MeV)	229.9	202.9	239.1
	M_0^*/M	0.697	0.697	0.637
a_t	(MeV)	32.03	31.12	28.42
	κ	0.250	0.660	0.884
	g_0	1.123	0.466	0.216
	g_0'	-0.141	0.631	1.007
	g_1'	1.043	0.610	0.146

NUCLEAR MATTER PROPERTIES

In Table 1, properties of the infinite nuclear matter are compared among several effective interactions, M3Y-P5$'$, the Skyrme SLy5 interaction [8] and the Gogny D1S interaction [9], within the Hartree-Fock (HF) theory [3]. All the interactions give close values to one another for the saturation density ($\rho_0 = 2k_{F0}^3/3\pi^2$), the saturation energy ($\mathcal{E}_0$), the incompressibility ($\mathcal{K}$), the effective k-mass (M_0^*) and the volume symmetry energy (a_t), because these quantities are more or less fitted to the data. On the contrary, we find significant interaction-dependence in the enhancement factor of the $E1$ energy-weighted sum (κ), and in the Landau-Migdal parameters relevant to the spin degrees of freedom (g_ℓ, g_ℓ').

Measurements have disclosed $\kappa \gtrsim 0.7$ [10] and $g_0' \approx 1$ [11]. Although experimental information is limited, g_0 seems relatively small. These properties are reproduced by M3Y-P5$'$ reasonably well. Not necessarily true for phenomenological interactions such as SLy5 and D1S, this may be an advantage of semi-realistic interactions. A typical case is found in g_0', for which $v_{\text{OPEP}}^{(C)}$ significantly contributes to the M3Y-P5$'$ result, as to the results with other M3Y-type interactions [3, 4].

MEAN-FIELD CALCULATIONS OF SPHERICAL NUCLEI

The semi-realistic interaction M3Y-P5$'$ is applied to MF calculations of spherical nuclei, and the results are compared with those of the D1S inteaction as well as with the available data. Computations are implemented by using the Gaussian expansion method [12, 13, 14]. Note that the results of M3Y-P5$'$ have no qualitative difference from those of M3Y-P5 [4].

It is shown in Table 2 that binding energies and rms matter radii of doubly magic nuclei, which are calculated in the spherical HF approximation, are reproduced to reasonable accuracy by M3Y-P5$'$. Although there remains certain discrepancy in the binding energies, we do not take this difference seriously, because correlations due to the residual interaction are ignored in the present calculations.

TABLE 2. Binding energies (MeV) and rms matter radii (fm) of doubly magic nuclei. Experimental data are taken from Refs. [15, 16, 17, 18].

		D1S	M3Y-P5$'$	Exp.
^{16}O	$-E$	129.5	124.1	127.6
	$\sqrt{\langle r^2 \rangle}$	2.61	2.60	2.61
^{40}Ca	$-E$	344.6	331.7	342.1
	$\sqrt{\langle r^2 \rangle}$	3.37	3.37	3.47
^{48}Ca	$-E$	416.8	411.5	416.0
	$\sqrt{\langle r^2 \rangle}$	3.51	3.51	3.57
^{90}Zr	$-E$	785.9	775.7	783.9
	$\sqrt{\langle r^2 \rangle}$	4.24	4.23	4.32
^{208}Pb	$-E$	1639.0	1635.7	1636.4
	$\sqrt{\langle r^2 \rangle}$	5.51	5.51	5.49

It is expected that ground-state properties of semi-magic nuclei are well described in the spherical Hartree-Fock-Bogolyubov (HFB) approximation. Difference between the HF and the HFB energies represents how strong the pair correlations are, and therefore gives a measure of degree of shell closure. There have been arguments whether $N = 32$ and/or 34 behave as magic numbers around neutron-rich Ca region [19, 20]. Both of the D1S and M3Y-P5$'$ give prediction that $N = 32$ is almost magic but $N = 34$ is not, for the Ca isotopes. On the contrary, predicted property of ^{60}Ca depends on the effective NN interactions. While ^{60}Ca is nearly a doubly magic nucleus by D1S, we have a significant pair correlation by M3Y-P5$'$, because of difference in the shell structure [4].

It has been argued that the tensor force significantly affects the single-particle (s.p.) levels [21], possibly accounting for the N-dependence of $\varepsilon_p(0h_{11/2}) - \varepsilon_p(0g_{7/2})$ on top of the Sn isotopes [22]. To draw a definite conclusion on this problem, it is desired to apply a realistic tensor force, together with a central force having reasonable characters. The semi-realistic interactions are suitable for this purpose. Using M3Y-P5$'$, we clearly view that the tensor force plays a crucial role in the N-dependence of the relative proton s.p. energies, as was already confirmed by M3Y-P5 [4].

RPA CALCULATIONS OF SPHERICAL NUCLEI

The semi-realistic interactions have also been applied to excited states of doubly magic nuclei in the self-consistent RPA [23]. Note that the residual interaction includes the c.m. Hamiltonian as well as the Coulomb force. Odd-parity lowest levels in ^{16}O are described reasonably well [24]. In the investigation of the $M1$ strengths in ^{208}Pb, the RPA caluclation with M3Y-P5 describes the excitation energy and transition strength of the low-lying isoscalar-dominant state to good accuracy [25]. The tensor force plays an important role in that result.

The excitation energy of the 2_1^+ state of ^{52}Ca is reproduced remarkably well, by the RPA caluclation with M3Y-P5$'$. This suggests reliability of the shell structure around $N = 32$ given by M3Y-P5$'$, and may support the result mentioned in the preceding section on the magicity of $N = 32$ and 34. This point will further be checked by the transition strength, for which M3Y-P5$'$ predicts $B(E2) \downarrow = 4.4\,e^2\mathrm{fm}^4$ within the RPA.

SUMMARY

Semi-realistic effective interactions have been developed, and been applied to self-consistent mean-field and RPA calculations. Selected results of the M3Y-P5$'$ parameter-set have been presented, which are in reasonable agreement with a good number of available data. Significance of the tensor force has been confirmed for the N-dependence of the proton single-particle energies of the Sn isotopes and for the low-lying $M1$ strength in ^{208}Pb. It is predicted that $N = 32$ is nearly magic but $N = 34$ is not in the neutron-rich Ca nuclei, consistent with the D1S result. Contradictory to D1S, a sizable pair correlation is predicted in ^{60}Ca, severely destroying magicity of $N = 40$.

Because of its relation to the microscopic theory, we may expect higher predictability in approaches with the semi-realistic interactions than those with conventional MF interactions. The present M3Y-type semi-realistic interaction keeps wide applicability at the same time, as demonstrated by implementing the MF and RPA calculations. It will be of interest to apply the semi-realistic interaction extensively, to deformed nuclei and unstable nuclei, as well as to theories that takes account of correlation effects.

REFERENCES

1. R. Roth, et al., *Phys. Rev. C* **73**, 044312 (2006).
2. G. Hagen, T. Papenbrock, D.J. Dean and M. Hjorth-Jensen, *Phys. Rev. Lett.* **101**, 092502 (2008).
3. H. Nakada, *Phys. Rev. C* **68**, 014316 (2003).
4. H. Nakada, *Phys. Rev. C* **78**, 054301 (2008).
5. G. Bertsch, J. Borysowicz, H. McManus and W.G. Love, *Nucl. Phys.* **A2 84**, 399 (1977).
6. N. Anantaraman, H. Toki and G.F. Bertsch, *Nucl. Phys.* **A 398**, 269 (1983).
7. H. Nakada, to be published.
8. E. Chabanat, P. Bonche, P. Haensel, J. Meyer and R. Schaeffer, *Nucl. Phys.* **A 635**, 231 (1998).
9. J.F. Berger, M. Girod and D. Gogny, *Comp. Phys. Comm.* **63**, 365 (1991).
10. A. Leprêtre, et al., *Nucl. Phys.* **A 367** 237, (1981).
11. C. Gaarde *et al.*, *Nucl. Phys.* **A 369**, 258 (1981); T. Suzuki, *Nucl. Phys.* **A 379**, 110 (1982); G. Bertsch, D. Cha and H. Toki, *Phys. Rev. C* **24**, 533 (1981); T. Suzuki and H. Sakai, *Phys. Lett.* **455B**, 25 (1999).
12. H. Nakada and M. Sato, *Nucl. Phys.* **A 699**, 511 (2002); *ibid.* **A 714**, 696 (2003).
13. H. Nakada, *Nucl. Phys.* **A 764**, 117 (2006); *ibid.* **A 801**, 169 (2008).
14. H. Nakada, *Nucl. Phys.* **A808**, 47 (2008).
15. G. Audi and A.H. Wapstra, *Nucl. Phys.* **A 595**, 409 (1995).
16. D.T. Khoa, H.S. Than and M. Grasso, *Nucl. Phys.* **A 722**, 92c (2003).
17. A. Ozawa *et al.*, *Nucl. Phys.* **A 691**, 599 (2001).
18. G.D. Alkhazov, S.L. Belostotsky and A.A. Vorobyov, *Phys. Rep.* **42**, 89 (1978).
19. J.I. Prisciandaro, et al., *Phys. Lett.* **510B**, 17 (2001).
20. M. Honma, T. Otsuka, B.A. Brown and T. Mizusaki, *Phys. Rev. C* **65**, 061301(R) (2002).
21. T. Otsuka, T. Suzuki, R. Fujimoto, H. Grawe and Y. Akaishi, *Phys. Rev. Lett.* **95**, 232502 (2005).
22. J.P. Schiffer, et al., *Phys. Rev. Lett.* **92**, 162501 (2004).
23. H. Nakada, K. Mizuyama, M. Yamagami and M. Matsuo, submitted to *Nucl. Phys. A*, arXiv:0904.4285[nucl-th].
24. H. Nakada, *Eur. Phys. J. A*, in press.
25. T. Shizuma, et al., *Phys. Rev. C* **78**, 061303(R) (2008).

Recent Developments about ΛN Spin-Orbit Interaction in Hypernuclei

Paolo Finelli

Physics Department, University of Bologna and INFN Section of Bologna, via Irnerio 46, I-40126 Bologna, Italy

Abstract. In the last years, a systematic study of Λ-hypernuclei has been initiated through a novel approach [1, 2] able to include contributions from the relevant degrees of freedom at low energies (π and kaon exchange interactions) into a (relativistic) energy density functional. In this contribution we will review the basic features of this approach and, in particular, we will report on some recent improvements [3] showing calculations for $^{16}_{\Lambda}$O.

Keywords: Chiral Dynamics, Λ-hypernuclei, Relativistic Mean Field Approximation
PACS: 21.10.Pc 21.60.Jz 21.80.+a

INTRODUCTION

Understanding the microscopic origin of the spin-orbit interaction is still one of the most interesting topics in nuclear physics. In particular, considering the $SU(3)$ flavor extension of the nuclear chart table, i.e. including Λ hypernuclei, no theoretical model can actually describe the large spin-orbit splittings observed in finite nuclei (i.e. $\Delta E \simeq$ 6 MeV for ^{16}O for p-states) and, at the same time, the quasi-degeneracy of spin-orbit partner levels in Λ-hypernuclei ($\Delta^{\Lambda} E(p) \simeq 100$ keV) [4, 5, 6]. In Refs. [2, 3] we introduced a relativistic framework in which the suppression of the ΛN spin-orbit interaction finds a very natural explanation in terms of an almost complete cancellation between short-range scalar/vector contributions and long-range terms generated by 2π exchange. In the following sections we briefly summarize the relevant features of this approach and show results for $^{16}_{\Lambda}$O.

THE MODEL

The relativistic energy density functional for single-Λ hypernuclei reads [2]:

$$E\left[\rho\right] = E^{N}[\rho] + E^{\Lambda}_{\text{free}}[\rho] + E^{\Lambda}_{\text{int}}[\rho], \qquad (1)$$

where $E^{N}[\rho]$ describes the core of protons and neutrons [1], $E^{\Lambda}_{\text{free}}$ is the free-Λ term and E^{Λ}_{int} collects the ΛN interaction in terms of density-dependent vector (G^{Λ}_{V}) and scalar (G^{Λ}_{S}) contact couplings and a surface term D^{Λ}_{S} (proportional to $\nabla \rho_{\Lambda} \cdot \nabla \rho$ in the equation of motion). Generally speaking, G^{Λ}_{V} and G^{Λ}_{S} contain mean-field contributions originating from short-distance dynamics (not resolved at the relevant energy scales) that can be generated by in-medium changes of the quark condensates (identified with

CP1165, *Nuclear Structure and Dynamics '09*
edited by M. Milin, T. Nikšić, D. Vretenar, and S. Szilner
© 2009 American Institute of Physics 978-0-7354-0702-2/09/$25.00

the superscript (0)) [7], and contributions from in-medium kaon- and 2π- exchange processes (with superscript (K, π)) [8]:

$$G_i^\Lambda(\rho) = G_i^{\Lambda(0)} + G_i^{\Lambda(K,\pi)}(\rho) \quad \text{with} \quad i = S, V \,. \tag{2}$$

In the next paragraphs, physics encoded in the couplings $G_i^\Lambda(\rho)$ will be shortly described.

$SU(3)$ in-medium chiral perturbation theory: π- and kaon-dynamics

Λ *Single particle potential.* In Ref. [8] the density dependent self-energy for a zero momentum Λ hyperon in isospin-symmetric nuclear matter has been calculated at two-loop order in the energy density. This calculation systematically includes kaon-exchange Fock terms and two-pion exchange with Σ hyperon and Pauli blocking effects in the intermediate state. This self-energy is translated into a mean field potential $U_\Lambda(k_f)$. A cutoff scale $\bar\Lambda$ (or equivalently, a contact term) represents short distance (high momentum) dynamics not resolved at scales characteristic of the Fermi momentum. In Fig. 1 (left) we show $U_\Lambda(k_f)$ as function of ρ for $\bar\Lambda = 700$ MeV. The density-dependent couplings $G_i^{\Lambda(K,\pi)}(\rho)$ are determined from $U_\Lambda(k_f)$ following the same procedures outlined in Refs.[1, 2].

ΛN *Spin-orbit potential.* In Ref. [8] the Λ-nucleus spin-orbit interaction generated by the in-medium two-pion exchange ΛN interaction has been evaluated. It depends only on known $SU(3)$ axial vector coupling constants and on the mass difference between Λ and Σ. The relevant momentum space loop integral is finite and hence model independent in the sense that no regularizing cutoff is required. The result $(U_{ls}^\Lambda(k_f^{(0)}) \simeq -15$ MeV fm^2 at $k_f^{(0)} \simeq 1.36$ fm^{-1}) is plotted in Fig. 1 (center) and has a sign *opposite* to the standard nuclear spin-orbit interaction. Evidently, this *wrong-sign* term should be (and will be) counterbalanced in some way. The inclusion of short-distance dynamics (in terms of scalar/vector background fields for example) is shown by the grey shaded area. For a more careful analysis about the derivation of this results and its consequences, we refer the reader to Refs. [8, 9]. In our approach, the effect of U_{ls}^Λ will be evaluated as a first order perturbative correction.

Λ *Gradient term.* This parameter can be evaluated from the spin-independent part of the self-energy of a Λ-hyperon interacting with weakly inhomogeneous nuclear matter [3]. In Fig. 1 (right) we plot the function $\rho D_\Lambda^{(\pi)}(\rho)$ as a function of ρ. For practical purposes it can be approximated by a constant value.

Short-distance dynamics

In addition to the mean field induced by kaon and two-pion exchange, there are also contributions from short distance dynamics that must be taken into account. They can

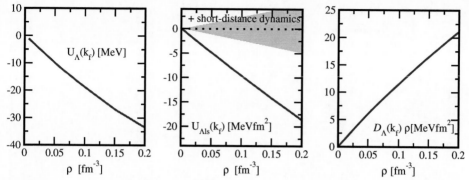

FIGURE 1. Left: Λ single particle potential as function of the density. Center: Λ*N* spin-orbit potential as function of the density (red colour) and the overall effect once the short-distance dynamics is included. Right: Λ gradient term [8] as function of the density.

be conveniently estimated introducing condensate background self-energies, as already shown in Refs.[1, 2]. They produce a sizeable spin-orbit potential with a *correct sign*. In general, one could estimate

$$G_{S,V}^{\Lambda(0)} = \zeta\, G_{S,V}^{(0)} \,, \tag{3}$$

where $G_V^{(0)}$ and $G_S^{(0)}$ are the vector and the scalar couplings to nucleons, arising from in-medium changes of the quark condensates, $\langle \bar{q}q \rangle$ and $\langle q^\dagger q \rangle$. ζ could vary from 2/3 (quark model prediction) to 0.4 (QCD sum rules predictions of Ref. [7]). In our approach ζ is taken as a free parameter.

Results

In Fig. 2 the Λ single particle energy levels of $^{16}_\Lambda$O are plotted (all theoretical calculations are performed for a simpler configuration, consisting of a closed core plus a single Λ hyperon). The parameters have been fixed ($\zeta = 0.5$, $\bar{\Lambda} = 721$ MeV and $D_S = -0.34$ fm^4) in order to reproduce known hypernuclei properties (see Ref. [3] for more details). In comparison we show experimental data [10] and other theoretical predictions [11]. The agreement with data is very good, in particular strengthening the hypothesis that Λ spin-orbit suppression can be easily explained once the correct low-energy degres of freedom are considered. We refer the reader to Ref. [3] for a complete investigation of the hypernuclear chart table.

ACKNOWLEDGMENTS

I would like to thank my collaborators N. Kaiser (TUM), W. Weise (TUM) and D. Vretenar (Zagreb). This work has been financed by INFN, MIUR and within the *Agreement of Scientific and Technological Cooperation between Italy and Croatia (2009-2010)*.

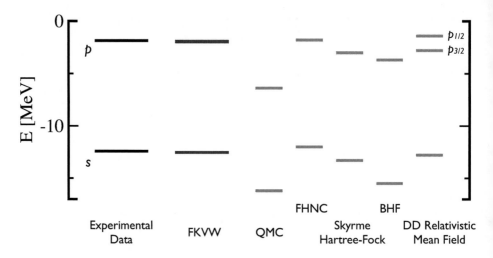

FIGURE 2. Λ single particle energy levels of $^{16}_{\Lambda}$O. Our results are compared with those of five different models: Quark-Meson coupling model (QMC), Fermi Hypernetted Chain (FHNC), Skyrme Hartree-Fock, Brueckner Hartree-Fock (BHF) and DD Relativistic Mean Field (DDRMF) [11].

REFERENCES

1. P. Finelli, N. Kaiser, D. Vretenar and W. Weise, *Nucl. Phys.* **A735**, 449 (2004). *Nucl. Phys.* **A770**, 1 (2006). *Nucl. Phys.* **A791**, 57 (2007).
2. P. Finelli, N. Kaiser, D. Vretenar and W. Weise, *Phys. Lett.* **B658**, 90 (2007).
3. P. Finelli, N. Kaiser, D. Vretenar and W. Weise, in preparation (2009).
4. O. Hashimoto and H. Tamura, it Prog. Part. Nucl. Phys. **57**, 564 (2006) and references therein.
5. *Topics in Strangeness Nuclear Physics, Lect. Not. in Phys.* **724** (2007).
6. *Special Issue on Strangeness Nuclear Physics, Nucl. Phys.* **A804** (2008).
7. T. D. Cohen, R. J. Furnstahl, D. K. Griegel and X. m. Jin, *Prog. Part. Nucl. Phys.* **35**, 221 (1995).
8. N. Kaiser and W. Weise, *Phys. Rev.* **C71**, 015203 (2005).
9. N. Kaiser and W. Weise, *Nucl. Phys.* **A804**, 60 (2008).
10. O. Hashimoto and H. Tamura, *Prog. Part. Nucl. Phys.* **57**, 564 (2006), and references therein.
11. **QMC**: P. A. M. Guichon, A. W. Thomas and K. Tsushima, *Nucl. Phys.* **A814**, 66 (2008); **FHNC**: F. Arias de Saavedra, G. Có and A. Fabrocini, *Phys. Rev.* **C63**, 64308 (2001); **Skyrme**: J. Cugnon, A. Lejeune and H.-J. Schulze, *Phys. Rev.* **C62**, 64308 (2000); **BHF**: I. Vidaña, A. Polls, A. Ramos and H.-J. Schulze, *Phys. Rev.* **C64**, 44301 (2001); **DDRMF**: C. M. Keil, F. Hofmann and H. Lenske, *Phys. Rev.* **C61**, 064309 (2000).

Imaginary Time Step Method to Solve the Dirac Equation with Nonlocal Potential

Ying Zhang*, Haozhao Liang*,† and Jie Meng *,**

*State Key Lab Nucl. Phys. & Tech., School of Physics, Peking University, Beijing 100871, China
†Institut de Physique Nucléaire, IN2P3-CNRS and Université Paris-Sud, F-91406 Orsay, France
**Department of Physics, University of Stellenbosch, Stellenbosch, South Africa

Abstract.
The imaginary time step (ITS) method is applied to solve the Dirac equation with nonlocal potentials in coordinate space. Taking the nucleus ^{12}C as an example, even with nonlocal potentials, the direct ITS evolution for the Dirac equation still meets the disaster of the Dirac sea. However, following the recipe in our former investigation, the disaster can be avoided by the ITS evolution for the corresponding Schrödinger-like equation without localization, which gives the convergent results exactly the same with those obtained iteratively by the shooting method with localized effective potentials.

Keywords: imaginary time step method, Dirac equation, nonlocal potential, coordinate space
PACS: 24.10.Jv, 21.60.-n, 02.60.Nm

To understand the phenomena in "exotic nuclei" [1, 2], especially the low density in the tails of nuclear matter distributions, one urgently need to solve the complex many-body problem in coordinate space. As one of the best candidates for the description of exotic nuclei, the density-dependent relativistic Hartree-Fock (DDRHF) theory [3] is developed and demonstrates the importance of the Fock terms in series of investigations [4, 5, 6, 7, 8]. More recently, the DDRHF theory is extended to the Bogoliubov representation (DDRHFB) for the description of the exotic nuclei [9].

For the future exploration of the deformed halo nuclei, a natural extension of the DDRHFB theory is to include the deformation degree of freedom in coordinate space. The imaginary time step (ITS) method [10] is just an effective approach to this end, which has achieved lots of success in the nonrelativistic mean field approach [11]. But, when applied in the relativistic system, the method will inevitably meet a great disaster due to the Dirac sea. However, taking the Dirac equation with scalar and vector potentials as an example, we demonstrated that this disaster can be avoided by the ITS evolution for the corresponding Schrödinger-like equation [12].

For the application of the ITS method in the relativistic Hartree-Fock theory, one has to solve the Dirac equation with nonlocal potentials due to the Fock terms. The corresponding Dirac equation will be a set of coupled integro-differential equations, which cannot be solved by the conventional shooting method directly unless with localized effective potentials. However, we demonstrated that the ITS evolution can be performed for the Schrödinger-like equation without localization [13].

In this paper, the ITS evolution will be applied to both the Dirac and Schrödinger-like equation with the nonlocal potentials. The results will be compared with those obtained iteratively by the shooting method with localized effective potentials.

CP1165, *Nuclear Structure and Dynamics '09*
edited by M. Milin, T. Nikšić, D. Vretenar, and S. Szilner
© 2009 American Institute of Physics 978-0-7354-0702-2/09/$25.00

For clarity and simplicity, a spherical symmetry is assumed for the nuclear system, and thus the radial Dirac equation with nonlocal potentials to be solved is

$$
\begin{pmatrix} V+S+M & -\frac{d}{dr}+\frac{\kappa_a}{r} \\ +\frac{d}{dr}+\frac{\kappa_a}{r} & V-(S+M) \end{pmatrix} \begin{pmatrix} F_a(r) \\ G_a(r) \end{pmatrix} + \begin{pmatrix} X_a(r) \\ Y_a(r) \end{pmatrix} = \varepsilon_a \begin{pmatrix} F_a(r) \\ G_a(r) \end{pmatrix}, \quad (1)
$$

where the nonlocal terms are

$$
X_a(r) = \int dr' U_X(r,r') F_a(r'), \quad U_X(r,r') \equiv u_{l_a}(r,r'), \tag{2}
$$

$$
Y_a(r) = \int dr' U_Y(r,r') G_a(r'), \quad U_Y(r,r') \equiv u_{l'_a}(r,r'). \tag{3}
$$

As demonstrated in Ref. [13], the ITS evolution will be performed without localization in the following.

Firstly, let us perform the ITS evolution directly for the radial Dirac equation with nonlocal potentials. Taking the nucleus ^{12}C as an example, the ITS evolutions of the neutron single-particle levels $s_{1/2}$ are shown in panel (a) of Fig. 1. The local and nonlocal potentials take the same form as in Ref. [13]. Only the local potentials are shown here to illustrate the location of the Fermi and Dirac sea. Since the evolution is performed in a spherical box with with $R = 10$ fm, and mesh size $dr = 0.5$ fm, one can evolve 19 orthogonal single-particle wave functions at most. The evolutions are shown for the lowest (1st) and the highest (19th) single-particle levels, and one can find that all the levels dive into the Dirac sea. This means that even with the nonlocal potentials, the ITS evolution for the Dirac equation still meets the disaster of the Dirac sea.

However, even in the disaster, one can still find some survivors such as the highest single-particle level (19th) in the Dirac sea. Its neutron radial wave functions $F(r)$, $G(r)$ and the corresponding nonlocal terms $X(r)$, $Y(r)$ are compared with the results of state $2s_{1/2}$ (labeled by the quantum number of the upper component) obtained by the shooting method in panel (b) of Fig. 1. The shooting method is performed iteratively with the localized effective potentials in a box with $R = 20$ fm and $dr = 0.1$ fm, which is supposed to give the exact solutions. The two results are in good agreement, which demonstrates that even with the nonlocal potentials, some physical solutions in the Dirac sea are accessible by the direct ITS evolution for the Dirac equation. However, it is at a great cost of precision, and more importantly, it still cannot give any solution in the Fermi sea.

To avoid the disaster of the Dirac sea, following the recipe in our former work [12], one should perform the ITS evolution for the corresponding Schrödinger-like equation. From the relation between the upper and lower components, the corresponding Schrödinger-like equation for the upper component reads

$$
-\frac{1}{M_+}\frac{d^2 F_a}{dr^2} + \frac{1}{M_+^2}\frac{dM_+}{dr}\frac{dF_a}{dr} + \left[(V+S+M) + \frac{1}{M_+^2}\frac{dM_+}{dr}\frac{\kappa_a}{r} + \frac{1}{M_+}\frac{\kappa_a(\kappa_a+1)}{r^2} \right] F_a
$$
$$
-\frac{1}{M_+}\frac{dY_a}{dr} + \frac{1}{M_+^2}\frac{dM_+}{dr}Y_a + \frac{1}{M_+}\frac{\kappa_a}{r}Y_a + X_a = \varepsilon_a F_a(r), \tag{4}
$$

with the effective mass $M_+ = M - (V - S) + \varepsilon_a$. The evolutions of the neutron single-particle energies in ^{12}C and the corresponding spectrum obtained by shooting method

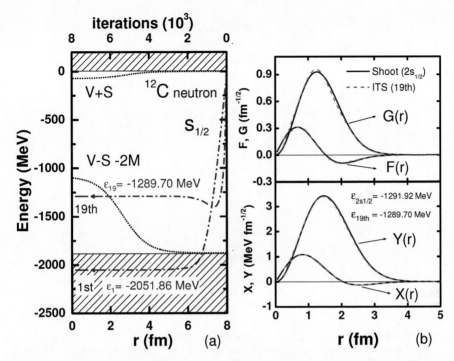

FIGURE 1. (Color online) Left: imaginary time step (ITS) evolutions of the neutron single-particle levels $s_{1/2}$ for the Dirac equation. Taking ^{12}C as an example, the local potentials (dotted line) are shown to illustrate the location of the Fermi and Dirac sea. The evolutions (dash-dotted line) are shown for the lowest (1st) and highest (19th) single-particle levels $s_{1/2}$, which are performed in a spherical box with $R = 10$ fm, and mesh size $dr = 0.5$ fm. The shaded areas are the continuum in the Fermi and Dirac sea. Right: neutron radial wave functions $F(r)$, $G(r)$ and the corresponding nonlocal terms $X(r)$, $Y(r)$ for the 19th level obtained by the ITS evolution for the Dirac equation (dashed line), in comparison with the state $2s_{1/2}$ obtained by the shooting method (solid line) with $R = 20$ fm and $dr = 0.1$ fm.

are shown in Fig. 2. Both the ITS evolution and shooting method are performed in a spherical box with $R = 20$ fm and $dr = 0.1$ fm. Due to the positive definite effective mass M_+, one can easily see that the evolutions are safely confined in the Fermi sea, and converge to exactly the same results with those obtained iteratively by the shooting method with localized effective potentials.

In conclusion, the ITS method is applied for solving the Dirac equation with nonlocal potentials. Similar as in Ref. [12], even with nonlocal potentials, the direct ITS evolution for the Dirac equation still meets the disaster of the Dirac sea. Although one could find some survivors in the Dirac sea, it is still not feasible for the poor precision, and none of the solutions can be obtained in the Fermi sea. Following the recipe as in Ref. [12] and [13], the ITS evolution is performed for the corresponding Schrödinger-like equation without localization. The evolution can be safely confined in the Fermi sea, and gives the convergent results exactly the same with those obtained iteratively by the shooting method with localized effective potentials.

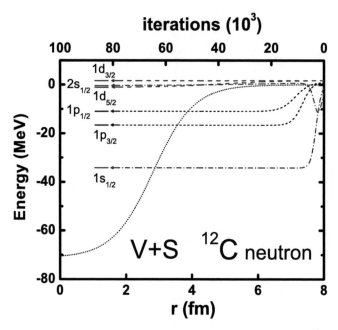

FIGURE 2. (Color online) Evolutions of the neutron single-particle energies in ^{12}C obtained by the ITS evolution for the Schrödinger-like equation, and the corresponding spectrum obtained by shooting method. Both the ITS evolution and shooting method are performed in a spherical box with $R = 20$ fm and $dr = 0.1$ fm.

ACKNOWLEDGMENTS

The authors are grateful to P. Ring, H. Sagawa, A. S. Umar, N. Van Giai, and D. Vretenar for helpful discussion. This work is partly supported by the Chinese-Croatian project "Nuclear structure far from stability", Major State 973 Program 2007CB815000, the NSFC under Grant Nos. 10435010, 10775004 and 10221003.

REFERENCES

1. I. Tanihata *et al.*, *Phys. Rev. Lett.* **55**, 2676 (1985).
2. B. Jonson, *Phys. Rep.* **389**, 1 (2004).
3. W. H. Long, N. Van Giai and J. Meng, *Phys. Lett.* **B640**, 150 (2006).
4. W. H. Long, H. Sagawa, N. Van Giai and J. Meng, *Phys. Rev.* **C76**, 034314 (2007).
5. W. H. Long, H. Sagawa, J. Meng and N. Van Giai, *Europhys. Lett.* **82**, 12001 (2008).
6. H. Z. Liang, N. Van Giai and J. Meng, *Phys. Rev. Lett.* **101**,122502 (2008).
7. B. Y. Sun, W. H. Long, J. Meng and U. Lombardo, *Phys. Rev.* **C78**, 065805 (2008).
8. H. Z. Liang, N. Van Giai and J. Meng, arXiv: 0904.3673[nucl-th].
9. W. H. Long, P. Ring, N. Van Giai and J. Meng, arXiv:0812.1103[nucl-th].
10. K. T. R. Davies, H. Flocard, S. Krieger, and M. S. Weiss, *Nucl. Phys.* **A342**, 111 (1980).
11. P. Bonche, H. Flocard, and P. H. Heenen, *Com. Phys. Com.* **171**, 49 (2005).
12. Y. Zhang, H. Z. Liang and J. Meng, arXiv:0905.2505[nucl-th].
13. Y. Zhang, H. Z. Liang and J. Meng, submitted to *Chinese Physics Letter*.

Covariant Energy Density Functionals: Time-Odd Channel Investigated

A.V. Afanasjev, H. Abusara

Department of Physics and Astronomy, Mississippi State University, Mississippi 39762, USA

Abstract. Time-odd mean fields (nuclear magnetism) are investigated in the framework of covariant density functional theory (CDFT). It is shown that they always provide additional binding to the binding energies of odd-mass nuclei. As a result, time-odd mean fields affect odd-even mass differences. They also have a profound effect on the properties of odd-proton nuclei in the vicinity of proton-drip line. Their presence can modify the half-lives of proton-emitters and considerably affect the possibilities of their experimental observation.

Keywords: Covariant density functional theory, time-odd mean fields
PACS: 21.60.Jz, 21.10.Dr, 21.10.Pc

While there was a dedicated effort to better understand time-odd (TO) mean fields in the framework of the Skyrme density functional (EDF) theory (see Refs. [1, 2] and references therein), much less attention has been paid to these fields in covariant density functional theory (CDFT) [3, 4, 5, 6, 7]. This is in part due to the fact that time-odd mean fields are defined through the Lorentz invariance in the CDFT (their coupling constants are the same as the ones for time-even mean fields) [7], while there are open problems with the definition of these fields in non-relativistic density functional theories [1, 2]. Our studies aim at better and systematic understanding of time-odd mean fields and their impact on physical observables in non-rotating [9] and rotating nuclei in the framework of the relativistic mean field (RMF) realization of the CDFT.

The presence of the magnetic potential $\mathbf{V}(\mathbf{r})$ in the Dirac equation as well as the currents $\mathbf{j}^{n,p}(\mathbf{r})$ in the Klein-Gordon equations in the systems with broken time-reversal symmetry leads to the appearance of time-odd mean fields in the CDFT. A magnetic potential $\mathbf{V}(\mathbf{r})$

$$\mathbf{V}(\mathbf{r}) = g_\omega \vec{\omega}(\mathbf{r}) + g_\rho \tau_3 \vec{\rho}(\mathbf{r}) + e\frac{1-\tau_3}{2}\mathbf{A}(\mathbf{r}), \qquad (1)$$

originates from the space-like components of the vector mesons. Note that in these equations, the four-vector components of the vector fields ω^μ, ρ^μ, and A^μ are separated into the time-like (ω_0, ρ_0 and A_0) and space-like [$\vec{\omega} = (\omega^x, \omega^y, \omega^z)$, $\vec{\rho} = (\rho^x, \rho^y, \rho^z)$, and $\mathbf{A} = (A^x, A^y, A^z)$] components. In the Dirac equation the magnetic potential has the structure of a magnetic field. Therefore, the effect produced by it is called *nuclear magnetism* (NM) [10].

The meson fields for the vector mesons are determined by the Klein-Gordon equations

$$\{-\Delta + m_\omega^2\} \,\omega_0(\mathbf{r}) \;=\; g_\omega[\rho_v^n(\mathbf{r}) + \rho_v^p(\mathbf{r})], \qquad (2)$$

$$\{-\Delta + m_\omega^2\} \,\vec{\omega}(\mathbf{r}) \;=\; g_\omega[\mathbf{j}^n(\mathbf{r}) + \mathbf{j}^p(\mathbf{r})] \qquad (3)$$

CP1165, *Nuclear Structure and Dynamics '09*
edited by M. Milin, T. Nikšić, D. Vretenar, and S. Szilner

$$\left\{-\Delta + m_\rho^2\right\} \rho_0(\mathbf{r}) = g_\rho [\rho_v^n(\mathbf{r}) - \rho_v^p(\mathbf{r})], \tag{4}$$

$$\left\{-\Delta + m_\rho^2\right\} \vec{\rho}(\mathbf{r}) = g_\rho [\mathbf{j}^n(\mathbf{r}) - \mathbf{j}^p(\mathbf{r})]. \tag{5}$$

with source terms involving the various nucleonic densities $\rho_{s,v}^{n,p}(\mathbf{r})$ and currents $\mathbf{j}^{p,n}(\mathbf{r})$ [7].

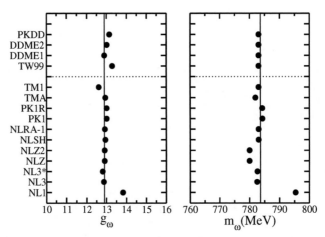

FIGURE 1. (Color online) The m_ω and g_ω parameters of the most frequently used non-linear and density-dependent parametrizations of the RMF Lagrangian.

The magnetic potential is the contribution to the mean field that breaks time-reversal symmetry in the intrinsic frame and induces non-vanishing currents $\mathbf{j}^{n,p}(\mathbf{r})$ in the Klein-Gordon equations (Eqs. (3, 5)), which are related to the space-like components of the vector mesons. In turn, the space-like components of the vector $\vec{\omega}$ and $\vec{\rho}$ fields form the magnetic potential (1) in the Dirac equation.

The currents are isoscalar and isovector for the ω and ρ mesons (Eqs. (3, 5)), respectively. As a consequence, the contribution of the ρ-meson to magnetic potential is marginal in the majority of the cases. Thus, time-odd mean fields in the RMF framework depend predominantly on the spatial components of the ω meson. Neglecting the contribution of the ρ meson, one can see that only two parameters, namely, the mass m_ω and the coupling constant g_ω of the ω meson define the properties of time-odd mean fields (Eqs. (1, 3, and 5)). Fig. 1 clearly indicates that these parameters are well localized in the parameter space for all modern non-linear and density-dependent parametrizations of the RMF Lagrangian. This suggests that the parameter dependence of the impact of time-odd mean fields on the physical observables should be weak. Indeed, the analysis of terminating states [6] and additional bindings due to NM in odd-mass nuclei [9] showed that time-odd mean fields are well defined in the CDFT framework.

Systematic analysis of rotational properties (moments of inertia, effective alignments i_{eff}) in different mass regions [7, 8] do not reveal substantial differences between theory and experiment. This suggests that time-odd mean fields are reasonably well defined in the CDFT as compared with experiment.

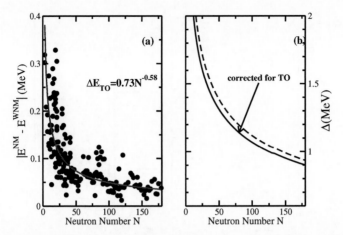

FIGURE 2. (Color online) (a) Neutron number dependence of additional binding due to NM. Solid circles show the results of the calculations for fixed single-particle configurations, while solid line shows the fit to these values using simple parametrization $\Delta E_{TO} = \frac{c}{N^\alpha}$. (b) Dashed line shows the global trend for OES extracted from experimental data (from Fig. 2 in Ref. [11]). Solid line shows the global trend corrected for the effects of time-odd mean fields (the ΔE_{TO} quantity) as extracted from the RMF calculations.

In odd-mass nuclei, time-odd mean fields always provide additional binding, the neutron number dependence of which (the $|E^{NM} - E^{WNM}|$ quantity) is shown in Fig. 2a. This additional binding only weakly depends on the RMF parametrization [9]. Fig. 2b shows that the effects of time-odd mean fields has to be taken into account when the strength of pairing is defined by the fit to odd-even mass staggerings (OES) since the binding energies of odd-mass nuclei are affected by these fields. Our calculations suggest that time-odd mean field contributions into OES can be as large as 10% in light systems and around 5-6% in heavy systems.

The presence of time-odd mean fields leads to the energy splitting $\Delta E_{split}(i)$ of the single-particle states which are time-reversal counterparts. This corresponds to the removal of the Kramer's degeneracy of these states. Unoccupied state moves up by $\approx \Delta E_{split}/2$ as compared with its position in the absence of NM, while occupied (blocked) one down by $\approx \Delta E_{split}/2$. This additional binding of the blocked state will affect the properties of the nuclei in the vicinity of the proton-drip line via two mechanisms discussed below. They are schematically illustrated in Fig. 3.

In the first mechanism, the nucleus, which is proton unbound (state A in Fig. 3) in the calculations without NM, becomes proton bound in the calculations with NM (state A' in Fig. 3). The necessary condition for this mechanism to be active is that the energy of the single-proton state in the absence of NM is smaller than $\Delta E_{split}/2$. This mechanism can be active both in the ground and excited states of the nuclei in the vicinity of the proton-drip line.

In the second mechanism, the energy of the single-particle state (state B' in Fig. 3) is lower in the presence of NM, but the state still remains unbound. This will affect the decay properties of proton emitters and the possibilities of their observation. Indeed, the lowering of the energy of the single-proton state will decrease the probability of emission

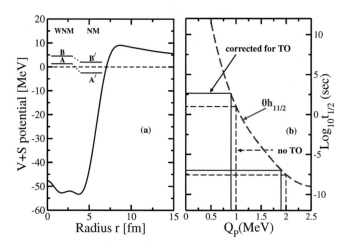

FIGURE 3. (Color online) Schematic illustration of the impact of time-odd mean fields on the properties of odd-proton nuclei in the vicinity of proton drip line. The single-proton states, involved in the mechanisms discussed in the text, and proton nucleonic potential (which also includes the Coulomb potential) are shown in panel (a). Dashed and dot-dashed lines in panel (b) show the results from Fig. 5 of Ref. [12].

of the proton through combined Coulomb and centrifugal barrier. This is illustrated in Fig. 3b in which the results of Ref. [12] for partial half-lives of the $0h_{11/2}$ state (dashed lines) are compared with the ones (solid lines) corrected for the presence of time-odd mean fields. The results shown by solid curves are drawn for the case in which time-odd mean fields lower the energy of the single-proton state by 100 keV: this is typical value for many states in rare-earth region. Such lowering of the single-proton state results in the increase of the half-lives of proton emitters by one order of magnitude at the upper end of the Q_p window and by almost two orders of magnitude at the bottom end of the Q_p window. The impact of NM will be even more dramatic on the half-lives of proton emitters in lighter nuclei (see Ref. [9] for details).

The material is based upon work supported by the Department of Energy under grant Number DE-FG02-07ER41459.

REFERENCES

1. M. Bender, J. Dobaczewski, J. Engel, and W. Nazarewicz, *Phys. Rev.* **C65**, 054322 (2002).
2. M. Bender, P.-H. Heenen, and P.-G. Reinhard, *Rev. Mod. Phys.* **75**, 121 (2003).
3. U. Hofmann and P. Ring, *Phys. Lett.* **B214**, 307 (1988).
4. J. König and P. Ring, *Phys. Rev. Lett.* **71**, 3079 (1993).
5. A. V. Afanasjev and P. Ring, *Phys. Rev.* **C62**, 031302(R) (2000).
6. A. V. Afanasjev, *Phys. Rev.* **C78**, 054303 (2008).
7. D. Vretenar, A. V. Afanasjev, G. A. Lalazissis, and P. Ring, *Phys. Rep.* **409**, 101 (2005).
8. A. V. Afanasjev and S. Frauendorf, *Phys. Rev.* **C71**, 064318 (2005).
9. A. V. Afanasjev and H. Abusara, submitted to *Phys. Rev.* C, preprint arXiv:0905.2445 [nucl-th].
10. W. Koepf and P. Ring, *Nucl. Phys.* **A493**, 61 (1989).
11. G. F. Bertsch et al., *Phys. Rev.* **C79**, 034306 (2009).
12. S. Åberg, P. B. Semmes, and W. Nazarewicz, *Phys. Rev.* **C56**, 1762 (1997).

Probing Time-Odd and Tensor Terms of The Skyrme Functional in Superdeformed Bands

V. Hellemans [*,†], M. Bender [**,‡] and P.-H. Heenen[†]

*University of Notre Dame, Department of Physics,
225 Nieuwland Science Hall, Notre Dame, IN 46556-5670, USA
†Université Libre de Bruxelles,Physique Nucléaire Théorique, CP229, B-1050 Bruxelles, Belgium
**Université Bordeaux, Centre d'Etudes Nucléaires de Bordeaux Gradignan,
UMR5797, F-33175 Gradignan, France
‡CNRS/IN2P3, Centre d'Etudes Nucléaires de Bordeaux Gradignan,
UMR5797, F-33175 Gradignan, France

Abstract. We briefly discuss Hartree Fock Bogoliubov (HFB) plus self consistent cranking calculations for the superdeformed band in ^{194}Hg, using the full Skyrme energy density functional including tensor terms in the p-h channel.

Keywords: Skyrme interaction, tensor interaction
PACS: 21.30.Fe, 21.60.Jz

Although a zero-range tensor force was already included in the original zero-range non-local interaction as proposed by Skyrme [1], its contribution to the Skyrme energy density functional (EDF) was seldomly taken into account in the fit of the parameterization of the EDF and its effect in calculations was only little explored until recently (see [2] for an extensive overview of the literature). The work of Otsuka *et al.* on the monopole effect of the tensor force [3] triggered the interest in the tensor interaction within the framework of self-consistent mean-field approaches.

In the context of the Skyrme EDF, the inclusion of the tensor force gives rise to additional terms in the EDF, some of which only contribute when time-reversal invariance is broken (the so-called *time-odd* terms). From the point of view of the parameterization of the EDF, the inclusion of the tensor terms has been treated in various ways. On the one hand, a strategy which focuses on the description of the $f_{5/2} - f_{7/2}$ spin-orbit splitting has been proposed [4] and band terminating states have been suggested to constrain the strength of the spin-orbit potential [5]. On the other hand, a set of parameterizations for the Skyrme EDF including tensor terms was constructed [2] using the Saclay-Lyon protocol which focuses on the simultaneous reproduction of nuclear bulk properties and the empirical characteristics of infinite nuclear matter. The latter work also provides a systematic study of the impact of the tensor terms in spherical nuclei.

In this contribution, we study the dynamical moments of inertia and the time-odd tensor terms for the superdeformed band in ^{194}Hg calculated with the full Skyrme EDF.

We carried out the calculation of the superdeformed band in ^{194}Hg in the HFB plus self-consistent cranking framework [6], determining the quasiparticle vacuum $|\Phi\rangle$ which minimizes

$$\mathscr{E}^{\omega} = \mathscr{E} - \langle\Phi|\lambda_N\hat{N} + \lambda_Z\hat{Z} + \omega\hat{J}_x|\Phi\rangle, \tag{1}$$

CP1165, *Nuclear Structure and Dynamics '09*
edited by M. Milin, T. Nikšić, D. Vretenar, and S. Szilner
© 2009 American Institute of Physics 978-0-7354-0702-2/09/$25.00

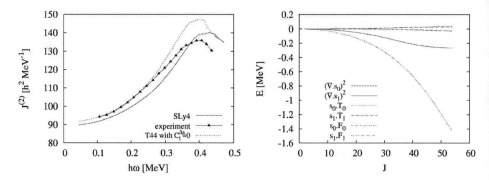

FIGURE 1. Left panel : Dynamical moments of inertia $\mathscr{J}^{(2)}$ as a function of the rotational frequency ω for the superdeformed band in ^{194}Hg for the SLy4 (no tensor terms!) and T44 parameterizations of the Skyrme force. Right panel : Energy contributions of the time-odd tensor terms in the Skyrme EDF for the superdeformed band in ^{194}Hg as a function of angular momentum J for the T44 parameterization.

for a given EDF $\mathscr{E}$. The constraints $\lambda_N \hat{N}$ and $\lambda_Z \hat{Z}$ have been added to obtain good average neutron- and proton-number, whereas $\omega \hat{J}_x$ imposes the angular momentum $J_0 = \langle \Phi | \hat{J}_x | \Phi \rangle$. We chose the Skyrme EDF (see e.g. [2] for the explicit expression) in the p-h channel and a density-dependent zero-range pairing interaction in the p-p channel. The Lipkin-Nogami prescription was used to prevent an abrupt collapse of pairing correlations.

In the left panel of Fig. 1, the dynamical moments of inertia $\mathscr{J}^{(2)}$ are displayed as a function of the rotational frequency ω for the SLy4 and the T44 [2, 7] parameterizations of the Skyrme force. In the right panel of Fig. 1, we display the evolution of the time-odd tensor contributions in the EDF as a function of angular momentum J. It is clear that $s_0 \cdot T_0$ and $(\nabla \cdot s_1)^2$ are the largest time-odd tensor contributions. Whereas, $(\nabla \cdot s_1)^2$ saturates with increasing J, the $s_0 \cdot T_0$ term becomes consistently more negative.

A systematic study of time-odd and tensor terms in the Skyrme EDF in superdeformed bands and odd (-odd) nuclei is in progress.

This work was supported by the US DOE under grant DE-FG02-95ER-40934 and the 'Interuniversity Attraction Pole' (IUAP) under project P6/23.

REFERENCES

1. T.H.R. Skyrme, *Phil. Mag.***1**, 1043 (1956) ; T.H.R. Skyrme, *Nucl. Phys.* **9**, 615 (1958) ; J.S. Bell and T.H.R. Skyrme, *Phil. Mag.***1**, 1055 (1956).
2. T. Lesinski, M. Bender, K. Bennaceur, T. Duguet, and J. Meyer *Phys. Rev.* **C76**, 014312 (2007).
3. T. Otsuka, T. Suzuki, R. Fujimoto, H. Grawe, and Y. Akaishi, *Phys. Rev. Lett.* **95**, 232502 (2005).
4. M. Zalewski, J. Dobaczewski, W. Satuła, and T. R. Werner, *Phys. Rev.* **C77**, 024316 (2008).
5. H. Zduńczuk, W. Satuła, and R. A. Wyss, *Phys. Rev.* **C71**, 024305 (2005).
6. P. Ring and P. Schuck, *The Nuclear Many-Body Problem*, Springer, Berlin, Heidelberg, New York, 1980 ; B. Gall, P. Bonche, J. Dobaczewski, H. Flocard, and P.-H. Heenen, *Z. Phys.* **A348**, 183 (1994).
7. The coefficient $C_t^{\Delta s}$ of the Laplacian of the spin density was put to zero to suppress finite-size instabilities (V. Hellemans, D. Davesne, et al., unpublished).

Relativistic Hartree-Fock-Bogoliubov Theory With Density Dependent Meson Couplings in Axial Symmetry

J.-P. Ebran*, E. Khan*, D. Peña Arteaga *, M. Grasso * and D. Vretenar †

*Institut de Physique Nucléaire, 15 rue Georges Clemenceau 91406 Orsay Cedex, France
†Physics Department, Faculty of Science, University of Zagreb, 10 000 Zagreb, Croatia

Abstract. Most nuclei on the nuclear chart are deformed, and the development of new RIB facilities allows the study of exotic nuclei near the drip lines where a successful theoretical description requires both realistic pairing and deformation approaches. Relativistic Hartree-Fock-Bogoliubov model taking into account axial deformation and pairing correlations is introduced. Preliminary illustrative results with density dependent meson-nucleon couplings in axial symmetry will be discussed.

Keywords: Mean Field, RHFB, RMF, Fock, Deformed
PACS: 21.10.Dr, 21.30.Fe, 21.60.Jz

We aim to describe the ground state properties of nuclei including pairing and deformation. For that purpose, a Relativistic Hartree-Fock-Bogoliubov (RHFB) code with density dependent meson couplings in axial symmetry is being developed. Let us mention that a RHFB code in spherical symmetry has been developed by W. H. Long et al. [1] and a RHB code in axial symmetry by T. Nikšić et al. [2] The RHFB theory treats the nucleons as point-like particles obeying a Dirac equation. They interact by the exchange of virtual mesons, characterized by their quantum numbers (spin J, parity π, isospin T). The theory is phenomenological, therefore as few effective meson fields as possible are introduced in order to properly describe nuclei. Moreover, the corresponding parameters, i.e. the meson masses and the coupling constants are neither taken from the meson properties in free space, nor derived from a more fundamental theory: they are fitted to experimental data. Based on such considerations one uses the following meson fields:

- $\vec{\pi}(0^-, 1)$: It mediates the long range (essentially tensor) attraction between nucleons. It does not contribute at the Hartree-level because it leads to a parity-breaking field, which has not been observed in nuclei. It only gives a non-zero contribution at the Fock-level and in the pairing channel.

- $\sigma(0^+, 0)$: It is a phenomenological meson which mediates the medium-range attraction between nucleons. Its origin lies in many complex effects, as two-pion resonances for example.

- $\omega^\mu(1^+, 0)$: It mediates the short-range repulsion between nucleons.

- $\vec{\rho}^\mu(0^+, 0)$: It provides an isospin dependence of the nuclear force.

- The photon γ is also introduced to take into account the Coulomb interaction.

CP1165, *Nuclear Structure and Dynamics '09*
edited by M. Milin, T. Nikšić, D. Vretenar, and S. Szilner
© 2009 American Institute of Physics 978-0-7354-0702-2/09/$25.00

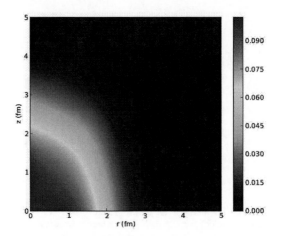

FIGURE 1. Proton density in ^{18}C

The hadronic degrees of freedom (nucleons and effective meson fields) are coupled in a phenomenological Lagrangian density:

$$\mathscr{L} = \bar{\psi}\left\{ i\gamma^\mu\partial_\mu - M - g_\sigma\sigma - g_\omega\gamma_\mu\omega^\mu - g_\rho\gamma_\mu\vec{\tau}.\vec{\rho}^\mu - \frac{f_\pi}{m_\pi}\gamma_5\gamma_\mu\vec{\tau}.\partial^\mu\vec{\pi} - e\gamma_\mu\mathscr{A}^\mu \right\}\psi$$
$$+ \tfrac{1}{2}\left(\partial_\mu\sigma\partial^\mu\sigma - m_\sigma^2\sigma^2\right) - \tfrac{1}{2}\left(\Omega_{\mu\nu}\Omega^{\mu\nu} - m_\omega^2\omega_\mu\omega^\mu\right)$$
$$- \tfrac{1}{2}\left(\vec{\mathscr{R}}_{\mu\nu}\vec{\mathscr{R}}^{\mu\nu} - m_\rho^2\vec{\rho}_\mu\vec{\rho}^\mu\right) + \tfrac{1}{2}\left(\partial_\mu\vec{\pi}\partial^\mu\vec{\pi} - m_\pi^2\vec{\pi}^2\right) - \tfrac{1}{2}\left(\mathscr{F}_{\mu\nu}\mathscr{F}^{\mu\nu}\right) \quad (1)$$

with

$$\Omega^{\mu\nu} = \partial^\mu\omega^\nu - \partial^\nu\omega^\mu$$
$$\vec{\mathscr{R}}^{\mu\nu} = \partial^\mu\vec{\rho}^\nu - \partial^\nu\vec{\rho}^\mu$$
$$\mathscr{F}^{\mu\nu} = \partial^\mu\mathscr{A}^\nu - \partial^\nu\mathscr{A}^\mu \quad (2)$$

The meson coupling constants are taken density dependent [3] The theory is formulated on the basis of two approximations: the Hartree-Fock approximation and the "no-sea" approximation. They lead to the Hartree-Fock equation.

We present an illustrative preliminary result in ^{18}C (Figure 1), corresponding to a relativistic Hartree-Fock calculation with no density dependence and where the pairing is treated in the BCS approximation. We use the effective interaction called set (e) in [4]. The calculation is based on a deformed harmonic oscillator expansion with 4 major shells for fermions and 20 major shells for bosons.

REFERENCES

1. W. H. Long et al. arXiv:0812.1103 (2009).
2. T. Nikšić et al. *Phys. Rev.* **C78**, 034318 (2008).
3. T. Nikšić et al. *Phys. Rev.* **C66**, 024306 (2002).
4. A. Bouyssy et al. Phys. Rev. **C36**, 380 (1987).

DYNAMICS OF LIGHT-ION REACTIONS

Formation of Degenerating Clusters in the $\alpha+{}^8$He Slow Scattering

Makoto Ito

Department of Pure and Applied Physics, Kansai University, 3-3-35 Yamate-cho, Suita, Osaka 564-8680, Japan

Abstract. The generalized two-center cluster model (GTCM), which can treat various single parti-cle configurations in general two center systems, is applied to the light neutron-rich system, ^{12}Be= $\alpha+\alpha+4N$. We discuss the change of the neutrons' configuration around two α-cores as a variation of an excitation energy. We found that the covalent, ionic and atomic configurations appear with a prominent degenerating feature above the $\alpha+{}^8$He$_{g.s.}$ particle-decay threshold.

Keywords: Cluster structures; Neutron-rich nuclei; Unbound systems.
PACS: 21.60.Gx,24.10.Eq,25.60.J

INTRODUCTION

In the last two decades, developments of experiments with secondary RI beam have extensively proceeded the studies on light neutron-rich nuclei. In particular, much efforts have been devoted to the investigation of molecular structure in Be isotopes. These isotopes can be considered as two-center superdeformed systems which build on an $\alpha+\alpha$ rotor of ^{8}Be. Theoretically, molecular orbital (MO), such as the π^- and σ^+ orbitals associated with the covalent bonding in atomic molecules, have been successful in describing the low-lying states of these isotopes [1].

The MO model can describe many kinds of characteristic properties of these isotopes, but they are mainly limited to the analysis on low-lying bound states, and theoretical studies on the highly excited states above the particle-decay threshold is still open area. In contrast to the situation of theoretical studies, recent experiments on nuclear breakup of Be isotopes by a light target revealed the existence of many resonant states, which strongly decay into ^{6}He$_{g.s.}$ or ^{8}He$_{g.s.}$. In the case of ^{12}Be, for instance, these results strongly suggest the formation of the ^{6}He$_{g.s.}$+^{6}He$_{g.s.}$ or $\alpha+{}^8$He$_{g.s.}$ molecular resonance (MR) in the unbound region of this nucleus [2, 3, 4]. They correspond to the ionic ($\alpha+{}^8$He) or atomic (^{6}He+^{6}He) configuration, where neutrons are trapped around one of α cores. This configuration is quite different from the covalent MO ones.

In previous experiments, breakup reactions by high-energy RI beam were mainly used to probe MRs in an unbound region of neutron-rich systems. However, there is a possibility that MRs can be directly excited by the slow scattering with low energy RI beam. The experimental techniques producing the slow RI beam are now extensively developed at many facilities. Recently, the low-energy beam of a neutrons' drip-line nucleus ^{8}He has just become available, and the scattering of $\alpha+{}^8$He$_{g.s}$ was measured at GANIL [5]. Since ^{8}He has the $\alpha+4N$ structure with four weakly bound neutrons and an α cluster, which is quite stable and inert, the valence four neutrons can easily be

CP1165, *Nuclear Structure and Dynamics '09*
edited by M. Milin, T. Nikšić, D. Vretenar, and S. Szilner

exchanged between two α particles during the collision. Therefore, it is quite interesting to study the unbound MRs in connection to the reaction mechanism, especially, the neutrons' transfer process.

In order to investigate the excitation of MRs in the slow $\alpha+{}^8\mathrm{He}_{g.s.}$ scattering, the reaction process should be treated consistently with the low-lying bound states of the compound system of $^{12}\mathrm{Be}$, because the resonances correspond to excited states, which are embedded in the continuum above the particle decay threshold. In Be isotopes including $^{12}\mathrm{Be}$, the MO model can successfully describe the low-lying states [1]. In the study of $\alpha+{}^8\mathrm{He}_{g.s.}$ scattering, therefore, the covalent MO configurations and the scattering process above the $\alpha+{}^8\mathrm{He}_{g.s.}$ threshold must be treated in a unified manner. For this purpose, we apply the generalized two-center cluster model (GTCM) [6, 7, 8, 9] to the $\alpha+{}^8\mathrm{He}_{g.s.}$ resonant scattering. The GTCM can cover the low-lying MO configurations [1] as well as the ionic and atomic ones excited as the $^X\mathrm{He}+{}^Y\mathrm{He}$ MRs ($X, Y = 4 \sim 8$) in the continuum; hence, treating resonant phenomena observed in slow scattering is possible. In this paper, we perform a unified study of the structural changes of $^{12}\mathrm{Be}$ and the $\alpha+{}^8\mathrm{He}_{g.s.}$ reaction by applying GTCM. In particular, we focus on the exotic phenomena appearing in an unbound continuum-states by investigating the excitation and the decay scheme of the resonances through the neutron-transfer reaction induced by the $\alpha+{}^8\mathrm{He}_{g.s.}$ slow collision.

FRAMEWORK

The unified treatment of the formation of the intrinsic states (MRs and MOs) and their coupling to the scattering continuum can be achieved by the combined method of GTCM [6, 7] and Kamimura's method [10, 11], which has been developed in Refs. [8, 9]. In this method, the total wave function of $^{12}\mathrm{Be}$ is given by the following linear combination

$$\Psi^{J^\pi(+)} = \sum_\beta \varphi_\beta \chi_\beta^{(+)} + \sum_\nu b_\nu \hat{\Psi}_\nu^{J^\pi} . \tag{1}$$

The first term of RHS stands for the "open channels" labeled by β, on which the scattering boundary condition is explicitly imposed [7, 11]. Here, φ_β and $\chi_\beta^{(+)}$ denote the internal and relative wave-functions of the two scattering nuclei, and the whole nucleons included in the nuclei are completely anti-symmetrized. We consider three rearrangement channels, $\alpha+{}^8\mathrm{He}_{g.s.}$, $^6\mathrm{He}_{g.s.}+{}^6\mathrm{He}_{g.s.}$ and $^5\mathrm{He}_{g.s.}+{}^7\mathrm{He}_{g.s.}$ as open channels.

The second term stands for the "intrinsic states" confined within the interaction region, which is a linear combination of the solutions obtained by diagonalizing the total Hamiltonian. The basis of the intrinsic states, $\hat{\Psi}_\nu^{J^\pi}$ labeled by an eigenvalue number ν, is damped in the asymptotic region. Therefore, the second term describes the compound states formed before decaying into the binary open channels. The $\hat{\Psi}_\nu^{J^\pi}$ is calculated by the GTCM, and its explicit form is written as

$$\hat{\Psi}_\nu^{J^\pi} = \int dS \sum_\mathbf{m} C_\mathbf{m}^\nu(S) \Phi_\mathbf{m}^{J^\pi}(S) \tag{2}$$

with the basis function given by

$$\Phi_{\mathbf{m}}^{J^\pi}(S) = \hat{P}_{K=0}^{J^\pi} \mathscr{A} \left\{ \psi_L(\alpha)\psi_R(\alpha) \prod_{j=1}^{4} \varphi_j(m_j) \right\}_S .$$ (3)

The detailed explanation of Eq. (3) is given in Refs. [8, 9], and we briefly explain these expressions. $\psi_n(\alpha)$ ($n=L,R$) denotes the α-cluster centered at the left (L) or right (R) side with relative distance S. The individual α core is expressed by the $(0s)^4$ configuration of the harmonic oscillator (HO). $\varphi_j(m_j)$ represents the atomic orbit (AO) state with the $0p$ orbital for the valence neutrons localized around one of the α clusters. Here, the index m is a set of quantum numbers needed to specify the AO state: the direction of the $0p$ orbital, the neutron's spin and its center (L or R). In Eqs. (2) and (3), $\mathbf{m}$ represents a set of AOs for the four neutrons, $\mathbf{m}=(m_1, m_2, m_3, m_4)$. The basis functions in Eq. (3) are fully anti-symmetrized by $\mathscr{A}$ and projected to the eigenstate of the total spin-parity J^π and its intrinsic angular projection K, which is restricted to the axial symmetric case ($K=0$) [12] by the projection operator $\hat{P}_K^{J^\pi}$. The $\hat{\Psi}_\nu^{J^\pi}$ describing the intrinsic state is finally given by taking the superposition of Eq. (3) over S and $\mathbf{m}$ as shown in Eq. (2).

As for the AO states of the four valence neutrons, $\mathbf{m}$ in Eq. (2), we include all the possible configurations, and hence, the model space of MO, where each valence neutron rotates around the two centers simultaneously, is also covered [6]. We use the Volkov No.2 and the G3RS for the central and spin-orbit parts of the nucleon-nucleon interaction, respectively. The parameters in the interactions and the size parameter of the HO are the same as those applied in Refs. [7, 8, 9], which successfully reproduce the structural properties of ^{10,12}Be. The adopted parameter set reasonably reproduces the threshold energies of $\alpha+^8$He$_{g.s.}$, ^{6}He$_{g.s.}+^6$He$_{g.s.}$ and ^{5}He$_{g.s.}+^7$He$_{g.s.}$. This is essential in the treatment of scattering phenomena.

RESULTS

The energy spectra of bound states are obtained by solving the eigenvalue problem with the total wave function in Eq. (2). In addition, the scattering matrices (S matrices) of the $\alpha+^8$He$_{g.s.}$ reaction is calculated with the total wave function of Eq. (1) when the system is excited above the threshold energy of this channel. By imposing the appropriate boundary conditions, we obtain whole energy spectra from the bound states to the unbound ones as shown in Fig. 1 ($J^\pi=0^+$). The obtained 0^+ states are classified into three categories:

(i) MO states. Two bound states appear below the $\alpha+^8$He$_{g.s.}$ threshold. The energy difference is about 2 MeV, and this difference agrees with a recent observation [13]. Here, the four neutrons are in MO motion around the two α cores $((\pi_{3/2}^-)^2(\sigma_{1/2}^+)^2$ and $(\pi_{3/2}^-)^2(\pi_{1/2}^-)^2$ for the 0_1^+ and 0_2^+ states, respectively).

(ii) MR states. The resonance states, 0_3^+, 0_4^+, and 0_6^+, have the developed cluster configurations of $\alpha+^8$He$_{g.s.}$, ^{6}He$_{g.s.}+^6$He$_{g.s.}$, and ^{5}He$_{g.s.}+^7$He$_{g.s.}$, respectively. Their energies are quite close to the corresponding cluster thresholds. The amplitude of the relative wave functions in these channels are enhanced around the ^{x}He$-^y$He barrier-top dis-

295

tance, although many kinds of other channels are strongly mixed in the inner region. Therefore, the intrinsic structure of these three resonances can be characterized as MR of the ^{x}He+^{y}He configurations.

(iii) Covalent SD state. The intrinsic structure of the 0^+_5 state is identified as a covalent superdeformed state (covalent SD), which has a hybrid configuration of covalent and ionic structures. In this configuration, two neutrons occupy the $0p$-wave AO, which is perpendicular to the $\alpha - \alpha$ axis ($L_z=\pm1$) around each α-cluster, like ^{5}He+^{5}He, while the remaining two neutrons form the $(\sigma^+_{1/2})^2$ bonding. Due to the formation of the σ^+-bonding, the $\alpha - \alpha$ distance is increased, which amounts to ~ 5 fm. Since the radius of the α particle is taken to be ~ 1.4 fm in the present model, this $\alpha - \alpha$ distance corresponds nearly to the hyperdeformation.

In the present model for ^{12}Be, we introduce the two degrees of freedom such as the $\alpha - \alpha$ relative motion and the single particle mo-

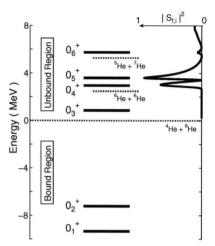

1: Energy spectra for $J^\pi=0^+$. The solid curve at the right side shows the squared magnitude of the S-matrix for $\alpha+^8$He$_{g.s.} \rightarrow$ ^{6}He$_{g.s.}+^6$He$_{g.s.}$ with a respective scale at the top-right corner. The dotted lines represent the threshold energy of the open channels considered in the calculation.

tions for the four valence neutrons, which can be activated as the total system is excited. By analyzing the distributions of the coefficient for the ν-th eigenstate, $C^\nu_m(S)$ in Eq. (2), all the excited states (0^+_ν, $\nu=2\sim6$) can be characterized in terms of the excitation degree of freedoms included in the ground states with the MO configuration of $(\pi^-_{3/2})^2(\sigma^+_{1/2})^2$. We summarized the identified excitation schemes in Fig. 2.

(1) Neutrons' Excitation Mode. The 0^+_2 state has the MO configuration, $(\pi^-_{3/2})^2(\pi^-_{1/2})^2$, while the 0^+_5 state has a hybrid MO-AO structure, which can be written in a symbolic form of $(^5$He+^{5}He$)\otimes(\sigma^+_{1/2})^2$. The neutrons' orbitals in these two states are different from the ground state with $(\pi^-_{3/2})^2(\sigma^+_{1/2})^2$. Therefore, they can be characterized as the single particles excitation modes of the excess neutrons from the ground state of $(\pi^-_{3/2})^2(\sigma^+_{1/2})^2$.

(2) Cluster Excitation Mode. The 0^+_3 and 0^+_6 states have a large component of the ionic structures of $\alpha+^8$He$_{g.s.}$ and ^{5}He$_{g.s.}+^7$He$_{g.s.}$, respectively. The mean distance between two clusters is more enhanced than that of the bound states. The 0^+_3 and 0^+_6 states are generated by growing the mean distance of two α-cores in the 0^+_1 and 0^+_2 states, respectively. Therefore, these ionic states can be considered as the excitation modes of the $\alpha-\alpha$ relative motion from the two bound states.

296

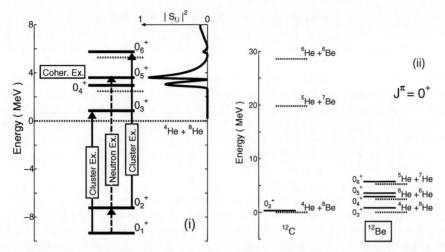

FIGURE 2. (i) Energy spectra classified by the excitation mode ($J^{\pi}=0^+$). See text for details. (ii) Threshold (dotted lines) and energy levels (solid lines) of ^{12}C and ^{12}Be ($J^{\pi}=0^+$). The threshold energies of the α emission is taken to be the origins.

(3) Coherent Excitation Mode. The 0_4^+ state has the atomic configuration, $^6\text{He}_{g.s.}+^6\text{He}_{g.s.}$. This configuration appear as a hybrid excitation mode between the single-particle type in *(1)* and the cluster type in *(2)*. Namely, the relative motion of the two α-cores and the single particle motions of the excess neutrons are coherently excited from the ground state.

The result shown in Fig. 2(i) means that the cluster structures can change from level to level, and various cluster configurations coexist in ^{12}Be. To show an anomalous property of the coexisting energy levels more clearly, in Fig. 2(ii), we show the threshold energies of ^{12}C and ^{12}Be. In ^{12}C with $N=Z$, there are large energy intervals in the threshold energy among the three channels of $^x\text{He}+^y\text{Be}$ with neutrons' rearrangements. The energy interval amounts to about 30 MeV ($\sim$ 20+10 MeV). Only the Hoyle state (0_2^+) is identified around the $\alpha+^8$Be threshold, and the existence of other rearrangement states remains unclear in experiments. However, these three cluster configurations must coexist with the energy interval of the respective threshold shown in Fig. 2(ii), according to the IKEDA's threshold rule [14], which makes the hypothesis that particular cluster structure structures will emerge for excitation energies near the corresponding threshold for decay.

The interval of the thresholds becomes small in ^{12}Be, and the energy interval of $^x\text{He}+^y\text{He}$ is only $\sim$ 5 MeV. This is smaller by about one order magnitude than that in ^{12}C. Therefore, we can clearly confirm that the different cluster structures appear with a strong degenerating feature in the neutron-rich system. This is due to the weakness of the binary potentials of the composite particles (There are no bound states in the α−neutron, neutron−neutron, and $\alpha - \alpha$ systems). As a result of this property in a weakly coupled system, the excess neutrons can easily changes their configurations around two α-clusters with a small energy difference.

SUMMARY AND DISCUSSIONS

In summary, we have studied the exotic structures of ^{12}Be in an unbound region and the two-neutron transfers in the $\alpha+^8$He$_{g.s.}$ slow-collision by applying GTCM. We explored the intrinsic structures of the excited states below and above the α decay threshold. Two important results were obtained from the present calculation as follows: Firstly, we confirmed that the excited states can be characterized according to the excitation degree of freedoms contained in the ground state. Due to the excitation of the relative motion of α clusters and single particle motion for excess neutrons, various structures appear as a variation of the excitation energy. Secondly, we found that there is a characteristic feature in the spacing of the energy levels above the particle decay threshold. The excited states appear with a strong degeneracy.

In particular, the second result should be stressed, because there is a possibility that similar degenerating features appear in neutrons' drip-line nuclei. In the region of neutron drip-line, neutrons' separation energy always becomes small, and neutron rearrangements among channels can be generated by a small energy increasing; hence, we can expect that the coexisting phenomena obtained in the present study is not specific only in ^{12}Be but quite general in neutrons' drip-line nuclei. Systematic studies on light neutron-rich systems are now proceeding.

ACKNOWLEDGMENTS

The authors thank Profs. N. Itagaki, K. Ikeda, and T. Nakatsukasa for useful discussions and encouragements. This work has been supported by the Grant-in-Aid for scientific Research in Japan (Nos. 18740129 and 21740211). The numerical calculations have been performed at the RSCC system, RIKEN.

REFERENCES

1. N. Itagaki and S. Okabe, *Phys. Rev.* C **61**, 044306 (2000).
2. A. A. Korsheninnikov, et al., *Phys. Lett.* **343B**, 53 (1995).
3. M. Freer, et al., *Phys. Rev.* C **63**, 034301 (2001).
4. A. Saito, et al., *Suppl. Prog. Theor. Phys.* **146**, 615 (2003); A. Saito, et al., AIP conference proceedings Vol. **891**, p205 (2006); A. Saito, Private communication (2008).
5. M. Freer, N. Orr and A. W. Ashwood, Private communication (2008).
6. M. Ito, K. Kato and K. Ikeda, *Phys. Lett.* **588B**, 43 (2004); *ibid, Mod. Phys. Lett.* A8, 178 (2003).
7. Makoto Ito, *Phys. Lett.* **636B**, 293 (2006); *ibid, Mod. Phys. Lett.* A **21**, 2429 (2006).
8. M. Ito, N. Itagaki, H. Sakurai, and K. Ikeda, *Phys. Rev. Lett.* **100**, 182502 (2008); *ibid, Jour. Phys. Con. Ser.* **111**, 012010 (2008).
9. M. Ito and N. Itagaki, *Phys. Rev.* C**78**, 011602(R) (2008); *ibid, Phys. Rev. Focus* Vol. 22, Story 4 (2008); *Int. Jour. Mod. Phys.* E**17**, 2055 (2008); *Mod. Phys. Lett.* A, inpress (2009).
10. M. Kamimura, *Prog. Theor. Phys. Suppl.* **62**, 236 (1977).
11. M. Ito and K. Yabana, *Prog. Theor. Phys.* **113**, 1047 (2005)
12. M. Ito, K. Yabana, T. Nakatsukasa and M. Ueda, *Phys. Lett.* **637B**, 53 (2006).
13. S. Shimoura, et al., *Phys. Lett.* **654B**, 87 (2007).
14. K. Ikeda, et al., *Suppl. Prog. Theor. Phys.* **68**, 1 (1980) and references therein.

Structure Effects in Collisions Induced by Weakly Bound Stable and Radioactive Nuclei

Alessia Di Pietro

INFN-Laboratori Nazionali del Sud, via S. Sofia 64, 95125 Catania, Italy

Abstract. In this paper I will discuss some results of experiments concerning the study of reaction mechanisms with halo and loosely bound stable nuclei at energies around the Coulomb barrier.

Keywords: radioactive beams, elastic scattering, fusion reactions
PACS: 25.60.-t, 25.60.Pj, 25.60.Bx

INTRODUCTION

It is well known that in order to reproduce the experimental behaviour of the cross-section for the different reaction processes at very low energies, dynamical effects, such as the coupling of the relative motion of projectile and target to their intrinsic excitations or to other reaction channels as for instance transfer, must be included in the calculations. Theoretically this is performed via Coupled Channel (CC) calculations. In the case of weakly bound nuclei, the ground state of the nucleus lies close to the continuum, therefore, in order to properly describe the reaction dynamics, the coupling not only to bound but also to continuum states (break-up states) must be included in CC calculations. CC calculations which include coupling to the continuum, predict enhanced sub-barrier total fusion cross-section with respect to the no-coupling case. In addition, due to the break-up, a suppression of the total fusion cross-section is predicted at energies above the barrier [1]. Conversely, different models as for example [2] based on a time-dependent wave-packet method predicts that the cross-section in the neutron halo case is slightly suppressed than in the non-halo case. In the case of halo nuclei were the low beam intensity hamper, in many cases, the possibility of measuring reaction processes, information concerning the effect of the coupling to break-up states can be attained from the study of elastic scattering. In fact, to properly describe the scattering, accurate description of the projectile structure, and the correct coupling to bound and unbound states of the halo nucleus must be considered. However, this requires high quality elastic scattering data which are not always available. In this paper an overview of the progress that have been done in reaction studies with halo and weakly bound stable nuclei is done.

ELASTIC SCATTERING OF HALO NUCLEI

Low energy reaction studies with halo nuclei have been performed so far mainly using ^{6}He. Elastic scattering and reaction data on several targets are nowadays available and

CP1165, *Nuclear Structure and Dynamics '09*
edited by M. Milin, T. Nikšić, D. Vretenar, and S. Szilner
© 2009 American Institute of Physics 978-0-7354-0702-2/09/$25.00

this has stimulated a lot of theoretical work as well. Some data are available also with [11]Be and, very recently, elastic scattering data of [8]B on medium mass [58]Ni have been published [3]. The common feature observed in [6]He induced elastic scattering, is a reduction of the elastic cross-section at the Coulomb rainbow see e.g. [4, 5]. In order to properly describe the reduced cross-section at the nuclear rainbow within the Optical Model (OM), large diffuseness parameter of the imaginary term of the optical potential must be used. Continuum Discretized CC calculations (CDCC) of [6]He elastic scattering on different target have been recently published [6, 7, 8]. All these calculations show that the reduction of the elastic cross-section in the rainbow region is originated by the coupling to states in the continuum. Since [6]He is a three body system, in order to simplify the very complicated CDCC calculations of a four body reaction, two body representations of [6]He are in many cases considered. In these representations, [6]He is described as a dineutron-alpha system [6]. Techniques to perform four-body (n-n-alpha+target) CDCC calculations have recently been developed [7, 8] and good agreement with the data is obtained in some cases. Again these calculations show that coupling to the continuum suppress the elastic cross-section at the rainbow. Also coupling to single neutron stripping can be of importance in describing the low-energy scattering of a halo nucleus. This has been investigated in [9] and according to the authors it has a significant effect on the elastic scattering of [6]He and should not be neglected. Also coupling to the 2n stripping could have a major role on the elastic scattering, owing to the large cross-section observed experimentally [10, 11], larger than 1n stripping and break-up.

Elastic scattering of [11]Be+[209]Bi have been measured by Mazzocco et al. [12]. This experiment was performed at RIKEN where a fragmentation beam degraded in energy was used. The beam energy resolution was insufficient to measure elastic scattering, but the quasi elastic (QE) cross-section was measured. The QE cross-section included the first excited state of [11]Be as well as the target excited states up to 2.6 MeV. The quality of the data does not allow to see whether elastic scattering with the halo [11]Be beams shows a similar behaviour as the scattering of [6]He. CC calculations were performed in order to extract information on the elastic cross-section. The extracted total reaction cross-section was compared to the one of [9]Be+[209]Bi, measured by the same group [13]. The total reaction cross-section is found to be very similar for the two Be isotopes. The authors concluded that since the absorption is due to fusion and break-up, and since the fusion for the two reactions [9,11]Be+[209]Bi has similar cross-section [14], the breakup process must also have comparable cross-section for both [9,11]Be nuclei. Therefore the differences in binding energies and radii do not play an important role.

Recently, we measured the elastic scattering angular distribution of [9,11]Be on a medium mass [64]Zn target, at the same cm energy (E_{cm}=25.4 MeV) using an ISOL [11]Be beam. The experiment with the stable [9]Be beam was performed at Laboratori Nazionali del Sud in Catania, the experiment with the radioactive [11]Be isotope was performed at REX-ISOLDE at CERN. In Fig. 1 the [9,11]Be+[64]Zn elastic scattering angular distribution is shown. In the case of [11]Be+[64]Zn the cross-section includes also the [11]Be first excited states. However, according to optical model calculations, the contribution of inelastic scattering is very small at all measured angles.

From Fig. 1 one can clearly see that for [11]Be induced scattering, a reduction of the elastic cross-section with respect to the weakly bound [9]Be induced scattering is present

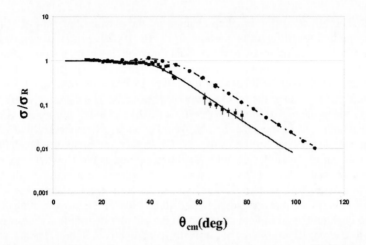

FIGURE 1. Elastic scattering angular distribution $^9Be+^{64}Zn$ (closed circle) and $^{11}Be+^{64}Zn$ (closed squares) at E_{cm}=25.4 MeV.

both in the rainbow region as well as at large angles. The total reaction cross-section deduced from OM analysis is σ_R= 1080 $\pm$ 50 mb for ^9Be and σ_R= 1980 $\pm$200 for ^{11}Be. Contrary to what found by [12], a large difference in the total reaction cross-section for the two Be isotopes is found. The break-up cross-section was measured in the case of $^{11}Be+^{64}Zn$. Preliminary results show that the transfer/break-up cross-section (the two processes cannot be distinguished in this experiment) is about 1b, i.e. about half of the total reaction cross-section. This conclusion disagree with the result of Mazzocco et al., but agrees with the results obtained with ^6He.

ELASTIC SCATTERING OF WEAKLY BOUND NUCLEI

Recently, much work has been devoted in studying threshold anomaly in elastic scattering of weakly bound nuclei $^{6,7,}Li$ and ^9Be on different targets. It is well known that the optical potential extracted from the analysis of the elastic scattering of heavy ions, involving tightly bound nuclei, shows a rapid variation with energy near the Coulomb barrier, so-called threshold anomaly, see e.g. [15]. The usual threshold anomaly has been ascribed to the coupling of the elastic scattering to the nonelastic channels. The decrease of the imaginary potential can be understood by the closure of nonelastic channels at energies near the Coulomb barrier. In weakly bound nuclei the breakup channel is expected to be important even at energies below the Coulomb barrier. The coupling to the breakup produces a repulsive polarisation potential and the usual threshold anomaly may disappear. It has recently been suggested that a new kind of anomaly may be present in the scattering of weakly bound nuclei, named as 'breakup threshold anomaly' (BTA) [16]. In this case the strength of the imaginary potential even increases as the incident energy decreases. The effects of the breakup channel on the elastic scattering of ^{6}Li and ^{7}Li have been studied for different target nuclei [17, 18, 19, 20, 21]. However, conclusions regarding the presence or absence of the usual threshold anomaly are contradictory. For ^{6}Li scattering on several targets, the results seem to indicate the presence of a break-up threshold anomaly, whereas this is not clear for ^{7}Li where in some analysis the optical potential seem to show the regular threshold anomaly. The different result has been justified as due to the higher break-up threshold of ^{7}Li and to the coupling to the first excited state which in ^{7}Li is bound and in ^{6}Li is unbound. The behaviour of the ^{9}Be elastic scattering, should be very similar to that of ^{6}Li. In fact, the break-up thresholds of the two nuclei are similar and they do not have bound excited states. However the results are rather controversial as for ^{7}Li [22, 23]. At low bombarding energies, below the Coulomb barrier, the elastic cross-section deviates very little from the Rutherford cross-section and the OM analysis is very sensitive to small variation of the cross-section. Therefore one has to be sure that all systematic errors in the angular distribution measurement are minimised. We recently measured at Laboratori Nazionali del Sud in Catania, the elastic scattering ^{6}Li+^{64}Zn in the energy range 10 MeV$\leq E_{cm} \leq$20 MeV. A very tight collimation system was used, in order to minimise the possible variation of the beam axis direction. The OM analysis was performed using a double folding potential. The same shape of the real and imaginary potential was used and the parameters of the fit were the two normalisation constants N_R and N_I. The results of the analysis are consistent with the so called BTA thus confirming the previous results of ^{6}Li measurements.

FUSION WITH HALO AND WEAKLY BOUND NUCLEI

In experiments with low intensity radioactive beams it is extremely difficult to measure at sub-barrier energies. The beam intensities, typically 10^5-10^7 pps, are several orders of magnitude smaller than in experiments with stable beams. Therefore, due to the low measured yield, fusion excitation function do not extend down to deep sub-barrier energies. Another experimental problem comes from the difficulty to discriminate complete fusion (CF) from incomplete fusion (ICF). An open issue is how to evidence effects due to the halo structure. The ideal would be having realistic CDCC calculations to be compared with the experimental data. These, especially in the case of ^{6}He, are still missing. The data have been compared with calculation of single barrier penetration, standard CC calculations, or with data involving reactions induced by well bound nuclei on the same target or by comparing the data of a reaction with different projectile and target but forming the same compound nucleus. However the result (enhancement/suppression of the cross-section) will depend upon the type of comparison. As an example, in the case of ^{6}He+^{64}Zn of [4] the data are compared with the reaction ^{4}He+^{64}Zn. From that comparison no differences of the two measured fusion excitation function are found in the energy range explored. In [24] the same ^{6}He+^{64}Zn data, are compared with CC calculation without coupling to the continuum. From that comparison, a suppression of the fusion cross-section of the order of 50% at energies above the barrier is claimed and no effect at energies below the barrier is found.

Experiments performed using weakly bound stable nuclei may help to investigate the specific role played by the break-up. Reactions induced by the least bound stable nuclei

302

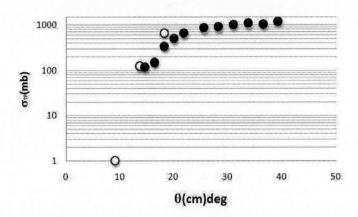

FIGURE 2. Total fusion excitation function for ^{6}Li+^{64}Zn. Closed circles this work, open circles data from [27].

(^{6,7}Li, ^{9}Be) have been studied on several targets see e.g. [25, 26, 27, 28]. The quality of the data is, of course, much better than the one obtained with the radioactive beams. In reactions on heavy targets a reduction of the CF cross-section is observed at energies above the barrier, whereas no clear effect is observed below the barrier. The reduction was attributed to the projectile break-up in the strong Coulomb field of the target nucleus, followed by ICF. No effect was found in the total fusion cross-section, which agrees with the calculations [25].

In the study of reactions between light-weakly-bound nuclei no effect has been observed. Reactions on medium mass targets have produced different results. For ^{6,7}Li and ^{9}Be on ^{64}Zn [27] the CF seems not to be affected by the break-up process, no reduction of the CF cross-section was observed within the experimental uncertainties. For ^{9}Be+^{64}Zn collision the total reaction cross-section was found almost equal to the total fusion cross-section. Conversely, the reaction induced by ^{6,7}Li on the same ^{64}Zn target have shown a total reaction cross-section much larger than the total fusion cross-section [27] which is not expected. However, in a recent publication by Gomes et al. [29] (see also the contribution of P.R.S. Gomes to this conference) where a universal fusion function (UFF) is derived in order to search for systematic trends in the total fusion excitation function, it is argued that there are problems with the ^{6,7}Li+^{64}Zn data. We recently measured the ^{6}Li+^{64}Zn total fusion excitation function at near and sub-barrier region. The experiment was performed by using the activation technique already used by us to measure ^{6}He+^{64}Zn [4]. In figure 2 our results are compared with the data of [27]. As one can see from the figure there is a disagreement between our data and the previously published data. This discrepancy could be originated by the time of flight technique used by Gomes at al. to obtain the cross-section. The low energy heavy fragment produced in fusion or incomplete fusion processes are partially stopped into the target, therefore threshold effect can be responsible for the lower cross-section measured.

CONCLUSIONS

Reaction studies with halo and weakly bound nuclei at energy around the barrier is currently being studied very extensively and a lot of data are becoming available. Elastic scattering of halo nuclei show a reduced cross-section at the rainbow region which is originated by the coupling to the break-up states. This coupling seems also responsible for the appearance of an "anomalous" threshold anomaly in the elastic scattering OM potential of weakly bound nuclei. Fusion studies with halo nuclei still shows controversy about possible enhancement/suppression effects, depending on how the data are compared to extract the searched effect. With weakly bound beams very detailed measurements exist on heavy targets, where a suppression of the complete fusion cross-section was found at energies above the barrier. On medium mass targets, the measurements do not extend to energies below the barrier and in some cases the measured cross-section is underestimated due to the experimental technique adopted.

REFERENCES

1. A. Diaz-Torres, et al., *Phys. Rev.* C **65**, 024606 (2002).
2. M. Ito, et al. *Phys. Lett.* B **37**, 53 (2006).
3. J.J. Kolata and E.F. Aguilera, *Phys. Rev.* C **79**, 027603 (2009).
4. A. Di Pietro, et al., *Phys. Rev.* C **69**, 044613 (2004).
5. A.M. Sánchez-Benítez, et al., *Nucl. Phys.* A **803**, 30 (2008).
6. A. Moro, et al., *Phys. Rev.* C **75**, 064607 (2007).
7. T. Matsumoto, et al., *Phys. Rev.* C **73**, 051602R (2006).
8. M. Rodríguez-Gallardo, et al., *Phys. Rev.* C **77**, 064609 (2008).
9. N. Keeley and N. Alamanos, *Phys. Rev.* C **77**, 054602 (2008).
10. A. Chatterjee, et al., *Phys. Rev. Lett.* **101**, 032701 (2008).
11. P.A. De Young, et al., *Phys. Rev.* C **71**, 051601(R) (2005).
12. M. Mazzocco, et al., *Eur. Phys. J.* A **28**, 295 (2006).
13. C. Signorini, et al., *Nucl. Phys.* A **701**, 23c (2002).
14. C. Signorini, et al., *Eur. Phys. J.* A **2** 227 (1998).
15. G.R. Satchler, *Phys. Rep.* **199**, 147 (1991).
16. M.S. Hussein, et al., *Phys. Rev.* C **73**, 044610 (2006).
17. N. Keeley, et al., *Nucl. Phys.* A **571**, 326 (1994).
18. A.M.M. Maciel, et al., *Phys. Rev.* C **59**, 2103 (1999).
19. A. Pakou, et al., *Phys. Lett.* B **556**, 21 (2003).
20. J.M. Figueira, et al., *Phys. Rev.* C **73**, 054603 (2006).
21. F.A. Souza, et al., *Phys. Rev.* C **75**, 044601 (2007).
22. C. Signorini, et al. *Phys. Rev.* C **61**, 061603 (2000).
23. R.J. Woolliscroft, et al., *Phys. Rev.* C **69** 044612 (2004).
24. E. Crema, et al., *Phys. Rev.* C **75**, 037601 (2007).
25. M. Dasgupta, et al., *Phys. Rev.* C **70**, 024606 (2004) and reference therein.
26. C. Beck, et al., *Phys. Rev.* C **67**, 054602 (2003).
27. P.R.S. Gomes, et al., *Phys. Rev.* C **71**, 034608 (2005).
28. A. Mukherjee, et al., *Phys. Lett.* B **636**, 91 (2006).
29. P.R.S. Gomes, et al., *Phys. Rev.* C **79**, 027606 (2009).

Treatment of Continuum in Weakly Bound Systems in Structure and Reactions

Francisco Pérez-Bernal* and Andrea Vitturi[†]

* Facultad de Ciencias Experimentales, Departamento de Física Aplicada, Universidad de Huelva, Huelva, Spain
[†] Dipartimento di Fisica and INFN, Padova, Italy

Abstract. We investigate different methods for the treatment of continuum states in nuclear structure calculations. We exploit a simple laboratory case: two particle moving in a one-dimensional mean field and interacting via a density-dependent short range residual interaction. We find that in procedures that involve continuum discretization a rather large basis has to be used in order to get convergence to the exact results, in particular for the radial dependence of the two-particle wave function. This may lead to unpracticable situations in the case of many interacting particles in the continuum.

Keywords: Two-particle correlations. Weakly-bound nuclei. Continuum states.
PACS: 21.10.Re, 21.10.Gv, 21.60.Cs

An enormous variety of models and methods has been advanced to describe the many facets of the nuclear many-body problems, both in nuclear structure and in nuclear dynamics. These approaches have normally modelized the nuclear many-body problem in a way where the hugh complexity of the solution of the full Schroedinger equation can be reduced to a solvable problem. Many of these approaches, normally named as microscopic, still keep the basic fermionic degrees of freedom, and usually involve an expansion of the full wave functions in basis states constructed from coupled single-particle states. These can be solutions of the self-consistent mean field, derived for example in Hartree-Fock, or, in more crude approaches, as solutions of a simple phenomenological harmonic or Woods-Saxon potential.

A typical example is provided by shell model calculations. For systems around the stability valley, the position of the Fermi surface in the single-particle potential is such that the wave functions of the ground state and of the low-lying states can be described within the model space that include the (still bound) particle levels above the Fermi surface. The situation is different as one moves from the stability valley towards the drip lines. In this case the Fermi surface tends to approach the particle threshold and the number of available (still bound) single particle levels decreases. At the drip lines, eventually, no unoccupied bound particle state is available, and only continuum states come into the game.

Similar features are also present in nuclear reaction calculations. Let us take as an example the description of grazing collisions in heavy-ion reactions. The master approach for the description of such processes is provided by the Coupled-Channel Formalism, where the total wave function is expanded in the eigenstates of the two colliding objects. In this case one is often only interested in the description of the soft part of the collision, where only a limited amount of energy is pumped into the internal

CP1165, *Nuclear Structure and Dynamics '09*
edited by M. Milin, T. Nikšić, D. Vretenar, and S. Szilner
© 2009 American Institute of Physics 978-0-7354-0702-2/09/$25.00

degrees of freedom, and, consequently, the expansion is normally limited to the bound excited states of projectile and target. Such an approach naturally encounters problems when weakly-bound projectiles are involved. In this case the most probable process is the excitation of the weakly-bound system to the continuum, i.e. the so-called break-up process, and the description in terms of discrete bound channels is clearly inadequate.

In all the above-mentioned situations the inclusion of continuum states becomes mandatory. Quite a variety of approaches have been developed to include continuum states in structure or reaction models. Some of them are intrinsically devised to only include in the many-body wave functions components with at maximum one particle in the continuum, others are based on the use of complex eigenstates of the hamiltonian [1]. Most of the approaches, on the contrary, introduce the so-called discretization procedure [2]. In this case the continuum part of the single-particle spectrum is replaced by an alternative discrete set of states with positive real energies, chosen according to different prescriptions. Examples are given by the expansion in Harmonic Oscillator basis, the Transformed Harmonic Oscillator basis, or the use of a large box where to impose vanishing boundary conditions.

All discretization procedures are equivalent, as long as a full complete basis is used. In practice, each procedure involves a number of approximations and truncations in the actual computations. Computational constraints may in fact become a severe problem, in particular when a large number of particles are allowed to move in the continuum. Comparing the positive and negative aspects of the different approaches is not an easy task, in view of the different sets of parameters that characterize each procedure. In addition, the problem used as a test case may be so complex that its use as a reference for checking convergence procedures is precluded.

For these reasons we have considered as a laboratory the simple case of particles moving in a one-dimentional attractive potential, with residual interactions that simulate simple conditions in nuclear reactions or nuclear structure. For the former case we have considered in ref. [3] the case of a single particle, initially bound by the potential, perturbed by an external time-dependent interaction. When the particle starts from a weakly-bound orbital the perturbation can excite the particle to the continuum, leading to a wave function that is no longer confined in the region of the potential (break-up process). The exact time-dependent vawe function has been compared with the results of the coupled-channel calculations using the continuum discretization method CDCC [4].

For the nuclear structure we consider here the simplest many-body problem: particles moving in the one-dimensional mean field and two particles outside the core interacting via an additional density-dependent short range residual interaction. We will take a case in which the particles of the core fully occupy all the bound states, and choose an interaction strength such as to produce a bound two-particle correlate state. The situation can therefore model the case of a two-particle halo nucleus (such as ^{6}He or ^{11}Li), which is bound in spite of the fact that the neighbour core+particle is unbound (as ^{5}He or ^{10}Li in the considered examples).

The one-body potential is assumed to be of Woods-Saxon form

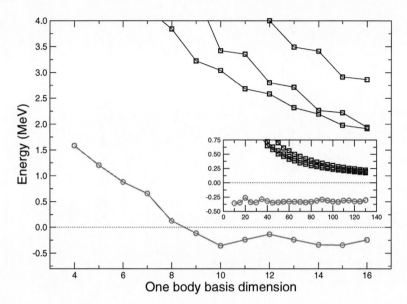

FIGURE 1. (Color online) Energies of the correlated two-particle states resulting from the diagonalization procedure. A Harmonic Oscillator has been as a basis to construct the WS single-particle states. In the abscissa is given the number of oscillator shells.

$$V(x) = -\frac{V_0}{1 + \exp\left(\frac{|x| - R}{a}\right)}, \tag{1}$$

with parameters $V_0 = 12$ MeV, $R=5$ fm and $a=0.8$ fm. This potential admits three bound states at energies $-10.22, -6.00$ and -1.39 MeV, all supposed to be fully occupied. The residual interaction among the two additional particles is assumed of the form

$$V_{int}(x_1, x_2) = -V_{int}\, \delta(x_1 - x_2)\frac{\rho(x)}{\rho_0}, \tag{2}$$

where $V_{int} = -60$ MeV and $\rho_0 = 0.15$. We have used different techniques to produce the (discrete) single-particle basis associated with our one-body potential. The complete analysis will be published in a forthcoming paper [5]. We report here only the results obtained by calculating the single-particle Woods-Saxon states using a harmonic oscillator basis, and then diagonalizing the residual interaction in the two-particle basis. Figure 1 shows the energies of the resulting correlated states as a function of the number of shells used in the HO basis. Positive energy states have no physical meaning (dependent as they are from the chosen basis) but, as apparent from the figure, the residual interaction has created a (physical) bound state with a converging value of the energy (compare the inset, where the number of shells oversizes one hundred). A decent value of the energy can be achieved already with a dozen of shells. However, this is not sufficient to ensure a correct description of the resulting wave function. To check this point we display in Figure 2 the correlated two-particle wave function (for $x_1 = x_2 = x$). As clear from the

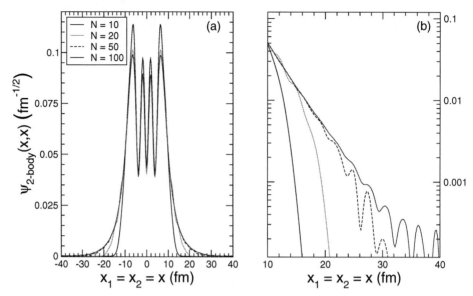

FIGURE 2. (Color online) Correlated two-particle wave function obtained with different numbers of HO shells. For a better view of the tail behaviour a log scale is used in panel (b).

figure (note the log scale used in the right frame) an extremely large number of shells has to be used to improve the behaviour in the tail region. Still, even with very large number of shells, unphysical oscillations appear in the far out region. Note that the behaviour on the tail is essential to describe properly the pairing field or the two-particle transfer processes in heavy-ion induced reactions.

The results of our analysis (only sketched here) shows that descriptions based on discretization of the continuum can be accurate only when a proper choice is made of the number of discrete states, of the energy mesh and of the energy cutoff. This may imply, even in simplified cases, the use of a rather large (and unpracticable) number of channels. The use of a more restricted number of channels may lead to rather misleading results.

REFERENCES

1. N. Michel, W. Nazarewicz, M. Ploszajczak, and J. Rotureau, *Phys. Rev.* **C74**, 54305(2006); G.Hagen, M.H. Jensen,and N.Michel, *Phys.Rev.* **C73**, 64307(2006).
2. K. Tsukiyama, T. Otsuka and R. Fujimoto, *Phys. Rev.* **C**, in press and in these proceedings.
3. C.H Dasso and A. Vitturi, *Nucl. Phys.* **A787**, 476C (2007); *Phys. Rev.* **C**, in press (2009).
4. M. Yahiro, N. Nakano, Y. Iseri, and M. Kamimura, *Prog. Theor. Phys.* **67**, 1464 (1982); *Prog. Theor. Phys. Suppl.* **89**, 32 (1986); J. A. Tostevin, F. M. Nunes, and I. J. Thompson, *Phys. Rev.* **C63**, 024617 (2001).
5. F. Perez-Bernal and A. Vitturi, to be published.

Study of Nuclei far From Stability by Using the CHIMERA 4π Detector and Radioactive Beams at LNS

G.Cardella[a], F.Amorini[b], A.Anzalone[b], L.Auditore[d], S.Cavallaro[b,c],
M.B.Chatterjee[h], E.De Filippo[a], E.Geraci[a,c], L.Grassi[a,c], A.Grzeszczuk[g],
P.Guazzoni[e], E.La Guidara[a,l], G.Lanzalone[b,i], I.Lombardo[b,c], S.Lo Nigro[a,c],
D.Loria[d], C.Maiolino[b], A.Pagano[a], M.Papa[a], S.Pirrone[a], G.Politi[a,c],
F.Porto[b,c], F.Rizzo[b,c], P.Russotto[b,c], A.Trifirò[d], M.Trimarchi[d], G.Verde[a],
M.Vigilante[f] and L.Zetta[e]

[a] INFN, Sez di Catania
[b] INFN Lab Naz del Sud, Catania, Italy
[c] Dip di Fisica e Astr. Univ. di Catania
[d] INFN & Dip.Fisica.Univ. di Messina, Italy
[e] INFN& Dip.Fisica. Univ. di Milano, Italy
[f] INFN& Dip. Fisica.Univ.Napoli, Italy
[g] Inst. of Phys., Univ. of Silesia, Katowice, Poland,
[h] Saha Inst. of Nucl. Phys., Kolkata, India
[i] Univ. "Kore", Enna, Italy
[l] CSFNSM Catania Italy

Abstract. At LNS are available radioactive beams at tandem and intermediate energies provided respectively by the EXCYT and by the fragmentation FRIBS facilities. Using these beams, and the 4π detector CHIMERA, we want to study excitation and decay of resonances in light exotic nuclei populated with pick-up stripping and other reaction mechanisms. Some preliminary results obtained with stable and unstable beams are reported.

Keywords: radioactive beams, kinematical coincidence, 4π detectors, transfer reactions.
PACS: 25.40.Ep; 25.40.Hs; 25.45.Hi; 25.60.Je.

INTRODUCTION

The availability at LNS of low [1] and intermediate energy [2] radioactive beams suggests to use CHIMERA [3] beside for dynamics studies [4-6], also for some nuclear structure measurements. To this aim, some experimental tests were performed with tandem beams. Various tagging systems were also tested in order to obtain the best performances. We report in this paper some preliminary results. The program of forthcoming experiments is also presented.

CP1165, *Nuclear Structure and Dynamics '09*
edited by M. Milin, T. Nikšić, D. Vretenar, and S. Szilner
© 2009 American Institute of Physics 978-0-7354-0702-2/09/$25.00

THE KINEMATICAL COINCIDENCE METHOD

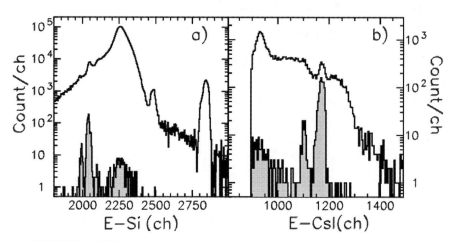

FIGURE 1. a) Silicon energy spectrum measured at $\theta=3.1°$ in single (empty histogram) and in coincidence (filled histogram) with the detector at 7.8° of fig.1.b. b) CsI(Tl) energy spectrum of a detector of at $\theta=7.8°$, single (empty histogram) and in coincidence with the detector of fig.1a (filled histogram).

A test of the use of CHIMERA to extract spectroscopic information from low energy reactions was performed using a 52 MeV ^{7}Li beam impinging on a polyethylene (H_4C_2) target, in order to study the ^{7}Li+p reaction. The main idea was to use kinematical coincidences to identify fragments produced in two-body reactions such as, for examples, elastic or inelastic scattering and transfer reactions to bound states. In fig.1 the energy spectra measured in a Silicon detector at $\theta=3.1°$ and in a CsI(Tl) detector at $\theta=7.8°$ are presented. In both spectra we see some structures superimposed to large background generated mainly by reactions on ^{12}C of the plastic target. The two considered detectors were selected with a difference in azimuthal angle of $\Delta\phi=180°$, in order to maximize the yield for kinematical coincidences. Such coincidence events, plotted as filled spectra in the figures, clearly show very clean peaks and reduced background. The strongest peak observed in coincidence is the second kinematical solution of the ^{7}Li+p elastic scattering, the other one, less intense, is produced by the ^{7}Li(p,d)^{6}Li reaction. Due to the background the (p,d) peak was hardly seen in the single spectra. On the contrary the kinematical coincidence allows a simple way to evaluate the reaction cross section. Apart background suppression, kinematical coincidences can also allow the discrimination of excited levels and can improve the angular resolution of the detector. As an example, in fig.2a we plot the energy measured in two silicon detectors at $\theta=6.4°$ and $\theta=58°$ in kinematical coincidence. We have selected events stopped in the silicon detector. We can note how the events are distributed along the kinematical lines relative to p,p elastic scattering and p,p' inelastic scattering, populating the first excited state of ^{7}Li at 0.47 MeV.

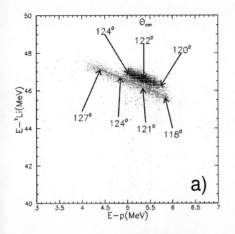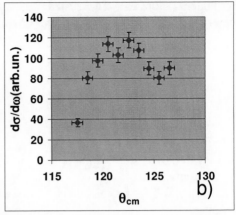

FIGURE 2. a) E_p-E_{7Li} plot measured at θ=6.4° and θ=58°. Kinematical lines for (p,p) and (p,p') scatterings are also plotted and CM angles are indicated. b) CM (p,p') angular distributions obtained from the same data.

The arrows, in the plot, correspond to the energy at different values of center of mass (CM) angles. It is easy to understand that by selecting an appropriate cut in the energy of the detected particles, one can select a definite angle in the CM system. In fig. 2b we plot the result that can be obtained with this selection. We underline that with the same method we are also able to detect and identify neutrons interacting with our CsI(Tl) detectors.

THE TAGGING DETECTORS

The beam provided by EXCYT is generally a quite pure beam, while the FRIBS beam is a fragmentation cocktail beam that need of an event by event identification. Moreover the beam is quite large and has a huge spread in energy. Therefore one has to measure its position and energy quite accurately, in order to allow a good reconstruction of kinematics. For this reason we have developed a tagging system constituted by a large surface microchannel plate detector (MCP) and by an XY Double Side Silicon Strip Detector (DSSSD). The MCP detector will also produce a good reference point to measure the time of flight of particles produced during the reactions in order to measure their mass. A similar time tagging can be also performed on the EXCYT beams. We have tested the tagging system with a calibration alpha source to measure efficiency, time and energy resolution of the assembly. 100% efficiency was measured using a mylar foil with some LiF deposition, to generate electrons. In fig.s 3 the results for the Time of flight measurements, obtained positioning the MCP at about 70 cm from the DSSSD, are shown. Selecting only one pixel of DSSSD we have the best time response, limited mainly by the used TDC resolution (250ps/ch). The system has also been tested using radioactive beams observing the production of ^{11}Be and surroundings beams.

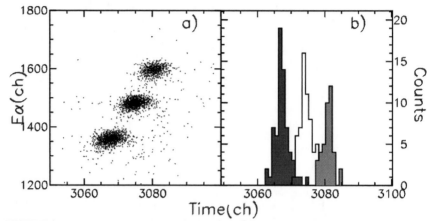

FIGURE 3. a) E_α-time scatter plot measured using MCP and DSSSD detectors. b) Time of flight spectrum measured between a pixel of DSSSD and the MCP: the three peaks correspond to the different energies of the calibration alpha source.

OUTLOOK AND CONCLUSIONS

We have tested the use of the CHIMERA 4π detector to extract spectroscopic information by using proton targets with radioactive light beams. Kinematical coincidences will provide clean enough data allowing the study of pick-up reactions. We are going to measure stripping (d,p) reactions using ^{9}Li beam delivered by EXCYT in order to study the ^{10}Li resonance [7]. Another experiment is already scheduled for July 2009 using FRIBS fragmentation beams. We will study transfer reactions in nuclei in the region of ^{11}Be. More in particular we will compare different reaction channels all populating ^{11}Be. Investigations on other nuclei, along both proton and neutron drip lines, will be possible in the near future also using the foreseen upgrading of the FRIBS facility [8] and new EXCYT beams.

REFERENCES

1. G. Cuttone, et al., *Nucl. Instrum. Meth. Phys. Res.* B **261**, 1040 (2007).
2. G. Raciti, et al., *Phys. Rev. Lett.* **100**, 19250 (2008).
3. A. Pagano, et al., *Nucl. Phys.* **A 734**, 504 (2004).
4. E. De Filippo, et al., *Phys. Rev.* C **71**, 044602 (2005).
5. I. Skwira-Chalot, *Phys. Rev. Lett* **101**, 262701 (2008).
6. F. Amorini, *et al.*, *Phys. Rev. Lett.* **102**, 112701 (2009).
7. H.B. Jeppesen, *et al.*, *Phys. Lett.* B **642**, 449 (2006).
8. L.Calabretta, G.Cardella and G.Raciti
   http://www.lns.infn.it/index.php?option=com_content&view=article&id=247&catid=31&Itemid=53

Resonant and Nonresonant Breakup of ^{11}Be and ^{19}C on a Proton Target

R. Crespo[*,†], E. Cravo[†], A. Deltuva[†] and A.C. Fonseca[†]

[*]Departamento de Física, Instituto Superior Técnico, Taguspark, Av. Prof. Cavaco Silva, Taguspark, 2780-990 Porto Salvo, Oeiras, Portugal
[†]Centro de Física Nuclear, Universidade de Lisboa, Av. Prof. Gama Pinto 2, 1649-003 Lisboa, Portugal

Abstract. Full Faddeev-type calculations are performed for the breakup of ^{11}Be and ^{19}C on proton target at 63.7 MeV/u and 70 MeV/u incident energy, respectively. We make use of a simplified two-body model for the one-neutron halo nucleus which involves an inert core and a valence neutron. Inclusive cross sections as a function of the center of mass angle of the n-core pair are calculated including both resonant and non-resonant contributions. The agreement between the calculated angular distributions and the data is discussed in each case.

Keywords: nuclear Reactions, halo nuclei, few-body structure
PACS: 21.45.-v,24.50.+g,25.60.Gc

INTRODUCTION

The Faddeev/Alt, Grassberger, Sandhas (Faddeev/AGS) formalism [1, 2, 3] is a non-relativistic three-body multiple scattering framework that can be used to calculate all relevant three-body observables (elastic, breakup and transfer) in equal footing.

This reaction framework makes a proper treatment of both the resonant and nonresonant contributions of the continuum and includes both nuclear and Coulomb interactions. We aim to study the breakup of the one-neutron halo nuclei ^{11}Be and ^{19}C on a proton target. The purpose is to unravel if the use of a numerically exact solution of the three-body Hamiltonian through the Faddeev/AGS framework brings new insight into the description of the breakup experimental data and estimate to what extent a nonlocal optical potential used for describing the interaction between the proton and the core modifies the calculated breakup observables.

Specifically, we calculate the angular distribution resulting from the breakup of ^{11}Be and ^{19}C on a proton target at 63.7 MeV/u and 70 MeV/u incident energy, respectively, and compare with experimental data

THE REACTION APPROACH

According to the Faddeev/AGS reaction framework, one calculates the operators $U^{\beta\alpha}$, whose on-shell matrix elements are the transition amplitudes. The AGS operators are

CP1165, *Nuclear Structure and Dynamics '09*
edited by M. Milin, T. Nikšić, D. Vretenar, and S. Szilner
© 2009 American Institute of Physics 978-0-7354-0702-2/09/$25.00

obtained by solving the integral equations [2, 3]

$$U^{\beta\alpha} = \bar{\delta}_{\beta\alpha}G_0^{-1} + \sum_{\gamma}\bar{\delta}_{\beta\gamma}t_{\gamma}G_0U^{\gamma\alpha} , \tag{1}$$

where $\bar{\delta}_{\beta\alpha} = 1 - \delta_{\beta\alpha}$, t_{γ} is the pair transition operator,

$$t_{\gamma} = v_{\gamma} + v_{\gamma}G_0t_{\gamma} , \tag{2}$$

and G_0 is the free resolvent

$$G_0 = (E + i0 - H_0)^{-1}, \tag{3}$$

with E being the total energy of the three-particle system in the center of mass (c.m.) frame. Equation (1) can only be solved for short-range potentials v_{γ}. Nevertheless, the Coulomb interaction is included using the method of screening and renormalization [4].

Before solving the Faddeev/AGS equations we have to specify the three pair interactions. Since we are considering the breakup of a one-neutron nucleus (assumed well described by a core and a neutron valence particle) by a proton target one needs then to specify the p-n, p-core and n-core interactions. For p-n we either use a realistic CD Bonn potential [5] or a simplified Gaussian interaction. For the potential between the proton and the core we use a phenomenological local optical model with parameters taken from the Watson global optical potential parametrization [6, 7]. In addition we also consider a nonlocal potential that significantly improves the description of the data in transfer reactions [8]. As for the n-core interaction we use an L-dependent interaction as described in [9] for ^{11}Be.

The interaction between the valence neutron and the ^{18}C core in ^{19}C is taken L-dependent as well,

$$V(r) = -V_c f(r, R_c, a_c), \tag{4}$$

where $f(r, R, a)$ is the usual Woods-Saxon form factor

$$f(r, R, a) = 1/[1 + \exp[(r - R)/a]], \tag{5}$$

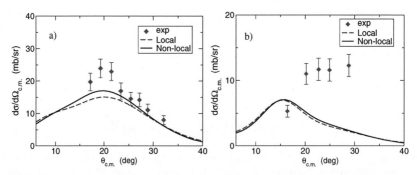

FIGURE 1. a) Angular distribution for the breakup p(^{11}Be,p)^{10}Be n at 63.7 MeV/u integrated over the energy range $E_{rel} = 0 - 2.5$ MeV (a) and $E_{rel} = 2.5 - 5.0$ MeV (b). The results obtained with nonlocal and local optical p-core potentials are given by solid and dashed lines, respectively. The data is from [10].

and $R_i = r_i A^{\frac{1}{3}}$. The depth of the central interaction is $V_c = 41.02$ MeV for the d partial wave and $V_c = 40.0325$ MeV for any other partial wave. The radius and difuseness are $r_c = 1.25$ fm and $a_c = 0.7$ fm. The s-wave interaction is adjusted to generate a 2s bound state with energy $E(2s) = -0.5794$MeV. This interaction also supports a Pauli forbidden 1s bound state with energy $E(1s) = -22.070$MeV which is projected out by moving it to large positive energy.

RESULTS

We first show the results for the breakup of ^{11}Be. The observables were calculated using a NN CD Bonn interaction in this case. In Fig. 1 we show the breakup angular distribution $d\sigma/d\Omega_{\text{c.m.}}$. We have not included very small angles $\theta_{\text{c.m.}} < 5°$ where there is no data and the convergence of the results with respect to the Coulomb screening radius is slow. In Fig. 1 a) due to the energy resolution of the experimental setup, the relative core-neutron energy is integrated around the resonance $E_r = 1.275$ MeV in the energy range $E_{\text{rel}} = 0 - 2.5$ MeV. Overall, the calculated observable using the AGS/Faddeev reaction framework and including both the resonant and nonresonant contributions reproduces fairly well the experimental data. Nevertheless it underpredicts the data for $\theta_{\text{c.m.}}$ around 20°. The angular distribution calculated using the nonlocal potential for the proton-core interaction [8] (represented by the solid line) slightly improves the description of the data reducing the discrepancy by about 30% compared to the observable calculated using a local interaction (dashed line).

In Fig. 1b) we show the breakup angular distribution, where the relative core-neutron energy is integrated over the energy range $E_{\text{rel}} = 2.5 - 5.0$ MeV. There is an evident discrepancy between the calculated observables and the data. The effect of the nonlocality of the p-core optical potential is minor in this case.

Next, in Fig. 2 we show the results for the breakup angular distribution of ^{19}C on a proton target where we have subtracted all the non-resonant continuum contribution. In this case, the results disagree considerably with the experimental data, indicating serious shortcomings in the used interaction models.

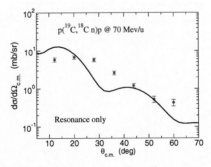

FIGURE 2. Angular distribution for the breakup p(^{19}C,p)^{18}C n at 70.0 MeV/u including only the resonant contribution. The data is from [11].

CONCLUSIONS

We have calculated de angular distribution for the breakup of ^{11}Be and ^{19}C on proton target at 63.7 MeV/u and 70 MeV/u incident energy using the full Faddeev/AGS scattering framework. In the case of the ^{11}Be projectile, due to the energy resolution of the experimental setup, the relative core-neutron energy was integrated around and away from the resonance, that is in the energy range $E_{rel} = 0 - 2.5$ MeV and $E_{rel} = 2.5 - 5.0$ MeV, respectively. In the former case, the Faddeev predicts reasonably well the trend of the breakup angular distribution although slightly underpredicting the data. In the later case there is an evident discrepancy between the calculated observables and the experimental results. We also found that the nonlocality of the proton-core interaction slightly improves the description of the $E_{rel} = 0 - 2.5$ MeV data.

The calculated resonant breakup angular distribution breakup of ^{19}C on a proton target disagrees considerably with the data.

More work is needed to understand the source of discrepancies between the calculated observables and the experimental data.

ACKNOWLEDGMENTS

This work was supported in part by the FCT under the Grant POCTI/ISFL/2/275 and PTDC/FIS/65736/2006. A.D. is supported the FCT grant SFRH/BPD/34628/2007.

REFERENCES

1. L. D. Faddeev, *Sov. Phys. JETP* **12**, 1014 (1961).
2. P. Alt, E. O.and Grassberger, and W. Sandhas, *Nucl. Phys.* **B 2**, 167 (1967).
3. W. Glöckle, *The Quantum Mechanical Few-Body Problem*, Springer-Verlag, Berlin/Heidelberg, 1983.
4. A. Deltuva, A. C. Fonseca, and P. U. Sauer, *Phys. Rev.* C **71**, 054005 (2005), **72**, 054004 (2005).
5. R. Machleidt, *Phys. Rev.* C **63**, 024001 (1991).
6. B. A. Watson, P. Singh, and R. Segel, *Phys. Rev.* **182**, 978 (1969).
7. R. Crespo, E. Cravo, A. Deltuva, M. Rodríguez-Gallardo, and A. C. Fonseca, *Phys. Rev.* C **76**, 014620 (1997).
8. A. Deltuva, *Phys. Rev.* C **79**, 021602(R) (2009), **79**, 054603 (2009).
9. E. Cravo, R. Crespo, A. Deltuva, and A. C. Fonseca, *Phys. Rev.* C (2009).
10. A. Shrivastava, et al, *Phys. Lett.* **596B**, 54 (2004).
11. Y. Satou, et al, *Phys. Lett.* **660B**, 320 (2004).

Scattering of ^{11}Be Around the Coulomb barrier

L. Acosta*, M.A.G. Álvarez[†,**], M.V. Andrés[†], M.J.G. Borge [‡], M. Cortés[‡], J.M. Espino[†], D. Galaviz[‡], J. Gómez-Camacho[†,**], A. Maira[‡], I. Martel*, A.M. Moro[†], I. Mukha[†], F. Pérez-Bernal*, E. Reillo[‡], K. Rusek[§], A.M. Sánchez-Benítez* and O. Tengblad[‡]

*Departamento de Física Aplicada, Universidad de Huelva, E-21071 Huelva, Spain
[†]Departamento de Física Atómica Molecular y Nuclear, Universidad de Sevilla, E-41080 Sevilla
**Centro Nacional de Aceleradores, Universidad de Sevilla-CSIC-Junta de Andalucía, E-41092 Sevilla, Spain
[‡]Instituto de Estructura de la Materia, CSIC, Madrid, E-28006 Madrid, Spain
[§]The Andrzej Sołtan Institute for Nuclear Studies, 00-681 Warsaw, Poland

Abstract.
Preliminary results on the ^{11}Be+^{120}Sn quasielastic scattering as well as the ^{11}Be $\rightarrow$ ^{10}Be + n breakup channel are presented in this work. The angular distributions of these channels were measured at REX-ISOLDE-CERN. The accuracy and angular range of the presented results provide stronger constrains to the theoretical interpretation than existing published results. We compare these new data with coupled-channel (CC) and continuum-discretized coupled-channel (CDCC) calculations. The role played by transfer and breakup channels in the elastic scattering is discussed.

Keywords: ^{11}Be, quasielastic, coupled-channel, halo nuclei, Coulomb barrier
PACS: 25.45.De 25.70.Mn 24.10.Eq 25.60.-t

INTRODUCTION

The ^{11}Be is a halo nucleus composed of a ^{10}Be core and a weakly bound neutron. This nuclide has a half life of 13.8 s and a separation energy for one neutron of 504(6) keV. The only bound excited state ($J^{\pi} = 1/2^-$) lies at 320 keV with a strong coupling to the ground state ($J^{\pi} = 1/2^+$) by the fastest known $E1$ transitions. Due to its loosely bound structure, coupling to the continuum should play an important role in near barrier scattering with heavy targets. Therefore the ^{11}Be nucleus is an interesting case to study the dynamics of nuclear haloes at Coulomb barrier energies [1, 2].

Another important issue is the role played by the highly deformed ^{10}Be core on the scattering cross sections [3]. Accurate data on ^{11}Be scattering are needed to study these effects. Presently the only results published are from one experiment performed in RIKEN [4, 5]. Unfortunately, the angular distributions presented in the mentioned analysis suffer of large experimental uncertainties, and elastic and other reaction channels could not be studied separately.

Aiming to improve the experimental situation we have recently performed measurements of ^{11}Be scattered on ^{120}Sn at 32 MeV (lab). The experiment was performed at the REX-ISOLDE facility at CERN (Geneva), using a detection system that covered a wide angular range.

CP1165, *Nuclear Structure and Dynamics '09*
edited by M. Milin, T. Nikšić, D. Vretenar, and S. Szilner
© 2009 American Institute of Physics 978-0-7354-0702-2/09/$25.00

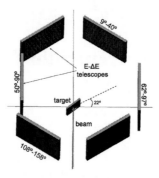

FIGURE 1. Schematic representation of the experimental setup. Downstream and upstream detectors are placed symmetrically with respect to the beam direction. The target was tilted 22° in order to cover angles around 90° with the lateral telescopes.

EXPERIMENT

The experiment IS444 was performed using a post-accelerated [11]Be beam at the energy of 2.91 MeV/u and a [120]Sn target, in order to study the scattering around the Coulomb barrier. The experimental setup consisted of an hexagonal configuration with 6 telescope detectors in E-ΔE configuration, surrounding the target. Each telescope was made up of two silicon detectors: a thin (ΔE) Double-Sided Silicon Strip Detector (DSSSD) with a thickness of 40 μm and divided in 16 strips in each side; and a PAD silicon detector (E) with a thickness of 500 μm. If we impose coincidences between strips in the front and back side (mutually perpendicular) of a DSSSD, we get what we call a "pixel" (16x16 in total). A schematic representation of the experimental setup is shown in Fig. 1. The target was tilted 68° with respect to the beam axis in order to allow the detection in the telescopes placed around 90°.

Due to the low intensity of the beam we used a thick (3.5 mg/cm^2) tin target. This fact limited the energy resolution of our detection system to about 350 keV, spoiling the possibility of resolving the excitation of [11]Be to the 1/2$^-$ first excited state (inelastic channel). However the resolution was good enough to identify [10]Be fragments resulting from breakup [11]Be → [10]Be + n for the angular range (15°-38°). Therefore, we could separate the quasielastic channel (elastic+inelastic) from the breakup channel in the mentioned range, by means of the analysis made pixel-by-pixel. Using pixels instead of full strips, it was possible to separate the breakup event from quasielastic ones, as the angular spread and kinematical effects are reduced. The difference between a full strip and one pixel is shown in the ΔE-E$_t$ spectra of the Fig. 2 at $\theta_{lab} \sim 34°$.

For the θ_{lab} range (52°-86°) it was not possible to separate the breakup from the quasielastic channel, because only part of the events had enough energy to go through the ΔE detector. In this case, we integrated the sum of quasielastic and breakup channels. The accumulated statistics registered in telescopes at θ_{lab} larger than 90° was too small to produce cross section data.

The measured angular distribution was normalized to Rutherford cross sections using the elastic scattering data of a ^{12}C beam impinging in the same [120]Sn target at 27 MeV.

318

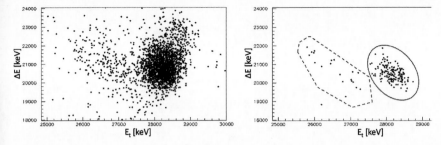

FIGURE 2. Differences between full strip and pixel. ΔE-E_t spectra around 34° (lab). The spectrum on the left shows the elastic+inelastic+breakup scattering for a full strip, the separation between channels seems not to be possible. On the right panel, one pixel from that strip is shown. In this case, we can observe the separation between quasielastic (solid polygon) and breakup (dashed polygon) channels.

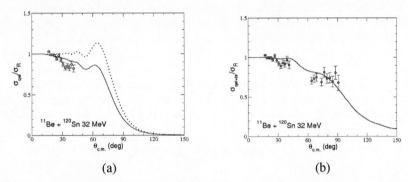

FIGURE 3. Comparison of experimental results and calculations. (a) Quasielastic scattering *vs.* $\theta_{c.m.}$ (dots) and CC calculations (dotted and solid lines). (b) Quasielastic+breakup channel *vs.* $\theta_{c.m.}$ (dots) and a CDCC calculation (solid line). See text for details.

RESULTS

The angular distribution obtained for quasielastic channel is shown in the Fig. 3a. The overall shape is similar to the angular distributions measured previously in the elastic scattering of the weakly bound ^{6}He on ^{208}Pb [6, 7]. However, the deviation from the Rutherford cross section at forward angles is much more pronounced in the case of ^{11}Be scattering. In order to reproduce the data shape we performed two CC calculations. The first (dotted line) includes the first excited state and two resonant states (1.78 MeV and 3.41 MeV). The strong absorption at very forward angles is not well reproduced. The second calculation includes, as proposed in [8, 9] two fictitious dipole states (*p*-states) located at excitation energy of 0.55 MeV, just above the breakup threshold, with spins $1/2^-$ and $3/2^-$. These states are intended to represent the low-lying dipole strength for the ^{11}Be continuum. With these parameters the agreement between calculation and data results is acceptable. Further details of this calculations can be consulted in [10].

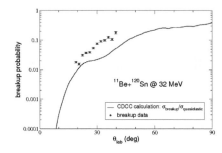

FIGURE 4. Breakup probability, defined as the ratio between the breakup and quasielastic events. Comparison between experimental results and CDCC calculation.

In Fig 3b we show preliminary results obtained for the quasielastic+breakup channel (dots). We include a CDCC calculation (solid line) generated with the potentials for the different channels taken from [11, 12, 13]. These new results show a good agreement with the calculation and even with the data presented in [5] at 46 MeV.

Fig. 4 shows the ratio between the breakup and the quasielastic events, shown in Fig. 3a. The solid line is the prediction of the CDCC calculation shown in Fig. 3b. In this case, the calculation underestimates the data. This discrepancy could be due to the contribution of higher partial waves of the ^{11}Be continuum, or the contribution of other channels not included in this calculation, such as the one-neutron transfer to the target.

ACKNOWLEDGMENTS

This work has been supported by the Spanish CICYT, under the project FPA2007-62170, the Spanish Council of Science and Education (MEC) under the research grants FINURA (FPA2007-63074), CPAN (CSD 2007-00042), the DGICYT under the project FPA2006-13807-c02-01 and Consolider-Ingenio CSD-2007-00042. L.A. acknowledges financial support by XI Plan Propio de Investigación from the Universidad de Huelva.

REFERENCES

1. L.F. Canto, et al., *Phys. Rev.* C **52**, R2848 (1995).
2. K. Yabana, Y. Ogawa and Y. Suzuki, *Phys. Rev.* C **58**, 2403 (1998).
3. N.C Summers, et al., *Phys. Rev.* **74**, 014606 (2006).
4. M. Mazzocco, et al., *Eur. Phys. J. A* **52**, 295 (2006).
5. M. Mazzocco, et al., *Eur. Phys. J. Special Topics* **150**, 37 (2007).
6. O.R. Kakuee, et al., *Nucl. Phys.* **A 765**, 294 (2006).
7. A.M. Sánchez-Benítez, et al., *Nucl. Phys.* **A 803**, 30 (2008).
8. M. Takashina, S. Takagi and Y. Sakuragi, *Phys. Rev.* C **67**, 037601 (2003).
9. D.J. Howell, J.A. Tostevin and J.S. Al-Khalili, *J. Phys. G* **31**, S1881 (2005).
10. L. Acosta, et al., *Eur. Phys. J. A* (*accepted*).
11. J.J. Kolata, et al., *Phys. Rev.* C **69**, 047601 (2004).
12. A.J. Koning and J.P. Delaroche, *Nucl. Phys.* **A 713**, 247 (2003).
13. P. Capel, G. Goldstein and D. Baye, *Phys. Rev.* C **70**, 064605 (2004).

Low Energy Transfer Reactions With ^{11}Be

Jacob Johansen,

on behalf of the IS430 collaboration

(Aarhus, Birmingham, CERN, Gothenburg, Madrid, Seville)

Department of physics and astronomy, Aarhus University, Denmark

Abstract. The low-energy transfer reaction ^{11}Be(d,p)^{12}Be gives us the opportunity to investigate single particle excitations in ^{12}Be. The breaking of the magic number N = 8 for ^{12}Be can be studied by comparing spectroscopic data with theoretical predictions.

Keywords: ^{11}Be beam. CD$_2$ target, DSSSD detector
PACS: 25.60.Je, 27.20.+n

MOTIVATION

The breaking of the N = 8 magic number in ^{12}Be makes the structure of ^{12}Be an interesting open question. Several beryllium isotopes including ^{11}Be and ^{12}Be are descriped by a cluster structure of α-particles and n [1]. The theory will be tested by comparing measured spectroscopic factors with theoretical predictions.

The REX-ISOLDE facility has made it possible to study ^{12}Be by low-energy transfer reaction. The low energy made it possible to only populate the bound states of ^{12}Be.

THE EXPERIMENT

The transfer experiment was done in autumn 2005 at REX-ISOLDE, CERN. The ^{11}Be from ISOLDE was bunched and accelerated to 2.25 MeV/u in the REX post accelerator. The beam was directed on to two targets: A deuteron target made of deuterated polyethylene ($(CD_2)_n$) and a proton target ($(CH_2)_n$).

Two double sided silicon strip detectors (dsssd) backed by two thick silicon detectors were used to detect the outgoing particles. The detectors were placed in forward angle ($\theta_{cm} = [20^o, 100^o]$), one on each side of the beam.

DATA AND RESULTS

With the deuteron target five different particles were detected (p, d, t, ^{4}He, ^{6}He). Deuterons arise from scattering of ^{11}Be. The one particle transfer reactions to ^{12}Be and ^{10}Be give p (d(^{11}Be,p)^{12}Be) and t (d(^{11}Be,t)^{10}Be). Break-up of ^{10}Be will produce ^{4}He and ^{6}He, but ^{4}He can also be produced in a compound reaction or directly in a transfer reaction to ^{9}Li.

CP1165, *Nuclear Structure and Dynamics '09*
edited by M. Milin, T. Nikšić, D. Vretenar, and S. Szilner
© 2009 American Institute of Physics 978-0-7354-0702-2/09/$25.00

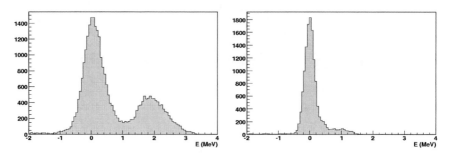

FIGURE 1. Excitation spectra for ^{11}Be on deuteron target and on proton target

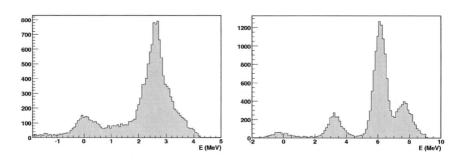

FIGURE 2. Excitation spectra for ^{12}Be and ^{10}Be

Excitation spectra for ^{11}Be were extracted for deuteron and proton targets, see Fig. 1. The resolution is clearly better for the proton target, where the reaction is mainly elastic scattering with some inelastic scattering (about 10 percent). The resolution for the deuteron target is not good enough to distinguish the two bound states in ^{11}Be, but the main contribution for this reaction is assumed also to be elastic scattering.

Figure 2 shows excitation spectra for ^{12}Be and ^{10}Be made from proton and triton data. The ^{12}Be spectrum contains the ground state as well as the excited 1^- state and may also contain some 2^+ and 0^+ excited states in the tail of the 1^--peak.

In the ^{10}Be spectrum three peaks are below the neutron emission threshold at 6.81 MeV. the ground state and the first excited state is clearly shown, while the peak at 6 MeV may contain 4 states. Spectroscopical factors will be derived from the differential cross sections.

A follow-up experiment will be done October 2009 at REX-ISOLDE including gamma detection in the MINIBALL array as well as particle detection.

REFERENCES

1. C. Romero-Redondo, et al., *Phys. Rev. C* **77**, 054313 (2008).

Factorization of the Cross Section for the ^{12}C(p,pα)^{8}Be(g.s.) Reaction at an Incident Energy of 100 MeV

A. A. Cowley[*,†], J. Mabiala[*,†], E. Z. Buthelezi[*], S. V. Förtsch[*],
R. Neveling[*], F. D. Smit[*], G. F. Steyn[*] and J. J. van Zyl[†]

[*]*iThemba LABS, PO Box 722, Somerset West 7129, South Africa*
[†]*Physics Department, Stellenbosch University, Private Bag X1, Matieland 7602, South Africa*

Abstract. Cross sections and analyzing powers for the reaction ^{12}C($p,p\alpha$)^{8}Be at an incident energy of 100 MeV, measured over a range of quasifree scattering angle pairs, are compared with elastic scattering of protons from ^{4}He. Remarkable agreement between angular distributions of the two sets of data, presented as a function of the scattering angle in the two-body centre-of mass, is found. Thus the α–cluster reacts to the projectile just like a free particle.

Keywords: Polarized proton interactions, Cluster knockout, Cross section, Analyzing power
PACS: 25.40.-h, 24.70.+s, 21.60.Gx

INTRODUCTION

Knockout reactions present a direct and convenient method for the study of cluster structure in the ground state of atomic nuclei. In α–cluster knockout by means of projectiles such as energetic protons and α–particles, the distorted-wave impulse approximation (DWIA) theory is able to provide a reasonably good representation of experimental data. However, there are indications that the reaction process may be viewed as being even simpler than implied by the full DWIA theory. For example, if spin-orbit interactions are ignored in a $(p,p\alpha)$ reaction, the coincidence cross section may be formulated to display its separate observable quantities as explicit factors. The resulting factorized expression of the cross section consists of the product of a two-body cross section, which represents the projectile-cluster scattering, and a distorted momentum distribution [1].

Previous results of Roos *et al.* [2] for ^{12}C($p,p\alpha$)^{8}Be at an incident energy of 100 MeV suggest a breakdown of the factorization approximation. However, in this work on the same target and incident energy, we explore a larger range of two-body scattering angles, which allows a better determination of the general trend. We select coincidence data measured at quasifree angle pairs, i.e. angle pairs at which zero recoil momentum of the residual nucleus is kinematically allowed. In addition, we investigate the analyzing power.

From the measured cross sections we extract knockout data corresponding to the quasifree projectile-cluster interaction. Theoretical guidance then indicates under which conditions these coincidence data would resemble free scattering.

CP1165, *Nuclear Structure and Dynamics '09*
edited by M. Milin, T. Nikšić, D. Vretenar, and S. Szilner
© 2009 American Institute of Physics 978-0-7354-0702-2/09/$25.00

THEORETICAL CONSIDERATIONS

Following the notation of Chant and Roos [1] we write the knockout reaction as $A(a, cd)B$. For a $(p, p\alpha)$ reaction we have $a = c$ and the bound knocked-out cluster $b = d$.

In a plane wave impulse approximation (PWIA) theory the cross section may be written as

$$\frac{d^3\sigma}{d\Omega_c d\Omega_d dE_c} = S_b F_k \frac{d\sigma^{p\alpha}}{d\Omega} |\psi|^2 , \tag{1}$$

where $d\Omega_c$ and $d\Omega_d$ are the solid angles of observation of the light ejectiles. E_c is a kinetic energy, S_b is a spectroscopic factor, and F_k is a kinematic factor. The quantity $d\sigma^{p\alpha}/d\Omega$ is a half-shell two-body cross section that describes the scattering of the projectile from the bound α–cluster, and ψ is the Fourier transform of the radial wave function of the bound particle.

The PWIA ignores not only the interaction of the projectile with the core of the target system, but also final state interactions between the outgoing light products with the residual nucleus. This is clearly unrealistic.

In a more realistic DWIA, the cross section [3] is expressed as

$$\frac{d^3\sigma}{d\Omega_c d\Omega_d dE_c} = S_b F_k \sum_{\rho_c' L\Lambda} \left| \sum_{\rho_a \sigma_a \sigma_c \sigma_c'} D^{s_a}_{\rho_a \rho_a'}(R_{ap}) \times D^{s_a^*}_{\sigma_c \sigma_c'}(R_{ac}) T^{L\Lambda}_{\sigma_a \sigma_c' \rho_a \rho_c'} \langle \sigma_c | t | \sigma_a \rangle \right|^2 , \tag{2}$$

where the D's are rotation matrices and the t-matrix for the two-body scattering is denoted by $\langle \sigma_c | t | \sigma_a \rangle$. The quantity $T^{L\Lambda}_{\sigma_a \sigma_c' \rho_p \rho_c'}$ contains the overlap of the various distorted waves with the target-structure.

Unfortunately, as is evident from Eq. 2, the cross section no longer factorizes. Only if we ignore spin-orbit forces in the distorted waves, do we retain an expression [3] which resembles the factorized form of Eq. 1. Under those conditions the quantity $|\psi|^2$ is a convolution of the distorted waves with the bound state wave function [1].

COMPARISON BETWEEN KNOCKOUT AND ELASTIC SCATTERING

As shown in Fig. 1, the angular distribution of the cross section $d\sigma^{p\alpha}/d\Omega$ for the two-body projectile-bound α–cluster system (shown as solid circles), extracted from the coincidence cross section $d^3\sigma/d\Omega_p d\Omega_\alpha dE_p$, is in remarkable agreement with elastic scattering of protons from ^{4}He (solid curve) at the same incident energy. However, we should keep in mind that this impressive demonstration of factorization of the knockout cross section, and its agreement with free scattering, relies on the assumption that the distorted momentum distribution $|\psi|^2$ is described accurately by the DWIA as implemented in this work. In other words, that the distorting potentials which are used, are appropriate.

Another, more sensitive and significant test of the factorization, is the analyzing power angular distribution. As shown in Fig. 2, we again see a remarkable correspondence

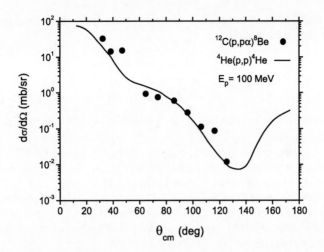

FIGURE 1. Comparison between the two-body cross section extracted from the experimental data for the reaction $^{12}C(p,p\alpha)^8$Be (filled circles) and elastic scattering of ^{4}He$(p,p)^4$He, represented as a smooth curve drawn through the experimental data of Ref. [4].

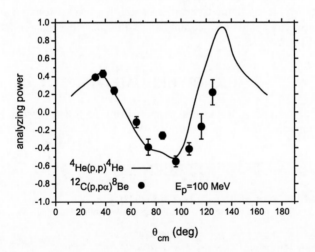

FIGURE 2. Comparison between the experimental analyzing power for the reaction $^{12}C(p,p\alpha)^8$Be (filled circles) and elastic scattering of ^{4}He$(p,p)^4$He, represented as a smooth curve drawn through the experimental data of Ref. [5].

between the knockout reaction and free scattering. Recall that the analyzing power for the knockout reaction is defined as

$$A_{p,p\alpha} = \frac{\frac{d^3\sigma}{d\Omega_p d\Omega_\alpha dE_p} \uparrow - \frac{d^3\sigma}{d\Omega_p d\Omega_\alpha dE_p} \downarrow}{\frac{d^3\sigma}{d\Omega_p d\Omega_\alpha dE_p} \uparrow + \frac{d^3\sigma}{d\Omega_p d\Omega_\alpha dE_p} \downarrow} , \tag{3}$$

(where arrows indicate spin directions of the projectile), which is only equal to the free p–^{4}He analyzing power A_{p-^4He} if $|\psi|^2 \uparrow = |\psi|^2 \downarrow$.

Thus, by comparing these cross sections and analyzing power angular distributions directly with those from free p–^{4}He elastic scattering, we have demonstrated that the factorization of the knockout cross section holds to a remarkable degree. This result is interpreted not only as compelling evidence for the existence of preformed α–clusters in ^{12}C, but also that the recoiling heavy residue acts merely as a spectator to the collision between the projectile and the cluster. Furthermore, the core of the target system is insensitive to the polarization of the projectile.

CONCLUSION

The fact that the cross section for the reaction ^{12}C$(p,p\alpha)^8$Be at 100 MeV has a much simpler structure than suggested by a proper DWIA theory has profound implications for the interpretation of the nuclear reaction mechanism.

Apart from general ideas about expectations regarding results from the DWIA theory, we have not presented details of a comparison with our experimental data here. This may be found [6] elsewhere.

ACKNOWLEDGMENTS

This work was performed with financial support from the South African National Research Foundation (NRF).

REFERENCES

1. N. S. Chant, and P. G. Roos, *Phys. Rev. C* **15**, 57-68 (1977).
2. P. G. Roos, N. S. Chant, A. A. Cowley, D. A. Goldberg, H. D. Holmgren, and R. Woody, *Phys. Rev. C* **15**, 69 (1977).
3. N. S. Chant, and P. G. Roos, *Phys. Rev. C* **27**, 1060 (1983).
4. N. P. Goldstein, A. Held, and D. G. Stairs, *Can. J. Phys.* **48**, 2629 (1970).
5. A. Nadasen, P. G. Roos, D. Mack, G. Ciangaru, L. Rees, P. Schwandt, K. Kwiatkowski, R. E. Warner, and A. A. Cowley, *Indiana University Cyclotron Facility, Scientific and Technical Report*, p.13 (1983).
6. J. Mabiala, A. A. Cowley, S. V. Förtsch, E. Z. Buthelezi, R. Neveling, F. D. Smit, G. F. Steyn, and J. J. van Zyl, *Phys. Rev. C* **79**, 054612 (2009).

Spectroscopy of ^{16}O Using α+^{12}C Resonant Scattering in Inverse Kinematics

N. I. Ashwood[a], M. Freer[a], N. L. Achouri[b], T. R. Bloxham[a],
W. N. Catford[c], N. Curtis[a], P. J. Haigh[a], C. W. Harlin[c], N. P. Patterson[c],
D. L. Price[a], N. Soić[d] and J. S. Thomas[c]

[a] *School of Physics and Astronomy, The University of Birmingham, Edgbaston, Birmingham, B15 2TT*
[b] *LPC, ISMRA and Université de Caen, IN2P3 CNRS, 14050 Caen Cedex, France*
[c] *School of Electronics and Physical Sciences, The University of Surrey, Guildford, Surrey, GU2 7XH*
[d] *Rudjer Bošković Institut, Bijenička 54, HR-1000, Zagreb, Croatia*

Abstract. A measurement of the α(^{12}C,α)^{12}C reaction has been performed using resonant scattering with a gas target. Beam energies of 46, 51, 56 and 63 MeV were used to populate resonances in the excitation energy range of 11.6 to 22.9 MeV in ^{16}O. The angular distributions of the elastic scattering were measured at zero degrees using an array of segmented silicon strip detectors with a minimum range of 0° to 30° in the centre of mass. The spins of 8 resonances between 14.1 and 18.5 MeV were obtained, confirming spin assignments made using elastic scattering in normal kinematics. An R-matrix analysis of the data was performed which indicates that the present understanding of ^{16}O in this region is good, but not complete.

Keywords: Resonant scattering, inverse kinematics, R-matrix, angular distributions, resonances.
PACS: 25.70.Ef. 25.70.Mn. 27.20+n.

INTRODUCTION

The α+^{12}C structure of ^{16}O has been extensively studied. A rotational band in ^{16}O, based on this configuration is predicted to lie close to the α-decay threshold (7.16MeV) [1]. Indeed the first excited state of ^{16}O lies only 1 MeV below this threshold, at 6.05 MeV, and is associated with a 4p-4h excitation i.e. an α-particle excited into the *sd* shell [2,3,4]. One of the most comprehensive of these studies was that of Ames [5] using the ^{12}C(α,α)^{12}C reaction. Here, the absolute cross-section was measured over the excitation energy range 15 to 22 MeV, using over 1000 bombarding energies in 10 keV steps. The elastically scattered α-particles were detected at 18 different angles, though not at 0 or 180 degrees in the centre of mass. Over 50 resonances were measured and the spins and decay widths determined using resonance line-shape fitting. The cluster structure of ^{16}O has also been studied via the (^{6}Li,d) reaction, using the coincident detection of α-particles from the subsequent α+^{12}C break-up of ^{16}O. Resonances were populated in the region of 10 to 35 MeV [6-11].

CP1165, *Nuclear Structure and Dynamics '09*
edited by M. Milin, T. Nikšić, D. Vretenar, and S. Szilner
© 2009 American Institute of Physics 978-0-7354-0702-2/09/$25.00

EXPERIMENAL DETAILS

This experiment was performed using the Van de Graaff accelerator at IPN Orsay, France. Fully stripped ^{12}C ions were accelerated at energies of 46, 51, 56 and 63 MeV, at intensities of ~10^9 pps. The target chamber was filled with ^{4}He gas, isolated from the beam line using a 5 μm Havar window. Gas pressures of between 200 and 510 mbar were used. As the ^{12}C ions traverse the gas they lose energy, tracing out the excitation function of ^{16}O. These resonances will subsequently decay via α-particle emission. The overall excitation energy range of the experiment was 11.6 to 22.9 MeV.

An array of silicon detectors, forming an arc of radius 70cm and angular range of 0 to 30 degrees in the centre of mass with respect to the chamber window, were used to detect the decaying α-particles. The two central detectors, centered at 0 and 5 degrees with respect to the window, had particle identification (PI), ΔE-E, capability. These consisted of a 65 μm silicon double sided detector (DSSD) 1.5 cm in front of a 500 μm DSSD. A 52 μm Al absorbing foil was placed in front of the telescope to stop any ^{12}C ions which were not fully stopped by the gas, from entering the detectors. Two further 500 μm DSSD detectors were centered at 9 and 13 degrees with respect to the window. The DSSD's comprised of 32 independent 3 mm strips, 16 in the horizontal direction and 16 in the vertical. In this way both the particle's energy and position could be determined. Steel baffles were placed 5 cm from the window to protect against the beam scattering from the window into the unprotected outer detectors.

RESULTS

The elastically scattered α-particles were detected and their energy and position calculated. A software gate was placed around the α-particles in the PI spectrum produced in the two central detectors, to select only ^{4}He. Where no PI was available it was assumed that the particle was a ^{4}He; the PI spectra from the central detectors suggested that the background contribution from other reactions is small.

Figure 1a shows typical excitation functions for the zero degree detector at each beam energy, normalised for detector efficiency. A background from inelastic scattering or maxima in the angular distributions for resonances close to the detector will contribute at low excitation energies in all but the 46 MeV data. As the α-particles are detected at 180 degrees in the centre of mass, any contribution from direct processes is at a minimum. In all cases, resonances seen in these spectra are in good agreement with resonances seen by both Ames [5] and Ophel et al. [12]. There is also good agreement for similar excitation energies between the resonance spectra for each beam energy, in the present data. The biggest difference is in the region of 15 to 16 MeV in the 51 and 56 MeV data. Here the resonances are very close to the detector and so there is an appreciable part of the angular distributions due to the large centre of mass coverage. The data is compared with that of Ames [5] at a similar centre of mass angle (close to 180 degrees in the ^{12}C(α,α)^{12}C reaction). Although the resolution is slightly better in the normal kinematics reaction, there is still good agreement between the two sets of data. The reduction resolution in the present measurement

(ΔE_x(FWHM) ~ 90 keV) can be attributed to a high count rate in the detector coupled with the intrinsic detector resolution.

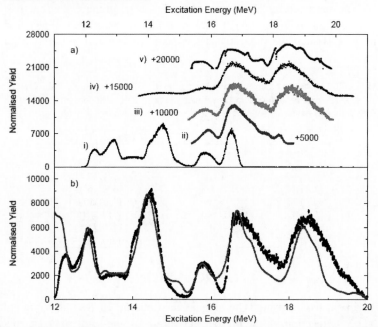

FIGURE 1. a) ^{16}O excitation energy spectra for the zero degree detector, normalised for detection efficiency, at beam energies and as pressures of i) 46 MeV and 360 mbar, ii) 51 MeV and 200 mbar, iii) 56 MeV and 200 mbar and iv) 63 MeV and 510 mbar. Spectrum v) shows the data taken from Ames [5] close to 180 degrees in the center of mass. The numbers next to each spectrum indicate the vertical displacement. b) Combined 46 and 63 MeV data with an R-matrix calculation folded with the experimental resolution (solid line).

The combination of the 46 and 63 MeV data is shown in Fig 1b. The excitation energy range observed is more extensive than that of Ames and overlaps with resonances observed by Ophel [12]. The solid line corresponds to a "global" R-matrix fit where we have attempted to reproduce the spins, widths and decay widths of ^{16}O resonances given in Table 16.15 of the compilation of Tilley *et al.* [13]. The R-matrix spectrum has been convolved with the experimental resolution. The R-matrix parameters are given in Table 1 of reference [14]. There is good agreement with the experimental data, although some noticeable differences do occur, such as in the region of 15 MeV, which shows that the present understanding of ^{16}O in this region is good, but not complete. The discrepancy above 17 MeV probably arises due to the need to compensate for resonance strength of high lying states cutting into this region.

Table 1 shows the spins and parities measured for the excitation energies listed. This was done by measuring the angular distributions associated with each resonance. The differential cross-section versus centre of mass scattering angle for these states can be found in Fig 3. and 4. of reference [14]. The spins of the states were determined from the order of the Legendre polynomial which best described the data. In all cases the spin and parity measured in the present data is in agreement with previously observed resonances [13]. The exception to this is the resonance at 17.6 MeV, which was measured by Ames as $(0^+,1^-)$. Reference [13] lists this state as being a 2^+ with T=1. This was measured by ^{15}N+p scattering. However, although the state is still listed as being measured by α-decay it is not possible to detect T=1 states from $\alpha+^{12}$C scattering and as such the present measurement would confirm l=1 strength in this region.

TABLE 1. Table showing spins measured using the $\alpha(^{12}C,\alpha)^{12}C$ reaction for the excitation energies listed. The data is compared to known natural parity states that decay by α emission [13].
* This state was measured by Ames [5] but has been recorded as a 2^+ with T=1 in [13].

Ex(MeV)	J^π	Ex(MeV) [J^π] [13]
14.1	3^-	14.1 [3^-]
14.6	4^+	14.62 [4^+]
15.5	$(3^-,2^+)$	15.26 [2^+], 15.41 [3^-]
16.7	6^+	16.28 [6^+]
17.1	2^+	16.94, 17.13, 17.20 [2^+]
17.6	1^-	15.51[1^-], 17.61 [$(0^+,1^-)$]*
17.8	4^+	17.78 [4^+]
18.5	5^-	18.40 [5^-]

CONCLUSION

This experiment has measured the angular distributions of resonances populated via the $\alpha(^{12}C,\alpha)^{12}C$ reaction. In the present measurement, the spins of 8 resonances were measured using the first minimum in the angular distributions to determine the spin of the state. The energies and spins of the states are in excellent agreement with previous measurements done in normal kinematics [5], which shows the advantage of doing this measurement in inverse kinematics, in being able to populate several resonances at a single beam energy, while there is little loss in the quality of the data. An R-matrix analysis has shown that although the data is well represented by the states listed in the compilation [13], differences in the calculations and resonant spectrum indicate that the knowledge of cluster states in ^{16}O is incomplete.

REFERENCES

1. K. Ikeda, N. Tawikawa and H. Horiuchi *Prog. Theor. Phys. Suppl.* **464** (1968).
2. S. J. Krieger, *Phys. Rev. Lett.* **22**, 97 (1962).
3. S. Aberg, I. Ragnarsson, T. Bengtsson and R. Sheline, *Nucl. Phys.* **A 391**, 327 (1982).
4. N. E. Reid, N. E. Davidson and J. P. Svenne, *Phys. Rev.* C **9**, 1882 (1974).
5. L. L. Ames, *Phys. Rev. C* **25**, 97 (1982).
6. K. P. Artemov, et al., *Phys. Lett. B.* **37**, 61 (1971).
7. K. P. Artemov, et al., *Sov. J. Nucl. Phys.* **20**, 368 (1975).
8. K. P. Artemov, et al., *Sov. J. Nucl. Phys.* **36**, 779 (1975).
9. K. P. Artemov, et al., *Sov. J. Nucl. Phys.* **37**, 643 (1975).
10. A. Cunsolo, et al., *Nuovo Cimento* **40**, 293 (1977)
11. A. Cunsolo, et al., *Phys. Rev. C* **21**, 2345 (1980).
12. T. R. Ophel, P. H. Martin, S. D. Cloud and J. M. Morris, *Nucl. Phys.* **A 173**, 609 (1971).
13. D. R. Tilley, H. R. Weller and C. M. Cheves, *Nucl. Phys.* **A 294**, 161 (1993).
14. N. I. Ashwood, et al., *J. Phys. G* **36**, 055105 (2009).

Radiative Capture Process at the Coulomb Barrier: the Resonant ^{12}C+^{16}O Case

D. Lebhertz*, S. Courtin*, F. Haas*, M.-D. Salsac†, C. Beck*, A. Michalon*, M. Rousseau *, P.L. Marley**, R.G. Glover**, P.E. Kent**, D.A. Hutcheon‡, C. Davis‡ and J.E. Pearson‡

*IPHC, CNRS/IN2P3, Université de Strasbourg, F-67037 Strasbourg Cedex 2, France
†CEA-Saclay, Service de Physique Nucléaire, 91191 Gif-sur-Yvette, France
**Department of Physics, University of York, Heslington, York YO10, UK
‡TRIUMF, 4004 Wesbrook Mall, Vancouver, B.C., V6T 2A3, Canada

Abstract. In a recent experiment performed at Triumf using the Dragon 0° spectrometer and its associated BGO array we have measured for the first time the full gamma decay of the radiative capture channel close to the Coulomb barrier. This measurement has been performed at 3 energies E_{cm}=8.5, 8.8 and 9 MeV. We have extracted a radiative capture cross section more than five times larger than what had been previously observed. A selective contribution of the entrance spins 5$^-$ and 6$^+$ has also been evidenced whereas 1$^-$ to 3$^-$ spins are predicted to be predominant by coupled-channel calculations. At E_{cm}= 9 MeV, stronger structural behaviour appears which is characterised by a larger total cross section and also by the particularly strong feeding of the ^{28}Si prolate 4$^+$ state at 9.16 MeV. This level is explained by several models in terms of ^{12}C-^{16}O cluster sub-structure. Our data is compared to such cluster-model predictions and the agreement is quite good.

Keywords: Radiative Capture,^{12}C+^{16}O, Resonance, 0° spectrometer, Cluster, Coulomb Barrier
PACS: 21.10.Re, 21.60.Gx, 25.70.Ef, 25.70.Gh

INTRODUCTION

The occurrence of resonances in collisions between ^{12}C+^{12}C and ^{12}C+^{16}O are well known phenomena at energies close to the Coulomb barrier (CB) and even below, as such resonances have been found recently in the ^{12}C+^{12}C Gamow energy window. The understanding of the existence of such narrow resonances ($\sim$ 200 keV), correlated in different reaction channels [1, 2], remains an open question. These resonances could be due to particular states of the compound nuclei, like molecular states. If this was the case, these states would preferentially decay to states of the compound nucleus with similar deformed structure. Several theoretical calculations [3, 4, 5] predict the existence of such ^{12}C-^{16}O states above 6 MeV in ^{28}Si. The best channel to observe such a selective γ-decay is the radiative capture (RC). The radiative capture between light heavy-ions has been scarcely studied, and both systems mentioned here show quite large RC cross-sections (up to 2μb) with narrow structures for the decay to the low lying states [6]. We have reopened the experimental study of the RC process in both systems, the status for ^{12}C+^{12}C has been presented by D.G Jenkins at this conference, and we will focus here on ^{12}C+^{16}O. The only known part of the RC decay for this system was the one feeding the lower states of ^{28}Si; we have measured the full γ-decay of the RC channel, in order to see eventual feeding of intermediate states with specific structure.

CP1165, *Nuclear Structure and Dynamics '09*
edited by M. Milin, T. Nikšić, D. Vretenar, and S. Szilner
© 2009 American Institute of Physics 978-0-7354-0702-2/09/$25.00

THE ^{12}C(^{16}O, γ)^{28}SI EXPERIMENT AT TRIUMF

The radiative capture between light heavy-ions represents a small part of the total fusion process ($\frac{\sigma_{RC}}{\sigma_F} \sim 10^{-5}$) close to CB. Therefore we need an accurate selection of the recoiling nuclei at 0°. A spectrometer with a very high rejection of the incident beam (10^{12}) is needed, as well as a high efficiency γ-array to detect the RC decay. Our experiment was therefore performed at Triumf (Vancouver) using the high rejection Dragon spectrometer and its associated BGO array at 3 beam energies E_{lab}= 19.8, 20.5 and 21 MeV on a thin (40 μg/cm^2) highly enriched ^{12}C target. The ^{28}Si recoils have been detected in a silicon detector at the spectrometer focal plan after a 21 m flight path trough Dragon. The γ-rays have been coincidently recorded in 30 BGO detectors placed as close as possible to the target to maximise the efficiency.

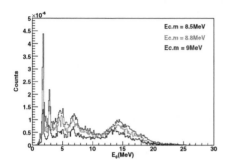 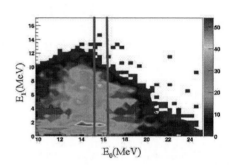

FIGURE 1. Left: spectrum of the highest energy γ-ray (E0) in coincidence with the recoils from the 12C+16O reaction at E_{cm}= 8.5, 8.8 and 9 MeV. Right: matrix of the energy of the second highest energy γ-ray (E1) versus the first one (E0) in coincidence with the recoils from the measurement at E_{cm}= 9 MeV.

The most striking feature of the γ-spectrum (Fig.1) is the prominent bump around 14 MeV. This corresponds to the feeding of states around 11 MeV in ^{28}Si. In the lower part of the spectrum we can identify transitions between the low lying levels of ^{28}Si, such as the first excited 2$^+$ state at 1.78 MeV and the 4$^+$ state at 4.62 MeV. This indicates that the main decay path involves a cascade and not a direct decay to the ground state, and this even after the intermediate states around 11 MeV. The cascade behaviour is confirmed by the measurement of the γ-fold which is around 2.7: this favours a spin larger than 4 for the entrance channel. Interestingly enough, a new decay path is revealed by a γ-peak around 7 MeV, which is in coincidence with γ-rays up to 19 MeV. This corresponds to the direct feeding of the 3$^-$ at 6.89 MeV. The feeding of this state is important in our study in that sense it is the first intruder state in ^{28}Si and results from a particle-hole excitation; the structure of this state was also discussed in terms of prolate or octupole deformation [7]. At E_{cm} = 9 MeV, a second component in the 7 MeV peak appears which is in coincidence with a 16 MeV γ-ray (see Fig.1). By gating on the 16 MeV peak, we can see that the new component is centred around 7.4 MeV. This corresponds to the feeding of the 4$^+$ (9.16 MeV) of the ^{28}Si prolate band. The feeding of this particular state is interesting because this band is predicted by cluster calculations to be the first band with ^{12}C-^{16}O substructure.

DISCUSSION OF THE RESULTS

To take into account the limited acceptance of Dragon for our experimental case we have performed GEANT simulations of the ^{28}Si flight pathway through Dragon and of the coincident γ-rays detected in the BGO array. We have compared the experimental data with three specific configurations of the entrance-channel and decay. The first one is a full statistical behaviour, the second one is a unique spin selection in case of an entrance resonance followed by the usual fragmentation of the γ-flux and the third a complete cluster behaviour in the entrance channel and decay mode. We will present the results for E_{cm}= 9.0 MeV.

Full statistical behaviour. In this scenario we introduce an entrance spin distribution of natural spin-parity (0^+ to 8^+) calculated in a coupled-channel scheme [8]. The γ-decay is then computed by including the 69 bound or quasi bound-states with $\frac{\Gamma_\gamma}{\Gamma} \sim 1$ reported in the literature, the branching ratios to all these states are estimated using the partial widths of the electromagnetic transitions deduced from their mean strengths tabulated in the literature [9]. Results are presented in Fig.2; the corresponding spin distribution is plotted in the upper right panel and shows a maximum around $J^\pi=2^+$. As seen in the Fig.2, this scenario is in disagreement with our data and a large χ^2/ndf of $\sim$ 37.4 is obtained.

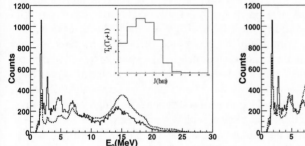 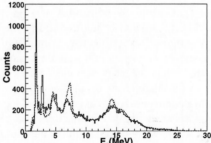

FIGURE 2. Left: Comparison of the experimental spectrum with simulated highest energy γ-ray spectrum (E0) in coincidence with the recoils for the statistical scenario (blue, dashed). The corresponding spin distribution is given in the upper right panel. Right: Comparison of the experimental spectrum with simulated highest energy γ-ray (E0) in coincidence with the recoils for the *cluster* scenario (blue, dashed).

Unique entrance spin. For a resonance behaviour, the expected spin distribution does not reproduce the data, because the main part of the flux is coming from one spin. We try then to compare our data with a unique entrance spin (between 0^+ and 8^+) scenario followed by a normal fragmentation of a specific entrance state as described before. The χ^2/ndf is minimum for a 6^+ entrance spin with χ^2/ndf$\sim$6.36. It should be noted that in comparison the spin distribution was centred at $J^\pi=2^+$. This χ^2 minimum for $J^\pi=6^+$ and the feeding of the 4^+ state of the prolate band incites us to compare the results at 9 MeV to cluster model predictions.

Comparison to cluster model calculations. To perform these calculations, we have used the branching ratio of a $6^+(2\hbar\omega)(14,0)$ at 25.75 MeV, calculated by the Debrecen group [3, 10, 11] with the semimicroscopic algebraic cluster model, to lower lying states of similar cluster structure. We have added such a scenario to a statistical background decay. This leads to a minimum χ^2/ndf of 9.2, in this case the cluster resonant part represents 60% of the decay. This scenario agrees with our data except for the 2.8 MeV peak corresponding to the ^{28}Si 4^+ state of the ground state band which is not included in the phase space of the cluster calculation, because of its oblate shape.

CONCLUSION

In our data we have identified a feeding of doorway states in ^{28}Si around 11 MeV which is ~ 1 MeV above the α threshold; similar results were found in the ^{12}C+^{12}C case [12]. This discovered new feeding increases by a factor of 5 the previously known radiative capture cross-section. We estimate $\sigma_{RC} \sim 11.6\pm 4$, 16.3 ± 6 and 23.4 ± 9 μb at 8.5, 8.8 and 9 MeV, respectively, whereas usual cross sections for the radiative capture process among light heavy-ions at the CB are in the few hundreds of nb range. We have measured a selective feeding of the 3_1^- at 6.88 MeV and of the 4_3^+ at 9.16 MeV. These states are both known to be deformed. We have payed special attention to the E_{cm}= 9 MeV results, because at this energy, the entrance spin seems to be a quasi pure J^π=6$^+$, a large part of the γ-flux feeds the prolate 4^+ state and the radiative capture cross-section is higher. These facts are good indications of a resonant cluster effect. Nevertheless the results should be completed to get a definitive signature of the resonant cluster effects. Our newly proposed experiment at Triumf will be performed below CB because at such low energies we expect less statistical background due to less spins involved and purer resonant behaviour. It is also clear that a γ-detector with better resolution would help a lot to disentangle all decay scenarii; it is the reason why we are also involved in Gammasphere+FMA RC experiments. Due to its low efficiency for high energy γ-rays, Gammasphere will not allow us to measure the complete γ spectrum from 1 to 25 MeV. In that sense, we hope to use in the near future a new type of scintillators array based on LaBr$_3$ like the PARIS calorimeter [13] for our RC experiments.

REFERENCES

1. E. C. Schloemer et al., *Phys. Rev. Lett.* **51**, 881 (1983).
2. E. Almquist ,D.A. Bromley and J.A. Kuehner, *Phys. Rev. Lett.* **4** 515 (1960).
3. J. Cseh, *Phys. Rev.* C **50**, 2240 (1994).
4. S. Ohkubo and K. Yamashita, *Phys. Lett.* **578 B**, 304 (2004).
5. Y. Kanada-En'yo, et al., *Nucl. Phys.* **A738**, 3 (2004) .
6. A.M. Sandorfi, Treatise on Heavy Ion Science, Vol.2, Sec. III, Ed. D.A. Bromley, p.52 (1984).
7. R. Sheline, et al., *Phys. Lett.* **119B**, 263 (1982).
8. N. Rowley, private communication.
9. P.M. Endt, *At. Data Nucl. Data Tables* **55**, 171 (1993) .
10. J. Cseh, G. Levai, *Ann. Phys.* (N.Y.) **230**, 165 (1994).
11. J. Cseh, J. Darai, G. Levai, private communication.
12. D.G. Jenkins et al., *Phys. Rev.* C **76**, 044310 (2007).
13. PARIS web site: http://paris.ifj.edu.pl/

Nuclear Proton-proton Elastic Scattering via the Trojan Horse Method

A. Tumino[*,†], C. Spitaleri[*], A. Mukhamedzhanov[**], G.G. Rapisarda[*],
L. Campajola[‡], S. Cherubini[*], V. Crucillá[*], Z. Elekes[§], Zs. Fülöp[§],
L. Gialanella[¶], M. Gulino[*], G. Gyürky[§], G.G. Kiss[§], M. La Cognata[*],
L. Lamia[*], A. Ordine[¶], R.G. Pizzone[*], S.M.R. Puglia[*], S. Romano[*],
M.L. Sergi[*] and E. Somorjai[§]

[*]*Laboratori Nazionali del Sud - INFN, via S. Sofia 62, 95123 Catania, Italy and Dipartimento di Metodologie Fisiche e Chimiche per l'Ingegneria, Universitá di Catania*
[†]*Università degli Studi di Enna "Kore", Enna, Italy*
[**]*Cyclotron Institute, Texas A&M University, College Station, USA*
[‡]*Dipartimento di Scienze Fisiche - Universitá Federico II, Napoli, Italy*
[§]*ATOMKI - Debrecen, Hungary*
[¶]*INFN - Sezione di Napoli, Italy*

Abstract. We present here an important test of the main feature of the Trojan Horse Method (THM), namely the suppression of Coulomb effects in the entrance channel due to off-energy-shell effects. This is done by measuring the THM $p - p$ elastic scattering via the $p+d \rightarrow p+p+n$ reaction at 4.7 and 5 MeV, corresponding to a $p - p$ relative energy ranging from 80 to 670 keV. In contrast to the on-energy-shell (OES) case, the extracted p-p cross section does not exhibit the Coulomb-nuclear interference minimum due to the suppression of the Coulomb amplitude. This is confirmed by the half-off-energy shell (HOES) calculations and strengthened by the agreement with the calculated OES nuclear cross sections.

Keywords: Quasi free mechanism, pole approximation, Plane Wave Impulse Approximation
PACS: 21.45.+v Few-body systems, 24.10.-i Nuclear reaction models and methods, 24.50.+g Direct reactions

INTRODUCTION

The THM [1, 2, 3] is a very powerful and known technique to study charged particle reactions at sub-Coulomb energies without experiencing Coulomb suppression. For this reason, in the last couple of decades it was successfully applied to rearrangement reactions of astrophysical interest [3, 4, 5, 6, 7, 8].

Here we investigate the suppression of the Coulomb amplitude when the THM is applied to scattering processes. This is done by considering the $p - p$ scattering at low energy whose features, shortly described below, can provide this important test for the THM. Proton-proton scattering cross section is well known since long time [9]. Its energy trend above the Coulomb barrier resembles that of $n - n$ or $p - n$ systems ($1/E$ behaviour, with E proton beam energy), while at ultra-low energies is dominated by the Coulomb field ($1/E^2$ behaviour). A deep minimum shows up in between ($p - p$ relative energy E_{pp}=191.2 keV, θ_{cm}= 90°), regarded as the signature of the interference between nuclear and Coulomb scattering amplitudes. Coulomb and nuclear amplitude nearly can-

CP1165, *Nuclear Structure and Dynamics '09*
edited by M. Milin, T. Nikšić, D. Vretenar, and S. Szilner
© 2009 American Institute of Physics 978-0-7354-0702-2/09/$25.00

cel each other in that region and the resulting cross section strongly deviates from pure Mott scattering. If one considers that a non sizeable Coulomb amplitude would make the minimum to disappear, the interference signature offers an important test of the THM Coulomb suppression also for scattering. This was done by measuring the $p - p$ elastic scattering through the quasi free (QF) ^{2}H(p, pp)n reaction. The low energy ^{2}H(p, pp)n reaction in the QF kinematics was measured several times before (see [10] and references therein), but not in the region where the $p - p$ Coulomb-nuclear interference takes place. The first measurements of this reaction at 6 MeV of beam energy approaching the interference area have been reported in [11, 12]. In order to validate those results and reach lower $p - p$ relative energies, the ^{2}H(p,pp)n experiment was performed at lower beam energies with more dedicated set-ups. The present paper reports on these experimental investigations performed at the ATOMKI in Debrecen [13, 14] and at the Dipartimento di Scienze Fisiche dell'Università Federico II, Naples (Italy) [14]. A 5 and 4.7 MeV proton beam was delivered onto a deuterated polyethylene target (98% of ^{2}H), 200 µg/cm^2. Two proton coincidences were measured by PSD's in the QF kinematics regime. For details see ref.[14]. These set-ups allowed us to investigate a wide E_{pp} region, ranging from 670 down to 80 keV.

EXPERIMENTAL RESULTS

The kinematics were reconstructed under the assumption of a neutron as third particle leading to a Q-value of -2.22 MeV, which refers to the $p + p + n$ channel of interest. A first selection of events was performed on the $p - p$ kinematic locus. Then, a flat contribution of less than 10% due to the $n - p$ Final State Interaction (FSI) [10] was subtracted from the selected three-body coincidence yield projected onto the E_{pp} axis. In order to investigate the reaction mechanism involved, a shape analysis of the experimental momentum distribution for the neutron $|\varphi(\mathbf{p}_n)|^2$ was carried out with the remaining data. This was performed following the procedure reported in [14], and employing the Plane Wave Impulse Approximation (PWIA) [10, 14]. The resulting momentum distribution is reported in Fig. 1 (full and open circles), together with the theoretical Hulthén shape for the $p - n$ system inside the deuteron (solid line). A quite good agreement shows up, making us confident that in the experimentally selected kinematical region, the quasi-free mechanism gives the main contribution. For further data analysis, only events with p_n values lower than 20 MeV/c were considered, since they give contributions in the $\theta_{c.m.}$ region close to 90°. We note that a rigorous analysis of the experimental data requires the full three-body Faddeev calculations with the Coulomb $p - p$ interaction [15]. However, it is impossible to extract the HOES $p - p$ scattering amplitude from such calculations. As mentioned before, and as it was successfully done in the previous investigations [10, 12] we employ a simple PWIA. It provides a straightforward expression of the TH $p - p$ cross section as given by the ratio between the cross section for the ^{2}H(p,pp)n three-body reaction and the product $KF |\varphi(\mathbf{p}_n)|^2$, with KF a kinematical factor containing the final state phase-space factor. Since $|\varphi(\mathbf{p}_n)|^2$ is known from experiment as well as from nuclear clustering studies, the product $KF |\varphi(\mathbf{p}_n)|^2$ can be calculated, either analytically (for fixed angles) or via a Monte Carlo simulation. Therefore it is possible to derive the TH cross section for the $p - p$ scattering (HOES) from

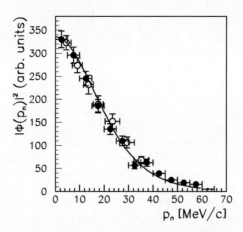

FIGURE 1. Experimental neutron momentum distribution(full and open circles). The solid line represents the shape of the theoretical Hulthén function in momentum space.

a measurement of ^{2}H(p,pp)n three-body cross section. The extracted $p - p$ HOES cross section from both runs is presented in Fig. 2 as weighted average of all sets of data (black dots) as a function of E_{pp}. The black solid line represents the free p-p cross sec-

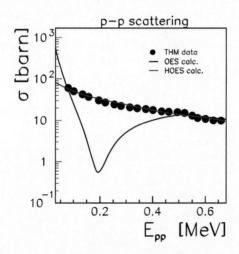

FIGURE 2. Weighted average of all the experimental THM data vs. E_{pp} The black and red solid lines represent the theoretical OES and HOES $p - p$ cross sections respectively, calculated as explained in [14].

tion [9, 14], whereas the red solid line is the calculated HOES $p - p$ cross section as reported in [14]. Experimental and calculated HOES cross-sections are normalized to the calculated OES one at E_{pp} close to the Coulomb barrier (500 keV). Vertical error bars in the figure include statistical and normalization errors as well as the error due to the subtraction of the FSI contribution. We observe a striking disagreement between the THM (HOES), which does not exhibit the minimum, and the free $p - p$ (OES) cross sections throughout the interference region. They come to the agreement right above the Coulomb barrier where the OES $p - p$ cross-section is given essentially by the nuclear part; in contrast, the THM data nicely fit the calculated HOES $p - p$ cross section. The THM $p - p$ scattering cross section has practically the same energy dependence as pure nuclear OES nucleon-nucleon cross sections throughout the whole E_{pp} range investigated (see [14]). The behaviour of the low-energy HOES cross section can be easily explained using Eq.22 of ref.[14]. In the HOES scattering, the initial $p - p$ relative momentum p and the final one k are different by definition. In particular, for the $p + d \rightarrow p + p + n$ reaction in the QF kinematics ($p_n = 0$), $p^2 = k^2 + 2\mu B_{np}$, where B_{np} is the deuteron binding energy. That is why p is always larger than k. Hence the transfer momentum in the $p - p$ elastic scattering $|\mathbf{p} - \mathbf{k}|$ at angles near 90° becomes large enough, compared to the OES case, to suppress the Coulomb HOES amplitude. For example, in the E_{pp} region where the OES cross section exhibits the interference minimum, $|\mathbf{p} - \mathbf{k}|$ is about of 0.3 fm^{-1}, making it possible to probe a distance between the two protons of the order of 3 fm, where the nuclear scattering dominates. In contrast, the corresponding transfer momentum in the on-shell $p - p$ scattering is around 0.1 fm^{-1}, matching with a distance of about 10 fm, where only the Coulomb interaction is present. The resulting HOES cross section at low energies is therefore dominated by the Coulomb modified nuclear amplitude (T_{CN} in eq.22 of ref.[14]), revealing the typical behaviour of the higher energy $p - p$ cross section, far from the interference region. In conclusion, through a mechanism different from that of nuclear rearrangement reactions, the present work strongly sustains the universality of the THM basic feature, namely the suppression of Coulomb effects in any two-body cross section at sub-Coulomb energies.

REFERENCES

1. G. Baur, *Phys. Lett.* B **178**, 135 (1986).
2. A. M. Mukhamedzhanov, C. Spitaleri and R.E. Tribble, *nucl-th/0602001* v1 (2006).
3. C. Spitaleri et al., *Phys. Rev.* C **63**, 055801 (2001).
4. A. Tumino et al., *Phys. Rev.* C **67**, 065803 (2003).
5. C. Spitaleri et al., *Phys. Rev.* C **63**, 055806 (2004) and references therein.
6. L. Lamia et al. *Nucl. Phys.* A **787**, 309 (2007)
7. M. La Cognata et al., *Phys. Rev.* C **76**, 065804 (2007) and references therein.
8. M. La Cognata et al., *Phys. Rev. Lett.* **101**, 152501 (2008).
9. J.D. Jackson and J.M. Blatt, *Rev. Mod. Phys.* **22**, 77 (1950).
10. V. Valković et al., *Nucl. Phys.* A **166**, 547 (1971).
11. M.G. Pellegriti et al., *Prog. Theor. Phys. Suppl.* **154**, 349 (2004).
12. A. Tumino et al., *Nucl. Phys.* A **787**, 337c (2007).
13. A. Tumino et al., *Phys. Rev. Lett.* **98**, 252502 (2007).
14. A. Tumino et al., *Phys. Rev.* C **78**, 064001 (2008) and references therein.
15. E. O. Alt and M. Rauh, *Phys. Rev.* C **49**, R2285 (1994).

ACTAR: the New Generation of Active Targets

R. Raabe for the ACTAR collaboration

GANIL, Bd Henri Becquerel, BP 55027, 14076 CAEN Cedex 05, France

Abstract. ACTAR is a new active target/time-projection chamber, designed for reaction and decay studies with nuclei far from stability. This class of instruments, initially developed for high-energy physics, has found profitable applications in medium- and low-energy nuclear physics as shown by a successful series of experiments. ACTAR builds on this experience to go beyond, incorporating developments in gas detector technology and a newly-designed electronic system of unprecedented scale in our research domain, to exploit the opportunities offered by the upcoming radioactive ion beams facilities.

Keywords: Time-projection chamber, nuclear reactions
PACS: 25.60.-t,29.40.Gx

INTRODUCTION

The use of beams of radioactive ions (RIBs) has revealed unexpected features in systems far from stability: from the peculiar structures in light nuclei (halos, cluster structures) to the change in shell closures in heavier nuclei [1]. Intensities of RIBs at the present facilities are much lower than in the case of stable beams, and their optical qualities (beam size and emittance) are also often much poorer. To overcome these difficulties, new detection systems need to be developed, having high geometrical efficiency and fine spatial resolution. In order to further increase the reaction yields, thick targets are also required, leading to a loss of resolution and a high threshold for the detection of low-energy recoils. These issues become even more important when using post-accelerated heavy secondary beams such as those that will become available at SPIRAL2, HIE-ISOLDE and ISAC2.

The idea of an active target, based on a gaseous ionization detector where the nuclei of the gas atoms are also the target nuclei, overcomes most of these difficulties. Gaseous detectors potentially have a very good geometric efficiency, a low detection threshold and excellent tracking capabilities; also, they have possibilities in particle identification. A large target thickness is possible without losing in resolution. The method also offers specific advantages: in case of short-leaving or unbound reaction products, the decay and its products can be detected in the gas volume itself; exploiting the energy loss of the incident beam in the gas, excitation functions can be obtained for selected reactions with a single tuning of the accelerator, thus optimizing the beam time. Several results have already been obtained in measurements covering different physics cases [2–5].

ACTAR builds on the experience gained with the present gaseous detectors to overcome their limitations, which concern especially the dynamic range, resolution of multiple tracks and counting rate capabilities.

CP1165, *Nuclear Structure and Dynamics '09*
edited by M. Milin, T. Nikšić, D. Vretenar, and S. Szilner
© 2009 American Institute of Physics 978-0-7354-0702-2/09/$25.00

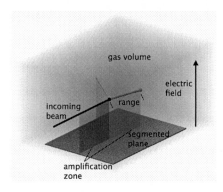

FIGURE 1. Working principle of a time-projection chamber as an active target

ACTAR DESIGN

The ACTAR detector uses the principle of a time-projection chamber (TPC) [6], a gaseous detector capable of tracking in three dimensions the charged particles traversing its volume. The working principle is illustrated in Fig. 1. Electrons produced by ionization by the energetic charged particles drift to an amplification zone, situated close to a segmented detection plane. The signals generated by the electrons create a a two-dimensional projection of the tracks; the third dimension is reconstructed by measuring the drift time of the electrons through the gas volume. In the active target mode the events of interest are those in which the incoming beam nuclei interact with the nuclei of the gas atoms. From the detection of the tracks, the reaction vertex can be directly reconstructed. This allows using a very large target thickness, with an increase in luminosity of a factor 5 to 10 with respect to solid targets, without losing in energy resolution. The length of the tracks (range) combined with the energy deposit provide an identification of the particles. Besides reactions, the detector can be used to measure charged particles (ions) emitted in the decay of unstable nuclei stopped in the detector volume.

To fulfill its physics goals, ACTAR will improve on the performances of the present active targets on several aspects. It should be capable of detecting signals in an energy interval of three orders of magnitude (dynamic range of 10^3); it will use beam intensities up to 10^5-10^6 pps (depending on the physics process) and have a rate of accepted events of 1 kHz. The spatial resolution should be better than 2 mm in each direction, in a volume of about $30\times30\times30$ cm^3. The instrument will be flexible, with parameters that can be adapted to optimize the detection for the events of interest.

A key element in the design is represented by the electronic system. In order to meet the requirements, a project for a General Electronics for TPCs (GET) has been initiated by GANIL, IRFU Saclay, CENBG and MSU. The electronics will be capable of handling signals (energy and time) from more than 20000 channels, selecting the events of interest thanks to a multi-level triggering system. GET will equip various TPC-type instruments (at GANIL, MSU, CENBG, RIKEN, GSI) with different physics goals.

Another essential element in the design is the amplification technology. Micro-pattern gaseous detectors such as Micromegas [7] and GEMs [8], developed for high-energy physics, present several advantages over the traditional wire-anode amplification: they are more robust with respect to discharge phenomena, allow higher signal rates because of the short ion feedback signal; and can be combined to achieve different amplification gains on different parts of the detection surface. Tests with these devices are in progress, where their performances in different conditions (pressure, gas mixtures, energy deposit of particles) are measured.

Finally, ACTAR will employ ancillary solid-state charged-particle detectors placed around the gas volume, to identify and measure the energy of escaping light particles.

PHYSICS CASES

A first category of physics cases for the active target are direct and resonant reactions induced by radioactive beams. Reactions of interest include elastic and inelastic scattering; transfer of one, two, or few nucleons; resonant reactions where states in the compound nucleus are investigated. We briefly present the possibilities of ACTAR in these cases.

One-nucleon transfer reactions are a powerful spectroscopic tool for the investigation of the single-particle structure of states in key nuclei. A reaction of the type ^{78}Ni(d,p), with a post-accelerated beam of ^{78}Ni at 5 MeV/nucleon, could be studied with ACTAR filled with deuterium gas at 1 bar presure. The backward-scattered protons would be stopped in the gas volume or in ancillary detectors with a high detection efficiency and low thresholds. With a total target thickness of about 10^{21} at/cm^2, the measurement would be feasible with a beam intensity as low as 10^3 pps, with a resolution of a few keV in the center-of-mass energy of the reaction.

Inelastic scattering can be used to populate giant resonances; isoscalar resonances in particular (the giant monopole resonance GMR, and the isoscalar giant dipole resonance ISGDR) are related to the incompressibility of nuclear matter, an important parameter of the equation of state. ACTAR is a unique instrument to measure the GMR and ISGDR in unstable nuclei: using He (with the addition of a quencher) as target gas at 1 bar pressure, the reaction ^{56}Ni(α,α') could be measured using a 10^6 pps beam at 50 MeV/nucleon produced by fragmentation and in-flight separation. The α particles, peaking at about 20 degrees in the laboratory frame, would be stopped in the gas, at a rate of about 1 event/sec, and the GRs could be measured with a resolution of ≈ 1 MeV. With this yield, even the fine structure of the GRs could be studied via their decay into proton-emission channels.

Resonant reactions are used to access and study unbound states, with implications relevant to nuclear structure (for example, states with a molecular structure) or nuclear astrophysics. With ACTAR, protons or He can be used as target nuclei, with a pressure set in order to stop the incoming beam in the gas volume. With respect to present techniques employing thin foils as target, in ACTAR the complete kinematic information would be available: energy and angle of the light recoil particles (possibly detected in forward-placed ancillary detectors), range (and thus energy) of the scattered beam particle, and position of the vertex of the interaction, from which the center-of-mass energy of the scattering can be directly obtained. For the measurement of the resonance, events in a large angular range can thus be used without losing in resolution, with a gain of a factor 50 to 100 with respect to the present techniques.

Another application where ACTAR present interesting features is the study of exotic radioactivities, such as two- and three-proton radioactivity, which require the precise determination of the energy and angle of the individual protons; or β-delayed multi-particle emission, where the nature and correlation of the emitted ions is of interest. The gas detector is in this case used a time-projection chamber and not as an active target stricto sensu, since the gas only acts as stopping medium for the nuclei of interest, however the technical requirements are very similar to those of an active target.

From the above list, it is clear that the physics cases considered for the active target cover a very broad range of nuclei, from light halo nuclei up to the relatively heavy beams which will be available from SPIRAL2. The energy range also goes from the lowest energies of an ISOL post-accelerated facility for resonant reactions studies up to several tens or hundreds of MeV/nucleon of the in-flight facilities for matter density determination, giant resonance studies and proton radioactivity. Active targets have the possibility of exploiting all these regimes, and ACTAR will be ready to profit from the opportunities offered by the upcoming radioactive ion beam facilities.

ACKNOWLEDGMENTS

The ACTAR collaboration: A. Chbihi, F. de Oliveira, B. Dominguez Fernandez, J. Pancin, R. Raabe, T. Roger, P. Roussel-Chomaz, F. Saillant, H. Savajols, G. Wittwer (GANIL); S. Anvar, P. Baron, F. Druillole, A. Gillibert, L. Nalpas, L. Pollacco (CEA/DSM/IRFU Saclay); B. Blank, J. Giovinazzo, J. L. Pedroza, J. Pibernat (CENBG); E. Khan, C. Petrache (IPN Orsay); J. Gibelin, M. Marques, N. Orr (LPC Caen); P. Egelhof (GSI); N. Keeley (Andrzej Sołtan Institute); K. Rusek (Heavy Ion Laboratory, Warsaw University); R. Wolski (Henryk Niewodniczański Institute); H. Alvarez-Pol, M. Caamaño, D. Cortina-Gil (University of Santiago de Compostela); I. Martel (University of Huelva); R. Lemmon (STFC Daresbury Laboratory); W. Catford (University of Surrey); M. Freer (University of Birmingham); A. Murphy (University of Edinburgh); M. Chartier (University of Liverpool); A. Laird (University of York); M. Huyse, P. Van Duppen (K.U. Leuven); Y. Blumenfeld, J. Van de Walle (ISOLDE, CERN); W. Mittig (NSCL, MSU).

REFERENCES

1. O. Sorlin and M. G. Porquet, *Prog. Part. Nucl. Phys.* **61**, 602 (2008).
2. M. Caamaño, D. Cortina-Gil, W. Mittig, H. Savajols, M. Chartier, C. E. Demonchy, B. Fernandez, M. B. Gomez Hornillos, A. Gillibert, B. Jurado, et al., *Phys. Rev. Lett.* **99**, 062502 (2007).
3. J. Giovinazzo, B. Blank, C. Borcea, G. Canchel, J.-C. Dalouzy, C. E. Demonchy, F. de Oliveira Santos, C. Dossat, S. Grevy, L. Hay, et al., *Phys. Rev. Lett.* **99**, 102501 (2007).
4. C. Monrozeau, E. Khan, Y. Blumenfeld, C. E. Demonchy, W. Mittig, P. Roussel-Chomaz, D. Beaumel, M. Caamaño, D. Cortina-Gil, J. P. Ebran, et al., *Phys. Rev. Lett.* **100**, 042501 (2008).
5. I. Tanihata, M. Alcorta, D. Bandyopadhyay, R. Bieri, L. Buchmann, B. Davids, N. Galinski, D. Howell, W. Mills, S. Mythili, et al., *Phys. Rev. Lett.* **100**, 192502 (2008).
6. J. A. MacDonald (ed.), *The Time Projection Chamber* (AIP Conf Proc. Vol. 108, New York, 1984).
7. Y. Giomataris, P. Rebourgeard, J. Robert, and G. Charpak, *Nucl. Instrum. Methods Phys. Res.* A **376**, 29 (1996).
8. F. Sauli, *Nucl. Instrum. Methods Phys. Res.* A **386**, 531 (1997).

Global Optical Potential for ^{6}He Interactions at Low Energies

Y. Kucuk, I. Boztosun and T. Topel

Department of Physics, Erciyes University, Kayseri, Turkey

Abstract.
   Within the framework of optical model, we present a set of global optical potential for the elastic scattering of ^{6}He halo nucleus from different target nuclei ranging from ^{12}C to ^{209}Bi at low energies. Consistent agreement with the experimental data has been obtained by using this global potential.

Keywords: ^{6}He, halo nuclei, optical model, elastic scattering
PACS: 24.10.Eq; 24.10.Ht; 24.50.+g; 25.60.-t; 25.70.-z

INTRODUCTION

The halo nuclei have become one of the main interests of nuclear physics and nuclear astrophysics since the discovery of their unusual structure [1–5]. The ^{6}He nucleus is a well-known halo type nucleus, which has been most studied. It has attracted enormous interest both theoretically and experimentally due to its Borromean structure and the large probability of its break-up near the Coulomb barrier [5–22]. In these works, the elastic scattering cross section and fusion cross section as well as the break-up/transfer cross section have been measured and studied theoretically for many systems at energies near the Coulomb barrier to investigate the behavior of the optical potential and the effect of break-up coupling to the reaction and scattering mechanism.

   The explanation of the measured elastic scattering angular distributions near the Coulomb barrier has been an important motivation for the theoretical calculations since elastic scattering bears significance in providing an idea about the nuclear optical potential of the system. ^{6}He interactions at energies around Coulomb barrier have crucial importance in the understanding of properties of exotic systems and a global potential set is required in the theoretical analysis of the reactions. So far, many potential sets have been derived either phenomenologically or of the folding type in order to describe the elastic scattering and other scattering observables of ^{6}He nucleus. The potentials derived for reactions of nuclei to describe ^{6}He interaction are very similar to those of the ^{6}Li potentials. Sometimes, ^{4}He potential has also been used by adjusting its radius to ^{6}He one. Although a good description of the observables by using these potentials has been obtained, there is no consistency between these potentials.

   Therefore, in this study, we aim to develop a global potential set to describe ^{6}He nucleus with light to heavy nuclei at low energies. In the next section, we present first the optical model and then the results of these theoretical analysis by using the derived potential set for many systems. The final section is devoted to our summary and conclusion.

CP1165, *Nuclear Structure and Dynamics '09*
edited by M. Milin, T. Nikšić, D. Vretenar, and S. Szilner
© 2009 American Institute of Physics 978-0-7354-0702-2/09/$25.00

OPTICAL MODEL CALCULATIONS

We have performed an extensive study for the elastic scattering of ^{6}He on different targets, from ^{12}C to ^{209}Bi, for a wide energy range. We have used the optical model for the theoretical calculations and the optical potential consists of the Coulomb, centrifugal and nuclear potentials. Nuclear part is the sum of the Woods-saxon square shaped real and Wood-Saxon shaped imaginary potentials given as

$$V_{nuclear}(r) = \frac{-V}{\left[1+e^{\frac{r-R_V}{a_V}}\right]^2} + i\frac{-W}{1+e^{\frac{r-R_W}{a_W}}} \tag{1}$$

Here, $R_i = r_i[A_P^{1/3} + A_T^{1/3}]$ $(i = V$ or $W)$, where $A_P^{1/3}$ and $A_T^{1/3}$ are the masses of projectile and target nuclei and r_V and r_W are the radius parameters of the real and imaginary parts of the nuclear potential respectively [23–26].

In order to obtain a global potential, we have taken the depth of the real and imaginary potentials as free parameters and have investigated the radius and diffuseness parameters for each parts that give the best fit for the elastic scattering cross section. We have obtained the radius parameters of real and imaginary part of the optical potential as 0.9 fm and 1.5 fm, respectively. The diffusion parameters have also been fixed as 0.7 fm for both parts of the potential. Coulomb radius has been taken as 1.2 fm for each of the calculations.

Having obtained the best fit for all data, we have investigated the change of the depth of the real and imaginary parts and we have derived the following equations for the variation of the real and imaginary parts of the nuclear potential.

$$V(r) = 90.2 + 0.85E + 0.067A_T e^{\frac{Z_T-E}{A_T}} \tag{2}$$

$$W(r) = 3E^{1/4} e^{\frac{Z_T-0.5E}{A_T}} \tag{3}$$

where E is the laboratory energy of the ^{6}He and Z_T and A_T are the charge and mass numbers of the target nuclei.

RESULTS

We have analyzed the elastic scattering of the ^{6}He from target nuclei of the ^{12}C, ^{27}Al, ^{58}Ni, ^{64}Zn, ^{65}Cu, ^{197}Au, ^{208}Pb and ^{209}Bi for a wide energy range below 50 MeV by using the derived new optical potential set given by Eqs. 2 and 3 within the framework of the optical model.

We present some of the results in Fig. 1. The analysis of ^{6}He nucleus from the above-mentioned reactions will be submitted for publication soon. The first system we have considered is the ^{6}He+^{12}C elastic scattering at 8.79, 9.18 and 18.0 MeV energies in the laboratory system. The experimental data for 8.79 and 9.18 have been measured by Smith et al. [27] and have been analyzed by using the potential parameters of ^{4}He, ^{6}Li and ^{7}Li. The elastic scattering data at these energies as well as the data at 18.0 MeV

344

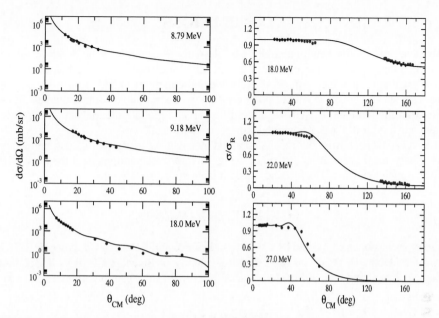

FIGURE 1. Left: Elastic scattering angular distributions for ^{6}He + ^{12}C. The solid lines show OM results while the circles show the experimental data taken from Refs. [27, 28]. Right: Elastic scattering angular distribution (ratio to Rutherford cross section) for ^{6}He + ^{208}Pb. The solid lines show OM results while the circles show the experimental data taken from Refs. [11, 30].

measured by Milin *et al.* [28] have been analyzed by using the new potential given by Eqs. 2 and 3 and better results have been obtained for all energies as presented in the left panel of Fig 1. When we compare the theoretical results with experimental data, we have obtained small χ^2/N values that are comparable with more sophisticated CDCC or similar approaches. We have also studied the elastic scattering of ^{6}He from heavy targets such as ^{208}Pb. The results obtained by using the global potential set given by Eqs. 2 and 3 are shown in the right panel of Fig. 1 at five different energies for the ^{6}He+^{208}Pb elastic scattering. The results are in good agreement with the experimental data except at 16 Mev where we have to increase the depth of the imaginary potential in order to get a better agreement with the measured data. We have also calculated the reaction cross sections for each system produced by this potential. These values are also in consistency with conducted values in literature. For theoretical calculations, the code Fresco [29] has been used.

The volume integrals and the dispersion relation between the real and imaginary potentials have been calculated and we have been unable to see a consistency between the volume integrals of the real and imaginary potentials. Further work to understand their behavior at low energies for different systems is under progress.

CONCLUSIONS

We have presented a new potential set by deriving a formula for the depth of the real and imaginary parts of the optical potential for ^{6}He elastic scattering below 50 MeV. Consistent agrement has been obtained between the theoretical and experimental results for different systems. We should point out that we do not aim to obtain the best fits for the experimental data. Rather, we attempt to derive a global potential set that produces the the behavior of the experimental data reasonably well.

As it can be seen from our results obtained by using the new potential parameters, one can easily use this potential instead of using the improved parameters of the most of the similar nuclei such as ^{4}He, ^{6}Li and ^{7}Li as it is most commonly done.

ACKNOWLEDGMENTS

This work has been supported by the Turkish Science and Research Council (TÜBİTAK) with Grant No:107T824 and the Turkish Academy of Sciences (TÜBA-GEBİP). Authors wish to thank Drs. A.M. Moro and N. Keeley for their useful comments.

REFERENCES

1. I. Tanihata, et al., *Phys. Rev. Lett.* **55**, 2676 (1985).
2. M.V. Zhukov, et al., *Phys. Rep.* **231**, 151 (1993).
3. P.G. Hansen, et al., *Ann. Rev. Nucl. Part. Sci.* **45**, 505 (1995).
4. B. Johnson and K. Riisager, *Phil. Trans. R. Soc.* (London) A **356**, 2063 (1998).
5. J.S. Al-Khalili, et al., *Phys. Lett.* **378B**, 45 (1996).
6. S. Karataglidis, et al., *Phys. Rev.* C **61**, 024319 (2000).
7. Y. Suzuki, *Nucl. Phys.* **528**, 395 (1991).
8. K. Varga, et al., *Phys. Rev.* C **50**, 189 (1994).
9. S. Funada, et al., *Nucl. Phys.* A **575**, 93 (1994).
10. P. Navratil, et al., Lawrence Livemore National Laboratory, UCRL-PROC-211912, (2005).
11. A.M. Sánchez Benítez, et al., *J. Phys. G: Nucl. Part. Phys.* **31**, S1953 (2005).
12. E.F. Aguilera, et al., *Phys. Rev.* C **63**, 061603(R) (2001).
13. I.J. Kolata, et al., *Phys. Rev. Lett.* **81**, 4580 (1998).
14. O.R. Kakuee, et al., *Nucl. Phys.* A **765**, 294 (2006).
15. R. Raabe, Katholieke Universiteit Leuven, PhD. Thesis, (2001).
16. L. Borowska, et al., *Phys. Rev.* C **76**, 034606 (2007).
17. A.A. Korsheninnikov, et al., *Nucl. Phys.* A **616**, 189 (1997).
18. L.R. Gasgues, et al., *Phys. Rev.* C **67**, 024602 (2003).
19. R. E. Warner, et al., *Phys. Rev.* C **51**, 178 (1995).
20. I. Boztosun, et al., *Phys. Rev.* C **77**, 064608 (2008).
21. N. Keeley and R.S. Machintosh, *Phys. Rev.* C **71**, 057601 (2004).
22. A.M. Moro, et al., *Phys. Rev.* C **75**, 064607 (2007).
23. I. Boztosun and W.D.M. Rae, *Phys. Lett.* **518B**, 229 (2001).
24. I. Boztosun and W.D.M. Rae, *Phys. Rev.* C **64**, 054607 (2001).
25. I. Boztosun and W.D.M. Rae, *Phys. Rev.* C **66**, 024610 (2002).
26. Y. Kucuk and I. Boztosun, *Nucl. Phys.* A **764**, 160 (2006).
27. R. J. Smith, et al., Phys. Rev. C **43**, 761 (1991).
28. M. Milin, et al., *Nucl. Phys.* A **730**, 285 (2004).
29. I. J. Thompson, Computer Phys. Rep. **7**, 167 (1988).
30. O.R. Kakuee, et al., *Nucl. Phys.* A **728**, 339 (2003).

A Microscopic Optical Potential Approach to ^{6,8}He+p Elastic Scattering

V. K. Lukyanov*, E. V. Zemlyanaya*, K. V. Lukyanov*, D. N. Kadrev†,
A. N. Antonov†, M. K. Gaidarov† and S. E. Massen**

*Joint Institute for Nuclear Research, Dubna 141980, Russia
†Institute for Nuclear Research and Nuclear Energy, Bulgarian Academy of Sciences, Sofia 1784,
Bulgaria
**Department of Theoretical Physics, Aristotle University of Thessaloniki, 54124 Thessaloniki,
Greece

Abstract. A microscopic approach to calculate the optical potential (OP) with the real part obtained by a folding procedure and with the imaginary part inherent in the high-energy approximation (HEA) is applied to study the ^{6,8}He+p elastic scattering data at energies of tens of MeV/N. The OP's and the cross sections are calculated using different models for the neutron and proton densities of ^{6,8}He. The role of the spin-orbit (SO) potential and effects of the energy and density dependence of the effective NN forces are studied. Comparison of the calculations with the available experimental data on the elastic scattering differential cross sections at beam energies < 100 MeV/N is performed and conclusions on the role of the aforesaid effects are made. It is shown that the present approach, which uses only parameters that renormalize the depths of the OP, can be applied along with other methods like that from the microscopic g-matrix description of the complex proton optical potential.

Keywords: light exotic nuclei, optical model, elastic scattering, nucleon density distributions
PACS: 24.10.Ht, 25.60.-t, 21.30.-x, 21.10.Gv

A widely used way to study the structure of exotic nuclei is to analyze their elastic scattering on protons or light targets at different energies. For example, proton elastic scattering angular distributions were measured at incident energies less than 100 MeV/N, namely, for ^{6}He, at energy 25.2, 38.3, 41.6 and 71 MeV/N, for ^{8}He at energy 15.7, 25.2, 32, 66 and 73 MeV/N and also at energy 700 MeV/N for He and Li isotopes. The main aim of this work is to calculate differential cross sections of elastic ^{6,8}He+p scattering at energies less than 100 MeV/N studying the possibility to describe the existing experimental data by calculating microscopically not only the real (in a double-folding procedure) but also the imaginary optical potential (instead of using phenomenological one) within the HEA and using a minimal number of fitting parameters.

We calculated the ^{8}He+p elastic scattering differential cross sections using the microscopically obtained real V^F and imaginary W^H contributions to the optical potential:

$$U_{opt}(r) = N_R V^F(r) + i N_I W^H(r) + 2\lambda_\pi^2 \left\{ N_R^{SO} V_0^F \frac{1}{r} \frac{df_R(r)}{dr} + i N_I^{SO} W_0^H \frac{1}{r} \frac{df_I(r)}{dr} \right\} (\mathbf{l.s}), \quad (1)$$

where V_0^F and W_0^H are the depths of the SO optical potential obtained simultaneously from the approximation of the volume real and imaginary OP's by Woods-Saxon form. We perform a fitting procedure, where the additionally introduced strength parameters N_R, N_I, N_R^{SO}, N_I^{SO} are varied step by step. Instead, three different combinations of V^F,

CP1165, *Nuclear Structure and Dynamics '09*
edited by M. Milin, T. Nikšić, D. Vretenar, and S. Szilner

V^H and W^H were used for the OP $U_{opt}(r)$ in calculations of the elastic ^{6}He+p cross sections [1].

As an example, in Figs. 1 and 2 we give the results of our calculations of ^{6}He+p and ^{8}He+p elastic scattering cross sections for various energies and the Large-Scale Shell Model (LSSM) density.

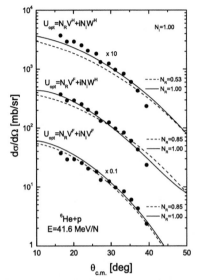

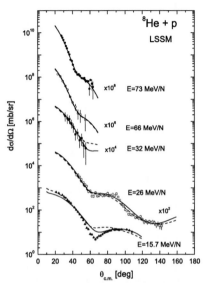

FIGURE 1. Elastic ^{6}He+p scattering cross section at energy $E = 41.6$ MeV/N calculated using different OP's for various values of the renormalization parameters N_R ($N_I = 1$). The LSSM density of ^{6}He is applied.

FIGURE 2. The ^{8}He+p elastic scattering cross sections at different energies calculated using U_{opt} [Eq. (1)] for various values of the renormalization parameters N_R, N_I, N_R^{SO} and N_I^{SO} giving the best agreement with the data.

The results show that the LSSM densities of ^{6}He and ^{8}He which have more diffuse tails at larger r than the densities based on Gaussians lead to a better agreement with the data for the ^{6,8}He+p elastic scattering at different energies. It was shown that, generally, at energies $E > 25$ MeV/N a good agreement with the experimental data for the differential cross sections can be achieved using OP with calculated both V^F and W^H varying mainly the volume part of the OP neglecting SO contribution. The explanation of the ^{6,8}He+p cross sections at lower energies ($E < 25$ MeV/N) needs accounting for the effects of the nuclear surface. A more successful explanation of the cross section at low energies could be given by inclusion of polarization contributions due to virtual excitations of inelastic and decay channels of the reactions.

Three of the authors (D.N.K., A.N.A. and M.K.G.) are grateful for the support of the Bulgarian Science Fund under Contracts Nos. DO02–285 and Ф–1501. The authors E.V.Z. and K.V.L. thank the Russian Foundation for Basic Research (Grant No. 06-01-00228) for the partial support.

[1] K. V. Lukyanov, V. K. Lukyanov, E. V. Zemlyanaya, A. N. Antonov, and M. K. Gaidarov, Eur. Phys. J. **A33**, 389 (2007).

Excitation functions of ^{6,7}Li+^{7}Li reactions at low energies

L. Prepolec*, N. Soić*, S. Blagus*, Đ. Miljanić*, Z. Siketić*, N. Skukan*, M. Uroić* and M. Milin†

*Ruđer Bošković Institute, Bijenička c. 54, HR-10000 Zagreb, Croatia
†Faculty of Science, University of Zagreb, Bijenička c. 32, HR-10000 Zagreb, Croatia

Abstract. Differential cross sections of ^{6,7}Li+^{7}Li nuclear reactions have been measured at forward angles (10° and 20°), using particle identification detector telescopes, over the energy range 2.75-10.00 MeV. Excitation functions have been obtained for low–lying residual–nucleus states. The well pronounced peak in the excitation function of ^{7}Li(^{7}Li,^{4}He)^{10}Be(3.37 MeV, 2$^+$) at beam energy about 8 MeV, first observed by Wyborny and Carlson in 1971 at 0°, has been observed at 10°, but is less evident at 20°. The cross section obtained for the ^{7}Li(^{7}Li,^{4}He)^{10}Be(g.s, 0$^+$) reaction is about ten times smaller. The well pronounced peak in the excitation function of ^{7}Li(^{7}Li,^{4}He)^{10}Be(3.37 MeV, 2$^+$) reaction could correspond to excited states in ^{14}C, at excitation energies around 30 MeV.

Keywords: ^{7}Li, ^{6}Li, excitation function, low energies
PACS: 20.21.25.10.+s

INTRODUCTION

Low-energy ^{6,7}Li+^{7}Li excitation function measurements have shown a pronounced peak for the ^{7}Li(^{7}Li,^{4}He)^{10}Be(3.37 MeV, 2$^+$) reaction at beam energy around 8.0 MeV, at angle of 0° [1]. The cross section of ^{7}Li(^{7}Li,^{4}He)^{10}Be(g.s., 0$^+$) reaction was reported to be much smaller [1].

Motivated to reproduce the peak observed in [1], we have performed a measurement at 10° and 20°. A small array of SSB detectors in particle-identification telescope configuration has been used (17.9 μm – 1000 μm and 15.9 μm – 1500 μm pairs), together with a monitor detector. Our setup has been optimised for energy resolution, while the acceptable trade-off has been lower statistics. Isotopically enriched ^{6}Li and ^{7}Li flouride targets have been used. The carbon backing of targets provided the reference cross sections for normalization of cross sections, using results from [2]. The measurement has been performed at 6MV EN tandem Van de Graaff accelerator at Ruđer Bošković Institute (RBI), beam energies ranging from 2.75 to 10.00 MeV in 200 keV steps.

RESULTS

Excitation functions have been obtained for a number of low-lying residual-nucleus states originating from the reactions: ^{7}Li(^{7}Li,^{4}He)^{10}Be, ^{7}Li(^{7}Li,^{3}H)^{11}B, ^{6}Li(^{7}Li,^{4}He)^{9}Be and ^{6}Li(^{7}Li,^{3}H)^{10}B.

^{7}Li(^{7}Li,^{4}He)^{10}Be reaction excitation functions, measured at 10° and 20° are shown in Fig. 1. The cross section for ^{10}Be in the ground state is ten times smaller than for ^{10}Be

CP1165, *Nuclear Structure and Dynamics '09*
edited by M. Milin, T. Nikšić, D. Vretenar, and S. Szilner
© 2009 American Institute of Physics 978-0-7354-0702-2/09/$25.00

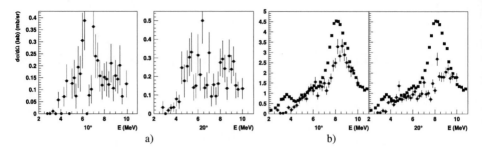

FIGURE 1. Excitation function of the $^7\text{Li}(^7\text{Li},^4\text{He})^{10}\text{Be}(\text{g.s.}, 0^+)$ reaction (a). Excitation function of the $^7\text{Li}(^7\text{Li},^4\text{He})^{10}\text{Be}(3.37\text{ MeV}, 2^+)$ reaction (filled circles) in comparison to results at $0°$(red squares) [1] (b). Error bars depict statistical errors only.

in the first excited state. Excitation functions shown in Fig. 1 have pronounced structure: in Fig. 1b) there is pronounced peak for measurement at $10°$, which is not so evident at $20°$, as well as some possible peaks in Fig. 1a) around 6 MeV. Similar structures have been observed in the spectra of the $^7\text{Li}(^7\text{Li},^3\text{H})^{11}\text{B}$ reaction, while excitation functions for the $^6\text{Li}+^7\text{Li}$ reactions are structureless.

OUTLOOK

Excitation functions for the $^{6,7}\text{Li}+^7\text{Li}$ reactions have been obtained in the beam energy range from 2.75 to 10.00 MeV. The pronounced peak for the $^7\text{Li}(^7\text{Li},^4\text{He})^{10}\text{Be}(3.37\text{ MeV}, 2^+)$ reaction could correspond to resonance in compound ^{14}C nucleus produced in a highly excited state (around 30 MeV) or may be introduced by direct nature of the reaction process. States at such a high excitation energies have been observed in recent experiments [3, 4]. We are planning to repeat the measurement at RBI in order to obtain better statistics and to cover wider range of angles, using large-area position-sensitive silicon strip detectors which provide high energy and angular resolution, indispensable to determine the nature of the observed peak.

REFERENCES

1. H.W. Wyborny, R.R. Carlson, *Phys. Rev.* C **3**, 2185 (1971).
2. J.E. Poling, E. Norbeck, R.R. Carlson, *Phys. Rev.* C **13**, 648 (1976).
3. M. Milin, et al. *Eur. Phys. J. Special Topics* **150**. 43 (2007).
4. N. Soić, et al., *Phys. Rev.* C **68**, 014321 (2003).

Microscopic Calculation of Absolute Values of Two–nucleon Transfer Cross Sections

G. Potel[*,†], B. F. Bayman[**], F. Barranco[‡], E. Vigezzi[†] and R. A. Broglia[*,§]

[*]Dipartimento di Fisica, Università di Milano, Via Celoria 16, 20133 Milano, Italy.
[†]INFN, Sezione di Milano Via Celoria 16, 20133 Milano, Italy.
[**]School of Physics and Astronomy, University of Minnesota, Minneapolis, Minnesota 55455
[‡]Departamento de Fisica Aplicada III, Universidad de Sevilla, Escuela Superior de Ingenieros, Sevilla, 41092 Camino de los Descubrimientos s/n, Spain.
[§]The Niels Bohr Institute, University of Copenhagen, Blegdamsvej 17, 2100 Copenhagen Ø, Denmark.

Abstract. Arguably, the greatest achievement of many–body physics in the fifties was that of developing the tools for a complete description and a thorough understanding of superconductivity in metals. At the basis of it one finds BCS theory and the Josephson effect. The first recognized the central role played by the appearance of a macroscopic coherent field usually viewed as a condensate of strongly overlapping Cooper pairs, the quasiparticle vacuum. The second made it clear that a true gap is not essential for such a state of matter to exist, but rather a finite expectation value of the pair field. Consequently, the specific probe to study the superconducting state is Cooper pair tunneling. Important progress in the understanding of pairing in atomic nuclei may arise from the systematic study of two–particle transfer reactions. Although this subject of research started about the time of the BCS papers, the quantitative calculation of absolute cross sections taking properly into account the full non–locality of the Cooper pairs (correlation length much larger than nuclear dimensions) is still an open question. In what follows we present results obtained, within a second order DWBA framework, of two–nucleon transfer reactions induced both by heavy and light ions. The calculations were carried out making use of software specifically developed for this purpose. It includes sequential, simultaneous and non–orthogonality contributions to the process. Microscopic form factors are used which take into account the relevant structure aspects of the process, such as the nature of the single–particle wavefunctions, the spectroscopic factors, and the interaction potential responsible for the transfer. Overall agreement with the experimental absolute values of the differential cross section is obtained without any free parameter.

Keywords: two–particle transfer, pairing correlations
PACS: 25.40.Hs

We first discuss the results obtained in the analysis of the heavy–ion reaction $^{208}Pb(^{16}O, ^{18}O)^{206}Pb$ at 86 MeV. Because the wavelength of the relative motion is relatively short at this energy ($\lambda \simeq 0.8$ fm), the semiclassical approximation is expected to be well suited. In Fig. 1 we display the results of such calculation for, together with those obtained within the framework of a full quantal theory, (second order DWBA formalism). Also shown are the experimental data. The basic ingredients of the calculations are optical potentials (typically fitted from elastic scattering data), single–particle wavefunctions of the transferred particles (obtained from a suitable central potential which reproduce experimental binding energies), spectroscopic amplitudes and the central single–particle potential responsible for the transfer, (taken equal to the one which generates the single–particle wavefunctions).

We now turn our attention to two–particle transfer reactions induced with light ions.

CP1165, Nuclear Structure and Dynamics '09
edited by M. Milin, T. Nikšić, D. Vretenar, and S. Szilner
© 2009 American Institute of Physics 978-0-7354-0702-2/09/$25.00

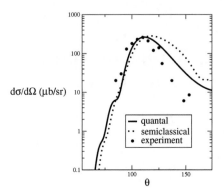

FIGURE 1. Differential cross section for the reaction $^{208}\text{Pb}(^{16}\text{O}, ^{18}\text{O})^{206}\text{Pb}$ at an energy of 86 MeV in the laboratory frame. We present the results of the quantal and semiclassical calculations, along with the experimental data, see also [1].

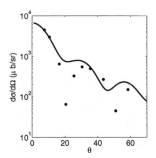

FIGURE 2. Differential cross section associated with the $^{112}\text{Sn}(p,t)^{110}\text{Sn}$ reaction in comparison with the experimental findings.

In this case, only a full quantal description is valid. In Fig. 2 the results for the $^{112}\text{Sn}(p,t)^{110}\text{Sn}$ reaction at $E_{lab} = 26$ MeV are displayed in comparison with the experimental data [2]. Of notice that in all the calculations of the absolute cross sections discussed above there are no free parameters adjusted to reproduce the data.

REFERENCES

1. B.F. Bayman, and Jongsheng Chen, *Phys. Rev.* C **26**, 1509 (1982).
2. P. Guazzoni, L. Zetta, A. Covello, A. Gargano, B.F. Bayman, G. Graw, R. Hertenberger, H.-F. Wirth, and M. Jaskola, *Phys. Rev.* C **74**, 054605 (2006).

Systematic Study of Tin Isotopes via High Resolution (p,t) Reaction Measurements

P. Guazzoni[a], L. Zetta[a], A. Covello[b], A. Gargano[b], B.F. Bayman[c], T. Faestermann[d], R. Hertenberger[e], H.-F. Wirth[e], M. Jaskola[f]

[a] Università degli Studi and I.N.F.N., Via Celoria 16, I-20133 Milano, Italy
[b] Università Federico II and I.N.F.N, Via Cintia, I-80126 Napoli, Italy
[c] Physics and Astronomy, University of Minnesota, Minneapolis MN 55455, Minnesota, USA
[d] Physik Department - Technische Universität München, D-85748 Garching, Germany
[e] Sektion Physik - Ludwig Maximilian Universität München, D-85748 Garching, Germany
[f] Soltan Institute for Nuclear Studies, 00-681 Warsaw, Poland

Abstract. The ^{118}Sn(p,t)^{116}Sn and ^{124}Sn(p,t)^{122}Sn reactions have been measured at an incident proton energies of 24.6 and 25 MeV respectively in the framework of a systematic study of the even tin isotopes using (p,t) reactions in high resolution experiments at the Munich HVEC Tandem, with the Q3D spectrograph.

Keywords: Transfer Reactions, Q3D Spectrograph, Cross Section Angular Distributions, J$^{\pi}$ Assignments.

PACS: 25.40.Hs, 21.10.Hw, 21.60.Cs, 27.60.+j

EXPERIMENTAL RESULTS AND CLUSTER DWBA ANALYSIS

The tin isotopes provide a very good opportunity for experimental and theoretical studies of variations in nuclear properties with changing neutron number. Transfer reactions play an important role in our understanding of tin isotopes since these reactions are very sensitive to the neutron structure of the isotopic sequence N=60 (^{110}Sn) to N=76 (^{126}Sn).

In the framework of a systematic study of the tin isotopes using (p,t) reactions in high resolution experiments at the Munich HVEC Tandem, with the Q3D spectrograph [1,2,3,4], the ^{118}Sn(p,t)^{116}Sn and ^{124}Sn(p,t)^{122}Sn reactions have been measured at an incident proton energies of 24.6 and 25 MeV, respectively. Angular distributions for 55 and 63 transitions to the final states of ^{116}Sn and ^{122}Sn, respectively, have been measured. A distorted-wave Born approximation (DWBA) analysis of the experimental differential cross sections, using conventional Woods-Saxon potentials and a semi-microscopic neutron pickup mechanism, has been carried out, allowing either the confirmation of previous spin and parity values or the assignment of new spin and parity to a large number of states.

In Fig.s 1-6 the angular distributions for the transitions to the observed 0^+, 2^+, 4^+, 1^-, 3^-, 5^-, 7^-, 9^-, 6^+ states and doublets of ^{116}Sn and ^{122}Sn residual nuclei (dots) are compared with the cluster DWBA calculations (solid lines). Zero-range one-step DWBA microscopic calculations for the ground state and some excited states using the two-neutron

CP1165, *Nuclear Structure and Dynamics '09*
edited by M. Milin, T. Nikšić, D. Vretenar, and S. Szilner
© 2009 American Institute of Physics 978-0-7354-0702-2/09/$25.00

spectroscopic amplitudes obtained from a shell model study of the [118,116]Sn and [124,122]Sn nuclei respectively, are in progress. The shell model calculations will be carried out within the framework of the seniority scheme, as in ref.s [3,4], using a realistic effective interaction derived from the CD-Bonn nucleon-nucleon potential [5].

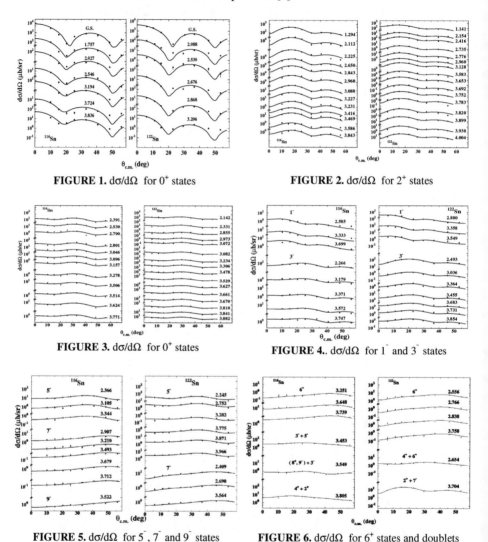

FIGURE 1. $d\sigma/d\Omega$ for 0^+ states

FIGURE 2. $d\sigma/d\Omega$ for 2^+ states

FIGURE 3. $d\sigma/d\Omega$ for 0^+ states

FIGURE 4. $d\sigma/d\Omega$ for 1^- and 3^- states

FIGURE 5. $d\sigma/d\Omega$ for 5^-, 7^- and 9^- states

FIGURE 6. $d\sigma/d\Omega$ for 6^+ states and doublets

REFERENCES

1. P. Guazzoni, et al., *Phys. Rev.* C **60**, 054603 (1999).
2. P. Guazzoni, et al., *Phys. Rev.* C **69**, 024619 (2004).
3. P. Guazzoni, et al., *Phys. Rev.* C **74**, 054605 (2006).
4. P. Guazzoni, et al., *Phys. Rev.* C **78**, 064608 (2008).
5. R. Machleidt, *Phys. Rev.* C **63**, 024001 (2001)

DYNAMICS OF HEAVY-ION REACTIONS

Aspects of Heavy Ion Transfer Reactions

L.Corradi

INFN, Laboratori Nazionali di Legnaro (Italy)

Abstract.
    With the large solid angle magnetic spectrometer PRISMA coupled to the γ array CLARA extensive investigations have been carried out for nuclear structure and reaction dynamics. In the present paper aspects of these studies will be presented, focusing more closely on the reaction mechanism, in particular on the properties of quasi-elastic and deep-inelastic processes and on measurements at energies far below the Coulomb barrier.

Keywords: Multinucleon transfer; quasi-elastic reactions; sub-barrier energies; magnetic spectrometer; γ-particle coincidences; coupled channels
PACS: 25.70.Hi, 29.30.Aj, 24.10.-i, 23.20.Lv

INTRODUCTION

With the development of high resolution and high efficiency experimental set-up's, one could unambiguously detect in mass and charge the nuclei produced in multineutron and multiproton transfer reactions (see [1, 2, 3] and references therein). The advent of the last generation large solid angle magnetic spectrometer PRISMA [4] pushed the detection limit more than an order of magnitude below previous limits, with a significant gain in mass resolution for very heavy ions. Further, the coupling of this spectrometer to the large gamma array CLARA [5] allowed to perform gamma-particle coincidences, thus detecting the transfer strength to the lowest excited levels of binary products and performing gamma spectroscopy for nuclei moderately far from stability produced via nucleon transfer or deep-inelastic reactions, especially in the neutron-rich region.

    In this paper I focus on specific aspects of reaction mechanism studies being performed with PRISMA. For the results concerning pure nuclear structure studies I refer to the contributions [6] to this conference.

ELASTIC SCATTERING

Elastic scattering is important to learn about the (outer part of) the nuclear potential and provides essential information on absorptive effects, to be accounted for in coupled channel calculations. The present set-up offers the possibility to separate elastic from inelastic scattering, at least for some nuclei. The pure elastic scattering can be determined by comparing the events with and without γ coincidences [7]. As an example, in the top panel of Fig. 1 are shown the total kinetic energy loss (TKEL) spectra for ^{90}Zr in the reaction ^{90}Zr+^{208}Pb with and without γ-coincidence, normalized in the tail (large TKEL) region. By subtraction, one obtains the contribution of pure elastic. This subtracted spectrum is characterized by a narrow peak centered at TKEL $\simeq 0$ MeV with

CP1165, *Nuclear Structure and Dynamics '09*
edited by M. Milin, T. Nikšić, D. Vretenar, and S. Szilner
© 2009 American Institute of Physics 978-0-7354-0702-2/09/$25.00

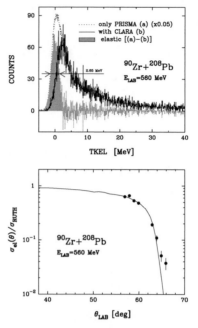

FIGURE 1. Top: Experimental angle integrated total kinetic energy loss distributions (TKEL) for ^{90}Zr in the ^{90}Zr+^{208}Pb reaction (a) without coincidence with γ rays and (b) with at least one γ ray detected in CLARA. The two spectra are normalized in such a way that the high TKEL tails match. The gray area corresponds to the subtraction between the two spectra [(a)-(b)], giving a peak whose width is ~ 2.65 MeV. Bottom: Experimental (points) and GRAZING calculated (curve) differential cross section for elastic scattering, normalized to Rutherford. Only statistical errors are included.

a FWHM of 2.65 MeV. Moreover, its centroid is separated by 2.15 MeV from the maximum of the TKEL spectrum in coincidence with CLARA, whose value is very close to the inelastic excitation of the first 2^+ state in ^{90}Zr. Such a procedure should be reliable, provided that the shape of the spectrum in coincidence with γ rays only weakly depends on the γ multiplicity. This fact is fulfilled for nuclei having low level density close to the ground state and rather narrow ($\simeq$ 2-3 MeV) TKEL distributions, as for the present near closed-shell nuclei.

By repeating this subtraction in steps of one degree over the entrance angular range ($\Delta\theta_{lab} = 12°$) of PRISMA one obtains the elastic angular distribution whose ratio to Rutherford is shown in the bottom panel of Fig. 1, in comparison with the results of GRAZING calculations [8]. The very pronounced fall-off of the elastic cross section for large angles clearly indicates that the elastic scattering for this system is dominated by strong absorption. The good agreement between theory and experiment gives us confidence on the used potential and on the fact that the included reaction channels correctly describe the depopulation of the entrance channel (absorption). A similar kind of anlysis has been successfully performed for the ^{48}Ca+^{64}Ni system [9].

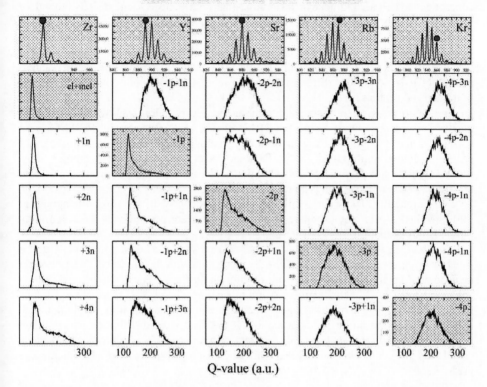

90Zr+208Pb E=560 MeV PRISMA

FIGURE 2. TKEL spectra obtained in the reaction ^{90}Zr+^{208}Pb for the indicated transfer channels. In the top row are shown the mass distribution associated to the different nuclear charges, while the circles indicate the specific masses corresponding to the spectra displayed along the upper-left/lower-right diagonal. The centroid of the elastic+inelastic channel corresponds to Q=0. The scale of the Q-value axis is 1 MeV/channel.

FROM QUASI-ELASTIC TO DEEP-INELASTIC REGIME

The Z and A identification capability and the large detection efficiency of PRISMA allows to follow the evolution of the reaction [10] from the quasi elastic (i.e. few nucleon transfer and low TKEL) to the deep inelastic regime (i.e. many nucleon transfer and large TKEL). Here, the challenging question is to what extent the fundamental degrees of freedom (single particle, surface and pair modes) used to describe few nucleon transfer processes, holds in the presence of large energy losses and/or large number of nucleons.

In Fig. 2 I show the TKEL spectra obtained in the ^{90}Zr+^{208}Pb reaction [11] for different transfer channels. One can follow the evolution pattern as function of the number of transferred neutrons and protons. For instance, in the case of pure neutron transfer one sees a quasi-elastic peak and an increasing strength on large energy loss components when adding neutrons. I remind that with PRISMA one detects secondary fragments and that the TKEL spectra are constructed assuming binary reactions. For channels which, due to optimum Q-values, are not directly populated, the shape of the corresponding TKEL differ a lot from the smooth behaviour just decribed. Look for instance at the comparison between the (-1p+1n) channel (mainly directly populated) and the (-1p-1n) one. This different behaviour tends to smooth out with larger number of transferred protons.

Large energy losses are associated with nucleon evaporation from the primary fragments. This can be clearly seen with PRISMA+CLARA. Gating with PRISMA on a specific Z and A (light partner) the velocity vector of the undetected heavy partner can be evaluated and applied for the Doppler correction of its corresponding γ rays. In those spectra not only the γ rays belonging to the primary binary partner are present but also the ones of the nuclei produced after evaporation takes place. For example in the ^{40}Ca+^{96}Zr reaction [7], for the $-2p+2n$ channel about 60% of the yield corresponds to ^{96}Mo, while the rest is equally shared between isotopes corresponding to the evaporation of one and two neutrons.

In general, for few nucleon transfer channels most of the yield corresponds to the true binary partner. This behavior is closely connected with the observed TKEL. For the neutron pick-up channels the major contribution in the TKEL is close to the optimum Q values ($Q_{opt} \simeq 0$), while in the proton stripping channels larger TKEL are observed, thus the neutron evaporation has a stronger effect on the final mass partition. The importance of neutron evaporation in the modification of the final yield distribution was outlined in inclusive measurements [1, 2, 3]. A direct signature of this effect was observed by correlating projectile-like and target-like fragment isotopic yields via $\gamma - \gamma$ coincidences [12].

SUB-BARRIER TRANSFER REACTIONS

In recent years there has been growing interest in studying dynamic processes at energies well below the Coulomb barrier, in particular sub-barrier fusion (see Refs. [13, 14] and references therein for the last conferences on the subject). This same energy range is also ideal to investigate transfer processes, which are strongly connected with fusion, as they probe different but complementary ranges of nuclear overlap.

To set the frame, one can write the transfer cross section as :

$$\sigma_{tr} \sim e^{-\frac{2}{\hbar}\int W(r(t))dt} \sum \left| \int F_{if}(r(t))e^{i\omega_{if}dt} \right|^2$$

where the first exponential term gives the probability to remain in the elastic channel and the second describes the direct population of the transfer channels being $F(r)$ the transfer form factor and $e^{i\omega_{if}}$ defining the Q-value window, with the sum running over all the final channels. The integrals are performed along the Coulomb trajectory. The imaginary potential $W(r)$, that describes the depopulation of the entrance channel, at very low energies is dominated by the single-nucleon transfer channels. Since the Q-

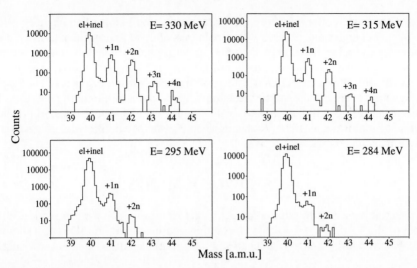

FIGURE 3. Mass distributions for pure neutron transfer channels obtained in the reaction ^{94}Zr+^{40}Ca at the indicated bombarding energies. Ca-like recoils have been detected at $\theta_{lab}=20°$ with the PRISMA spectrometer.

value distributions get narrower at low bombarding energies these subbarrier studies may provide important information on the nuclear correlation close to the ground state. In this energy region the multinucleon transfer channels should be dominated by a successive mechanism with negligible contribution from a cluster-like transfer [15]. This fact should provide a simpler analysis of the data.

From the experimental point of view, measurements of heavy-ion transfer reactions at far sub-barrier energies have significant technical difficulties. At low bombarding energies angular distributions result, in the center of mass frame, in a strong backward peaking, with a maximum at $\theta_{cm} \simeq 180°$. The absolute yield gets very small, therefore high efficiency is needed. At the same time, mass and nuclear charge resolutions must be maintained at a level sufficient to distinguish the different reaction channels. For situations where the projectile has a significant fraction of the target mass, as it is in most cases, the backscattered projectile-like fragment has such a low energy that usual identification techniques become invalid. A suitable way to overcome these limitations is by means of inverse kinematics, thus we recently detected multinucleon transfer channels in the reactions ^{94,96}Zr+^{40}Ca at different bombarding energies below the Coulomb barrier [16], making use of the PRISMA+CLARA set-up. The use of inverse kinematics and the detection at very forward angles, allowed to have, at the same time, enough kinetic energy of the outgoing recoils (for energy and therefore mass resolution) and forward focused angular distribution (high efficiency). Sub barrier fusion cross sections for the same system had been previosuly measured with high precision [17] and a complete set of data for both multinucleon transfer and fusion reactions would provide an excellent basis for coupled channel calculations.

In Fig. 3 I show the mass spectra for pure neutron transfer channels in the system

361

^{94}Zr+^{40}Ca obtained after trajectory reconstruction at four bombarding energies. While at the higher energies one observes the populations of up to four nucleon transfer, at the lower energies (below the Coulomb barrier) only one and two neutron transfer survive. The mention that the Q-value distributions for the +2n channel at the lowest energies are very narrow and close to the ground state to ground state transition, as a result of the very low excitation energy of the transfer reaction products at these sub-barrier energies. The experimental results will be compared with coupled channel calculations, in particular the comparison of two nucleon vs. one nucleon transfer should provide information on nucleon-nucleon correlation effects.

ACKNOWLEDGMENTS

The material presented in this conference is the result of a cooperative work of many people of the PRISMA+CLARA collaboration and belonging to different institutions (LNL, Padova, Torino, Zagreb, Bucharest, Strasbourg) which I wish to acknowledge.

REFERENCES

1. L. Corradi and G. Pollarolo, *Nucl. Phys. News*, Vol. **15**, N.4 (2006).
2. S. Szilner, et al., *Phys. Rev.* C **71**, 044610 (2005).
3. L. Corradi, G. Pollarolo and S. Szilner, *J. Phys. G (ST)*, in preparation.
4. A. M. Stefanini, et al., *Proposta di esperimento PRISMA, LNL-INFN (Rep)-120/97* (1997); *Nucl. Phys.* A **701c**, 217 (2002); http://www.lnl.infn.it/~prisma.
5. A. Gadea, et al., *Eur. Phys. J.* A **20**, 193 (2004).
6. Contributions to this conference by E. Farnea, S. Lunardi, D. Mengoni, et al., J.J. Valiente-Dobon, et al. and references therein.
7. S. Szilner, et al., *Phys. Rev.* C **76**, 024604 (2007).
8. A. Winther, *Nucl. Phys.* A **572**, 191 (1994); *Nucl. Phys.* A **594**, 203 (1995); program GRAZING, htpp:/www.to.infn.it/~nanni/grazing.
9. D. Montanari et al, contribution to this conference.
10. K. E. Rehm, *Annu. Rev. Nucl. Part. Sci.* **41**, 429 (1991).
11. L. Corradi, *Nucl. Phys.* A **787c**, 134 (2007).
12. R. Broda, *J. of Phys. G* **32**, R151 (2006).
13. *Fusion06: Int. Conf. on Reaction Mechanisms and Nuclear Structure at the Coulomb barrier*, S.Servolo (Venezia), Italy,19-23 March 2006, AIP Proceedings Series, Vol. 853, Melville (New York), L. Corradi et al. eds.
14. *Fusion08 : Int. Conf. on New Aspects of Heavy Ion Collisions Near the Coulomb Barrier*, Chicago (USA), September 22-26, 2008, AIP Proceedings Series, Vol. N. 1098 (2009), Melville, New York, K. E. Rehm et al eds.
15. B. F. Bayman and J. Chen, *Phys. Rev.* C **26**, 1509 (1982).
16. L. Corradi, S. Szilner, et al., *LNL-INFN PAC Proposal July 2007 and July 2008*.
17. A. M. Stefanini, et al., *Phys. Rev.* C **76**, 014610 (2007).

Transfer Reactions on Neutron-rich Nuclei at REX-ISOLDE

Th. Kröll[*,†], V. Bildstein[†], K. Wimmer[†], R. Krücken[†], R. Gernhäuser[†],
R. Lutter[**], W. Schwerdtfeger[**], P. Thirolf[**], B. Bastin[‡], N. Bree[‡],
J. Diriken[‡], M. Huyse[‡], N. Patronis[‡], R. Raabe[‡], P. Van Duppen[‡],
P. Vermaelen[‡], J. Cederkäll[§], E. Clément[§], J. Van de Walle[§], D. Voulot[§],
F. Wenander[§], A. Blazhev[¶], M. Kalkühler[¶], P. Reiter[¶], M. Seidlitz[¶],
N. Warr[¶], A. Deacon[‖], C. Fitzpatrick[‖], S. Freeman[‖], S. Das Gupta[††],
G. Lo Bianco[††], S. Nardelli[††], E. Fiori[‡‡], G. Georgiev[‡‡], M. Scheck[‖],
L. M. Fraile[§§], D. Balabanski[¶¶], T. Nilsson[***], E. Tengborn[***],
J. Butterworth[‖], B. S. Nara Singh[‖], L. Angus[†††], R. Chapman[†††],
B. Hadinia[†††], R. Orlandi[†††], J. F. Smith[†††], P. Wady[†††], G. Schrieder[*],
M. Labiche[‡‡‡], J. Johansen[§§§], K. Riisager[§§§], H. B. Jeppesen[¶¶¶],
A. O. Macchiavelli[¶¶¶], T. Davinson[†††] and the REX-ISOLDE and
MINIBALL collaborations[§]

[*]Institut für Kernphysik, Technische Universität Darmstadt, Germany
[†]Physik-Department E12, Technische Universität München, Garching, Germany
[**]Fakultät für Physik, Ludwig-Maximilians-Universität München, Garching, Germany
[‡]Instituut voor Kern- en Stralingsfysica, Katholieke Universiteit Leuven, Belgium
[§]CERN, Genève, Switzerland
[¶]Institut für Kernphysik, Universität zu Köln, Germany
[‖]Nuclear Physics Groups, Universities of Manchester, York, and Liverpool, United Kingdom
[††]INFN and Dipartimento di Fisica, Università di Camerino, Italy
[‡‡]Centre de Spectrométrie Nucléaire et de Spectrométrie de Masse, Orsay, France
[§§]Facultad de Sciencias Físicas, Universidad Complutense, Madrid, Spain
[¶¶]INRNE, Bulgarian Acadamy of Sciences, Sofia, Bulgaria
[***]Fundamental Fysik, Chalmers Tekniska Högskola, Göteborg, Sweden
[†††]Nuclear Physics Groups, Universities of the West of Scotland and Edinburgh, United Kingdom
[‡‡‡]Daresbury Laboratory, Warrington, United Kingdom
[§§§]Institut for Fysik og Astronomi, Århus Universitet, Denmark
[¶¶¶]Lawrence Berkeley National Laboratory, USA

Abstract. We report on one- and two-neutron transfer reactions to study the single-particle properties of nuclei at the border of the "island of inversion". The (d,p)- and (t,p)-reactions in inverse kinematics on the neutron-rich isotope ^{30}Mg, delivered as radioactive beam by the REX-ISOLDE facility, have been investigated. The outgoing protons have been detected and identified by a newly built array of Si detectors. The γ-decay of excited states has been detected in coincidence by the MINIBALL array. First results for ^{31}Mg and from the search for the second, spherical, 0^+ state in ^{32}Mg are presented.

Keywords: transfer reaction, gamma-ray spectroscopy, neutron-rich nuclei
PACS: 25.60.Je, 23.20.-g, 27.30.+t

CP1165, *Nuclear Structure and Dynamics '09*
edited by M. Milin, T. Nikšić, D. Vretenar, and S. Szilner
© 2009 American Institute of Physics 978-0-7354-0702-2/09/$25.00

MOTIVATION

Light-ion induced transfer reactions, like (d,p) or (t,p), offering selectivity to both kinematical matching conditions and nuclear structure are a well-established spectroscopic tool to study single-particle properties of nuclear states. For radioactive ions, the reactions have to performed in inverse kinematics. From the energies of the outgoing particles the (single-particle) level energies can be determined. Their angular distributions lead to spin and parity assignments. Both are assisted by the observation of coincident γ-rays. Eventually, the (relative) spectroscopic factors deduced from the cross sections allow to extract information on the configurations forming the populated state.

However, there are conceptual limitations. Transfer reactions probe only the wave functions on the nuclear surface. The analysis is based on phenomenological approaches (optical model, DWBA) and, therefore, less well under control compared e.g. to Coulomb excitation. Finally, the observed quenching of spectroscopic factors is an open problem but on the other hand also an opportunity to probe the validity of single-particle models. Despite these facts, transfer reactions have been a successful tool in nuclear spectroscopy of stable nuclei for many decades.

Concerning the study of radioactive nuclei, three points are worth to be mentioned. As population mechanism for neutron-rich nuclei a neutron-transfer reaction starting from a less neutron-rich nucleus may be competitive to a direct production in certain cases. Due to its selectivity, states may be populated which are not accessible by other methods, like Coulomb excitation, β-decay, or fragmentation. In particular, the pair transfer is a valuable tool to study phenomena like shape coexistence and pairing correlations.

In practice, there are also challenges and (open) questions. The parameters of the optical potentials are not known and have to extracted from elastic scattering or extrapolated from stable systems. At REX-ISOLDE, the beam energy is limited to 3 MeV/u. This, together with the low Q values for neutron transfer on neutron-rich nuclei, causes low energies of the outgoing protons, smooth angular distributions, and possible inconsistencies in the determination of spectroscopic factors. These points have been investigated in detail for a stable system [1]. Despite the low beam energies, the observed protons still originate from direct reactions rather than from fusion-evaporation reactions because thermal energy spectra and isotropic angular distributions have not been observed experimentally. As statistical process, only neutrons will be evaporated because their separation energy in neutron-rich nuclei is much smaller compared to that of protons. This simple argument has been confirmed by Hauser-Feshbach calculations [2].

EXPERIMENTAL SET-UP

ISOLDE is the radioactive beam facility at CERN having more than 40 years of experience in producing RIBs by a 1.4 GeV proton beam impinging on a solid target, in our case UC_x. The exotic species are extracted as singly-charged ions and mass separated. In REX-ISOLDE, this low-energy beam is cooled and bunched in a penning trap, charge bread to $A/q \approx 4$ in an EBIS, and finally post-accelerated by a linac to energies up to 3 MeV/u. REX-ISOLDE has delivered since 2002 more than 60 RIBs to experiments.

The experiments reported in this contribution require the coincident measurement of

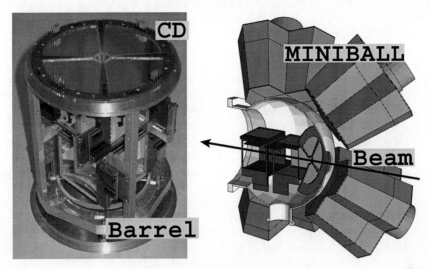

FIGURE 1. Photo (left) and drawing mounted inside of MINIBALL (right) of the new Si array.

the light target-like particles and the γ-rays. The latter are detected by the MINIBALL array consisting of 24 6-fold segmented HPGe detectors [3]. The photopeak efficiency is around 3% at 1333 keV with the particle detector inside. This Si detector array has been newly built and offers a large solid angle, position sensitivity, and the capability to identify light particles by the $\Delta E - E$ method. The array consists of a DSSSD detector (500 µm), so-called CD detector, in backward direction and a barrel formed by 8 quadratic detectors: stacks of 140 µm and 1000 µm detectors in forward direction or a single 500 µm detector in backward direction [4] (Fig. 1, left). The 140 µm and 500 µm detectors are read out via resistive strips. The forward DSSSD detector is still missing. The particle identification seemed to be important only in forward direction as backwards only protons originating from transfer reactions are expected. However, it turned out that electrons from the β-decay of scattered beam particles are badly separated from the low-energy protons. Therefore, the first version of the set-up built and used in 2007 has been upgraded in 2008 by using in forward and backward direction for the barrel the same type of detector stacks and by backing the CD detector with an unsegmented Si detector. As all light ions in backward direction are stopped in the first Si layer, the second only serves to discriminate electrons. The drawing in Fig. 1 (right) shows the arrangement of the Si array inside of MINIBALL as it has been implemented into GEANT4 [5] for simulation.

Additionally, a segmented diamond detector on the target ladder which can be moved at target position and an active collimator with four PIN diodes in front of the chamber for beam focusing have been included.

This set-up was a major upgrade of the instrumentation available at REX-ISOLDE and has been funded by TU München, KU Leuven, University of Edinburgh, and CSNSM Orsay. For 2009, the upgrade to the final set-up is planned.

ISLAND OF INVERSION

Although it has been discovered already more than 30 years ago, the structure of nuclei within the "island of inversion" [6, 7], a region of nuclei with deformed ground states in the sea of spherical sd-nuclei, is still not fully understood. The gross features of this inversion are interpreted as effect of a residual interaction derived from the tensor part of the nuclear force [8, 9]. The intruder fp-orbitals are lowered with respect to the normal sd-orbitals and excitations of neutrons to the fp-orbitals are energetically favoured. This breaking of the shell closure at $N = 20$ leads to deformed states.

At the north-west coast of the island, the even Mg isotopes have been investigated by lifetime measurements at ISOLDE [10] and by Coulomb excitation at "safe" energies at REX-ISOLDE [11, 12] as well as at intermediate energies, e.g. Refs. [13, 14]. The results establish consistently a nearly spherical shape of the ground state and the first 2^+ state in ^{30}Mg but deformed shapes for the corresponding states in ^{32}Mg. Hence, the former is outside whereas the latter is inside of the "island of inversion". This interpretation is supported by theory. A good agreement has been achieved by both modern large scale shell model calculations, e.g. Refs. [15, 16], and a "beyond mean field" approach [17]. Therefore, the underlying physics seems to be well understood.

It was a surprise that in an experiment at ISOLDE for the ground state of ^{31}Mg, the nucleus right on the shore of the island, a $1/2^+$ assignment was found [18, 19] which is in contrast to all theoretical predictions. A possible solution was an adjustment of a few matrix elements of the monopole interaction used in the shell model calculations [20]. In this approach, the ground state and the first excited state in ^{31}Mg are largely dominated by 2p-2h intruder configurations (93% and 95% of the wave function, respectively) with a fp-shell occupancy close to 2. Hence, the transition into the island is much steeper than e.g. in Na. However, there is no consistent microscopic description of this region.

Aiming to study the single-particle properties also of excited states in ^{31}Mg, we investigated the d(^{30}Mg, ^{31}Mg)p reaction. A ^{30}Mg at 2.86 MeV/u with a, due to technical problems, low intensity of 10^4 part/s impinged on a 1 mg/cm^2 deuterated PE target. The experiment has been performed in 2007 and the analysis is still in progress [21].

In Fig. 2 (left), the $\Delta E - E$ particle identification plot is shown. Clearly, the different particles, i.e. p, d, t, and, β, are separated. As the energy resolution is not sufficient to identify the populated states, the γ-rays are needed. Fig. 2 (right) shows part of the γ-ray spectrum in coincidence with the protons. Gating on the 170 keV transition, the angular distribution of the respective protons is consistent with a $\Delta\ell = 1$ orbital momentum transfer, giving support for the tentative $3/2^-$ assignment for the state at 221 keV.

An alternative look into the evolution of the single-particle structure can be obtained from the study of the excited 0^+ states. The configuration which forms the ground state in ^{30}Mg becomes an excited 0^+ state in ^{32}Mg and vice versa. The coexistence of 0^+ states with different shapes and the inversion of their order is a direct consequence of the migration of the single-particle levels.

In ^{30}Mg, the second 0^+ state at 1789 keV is quite well known as it is populated in the β-decay of ^{30}Na. Its γ-decay to the 2^+ state and its lifetime [10] as well as its E0 decay to the ground state [22] have been measured at ISOLDE. The results are consistent with a deformed shape and a small mixing with the ground state [22].

The excited 0^+ state in ^{32}Mg has not been observed experimentally so far. It is not

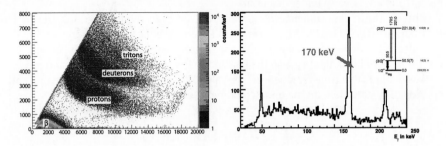

FIGURE 2. PID plot (left) and part of the γ-ray spectrum in coincidence with the protons (right).

populated in the β-decay of ^{32}Na [10, 23], in Coulomb excitation, or in fragmentation reactions [24]. The theoretical predictions range from 1.4 MeV to 3 MeV [25, 26].

As its particle-hole structure is assumed to be similar to that of the ground state of ^{30}Mg, hence a spherical 0p-0h configuration, it should be selectively populated in a two-neutron transfer reaction because an large overlap of the wave functions means a large spectroscopic factor. The population of other states in ^{32}Mg with 2p-2h intruder configuration, e.g. the first 2^+ state, should be disfavoured.

In order to search for the second 0^+ state in ^{32}Mg, the t(^{30}Mg, ^{32}Mg)p reaction has been employed. A ^{30}Mg beam at 1.83 MeV/u with an intensity of $4 \cdot 10^4$ part/s impinged on a 0.5 mg/cm^2 Ti target loaded with tritium (^{3}H/Ti=1.5). The experiment has been performed in 2008 and the analysis is in progress [27].

From the first preliminary results several interesting points can be mentioned. Protons from the transfer reaction are seen, kinematically clearly separated from elastically scattered protons (and tritons). Hence, the two-neutron transfer works. No deuterons are observed, as the $Q = -3.9$ MeV disfavours strongly the (t,d) reaction because of kinematic mismatching. There is also no evidence for the population of the 2^+ state, consistent with the different structure and the negative $Q = -1.2$ MeV. Reconstructing the excitation energies from the proton energies, the population of two states can be identified: the ground state and a new state at around 1.1 MeV. The first can be understood by the small $Q = -0.3$ MeV, near to the optimal value of $Q = 0$, and the small configuration mixing in the wave functions. The angular distributions of the respective protons indicate a $\Delta \ell = 0$ orbital momentum transfer for both states. Therefore, the new state is a promising candidate for the long-searched second 0^+ state in ^{32}Mg.

Summary and outlook

Further experiments using the described set-up have already been approved: d(^{11}Be, ^{12}Be)p [28], d(^{66}Ni, ^{67}Ni)p [29], and d(^{78}Zn, ^{79}Zn)p [30]. The reported studies at the shore of the "island of inversion" will be extended to neighbouring Al or Na isotopes.

For the farer future experiments addressing many interesting topics like the quenching of spectroscopic factors, shape coexistence in the region of neutron-deficient Pb isotopes, paring correlations in exotic nuclei (surface pairing, $T = 0$ np-paring in $N = Z$

nuclei) etc. are already under discussion.

Most of these experiments will profit from or even require the upgrade of REX-ISOLDE to HIE-ISOLDE [31]. Apart from a step-wise increase in energy to 5.5 MeV/u and later to 10 MeV/u, larger intensities and better beam qualities are envisioned. Studies for a 0° separator to identify also the heavy transfer products have been started.

We conclude that transfer reactions are valuable tool for nuclear spectroscopy and will considerably contribute to the understanding of exotic nuclei.

ACKNOWLEDGMENTS

This work is supported by the German BMBF under grant No. 06MT238, by the MLL, by the EU through EURONS (contract No. 506065), by the state of Hessen within the HIC for FAIR programme and the LOEWE excellence initiative, and by the DFG cluster of excellence Universe (www.universe-cluster.de). REX-ISOLDE and MINIBALL are supported by the German BMBF, the Belgian FWO-Vlaanderen and IAP, the U.K. EPRSC, and the EU, as well as the ISOLDE collaboration.

REFERENCES

1. M. Mahgoub, PhD thesis (TU München, 2008).
2. M. Pantea, PhD thesis (TU Darmstadt, 2005).
3. J. Eberth, et al., *Prog. Part. Nucl. Phys.* **46**, 389 (2001).
4. V. Bildstein, et al., *Prog. Part. Nucl. Phys.* **59**, 386 (2007).
5. GEANT4, http://geant4.web.cern.ch/.
6. C. Thibault, et al., *Phys. Rev.* C **12**, 644 (1975).
7. E. K. Warburton, et al., *Phys. Rev.* C **41**, 1147 (1990).
8. T. Otsuka, et al., *Eur. Phys. J.* A **15**, 151 (2002).
9. T. Otsuka, et al., *Phys. Rev. Lett.* **95**, 232502 (2005).
10. H. Mach, et al., *Eur. Phys. J.* A **25**, 105 (2005).
11. O. Niedermaier, et al., *Phys. Rev. Lett.* **94**, 172501 (2005).
12. O. Niedermaier, PhD thesis (Universität Heidelberg, 2005); to be published.
13. T. Motobayashi, et al., *Phys. Lett.* **346B**, 9 (1995).
14. J. A. Church, et al., *Phys. Rev.* C **72**, 054320 (2005).
15. Y. Utsuno, et al., *Phys. Rev.* C **60**, 054315 (1999).
16. E. Caurier, et al., *Nucl. Phys.* **A 693**, 374 (2001).
17. R. Rodríguez-Guzmán, et al., *Nucl. Phys.* **A 709**, 201 (2002).
18. G. Neyens, et al., *Phys. Rev. Lett.* **94**, 022501 (2005).
19. M. Kowalska, PhD thesis (Universität Mainz, 2006).
20. F. Maréchal, et al., *Phys. Rev.* C **72**, 044314 (2005).
21. V. Bildstein, PhD thesis (TU München, 2009).
22. W. Schwerdtfeger, et al., *Phys. Rev. Lett.* (in press).
23. C. M. Mattoon, et al., *Phys. Rev.* C **75**, 017302 (2007).
24. M. Gelin, PhD thesis (Université de Caen, 2007).
25. D. Guillemaud-Mueller, et al., *Eur. Phys. J.* A **13**, 63 (2002).
26. T. Otsuka, et al., *Eur. Phys. J.* A **20**, 69 (2004).
27. K. Wimmer, PhD thesis (TU München, expected for 2010).
28. J. Johansen, et al., contribution to this conference.
29. N. Patronis, et al., Proposal CERN-INTC-2008-007, INTC-P-238 (2008).
30. R. Orlandi, et al., Proposal CERN-INTC-2009-017, INTC-P-264 (2009).
31. *HIE-ISOLDE - the scientific opportunities*, Eds. K. Riisager, P. Butler, R. Krücken, CERN-2007-008.

Role of Transfer Channels in Heavy-ion Reactions

Giovanni Pollarolo

Dipartimento di Fisica Teorica, Università di Torino and INFN, Sez. di Torino,
Via Pietro Giuria 1, I-10125 Torino, Italy

Abstract. Transfer reactions constitute the dominant contribution to the back-angles quasi-elastic excitation functions measured in collisions between heavy-ions. This is shown by using a semiclassical model that incorporates both the excitation of surface modes and the particle transfer degrees of freedom.

Keywords: Heavy-ion reactions, transfer reactions
PACS: 25.70.Hi, 21.30.Fe,24.10.-i,25.70.Jj

INTRODUCTION

Exploiting the very short wave length of the relative motion, one can use simple classical arguments to understand the main characteristics of a heavy ion reaction by introducing a potential that is function of the center-of-mass distance. With this simple ingredient it is, for example, possible to provide a reasonable estimation of the total reaction cross section and to predict at which angle the yields are mostly concentrated. The ion-ion potential has as its most conspicuous feature a barrier originating from the balance between a long-range repulsive Coulomb and a short-range attractive nuclear components.

Despite its merits this simple potential description has been readily recognised as leading to important shortcomings. From elastic scatterings one learned [1] that the potential must be energy dependent and must have an imaginary part. From fusion reactions one learned [2] that the simple potential description strongly under predicts fusion cross sections at very low energies.

To arrive at a consistent description of the data one has to include in the reaction mechanism couplings that take into account the intrinsic states of the two nuclei. These variables are associated to single particles and collective modes, surface vibrations and rotations. In the case of fusion reaction it has been shown that the couplings to surface modes [3] account for most of the missing cross section. For these reactions the effect of the couplings preclude us from talking about a single barrier but it is more convenient to talk about a distribution of barriers (several Mev wide) around the nominal Coulomb barrier of the ion-ion potential. These barrier distributions can be extracted directly from the fusion excitation functions by taking the second energy derivative of the energy weighted fusion cross section [4]. More ricently, it has been suggested that the same information on the barrier may be extracted from the energy dependence of the quasi-elastic cross section at backward angles [5]. In this case the barrier distributions are obtained by the energy derivative of the quasi-elastic excitation functions.

CP1165, *Nuclear Structure and Dynamics '09*
edited by M. Milin, T. Nikšić, D. Vretenar, and S. Szilner
© 2009 American Institute of Physics 978-0-7354-0702-2/09/$25.00

The importance of transfer reactions, i.e. of couplings to single-particle degrees of freedom, in the description of a heavy-ion reaction has been underlined in several papers [6, 7, 8]. These transfer degrees of freedom are weak, very numerous and span a wide range of Q-values. They are governed by long range formfactors and are providing the main contribution to the absorptive and polarization potential. Unfortunately fusion reactions have been very elusive in pinning down the role of particle transfer, many good fits of the data could, in fact, be obtained by including only surface modes.

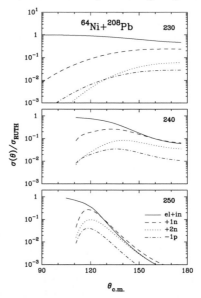

FIGURE 1. Center-of-mass angular distributions for elastic plus inelastic and some transfer channels. The cross section are plotted as ratio to the Rutherford cross section. The label in each frame indicates center-of-mass bombarding energy in MeV.

Quite recently quasi-elastic excitation functions have been measured [9] for several systems, whose use have been proposed for cold-fusion production of super-heavy elements, and the corresponding barrier distributions extracted. Because of the final energy, mass and charge resolution of the experiment, the quasi-elastic reactions, beside elastic and inelastic channels, receive contributions also from transfer channels, both neutrons and protons. As a consequence these reactions are providing a very interesting tool to investigate the role of transfer reactions at near barrier energies.

THE MODEL

To analyse the quasi-elastic reactions we use a semiclassical model, GRAZING [12, 13, 14, 15], that generalizes the well known theory for Coulomb excitation by incorporating the effects of the nuclear interaction in the trajectory and in the excitation process and by including the exchange of nucleons between the two partner of the reaction. The model solves, in an approximate way, the system of semi-classical coupled equations

$$i\hbar\dot{c}_\beta(t) = \sum_\alpha <\beta|V_{int}|\alpha> c_\alpha(t)e^{\frac{i}{\hbar}(E_\beta - E_\alpha)t + i(\delta_\beta - \delta_\alpha)} \qquad (1)$$

where c_β gives the amplitude for the system to be, at time t, in channels β. This system of coupled equation derives from the Schrödinger equation by expanding the the total wave function in term of channels wave functions describing the states belonging to the different asymptotic mass partitions. The interaction V_{int} is responsible for the excitation of the surface modes and for the transfer of nucleons. The time-dependence of the matrix elements is obtained by solving the Newtonian equations of motion for the relative motion that develops in a nuclear plus Coulomb field. For the nuclear part the model uses the empirical potential of Ref. [16] and for the Coulomb component the two point charges expression is used.

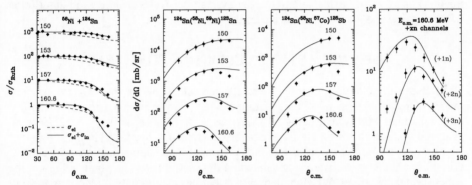

FIGURE 2. In the first column is shown the ratio to Rutherford of the elastic plus inelastic scattering (full line) in comparison with the experimental data. The pure elastic scattering (dash-line) is also shown. The following two columns display the calculated angular distributions of the inclusive one-neutron pickup and one-proton stripping channels. The last column displays the angular distributions for some multi-neutron transfer channels at the indicated bombarding energy.

The model is able to calculate the distribution of the total reaction cross section among all binary final states but it is not able to follow the evolution of the di-nuclear complex up to the formation of the compound nucleus. All the flux that reaches the inner pocket of the potential is considered to lead to capture. The component of the interaction V_{int}, responsible for the exchange of nucleons, is constructed from the one-particle transfer formfactors calculated by using the parametrization of ref. [17]. This parametrization has been tested for several target and projectile combinations and it has been found to provide a quite good description of one-nucleon transfer reactions. The component of the interaction V_{int}, responsible for the excitation of the surface modes, is constructed by using formfactors that are proportional to the r-derivative of the ion-ion potential. In Fig. 1, for the ^{58}Ni plus ^{208}Pb system, are shown the calculated angular distributions for the elastic plus inelastic channels in comparison with the angular distributions of several transfer channels. It is clear from the figure that at large angles the quasi-elastic angular distribution (that is a sum over elastic, inelastic and transfer channels) receive sizable contributions from transfer channels. These transfer channels are the dominant one at the higher bombarding energies.

Before drawing conclusions one has to demonstrate that the above calculations describe adequately the main properties of grazing reactions. To this purpose we analyse the ^{58}Ni + ^{124}Sn system. This is one of the few systems for which we have complete measurements of all reaction channels in a wide energy range [18, 19, 20, 21, 22, 23, 24, 25, 26] and for which a coupled channels analysis [27], that includes inelastic and transfer channels, has been performed. From Fig. 2 is clear that the model gives, for all energies, a good description of the elastic angular distributions (first column), of the angular distributions for the inclusive cross sections of one-neutron pick-up, of one-proton stripping channels and of some multi-neutrons transfer channels (last three columns). In the case of elastic scattering the good description shown by the full line is obtained by adding to the true elastic (shown with a dash-line) all the inelastic channels. The shown results are very similar to the one obtained in ref. [27] where a quantum mechanical coupled-channels formalism has been used. This indicates that the semiclassical approximation (that is easily extensible to heavier systems) provides a quite good description of the reaction and gives reassurance over the present results.

QUASIELASTIC EXCITATION FUNCTION

To produce the excitation function of ref. [9] one calculates, for the different systems, the angular distributions of all the reaction channels shown in Fig. 1 in step of 1 MeV of bombarding energy and sums all the cross section taken at $\theta_{lab} = 172^o$. For all analyzed systems the quasi-elastic excitation functions are displayed in the top row of Fig. 3. The barrier distributions $B(E)$ obtained from the excitation functions with a three-point formula energy derivative, are shown in the central row. The points represent the experimental data of ref. [9]. Both barrier distributions and excitation functions are very well described by the theory. Interpreting the centroid of the barrier distributions as the position of the effective barrier E_B^{eff} one sees that the couplings give rise to a lowering of the Coulomb barrier of the entrance channels by 4∼7 MeV depending on the systems. The full width-half-maximum of the barrier distributions, all of Gaussian-like shape, is of the order of 10∼12 MeV and is almost constant for all the systems.

The contribution of the particle transfer channels is shown in the bottom row of Fig. 3 as the ratio of the transfer cross section to the total quasi-elastic one. It is clear that transfer channels give sizable contributions in all the energy range and are the dominant processes at the higher energies. The contribution of more massive transfer channels is at this angle negligible. The last column of Fig. 3 shows the prediction of the model for the collision of ^{76}Ge plus ^{208}Pb system that, in ref. [11], has been proposed for cold fusion production of superheavy elements.

An alternative illustration of the role of particle transfer channels is obtained by looking at the evolution of the barrier distribution as a function of the channels that are contributing to the quasi-elastic cross section. If for quasi-elastic we consider all the final states that belong to the entrance channel mass partition (i.e. only elastic plus inelastic channels) we obtain the quasi-elastic excitation functions and barrier distributions shown with dash-lines in Fig. 3. It is clear from the figure that the quasi elastic barrier distribution depends on what we consider quasi-elastic. It is thus difficult to have a direct comparison between quasi elastic and fusion barrier distributions, differences

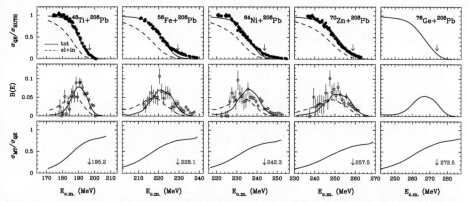

FIGURE 3. Quasi-elastic excitation function (top), barrier distribution (middle), ratio of transfer channels to the total quasi-elastic cross section (bottom). All the cross sections have been calculated at $\theta_{lab} = 172^o$. The down-arrows represent the Coulomb barrier for the entrance channels calculated with the empirical potential of ref. [16] and using a two points-charge Coulomb potential. The dash-lines are the results considering as quasi-elastic all the final channels belonging to the entrance channel mass partition. The data are from ref. [9]

may appear due to the different definition of what it is quasi-elastic. In pursuing these comparisons one should keep in mind that, while the barrier distribution extracted from fusion reactions gives an illustration of how the couplings modify the transmission coefficient through the barrier, the barrier distribution extracted from quasi-elastic scattering illustrates the modification of the reflection coefficient. For systems where fusion and quasi-elastic scattering exhaust most of the total reaction cross section, it is reasonable to expect equivalence between the two barrier distributions. This may not be the case for heavy system where the reaction is dominated by more complicated processes where the two reactants may overcome the Coulomb barrier but separate again with large energy losses and substantial exchange of mass and charge.

CONCLUSIONS

We have shown that the semi-classical theory offers a very powerful tool for the analysis of heavy-ion reactions. It allows a clear separation between relative motion variables and intrinsic degrees of freedom, surface vibrations and particle transfers. This separation is essential to pin down the relative role played by the different degrees of freedom in the large variety of nuclear processes. In this contribution we have seen that particle transfer channels give sizable contributions to the quasi-elastic cross sections in all the energy range. It has also been shown that the shape of the barrier distributions are related to the processes that are contributing to the quasi-elastic scattering.

REFERENCES

1. J.S. Lilley, B.R. Fulton, M.A. Nagarajan, I.J. Thompson and D.W. Banes, *Phys. Lett.* **151B**, 181 (1985).
2. M. Beckerman, *Phys. Rep.* **129**, 145 (1985).
3. H. Esbensen, *Nucl. Phys.* **A 352**, 147 (1981).
4. N. Rowley, G.R. Satchler and P.H. Stelson, *Phys. Lett.* **254B**, 25 (1991).
5. H. Timmers, J.R. Leigh, M. Dasgupta, D.J. Hinde, R.C. Lemmon, J.C. Mein, C.R. Morton, J.O. Newton, N. Rowley, *Nucl. Phys.* **438**, 190 (1995).
6. R. A. Broglia, G. Pollarolo and Aa. Winther, *Nucl. Phys.* **A 361**, 307 (1981).
7. G. Pollarolo, R. A. Broglia and Aa. Winther, *Nucl. Phys.* **A 406**, 369 (1983).
8. I.J. Thompson, M.A. Nagarajan, J.S. Lilley, B.R. Fulton *Phys. Lett.* **157B**, 250 (1985).
9. S. Mitsuoka, H. Ikezoe, K. Nishio, K. Tsuruta, S. C. Jeong, and Y. Watanabe, *Phys. Rev. Lett.* **99**, 182701 (2007).
10. R. A. Broglia, C. H. Dasso, G. Pollarolo and Aa. Winther. *Phys. Rep.* **48**, 351 (1978).
11. T. Ichikawa, A. Iwamoto, P. Møller and A.J. Sierk *Phys. Rev.* C **71**, 044608 (2005).
12. A. Winther, *Nucl. Phys.* **A 572**, 191 (1994).
13. A. Winther, *Nucl. Phys.* **A 594**, 203 (1995).
14. G. Pollarolo and A. Winther, *Phys. Rev.* C **62**, 054611 (2000).
15. A. Winther, GRAZING, computer program. $(http://www.to.infn.it/ \sim nanni/grazing)$
16. R. Broglia and A. Winther, *Heavy Ion Reactions*, Addison-Wesley Pub. Co., Redwood City CA, 1991
17. J.M. Quesada, G.Pollarolo, R.A. Broglia and A. Winther, *Nucl. Phys.* **A 442**, 381 (1985).
18. W.S. Freeman, et al., *Phys. Rev. Lett.* **50**, 1563 (1983).
19. K.T. Lesko, et al., *Phys. Rev. Lett.* **55**, 803 (1985) and *Phys. Rev.* C **34**, 2155 (1986).
20. F.L.H. Wolf, et al. *Phys. Lett.* **196B**, 113 (1987).
21. F.L.H. Wolf, *Phys. Rev.* C **36**, 1379 (1987).
22. R.R. Betts, et al., *Phys. Rev. Lett.* **59**, 978 (1987).
23. C.N. Pass, et al., *Nucl. Phys.* **A 499**, 173 (1989).
24. W. Henning, et al., *Phys. Rev. Lett.* **58**, 318 (1987).
25. A.M. Van den Berg, et al., *Phys. Rev. Lett.* **56**, 572 (1986) and *Phys. Rev.* C **37**, 178 (1988).
26. C.L. Jiang, et al., *Phys. Rev.* C **57**, 2383 (1998).
27. H. Esbensen, C.L.Jiang and K.E. Rehm *Phys. Rev.* C **57**, 2401 (1998).

Fusion of ^{48}Ca + ^{48}Ca Far Below the Barrier

F.Scarlassara[a], A.M.Stefanini[b], G.Montagnoli[a], R.Silvestri[b], L.Corradi[b],
S.Courtin[c], E.Fioretto[b], B.Guiot[b], F.Haas[c], D.Lebhertz[c], P.Mason[a],
S.Szilner[d]

[a] Dipartimento di Fisica "G.Galilei", Università di Padova and INFN Sezione di Padova, via Marzolo 8, I-35231 Padova, Italy.
[b] INFN, Laboratori Nazionali di Legnaro, I-35020 Legnaro (Padova), Italy
[c] IPHC, CNRS-IN2P3, Université de Strasbourg, F-67037 Strasbourg Cedex 2, France
[d] Ruder Boskovic Institute, HR-10002 Zagreb, Croatia

Abstract. In recent years, a puzzling pattern has been observed in fusion cross sections well below the Coulomb barrier, characterized as a departure from the exponential-like behavior predicted by standard coupled-channels models, known as fusion hindrance. We report on recent fusion measurements performed at the Laboratori Nazionali di Legnaro, in particular the ^{48}Ca + ^{48}Ca reaction down to the level of 0.6 µb. Unlike most recent results in this field, we do not observe the typical divergent behavior of the logarithmic derivative; but rather a sort of saturation, albeit at a larger value than predicted with a standard nucleus-nucleus potential.

Keywords: Heavy-ion fusion, sub-barrier cross sections, coupled-channels models.
PACS: 25.70.Jj, 24.10.Eq

INTRODUCTION

Up to a few years ago, studies in heavy-ion fusion cross sections concentrated on "fusion enhancement" at energies around and slightly below the Coulomb barrier, and a consensus arose on the role of channel-coupling between relative motion and intrinsic degrees of freedom, investigated in precise experiments (see e.g. [1] and references therein).

Calculations that reproduced the enhancement predicted an exponential-like behavior of the cross section at energies well below the barrier i.e. below the lowest coupled-channels barrier, irrespective of the coupling scheme or the nucleus-nucleus potential. However, the experimental data systematically drop well below such predictions for energies sufficiently below the barrier, hence the name "fusion hindrance" [2, 3] a topic of broad experimental and theoretical interest in recent years Experimentally it is a challenging task as one has to measure very low cross sections at (close to) zero degrees.

The two most popular ways to represent the hindrance: are the logarithmic slope $L(E) = d(E\sigma)/dE$ or the astrophysical S-factor, a concept borrowed from astrophysics, namely $S(E) = E\sigma \cdot \exp(2\pi\eta)$, where η is the Sommerfeld parameter. In several experimental data the S-factor shows a maximum, at variance with standard

CP1165, *Nuclear Structure and Dynamics '09*
edited by M. Milin, T. Nikšić, D. Vretenar, and S. Szilner
© 2009 American Institute of Physics 978-0-7354-0702-2/09/$25.00

expectations, corresponding to a logarithmic derivative $L(E) = L_s(E) = \pi\eta / E$. At the same energy E_S corresponding to a maximum of S(E), the experimental L(E) crosses the "constant S-factor" line $L_s(E) = \pi\eta / E$. A maximum in the S-factor or a crossing of the logarithmic derivative are the clearest indication of low-energy fusion hindrance.

The best interpretation of those results invokes a shallow inner nucleus-nucleus potential, that emerges naturally through the incompressibility of nuclear matter and the Pauli exclusion principle of the overlapping nuclear distributions [4,5]. Ichikawa, Hagino and Iwamoto have shown recently that below the threshold energy E_S, the inner turning point is more inside than the touching point [6], suggesting a link between the fusion process in heavier systems, where fusion is strongly suppressed and can only proceed adiabatically, and lighter systems where fusion is well described in the "sudden" approximation.

A threshold energy is naturally expected due to the negative fusion Q-value of most medium-heavy ion combinations, given by $E_{THR} = -Q$; however, the observed threshold energies (corresponding to the maximum S-factor) are much larger than expected from Q-value arguments: 20-40 MeV higher according to systematics. Medium-light systems play an important role in this context as they bridge the gap between heavier systems where clear cut evidence of fusion hindrance exists and the lighter systems of astrophysical interest, with large positive Q-values and no clear evidence of a maximum in the S-factor exists [6].

In the fusion of ^{28}Si + ^{30}Si, low-energy hindrance has been observed [8] despite the large positive Q-value (Q = +14.3 MeV) but the phenomenon of hindrance appears to be more nuanced: in ^{16}O + ^{208}Pb the logarithmic derivative is very steep but, unlike in the Ni-induced reactions, seems to saturate around L_{CS}, a behavior that can be reproduced within the ingoing-wave boundary condition only by using an unphysical diffuseness value $a = 1.65 \, \text{fm}$, incompatible with the higher energy data.

Slope saturation at low energy has also been observed in our recent measurement of ^{36}S + ^{48}Ca (Q = +7.6 MeV) [9]. The results are reminiscent of the light systems with no pronounced change of slope below the barrier, and show no evidence of a maximum in the S-factor or L_{CS} crossing.

A rather good fit to the data, from low to high energy, has been possible within the coupled-channels model [10] using a diffuseness parameter $a = 0.95 \, \text{fm}$.

EXPERIMENT AND RESULTS

In view of the above results, a comparison with ^{48}Ca + ^{48}Ca looked very interesting. This system has a slightly negative Q-value (Q= -3.0 MeV) but is otherwise very similar to ^{36}S + ^{48}Ca: both of them involve magic (or semi-magic) neutron-rich nuclei. Existing data for ^{48}Ca + ^{48}Ca [11] actually displayed a similar behavior, but the lowest measured cross section (150 µb) was too large to extrapolate into the low-energy region of interest. The aim of the experiment was extending the excitation function to lower energies, below 1 µb, but several high-energy points have also been carefully re-measured and resulted in a 20% reduction of the previously measured cross sections, a result that brings them in line with the systematics.

The measurement was performed at the INFN, Laboratori Nazionali di Legnaro. ^{48}Ca ions from the XTU Tandem were accelerated between 93 and 122 MeV, onto thin, 50 μg/cm^2 thick, ^{48}CaF$_2$ targets evaporated on 10 μg/cm^2 carbon backing.

The recoiling evaporation residues (ER) were detected at 0° (some of the lower energy points at 2° to reduce the background) after beam rejection by means of an electrostatic deflector. The ER were further discriminated against scattered beam particles by means of an Energy - Time of Flight telescope. Discrimination of the scattered beam particles is crucial as it determines the lower measurable cross sections. The telescope consisted of two micro-channel plate (MCP) time detectors and a final silicon detector (450 mm^2) for energy and timing. The redundancy of the time-of-flight information was very useful at the lowest energies as it helped discriminate against the background, due to random coincidence with scattered beam particles. Four silicon detectors, placed at about 16° to the beam, were used for beam monitoring and normalization purposes.

The fusion excitation function is shown in Fig. 1 (left, full circles), together with the ^{36}S + ^{48}Ca [9] (open squares), after scaling the energy in inverse proportion to the fusion barrier. The gray circles in the left panel refer to the previously measured data [11] scaled by a factor 0.8. The right panel represents the logarithmic slope, obtained from the experimental data by applying a gaussian smoothing of 1.5 MeV fwhm. Despite the large error bars of the lower energy points, the data seem to saturate around 2.5 MeV^{-1} and do not reach the L_{CS} limit (dotted line). Coupled-channels cal-

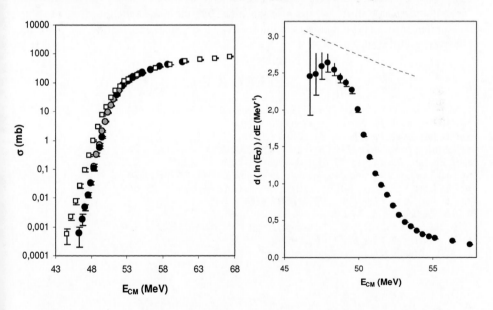

FIGURE 1. Fusion excitation functions (left) and logarithmic derivative (right) of .^{48}Ca + ^{48}Ca (full circles) and ^{36}S + ^{48}Ca (open squares). The ^{36}S + ^{48}Ca energies have been scaled to the ^{48}Ca + ^{48}Ca. In the left panel, the gray circles represent the previously published data [11] scaled by a factor 0.8.

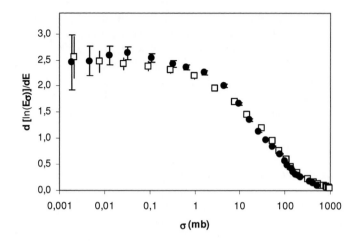

FIGURE 2. Logarithmic slope versus cross section: an unusual representation of the data. The data points for ^{48}Ca + ^{48}Ca (full circles) and ^{36}S + ^{48}Ca (open squares). scale nicely together.

-culations (not discussed here) can reasonably reproduce the experimental data provided a large diffuseness $a = 0{,}90$ fm is used. For more details see [12].

The ^{36}S + ^{48}Ca and the ^{48}Ca + ^{48}Ca systems seem to behave in a similar way, as is nicely demonstrated in Fig:2, which plots the logarithmic slope not as a function of energy but of cross section, a representation that seems to correct for the different "asymptotic energy shifts" of the reactions.. Despite the different sign of the Q-value, the similarity of the two data sets is impressive. A physical explanation of this result, however, is still wanting.

REFERENCES

1. M. Dasgupta, D.J. Hinde, N. Rowley and A.M. Stefanini, *Ann. Rev. Nucl. Part. Sci.* **48**, 401 (1998).
2. C.L: Jiang et al., *Phys. Rev. Lett.* **89**, 052704 (2002).
3. C.L. Jiang et al., *Phys. Rev.* C **73**, 014613 (2006).
4. S. Misicu, H. Esbensen, *Phys. Rev. Lett.* **96**, 112701 (2006).
5. H. Ebensen, *Phys. Rev.* C **77**, 054608 (2008).
6. T. Ichikawa, K. Hagino and A. Iwamoto, *Phys. Rev. Lett.* **99**, 192701 (2007).
7. C.L. Jiang, K.E. Rehm, B.B.Back and R.V.F. Janssens, *Phys. Rev.* C **75**, 01583 (2006).
8. C.L. Jiang et al., *Phys. Rev.* C **78**, 017691 (2008).
9. A.M. Stefanini et al., *Phys. Rev.* C **78**, 044607 (2008).
10. K. Hagino, N. Rowley and A.T. Kruppa, *Comp. Phys. Comm.* **123**, 143 (1999).
11. M. Trotta et al., *Phys. Rev.* C **65**, 011601 (2001).
12. A.M. Stefanini et al., *Phys. Lett.* B, accepted

Survivability and Fusability in Reactions Leading to Heavy Nuclei in the Vicinity of the N=126 Closed Shell

Roman N. Sagaidak

Flerov Laboratory of Nuclear Reactions, Joint Institute for Nuclear Research, Dubna 141980, Russia

Abstract. The macroscopic component of fission barriers for Po to Th nuclei around the N=126 closed neutron shell has been derived within the framework of the analysis of available fission and evaporation residues excitation functions using the conventional barrier passing (fusion) model coupled with the standard statistical model and compared with the predictions of various theoretical models.

Keywords: Fusion-fission and fusion-evaporation reactions, fusion probability, fission barriers.
PACS: 25.70.Jj, 24.60.Dr, 25.70.Gh, 27.80.+w

INTRODUCTION

Nuclear fission is well suited to study the dynamic properties and dissipative processes in cold and moderately excited nuclei. It is also a unique tool to explore level density and shell effects at an extreme deformation. Despite the significant progress in the fission studies, the isospin dependence of fission properties and, in particular, of fission barrier heights still remains an open problem. Using the adopted barrier data close to stability [1], theoretical fission model parameters are tuned, which provides a reasonable description of the fission barriers close to the stability line. However, a large spread is observed in the predictions of fission barrier heights in different model calculations (see, e.g. [2]). These discrepancies become especially undesirable in the r-process calculations for extremely neutron-rich nuclei, whose fission barriers determine the termination of the r-process by fission. Unfortunately, such neutron-rich nuclei will be not accessible for studies in the nearest experiments. Therefore, fission properties of exotic nuclei, e.g., their isospin dependence can be studied in an alternative (accessible) neutron-deficient region of the Nuclear Chart.

In the framework of the standard statistical model (SSM) evaporation residues (ER) and fission cross sections obtained in heavy ion induced reactions are determined by statistical properties of nuclei involved in the compound nucleus (CN) de-excitation and depend strongly on their fission-barrier heights. Early experiments on Th nuclei production in ^{40}Ar induced reactions [3] did not show any enhancement due to the shell stabilization around $N=126$, although a strong shell effect ~5 MeV exists in the ground-state masses of the nuclei. This result was in a sharp contrast to expectations

CP1165, *Nuclear Structure and Dynamics '09*
edited by M. Milin, T. Nikšić, D. Vretenar, and S. Szilner
© 2009 American Institute of Physics 978-0-7354-0702-2/09/$25.00

based on the SSM calculations using intrinsic level densities [4]. The lack of stabilization against fission around N=126 was explained by the effect of collective enhancement in the level density (CELD), which is different for spherical and deformed nuclei [5]. Later the expected stabilization against fission for spherical nuclei was not found again in the cross-sections data analysis for heavy nuclei produced in the relativistic U-projectile fragmentation. With the inclusion of CELD, the experimental data could be well described [6]. In addition, significant limitations for fusion in heavy systems were revealed in fission studies in reactions with massive nuclei. These limitations are caused by quasifission (QF), the process of reseparation of two fission-like fragments in a dinuclear system, which proceeds besides the CN formation. The quasifission effect reducing a complete fusion cross section was observed even in the asymmetric ^{32}S+^{182}W combination leading to the ^{214}Th* CN [7]. One should take into account both effects because they may distort the fission barrier estimates derived with the cross-sections data analysis using SSM approximations [4].

DATA ANALYSIS AND RESULTS

In the framework of the SSM [4] used for the analysis the fission barrier height is expressed as $B_{f}(L) = B_{f}^{m} - \Delta W_{gs}$, where the macroscopic component $B_{f}^{m} = k_{f} B_{f}^{LD}(L)$ could be varied with the scaling parameter k_{f} at the liquid-drop barrier B_{f}^{LD} depending on the angular momentum L [8]. In Fig.1, the results of the SSM analysis of ER and fission excitation functions obtained in the very asymmetric (free of QF) ^{16}O+^{204}Pb reaction [9] are applied to the similar data obtained in more symmetric ^{40}Ar and ^{124}Sn induced reactions leading to the same ^{220}Th* CN. As one can see, ER cross sections obtained in more symmetric combinations can be described either independently using the LD fission barrier scaling obtained with the data fit or using the same LD fission barriers as obtained for ^{16}O+^{204}Pb and introducing some fusion probability value P_{CN}.

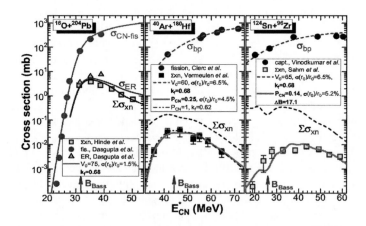

FIGURE 1. The application of the results of the SSM analysis (lines) of the ^{16}O+^{204}Pb cross sections obtained in [9] (left panel) to the analysis of similar data (symbols) obtained in more symmetric ^{40}Ar+^{180}Hf and ^{124}Sn+^{96}Zr combinations [4,10,11] (central and right panels) leading to the same ^{220}Th* CN.

380

The cross-section data analysis of the ^{16}O+^{208}Pb reaction leading to the ^{224}Th* CN shows noticeably larger value of k_f=0.78 [12]. Excitation functions obtained in other very asymmetric combinations leading to compound nuclei with masses $A_{CN}\geq$223 [13,14] are reproduced with about the same scaling parameter value, whereas the ^{4}He+ ^{226}Ra data [15] corresponding to the ^{230}Th* CN shows k_f=0.84 (see Fig.2).

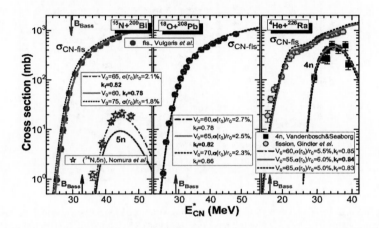

FIGURE 2. The results of the SSM analysis (lines) of the ^{14,15}N+^{209}Bi [13,14] (left panel), ^{18}O+^{208}Pb [14] (central panel) and ^{4}He+^{226}Ra [15] (right panel) evaporation and fission cross sections (symbols).

Going to the analysis of the neutron-deficient nuclei production in less asymmetric combinations, we encounter the effect of fusion suppression caused by QF, as observed in reactions induced by Ar. More severe limitations for fusion are observed in nearly symmetric reactions. Besides smaller values of the fusion probability at energies well above the nominal fusion (Bass) barrier, some additive ΔB to the potential barrier has to be applied in calculations to reproduce the ER cross sections at energies around the barrier, as shown for the ^{124}Sn+^{96}Zr data (right panel in Fig.1).

The macroscopic fission barriers derived with the SSM analysis of the cross-section data obtained for very asymmetric combinations leading to Rn, Ra and Th compound nuclei are shown in Fig.3. They are compared with the adopted ones [1] and with the model calculations [8,16-18]. The LD fission barrier scaling for Th nuclei shows that the transition from CN evaporation chains corresponding to deformed nuclei to the one of ^{220}Th* seems to be very sharp, whereas the macroscopic fission barriers should be a smooth function of N. An effective sharp decrease in fission barriers can be caused by the manifestation of the CELD effect for spherical nuclei, as proposed earlier [5,6]. The estimate of the effect strength shows the factor of 3–4 in the decrease in the production cross section for spherical Th nuclei. It is obtained with the ER excitation function calculation using the scaling parameter obtained by the extrapolation of k_f-values for deformed nuclei. Such estimate of the CELD effect is much smaller than one could expect according to the model prediction [6] that gives the factor of 7–12.

For Rn nuclei a stepwise discontinuity in the barriers at N=126, which follows from the analysis of the cross-section data obtained in Refs. [19-21], might have the CELD reason, but it requires experimental confirmations. Cross-section data obtained in reactions leading to Ra nuclei were earlier analyzed within the similar approach [22]. More detailed SSM analysis of reactions leading to Po nuclei is given in Ref. [23].

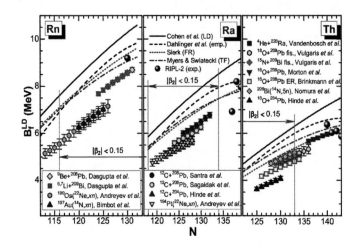

FIGURE 3. Macroscopic fission barriers derived with the SSM analysis (symbols) of the cross-section data obtained in very asymmetric reactions leading to Rn (left panel), Ra (central panel) and Th nuclei (right panel) in comparison to the adopted ones [1] and the model calculations [8,17-19] (lines).

REFERENCES

1. "Handbook for calculations of nuclear reaction data", IAEA (2006), http://www-nds.iaea.org/ripl-2/.
2. A. Mamdouh et al., *Nucl. Phys.* **A679**, 337 (2001).
3. D. Vermeulen, et al., *Z. Phys. A* **318**, 157 (1984).
4. W. Reisdorf, *Z. Phys. A* **300**, 227 (1981); W. Reisdorf and M. Schädel, *ibid.* **343**, 47 (1992).
5. A. V. Ignatyuk, et al., *Yad. Fiz.* **37**, 831 (1983) [*Sov. J. Nucl. Phys.* **37**, 495 (1983)].
6. A. R. Junghans, et al., *Nucl. Phys.* **A629**, 635 (1998).
7. J. G. Keller, et al., *Phys. Rev. C* **36**, 1364 (1987).
8. S. Cohen, F. Plasil, and W. J. Swiatecki, *Ann. Phys. (NY)* **82**, 557-596 (1974).
9. D.J. Hinde et al., *Phys. Rev. Lett.* **89**, 282701 (2002); M. Dasgupta et al., *Phys. Rev. Lett.* **99**, 192701 (2007).
10. H.-G. Clerc et al., *Nucl. Phys.* **A419**, 571 (1984); C.-C. Sahm et al., *ibid.* **A441**, 316 (1985).
11. A. M. Vinodkumar et al., *Phys. Rev. C* **74**, 064612 (2006).
12. R. N. Sagaidak et al., "Sub-Barrier Fusion in the HI+^{208}Pb Systems and Nuclear Potentials for Cluster Decay" in *Exotic Nuclear Systems*, edited by Z. Gácsi et al., AIP 802, NY, 2005, p. 61.
13. T. Nomura et al., *Nucl. Phys.* **A217**, 253 (1973).
14. E. Vulgaris et al., *Phys. Rev. C* **33**, 2017 (1987).
15. R. Vandenbosch and G. Seaborg, *Phys. Rev.* **110**, 507 (1958); J. E. Gindler et al., *Phys. Rev.* **136**, B1333 (1964).
16. M. Dahlinger, D. Vermeulen, and K.-H. Schmidt, *Nucl. Phys.* **A376**, 94 (1982).
17. A. J. Sierk, *Phys. Rev. C* **33**, 2039 (1986).
18. W. D. Myers and W. J. Świątecki, *Phys. Rev. C* **60**, 014606 (1999).
19. R. Bimbot et al., *J. Phys. (Paris)* **29**, 563 (1968).
20. A. N. Andreyev et al., *Nucl. Phys.* **A626**, 857 (1997).
21. M. Dasgupta et al., *Phys. Rev. C* **70**, 024606 (2004).
22. R. N. Sagaidak et al., *Phys. Rev. C* **68**, 014603 (2003).
23. R. N. Sagaidak and A. N. Andreyev, *Phys. Rev. C* **79**, 054613 (2009).

Density-Constrained TDHF Calculation of Fusion and Fission Barriers

A.S. Umar*, V.E. Oberacker*, J.A. Maruhn[†] and P.-G. Reinhard**

*Department of Physics and Astronomy, Vanderbilt University, Nashville, Tennessee 37235, USA
[†]Institut für Theoretische Physik, Goethe-Universität, 60438 Frankfurt am Main, Germany
**Institut für Theoretische Physik, Universitat Erlangen, D-91054 Erlangen, Germany

Abstract. The density-constrained time-dependent Hartree-Fock (DC-TDHF) theory is a fully microscopic approach for calculating heavy-ion interaction potentials and fusion cross sections below and above the fusion barrier. We discuss applications of DC-TDHF method to fusion, the effects of neutron transfer, and potential application to fission barrier calculations.

Keywords: TDHF, density constraint, fusion barrier, fission barrier
PACS: 21.60.Jz, 24.10.Cn

INTRODUCTION

In recent years major strides have been made in microscopic calculation of fusion barriers for heavy-ion reactions. The underlying theory for these calculations is the time-dependent Hartree-Fock (TDHF) approach coupled with a novel method of using a density constraint to obtain the corresponding TDHF trajectory on the multi-dimensional potential energy surface of the combined nuclear system [1, 2, 3, 4, 5]. Thus, TDHF provides the evolution of the nuclear shape (density) including all of the self-consistent dynamical effects present in the mean-field limit. The success of this approach together with a more in depth understanding of neutron transfer in TDHF collisions [6] leads us to believe that a similar approach may be used in calculating certain types of fission phenomena.

Fission following a heavy-ion collision is of great interest since this will be an important mechanism influencing the formation of super-heavy nuclei and fusion of heavy neutron-rich systems. While the initial configuration for prompt, isomeric, or fission induced by relatively low energy neutrons, is most likely the ground state or isomeric states in the vicinity of the ground state, the fission following a heavy-ion reaction may result from very different initial state. Recently, it has been shown that as we increase the temperature of the system the potential energy surfaces are completely changed in comparison to $T = 0$ case [7]. This is certainly the case in fission from a compound nucleus formed in heavy-ion fusion. However, in heavy-ion collisions the conversion of the relative kinetic energy into internal excitations is also likely to excite certain collective modes, thus, at least for low energies, not all of the available energy may be channeled into heating a compound nucleus but instead we may have a combined nuclear system with large collective excitations. A similar phenomenon is seen for light systems and leads to the well known nuclear molecular resonances. This type of *transition state* is an excellent candidate as an initial state for the quasi-fission process.

CP1165, *Nuclear Structure and Dynamics '09*
edited by M. Milin, T. Nikšić, D. Vretenar, and S. Szilner
© 2009 American Institute of Physics 978-0-7354-0702-2/09/$25.00

We believe that TDHF theory can be used to study the evolution of the colliding system in the light of the above discussion.

FUSION BARRIERS FROM TDHF

The DC-TDHF theory provides a comprehensive approach to calculating fusion barriers in the mean-field limit. As a fully microscopic, time-dependent, and self-consistent theory the only input is the effective interaction. Here, we only would like to emphasize that the theory has been successfully applied to calculating fusion cross-sections for ^{64}Ni+^{132}Sn [2, 3], ^{64}Ni+^{64}Ni [4], and ^{16}O+^{208}Pb [5] systems. For systems involving deformed nuclei the barriers for all orientations of the deformed system were averaged over, including a separate calculation of orientation probability due to Coulomb excitation. These calculations show that if the participating nuclei are well described by the effective interaction the resulting barriers yield a very good description of the fusion cross-sections.

Furthermore, we have also shown that even though TDHF is essentially a classical limit of a quantal system, neutron transfer present in the theory is in agreement with fully quantum mechanical model calculations. This is due to the fact that TDHF is a many-body theory and each single-particle state sees a different potential barrier, some ending up above their effective ion-ion barrier [6].

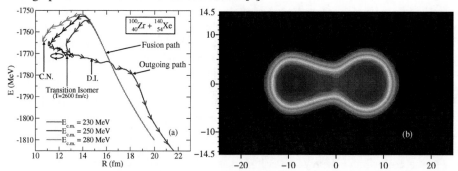

FIGURE 1. (a) Ion-Ion potential for the ^{100}Zr+^{140}Xe system at three different energies. (b) New shape isomer of ^{240}Pu found from the transition state.

FISSION AND TDHF

As we have discussed above, fission following a heavy-ion collision may be studied using the TDHF approach. We have first performed collisions of ^{100}Zr+^{140}Xe, which results in the extensively studied ^{240}Pu composite system. The choice of these nuclei were motivated by the most probable fragment mass and charge for the induced fission of ^{240}Pu. Along the TDHF trajectory density-constraint was applied to obtain the ion-ion potentials. In Fig. 1a we plot the result at three different bombarding energies. As we see at the lowest energy the system undergoes a deep-inelastic collision, while at the highest energy we observe complete fusion, as seen from the fact that the system is void of any collective motion for a long time (2500 fm/c). On the other hand, for the intermediate energy the system executes large collective oscillations with very little damping for the same time interval. This is what we call the *transition state* and indicates that the system is trapped in an isomeric minimum and executes collective motion. Following this idea we have started from a density along this path and performed an *unconstrained* minimization. The result was an isomer of the ^{240}Pu system shown in Fig 1b ($\beta_2 = 2.27$, $Q_{20} = 230b$, and $Q_{30} = -28b^{3/2}$). This isomer is far from the well known isomer of ^{240}Pu, which cannot be reached in such reactions. One feature of these transition states is that they fission with a very small collective boost, e.g. $e^{\pm\alpha q_{20}(\mathbf{r})}$. In this case the isomer fissions for boost energies of just a few MeV. By using the above boost and performing DC-TDHF calculations for the fissioning nuclei one can investigate the potential barriers in the vicinity of the isomeric minimum. This is shown in Fig. 2. In conclusion, we have shown that many interesting phenomena related to heavy-ion fusion and fission can be investigated using the methods developed based on the TDHF theory.

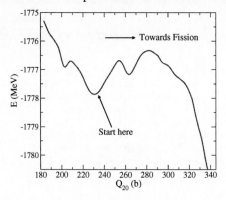

FIGURE 2. Potential barrier in the vicinity of the new shape isomer.

ACKNOWLEDGMENTS

This work has been supported by the U.S. Department of Energy under grant No. DE-FG02-96ER40963 with Vanderbilt University, and by the German BMBF under contracts No. 06 F 131 and No. 06 ER 124.

REFERENCES

1. A. S. Umar and V. E. Oberacker, *Phys. Rev.* C **74**, 021601(R) (2006).
2. A. S. Umar and V. E. Oberacker, *Phys. Rev.* C **74**, 061601(R) (2006).
3. A. S. Umar and V. E. Oberacker, *Phys. Rev.* C **76**, 014614 (2007).
4. A. S. Umar and V. E. Oberacker, *Phys. Rev.* C **77**, 064605 (2008).
5. A. S. Umar and V. E. Oberacker, *Eur. Phys. J.* A **39**, 243 (2009).
6. A. S. Umar and V. E. Oberacker, and J. A. Maruhn, *Eur. Phys. J.* A **37**, 245 (2008).
7. J. C. Pei, W. Nazarewicz, J. A. Sheikh, and A. K. Kerman, *Phys. Rev. Lett.* **102**, 192501 (2009).

Population Of Neutron Rich Nuclei Around ^{48}Ca With Deep Inelastic Collisions

D.Montanari[a], S.Leoni[a], G.Benzoni[a], N.Blasi[a], A.Bracco[a], S.Brambilla[a], F.Camera[a], A.Corsi[a], F.C.L.Crespi[a], B.Million[a], R.Nicolini[a], O.Wieland[a], L.Corradi[b], G.de Angelis[b], F.Della Vedova[b], E.Fioretto[b], A.Gadea[b, c], B.Guiot[b], D.R. Napoli[b], R.Orlandi[b, †], F.Recchia[b], R. Silvestri[b], A.M.Stefanini[b], R.P.Singh[b], S.Szilner[b,d], J.J.Valiente-Dobon[b], D.Bazzacco[e], E.Farnea[e], S.M.Lenzi[e], S.Lunardi[e], P.Mason[e], D.Mengon[e], G.Montagnoli[e], F.Scarlassara[e], C.Ur[e], G.Lo Bianco[f], A.Zucchiatti[g], M.Kmiecik[h], A.Maj[h], W.Meczynski[h] and G.Pollarolo[i]

[a] Università degli Studi di Milano and INFN, Sezione di Milano, Milano, Italy
[b] Laboratori Nazionali diLegnaro Padova, Italy
[c] IFIC, CSIC-University of Valencia, Spain
[d] RBI, Zagreb, Croatia
[e] Università degli Studi di Padova and INFN, Sezione di Padova, Padova, Italy
[f] Università degli Studi di Camerino and INFN, Sezione di Perugia, Camerino (Pg), Italy
[g] INFN, Sezione di Genova, Genova, Italy
[h] The Niewodniczanski Institute of Nuclear Physics, PAN, Krakow, Poland
[i] Università degli Studi di Torino and INFN, Sezione di Torino, Torino

Abstract. The reaction ^{48}Ca+^{64}Ni has been studied in the deep inelastic regime at approximately 6 MeV/A, by the CLARA-PRISMA setup. Angular distributions of the reaction cross sections have been obtained for the most intense reaction products, taking into account the response function of the magnetic spectrometer. The response of PRISMA has been calculated making use of a MonteCarlo simulation of the transport of the ions in the spectrometer, starting from known events distributions. For the one-nucleon transfer channels, the experimental data are in good agreement with predictions from a semiclassical multi-nucleon transfer model.

Keywords: Nuclear Transfer Reactions.
PACS: 21.1.Re, 21.10.Gv, 25.85.Ge, 27.90.+b, 25.70.Hi, 25.70.Bc, 29.30.Aj

INTRODUCTION

One of the most interesting issues in nuclear physics is the study of nuclei far from stability. High energy collisions between heavy ions have been proved to be a valuable tool to populate neutron rich nuclei. Furthermore, the knowledge of these reaction mechanisms provides information on nuclear potentials, spectroscopic factors, pair transfer and particle-vibration couplings, which are the starting points toward the study of the nuclear structure of exotic nuclei.

In this contribution we present a study, with the PRISMA-CLARA setup, of the population of moderately neutron rich nuclei around A = 50 via the deep inelastic reaction ^{48}Ca+^{64}Ni, at ~ 2.5 times above the Coulomb barrier. Particular emphasis is given to the study of the transport of the ions in the magnetic spectrometer (i.e. to the response function of PRISMA), which is crucial for a proper analysis of the experimental data. Experimental results have been obtained for the energy integrated angular distributions of the most intense reaction products, which have been compared with predictions from the semiclassical model GRAZING [1].

Experimental Setup and Data Analysis

The experiment has been performed at the Laboratori Nazionali di Legnaro, (Padova, Italy). The ^{48}Ca beam has been provided by the Tandem-Alpi accelerators at 270 MeV and impinged on a 0.98 mg/cm^2 thick ^{64}Ni target. The reaction products have been measured by the magnetic spectrometer PRISMA and the coincident γ-rays by the CLARA Ge-array [2]. The large acceptance magnetic spectrometer (≈ 80 msr) was placed at the grazing angle for this reaction, i.e. 20°, with an angular acceptance of ± 6°. Atomic species from -4p to +4p have been populated and many nucleons transfer channels, pick-up or stripping, have been observed.

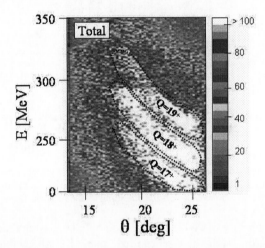

FIGURE 1. Bidimensional plots showing the (E_{kin},θ) distribution of an input uniform distribution in (E_{kin},θ,ϕ) of ^{48}Ca. The contributions of the three different charge states are visibile. A considerable suppression of the yield is present at small and large angles, due tio the different transport of the ions in the spectrometer.

To obtain a proper evaluation of the cross sections, one must know how the transport of the ions in PRISMA affects angular and energy distributions of the reaction products. This study has been performed by a MonteCarlo simulation of the transport in PRISMA using a uniform distribution in θ, φ and kinetic energy (E_{kin}) of ^{48}Ca ions as input. The trajectories of the ions into the spectrometer are calculated, event by event, transporting the ions up to the focal plane. This is done on basis of a detailed knowledge of the magnetic fields (including the fringing fields) and of the geometry of the instruments. The procedure employs a ray tracing code, which uses numerical integrators to determine the trajectories of individual ions through the electromagnetic fields, the latter being calculated by means of the Finite Element Method [3,4]. As a first step, the optics of the magnets has to be adjusted according to the experimental conditions under analysis. This requires a tuning of the intensity of both the quadrupole and dipole magnetic fields on basis of the position of the different charge states at the focal plane for ions with a given kinetic energy. After this, one million events, uniformously distributed in E_{kin}, θ and ϕ, have been generated in such a way to cover the entire experimental ranges, namely, E_{kin} between 150 and 400 MeV, θ between 10° and 40° and ϕ between -40° and 40° with respect to the beam axis. The charge states distributions have been first studied individually and then as a whole. Figure 1 shows the two dimensional spectrum, integrated over ϕ, of E_{kin} versus the polar angle θ for the events after the transport in the spectrometer.

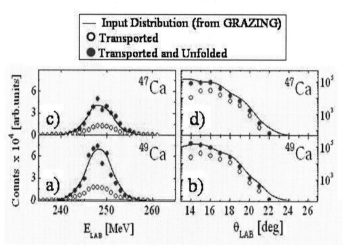

FIGURE 2. Monodimensional plots showing E_{kin}, left panels, and θ, right panels, distributions of ±1n channels. Solid lines are theoretical calculations, open circles are the transported events and dots are transported events after the unfolding by the response function.

The response function of PRISMA has been obtained as the ratio between the input uniform distribution and the transported one. Before being applyied to the experimental data, the response function has been tested on theoretical input distributions for the one-nucleon transfer channels, calculated with the semiclassical model GRAZING. Results for the unfolding procedures for the ±1n channels are reported in Figure 2. The left (right) panels represent projections on the kinetic energy (polar angle θ) for these two reaction channels. The solid line is the calculation given as input to the simulation, the open circles represent the distribution of the transported events and the full dots are the events unfolded by the response function previously obtained. The satisfactory agreement between the input calculations and the unfolded distributions gives us confidence to the quality of the unfolding procedure, which therefore has been applied to the data.

Figure 3 shows the experimental results for the energy integrated angular distributions of the most intense reaction channels in comparison with the theoretical calculations of the GRAZING code. Theoretical values are represented by the solid red line, while black dots are the experimental results. The central panel shows the experimental elastic cross section for ^{48}Ca obtained as described in ref. [5], i.e as a difference between the total kinetic energy loss (TKEL) measured in PRISMA and the one requiring a γ-coincidence. In this way, the elastic scattering distribution provides an absolute normalization for all the reaction products. A similar experimental analysis has been carried out on a number of transfer channels between -4p to +2p. This provides valuable experimental information for theoretical modelling, including additional reaction mechanisms such as evaporation and pair transfer.

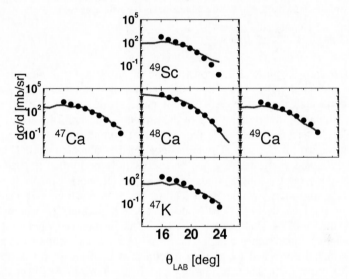

FIGURE 3. Angular distributions for the elastic channel, the ±1n channels and the ±1p channels. Experimental data, full dots, and theoretical calculation of the semiclassical model grazing, solid line, are in good agreement.

REFERENCES

1. A. Winther, *Nucl. Phys.* **A594**, 203(1995).
2. A. Gadea et al., *Eur. Phys. J.* **A20**, 193 (2004).
3. A. Latina, *PhD Thesis*, University of Torino (2004).
4. D.S. Burnett, *Finite Elements Analysis: From Concepts to Applications*, Addison-Wesley, New York, 1988.
5. S.Szilner et al., *Eur. Phys. J.* **C76**, 193 (2004).

The AGATA Demonstrator Array at Laboratori Nazionali di Legnaro: Status of the Project

E. Farnea

INFN Sezione di Padova, Padova, Italy
on behalf of the AGATA and the PRISMA Collaborations

Abstract. The AGATA Demonstrator Array is presently under installation at Laboratori Nazionali di Legnaro, Italy, where it will replace the CLARA array at the target position of the PRISMA magnetic spectrometer. In the present contribution, the details of the installation will be reviewed. Preliminary results from the first in-beam commissioning test will be given.

Keywords: γ-ray tracking, pulse shape analysis
PACS: 29.30.Aj, 29.30.Kv, 25.70.Hi

The AGATA project aims at the construction of an array of germanium detectors where a photopeak efficiency larger that 40% and a peak-to-total ratio larger than 50% are obtained through the use in real time of pulse shape analysis (PSA) and γ-ray tracking algorithms. In the initial phase of the project, a subset of the array, composed of five triple clusters and known as the AGATA Demonstrator, will operate at the Laboratori Nazionali di Legnaro. The initial goal of the campaign at LNL is to prove that indeed PSA and γ-ray tracking can be successfully performed in real time. The validation of the γ-ray tracking at LNL will occur on the most demanding conditions achievable in a low-energy stable-beam facility, i.e. with reactions with velocities of the γ-emitting products up to $\beta \approx 10\%$ and relatively high intensity beams. Once this is achieved, the AGATA Demonstrator will be used in coupled operation with the PRISMA magnetic spectrometer [1] to perform spectroscopic studies of moderately neutron-rich nuclei populated by grazing reactions as multi-nucleon transfer or deep inelastic collisions with the stable beams delivered by the Tandem-ALPI and the PIAVE-ALPI accelerator complex. Nevertheless, the coupling of the AGATA Demonstrator with complementary detectors, other than PRISMA, opens experimental possibilities beyond the aforementioned reactions, with direct, Coulomb excitation as well as fusion-evaporation reactions.

The compact arrangement of five triple clusters of AGATA is the optimal geometry of the array for the experimental activity of the Demonstrator foreseen at LNL. The detectors will be placed in front of the PRISMA spectrometer input aperture, therefore distributing the active volume in the best positions regarding Doppler broadening, considering that the nuclei of interest will be detected by PRISMA. The photopeak efficiency of the AGATA Demonstrator placed at the nominal 23.5 cm target-to-detector distance is roughly 3%. Given the low solid angle coverage, the Demonstrator can be also used at shorter target-to-detector distances, with an increase in photopeak efficiency (the value is approximately 7% when the detectors are moved by 10 cm closer to the target position) and without significant losses in the resolution and peak-to-total performance [2].

CP1165, *Nuclear Structure and Dynamics '09*
edited by M. Milin, T. Nikšić, D. Vretenar, and S. Szilner
© 2009 American Institute of Physics 978-0-7354-0702-2/09/$25.00

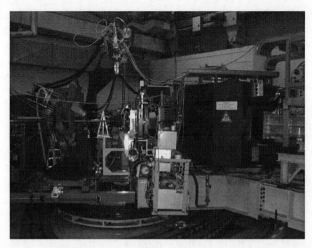

FIGURE 1.   Photo of the AGATA support structure coupled to the PRISMA magnetic spectrometer.

PRISMA is a large acceptance magnetic spectrometer designed to work with grazing reactions with the heavy ion beams provided by the LNL accelerator complex. The basic characteristics of PRISMA are described in ref.[1]. For the following discussion it is relevant to mention that PRISMA uses ion-tracking position-sensitive detectors to achieve the mass resolution. The tracking detectors provide the basic information to obtain the trajectory and velocity of the reaction products. According to the Monte Carlo simulations [3], up to velocities of approximately $v/c=10\%$, the intrinsic AGATA detector resolution is almost fully recovered if the recoil velocity module is measured with a relative precision better than 1%, and if the recoil velocity direction is measured with a precision better than $1°$. These values are actually well within the possibilities of PRISMA.

As mentioned before, the AGATA Demonstrator at LNL is strongly constrained by the experimental campaign coupled to PRISMA. The different elements of the mechanics and infrastructures are such that the coupling and experimentation with both setups are possible. In the following, the different elements of the infrastructure will be described in more detail.

The setup is intended to measure coincidences between the γ-rays detected by the AGATA Demonstrator and the reaction products detected by PRISMA. The AGATA Demonstrator is installed on a mobile platform, shown in Fig. 1, that will rotate together with PRISMA in such a way that reaction products, detected in the spectrometer focal plane in coincidence with the γ-rays, will have a forward trajectory with respect to the array in order to benefit from the lowest Doppler broadening. The detectors are hosted into a shell made out of 15 elementary AGATA flanges. The shell is positioned on a trolley which can slide on the same platform, rigidly linked to PRISMA. It is thus possible to easily modify the target-detector distance and to access the scattering chamber and to the instrumentation placed closed to the target. The whole support structure has minimal impact on the rotation of PRISMA. Taking into account the rest of the mechanical structure (beam line, scattering chamber), the angular range 41° to 110° is possible (with the Demonstrator placed at the closest distance from the target), while for the largest distance from the target the possible range is 37° to 110°. If one or more of the detectors are removed, PRISMA can be positioned at smaller angles, respectively $15° - 32°$ and $14° - 29°$.

The sliding seal scattering chamber previously used with CLARA was replaced with a new lower absorption chamber. The final part of the beam line as well was replaced by a specially designed telescopic beam line.

The digitizers and the Detector Support System (autofill and power supplies) will be hosted in racks mounted on the same platform used previously for the front-end electronics and power supplies of CLARA, which is rigidly linked to the structure of PRISMA. The 75 m long optical fibres connecting the digitizers to the pre-processing electronics are taken through the basement up to the Pre-Processing racks sitting in the AGATA-PRISMA control room. The racks for the pre-processing electronics are water cooled and fully protected for thermal and acoustic noise insulation. The pre-processing racks are connected to the PSA computer farm, placed in the main computer room of the Tandem building, via 15 m long fibres. The disk storage for AGATA is instead placed in another building, at the location of the TIER-2 centre for the CMS and ALICE experiments, and is connected to the computing farm via optical fibres.

THE FIRST IN-BEAM COMMISSIONING EXPERIMENT

Following the installation of the required infrastructures, an in-beam commissioning test was performed early in 2009. Besides providing the opportunity to verify the correct functioning of the several parts composing the front-end electronics and data acquisition chain of the AGATA Demonstrator, the main goal of the experiment was to measure the overall position resolution provided by the pulse shape analysis algorithms currently implemented in the system. As discussed thoroughly in [4], this parameter, which is critical in determining the overall performance of the array, can be determined with an in-beam measurement by comparing the effective energy resolution of the detectors (taking into account the proper Doppler correction) at several target-detector distances.

During this test, a beam of ^{30}Si with an energy of 70 MeV was fired onto a ^{12}C target, 200 μg/cm^2 thick. The prompt γ radiation was detected with the first AGATA asymmetric triple-cluster detector (ATC1), positioned as close as possible to 90° with respect to the beam direction. The detector was operated with a partial version of the AGATA Detector Support System, including the autofill system and the low-voltage power supply. The high-voltage was provided instead with a standard SY527 system by CAEN.

A full AGATA system was used to collect data to disk, including AGATA digitizers, Global Trigger and Synchronization, pre-processing electronics and a Narval system [5] running on several nodes. The DAQ chain was started using the Cracow GUI, communicating with the Narval system via a Run Control server. Dedicated Narval actors were performing pulse shape analysis and γ-ray tracking in real time. As this was the first time in which the full system was running, the original digitized signals were

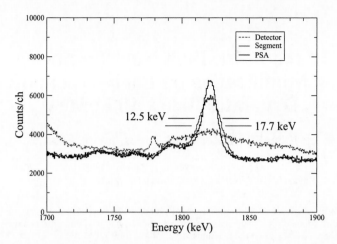

FIGURE 2. Comparison of Doppler-corrected spectra obtained under different conditions. See text for details.

stored to disk for later replay. Because of this, the overall counting rate was limited to approximately 1.5 kHz per crystal, corresponding to approximately 1 GB/min per crystal.

Through a preliminary data analysis it was possible to spot and to correct for some inconsistencies in the cabling and in the positioning of the detector. The Doppler-corrected spectra for the 1823 keV line of ^{40}K, namely the pn evaporation channel, are shown in Fig. 2. Since no ancillary devices to measure the recoil vector velocity on an event-by-event basis were used, Doppler correction was performed by assuming an average recoil velocity. Using the crystal centre-of-gravity to perform Doppler correction, no peak can be clearly identified corresponding to the 1823 keV transition. This case corresponds to discarding the information from the segments, in other words, to treat the detectors as "conventional" ones. The peak FWHM is instead 17.7 keV using the segment centre-of-gravity to perform Doppler correction. A peak FWHM of 12.5 keV is obtained when a full PSA algorithm is applied, in this case a grid search algorithm [6] using a signal basis calculated with the JASS code [7]. This last value is compatible with estimates considering realistic values for the dispersion in recoil velocity module and direction.

REFERENCES

1. A. Latina, et al., *Nucl. Phys.* **A 734**, E1 (2004).
2. http://agata.pd.infn.it/documents/simulations/demonstrator.html
3. F. Recchia, et al., LNL-INFN(REP)-202/2004, 160 (2004).
4. F. Recchia, In-beam test and imaging capabilities of the AGATA prototype detector, PhD Thesis, University of Padova (2008).
5. X. Grave, et al., 14th IEEE-NPSS 1 (2005).
6. R. Venturelli and D. Bazzacco, LNL-INFN(REP)-204/2005, 220 (2005).
7. M. Schlarb, et al., GSI Report 2009-1, 232 (2009).

Impact of the In-medium Nucleon-nucleon Cross Section Modification on Early-reaction-phase Dynamics Below 100*A* MeV

Z. Basrak*, M. Zorić*, P. Eudes† and F. Sébille†

*Ruđer Bošković Institute, Zagreb, Croatia
†Subatech, EMN-IN2P3/CNRS-Université de Nantes, Nantes, France

Abstract. With a semi-classical transport model studied is the impact of the in-medium *NN* cross section modifications on the early energy transformation, dynamical emission and quasiprojectile properties of the Ar+Ni and Ni+Ni reactions at 52, 74 and 95(90)*A* MeV.

Keywords: heavy-ion reactions, NN scattering in-medium; medium modifications
PACS: 24.10.-i; 25.70.-z; 21.65.-f

In this work we extend our study of the early dynamical and compact phase of heavy ion reactions (HIR) [1] by complementing it with an investigation of a possible influence of the in-medium modifications of nucleon-nucleon (*NN*) cross section. In our earlier investigations we have shown that HIR from the Fermi energy to about 100*A* MeV are strongly dominated by the mid-rapidity emission, especially in central collisions, a component which is emitted early during the dynamical reaction phase. This prompt and copious dynamical emission (DE) is proportional to the impact parameter *b* and evacuates a large amount of available system energy, a result confirmed experimentally [2]. We have also studied the early transformation of the initial relative motion of the entrance reaction channel into other forms of energy in particular to its main components which are responsible for DE, namely, heat E_{th} and compression E_{compr}. Above HIR simulations have been performed with the semi-classical Landau-Vlasov model and the momentum-dependent Gogny interaction D1-G1 has been used [3]. In this model, σ_{NN} is the free *NN* cross section σ^f with its usual energy and isospin dependence. In our earlier work [1] the σ^f was considered only. Here, we investigate the in-medium effects, i.e. how the modification of σ_{NN} influences the early compact and dominantly dynamical phase of HIR. The change is taken into account by multiplying σ^f by a corrective constant factor *F*, $\sigma_{NN} = F\sigma^f$. In other words, we examine how the E_{th} and E_{compr} evolve with the reaction time and how they and the DE and quasiprojectile (QP) behave as a function of the factor *F*. In a fine step in *b* from central to peripheral collisions we investigate the $^{36}Ar + ^{58}Ni$ and $^{58}Ni + ^{58}Ni$ reactions at 52, 74, and 95*A* MeV (52, 74, and 90*A* MeV for the latter reaction). The factor *F* has been varied between 0.5 and 2.

The results of the simulations show that the time evolution of heat E_{th} and compression E_{compr} during the early dynamical reaction phase displays maxima at all incident energies. A maximum is a rather robust observable. These maxima are function of the factor *F*. Their dependence on σ_{NN} is the most important in the head-on regime. For other impact parameters, it is weaker and vanishes for grazing collisions. It is worth not-

CP1165, *Nuclear Structure and Dynamics '09*
edited by M. Milin, T. Nikšić, D. Vretenar, and S. Szilner
© 2009 American Institute of Physics 978-0-7354-0702-2/09/$25.00

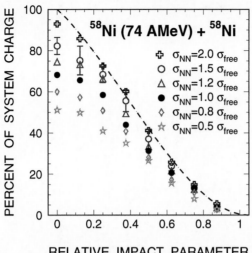

FIGURE 1. Simulation results for the evolution of dynamical emission as a function of the reduced impact parameter for the Ni+Ni reaction at $74A$ MeV as a function of σ_{NN}. Different symbols used denote different values of F: open crosses for $F=2$, circles $F=1.5$, triangles $F=1.2$, filled circles $F=1$, diamonds $F=0,8$, and open stars $F=0,5$. The curve is due to a simple estimate of the contribution of participant matter assuming a purely geometrical overlap of the two interpenetrating spheres.

ing that E_{th} increases with σ_{NN}, whereas E_{compr} decreases. As expected, the value of the energy maxima increases with the entrance channel energy. Maxima are larger for E_{th} showing that the compression is more prompt in developing and fading than the heat is. One also observes that E_{th} reaches its maximal value a few fm/c before E_{compr} reaches its. It should be noticed that the instant at which these maxima are reached is almost independent of σ_{NN} and decreases with increasing relative velocity of colliding nuclei.

The amount of dynamically emitted particles also displays a maximum which regularly depends on the factor F. The DE is proportional to the value of σ_{NN}, while mass of the primary QP is inversely proportional to F. Also, a regular function of F is the QP maximal center-of-mass angle and the QP velocity. The nice regularity of these observables on F and its experimental accessibility may be used to constrain the factor by which the free NN cross section σ^f has to be modified in order to account for the modifications of the elementary NN scattering process caused by the presence of nuclear surroundings. This is corroborated in Fig. 1 for DE and the Ni+Ni reaction at $74A$ MeV.

REFERENCES

1. P. Eudes, Z. Basrak, and F. Sébille, *Phys. Rev.* C **56**, 2003 (1997); F. Haddad, P. Eudes, Z. Basrak, and F. Sébille, *Phys. Rev.* C **60**, 031603 (1999); P. Eudes, and Z. Basrak, *Eur. Phys. J.* A **9**, 207 (2000); I. Novosel, Z. Basrak, P. Eudes, F. Haddad, and F. Sébille, *Phys. Lett.* **625B**, 26 (2005).
2. D. Doré et al., *Phys. Lett.* **491B**, 15 (2000); T. Lefort et al., *Nucl. Phys.* A **662**, 397 (2000).
3. F. Sébille, G. Royer, C. Grégoire, B. Remaud, and P. Schuck, *Nucl. Phys.* A **501**, 137 (1989).

WEAK INTERACTIONS AND ASTROPHYSICS

The Influence of Nuclear Structure and Reactions in Astrophysics

K. E. Rehm

Physics Division
Argonne National Laboratory
9700 South Cass Av.
Argonne, IL, 60439, USA

Abstract. Nuclear structure and nuclear dynamics play an important role in astrophysics. The occurrence (or non-occurrence) of a certain state at a given excitation energy can change the reaction rate, and thus the abundance of the elements, by many orders of magnitude. An example is the existence of an excited 0^+ state in ^{12}C at 7.654 MeV, right in the middle of the Gamow window. This is a crucial step for the formation of ^{12}C via the triple-α reaction. Similarly the lack of a state with natural parity in ^{20}Ne at $E_x \sim 5$ MeV prevents the conversion of chemically active ^{16}O into chemically inert ^{20}Ne. In this contribution some recent accomplishments relevant to studies of nuclear structure and dynamics in astrophysics will be discussed.

Keywords: Nuclear Astrophysics
PACS: 26.20.-7, 26.30.-k, 26.50.+x

INTRODUCTION

The year 2009 has been declared by UNESCO as the 'Year of Astronomy' because of two events that happened 400 years ago. It was in 1609 that Galileo Galilei pointed his newly built telescope to the stars and saw the movements of Jupiter's moons, which gave a severe blow to the heliocentric picture of our world. In the same year, Johannes Kepler published his book "Astronomia Nova", where, based on observations of the planet Mars, he discovered his first two laws. This attempt to describe astronomical observations through the laws of physics can be seen as the beginning of astrophysics.

While many subfields of physics contribute to the understanding of astronomical phenomena, it is nuclear physics which is needed to explain the energy production as well as the generation of the elements in the universe. The lightest elements, hydrogen and helium, together with small amounts of lithium were produced in the Big Bang. The other heavier elements which make life on Earth possible are 'cooked' in the stellar cauldrons, known as stars.

Nuclear astrophysics has seen a renaissance in the last decades. Reactions with small cross sections, which have been strongly hampering progress in the past, are now measured using high-efficiency detector arrays and beams from high-intensity accelerators. Background reactions have been suppressed by studying the important reactions in laboratories deep underground; reactions between nuclei that exist only for a fleeting moment on earth can now be studied with beams from the new radioactive

CP1165, *Nuclear Structure and Dynamics '09*
edited by M. Milin, T. Nikšić, D. Vretenar, and S. Szilner
© 2009 American Institute of Physics 978-0-7354-0702-2/09/$25.00

beam facilities. There are thousands of reactions occurring in stellar explosions. Most of their rates have not been experimentally determined, leaving this a fruitful field for future studies.

ALPHA BURNING IN RED GIANT STARS

Nuclear structure and dynamics influence astrophysics in many ways, and it is impossible to pay tribute to all recent experiments and calculations. Some of these are discussed in other contributions to this conference [1-10]. In this contribution I want to discuss two reactions, which, despite their long history, are still not known with sufficient accuracies and are therefore at the center of present-day theoretical and experimental nuclear astrophysics.

One of them, the formation of ^{12}C through the so-called triple-α reaction occurring in red giant stars, has been studied for more than 50 years. It led to the prediction [11] and later to the discovery [12] of the first excited 0^+ state (the 'Hoyle' state) in ^{12}C. The structure of this state is presently investigated by various theoretical calculations [13]. A lingering question in this field is the possible existence of a rotational 2^+ state built on the 0^+ 'Hoyle' state. The predictions for the excitation energy of this state vary widely from about 8.4 MeV [14] to 11.5 MeV [15]. (see also the contributions by H. Fynbo and M.'Freer [8, 10] to this conference). In astrophysics, such a 2^+ state, which is assumed to exist in the European Nuclear Astrophysics compilation NACRE [16], will increase the reaction rate at higher temperature ($T_9 \sim 1$) by up to an order of magnitude. The history of the search for this state is covered in more detail in H. Fynbo's contribution [8].

We have performed a new search for this state through the study of the β-delayed α-decay of ^{12}B and ^{12}N using a position-sensitive gas-filled calorimeter. Compared to other techniques (e.g. inelastic scattering), fewer states in ^{12}C are populated in the β-decay.

The detector in the experiment is based on a twin-ionization chamber, which has been extensively used for the study of fission fragments [17]. From the anode and Frisch-grid signals, the energy and angle of a binary fission event can be determined. The setup used in our studies is shown in Fig.1. A ^{12}B (E=70 MeV) or ^{12}N (E=90 MeV) beam produced with the In-Flight technique [18] via the d(^{11}B,^{12}B)p or ^{3}He(^{10}B,^{12}N)n reaction, respectively, is slowed down in a gas-filled attenuator such that the particles of interest are stopped in the center of the twin-ionization chamber. Because of the short half-life of 20 ms (^{12}B) or 10ms (^{12}N), the stopped particles diffuse only a few mm before they decay through the emission of an electron (or positron) and, for higher excited states (E_x=7.275 MeV), into three α's. In the experiment, the ^{12}B and ^{12}N particles enter the calorimeter for a time period of about 2 half lives. The beam is then stopped for another two half lives, and the decay of the ^{12}C particles in the calorimeter is measured during the beam-off period.

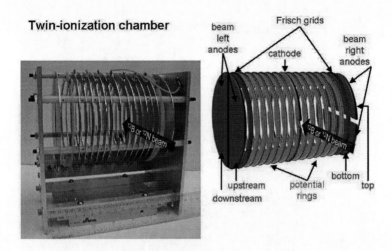

FIGURE 1. Schematic (right) and photograph (left) of the twin-ionization chamber used for the measurement of the β-delayed α-decay of ^{12}B or ^{12}N. The total energy of the decay is obtained from the sum of the beam-left and beam-right anodes, while the emission angle is obtained from the Frisch-grid signal.

The easiest signal that can be analyzed is the total decay energy obtained from the beam-left and beam-right anode signals. A two-dimensional plot of these events obtained with a ^{12}B beam is shown in Fig.2. Events originating from the decays of the 'Hoyle' state at E_x=7.654 MeV, the 0^+ state at E_x=10.3 MeV and the 1^+ state at E_x=12.71 MeV are clearly visible.

Since the anodes of the ionization chambers are subdivided into four individual sections, the decay of ^{12}C into three α-particles can be identified. For this purpose Monte Carlo simulations are presently being performed [19]. At the present stage of the analysis no indication of a 2^+ state in the excitation energy region between 8.5 and 11.5 MeV is observed.

Once ^{12}C has been formed, the capture of another α-particle via the ^{12}C$(\alpha,\gamma)^{16}$O reaction leads to the formation of ^{16}O, another nuclide that is crucial for life on Earth. This reaction, sometimes called the 'holy grail of nuclear astrophysics', has been studied for many years. Despite the use of high-intensity beams and elaborate detector systems [20], the small cross sections (~ 10 fb) at astrophysical energies ($\sim$300 keV) have not allowed an accurate measurement of this important reaction rate.

Recently, a new approach has been suggested that makes use of the favorable phase space conditions in the time-inverse ^{16}O$(\gamma,\alpha)^{12}$C reaction [21]. As with (p,γ) reactions

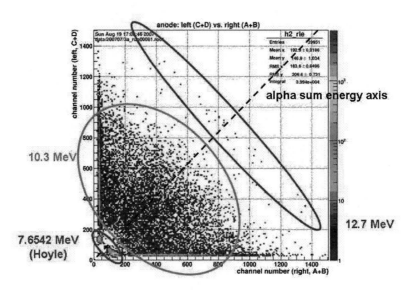

FIGURE 2. Spectrum of the beam-left and beam-right signals obtained from the decay of [12]B into three α-particles.

and Coulomb dissociation a measurement of the time-inverse reaction leads to an increase in cross section by a factor of ~100. In this new approach gamma rays from a Free Electron Laser (FEL) bombard a CO_2-filled time-projection chamber, which is used to identify the outgoing α and ^{12}C particles. First test experiments with this technique have been performed [22]. The intensities at existing FEL's, however, so far restrict the lowest energy that can be reliably studied to about E_x=8.7 MeV, i.e. 500 keV higher than what has been measured in the standard $^{12}C(α,γ)^{16}O$ experiments performed so far [20]. While such an experiment will give valuable information about the systematic uncertainties, it is very important to push these measurements further towards even lower energies [23].

An increase in luminosity by another factor of 10^3 to 10^4 can be achieved by replacing the ^{16}O gas with a ^{16}O-containing liquid. For this a 'supercritical bubble detector' is presently being developed at Argonne. Detectors of this type have been successfully used for Dark Matter searches [24], where they detect the recoil particles originating from collision with weakly interacting massive particles. Besides their higher density, these detectors have a very low detection efficiency for minimum-ionizing particles. For the COUPP dark matter detector a γ suppression of more than 10^{10} has been achieved

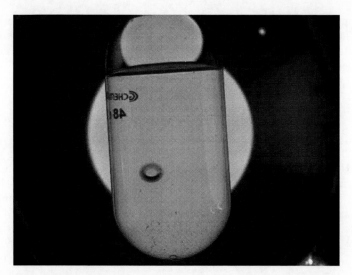

FIGURE 3. Photograph of a neutron-induced bubble in a 48 ml supercritical detector.

[24]. Furthermore, the operating conditions can be chosen so that only heavier, strongly ionizing particles like ^{4}He and ^{12}C are detected.

For the study of the ^{16}O$(\gamma, \alpha)^{12}$C reaction, a $\sim$30-cm-long vessel filled with water at supercritical conditions (T$\sim$190 C and P$\sim$1 atm) will be bombarded by gamma rays from a FEL. Depending on the energy of the incoming gamma rays, the breakup of ^{16}O leads to the formation of 0.5-1 MeV α's and 0.15-0.4 MeV ^{12}C ions. The energy loss of these particles in the supercritical liquid leads to formation of a bubble, which is detected by a high-speed video camera. A picture of a bubble induced by recoil particles generated by neutrons from a Cf fission source is shown in Fig.3. After the detection of a bubble the pressure in the detector is immediately increased to avoid an uncontrolled boiling of the supercritical liquid. Details regarding supercritical bubble detectors can be found in Ref.[24]. In order to eliminate reactions induced by neutrons or gamma rays on deuterium, ^{17}O and ^{18}O the abundance of these target contaminants needs to be reduced by about a factor of $\sim$10^6. Despite this requirement the increase in luminosity and the high suppression of gamma rays make supercritical bubble detectors ideally suited for extending the measurements of the ^{16}O$(\gamma, \alpha)^{12}$C reaction to lower energies.

SUMMARY

Nuclear astrophysics has seen a strong revival in recent years. This revival is triggered by advances in observational astronomy, covering all wavelengths from radio waves to high energy γ-rays. Making use of high-intensity stable beam accelerators, sometimes located deep underground to reduce background from cosmic rays, or using beams of short-lived radioactive nuclei together with high-efficiency detector systems, we have started to unravel many of the processes occurring in quiescent and explosive stellar burning. These experiments have provided us with a first glimpse into this exciting area of nuclear physics. Next-generation facilities and detectors are needed to give us a quantitative understanding of these phenomena.

ACKNOWLEDGMENTS

I want to thank my collaborators on the ^{12}B,^{12}N (N. Patel, C. Deibel, J. Greene, U. Greife, D. Henderson, C. L. Jiang, B. Kay, S. Marley, M. Notani, R. Pardo, X. D. Tang) and ^{16}O(γ,α) experiments (C. Ugalde, A. Champagne, B. Digioveni, D. Henderson, R. Holt, A. Sonnenschein and A. Tonchev) for their help with these experiments. This work was supported by the US Department of Energy, Office of Nuclear Physics, under contract No. DE-AC02-06CHl1357 and by the NSF Grant No. PHY -02-16783 (Joint Institute for Nuclear Astrophysics).

REFERENCES

1. N. Aoi, contribution to this conference.
2. A. Jokinen, contribution to this conference.
3. C. Broggini, contribution to this conference.
4. E. B. Norman, contribution to this conference.
5. M. Heil, contribution to this conference.
6. M. Kowalska, contribution to this conference.
7. R.G. Pizzone, contribution to this conference.
8. H.O.U. Fynbo, contribution to this conference.
9. M. Alcorta, contribution to this conference.
10. M. Freer, contribution to this conference.
11. F. Hoyle, *Astrophys. J. Suppl.* **1**, 121 (1954).
12. D.N.F. Dunbar, et al., *Phys. Rev.* **92**, 649 (1953).
13. M. Chernyk, et al., *Phys. Rev. Lett.* **98**, 032501 (2007) and references therein.
14. R.R. Betts, *Nuovo Cimento* A**110**, 975 (1997).
15. B. John, et al., *Phys. Rev.* C **68**, 014305 (2003).
16. C. Angulo, et al., *Nucl. Phys.* **A 656**, 3 (1999).
17. C. Budtz-Jorgenson, et al., *Nucl. Instrum. Meth.* A**258**, 209 (1987).
18. B. Harss, et al., *Rev. Sci. Instrum.* **71**, 380 (2000).
19. N. Patel, PhD thesis, 2009, Colorado School of Mines, and to be published.
20. M. Assuncao, et al., *Phys. Rev.* C **73**, 055801 (2006) and references therein.
21. M. Gai, et al., arXiv:nucl-ex/0504003.
22. M. Gai, contribution to the SARAF workshop 2008, http://www.phys.huji.ac.il/SARAF/index.htm
23. Y. Xu, et al., *Nucl. Instrum. Meth.* A**581**, 866 (2007).
24. J. Collar, et al., COUPP project proposal, http://www-coupp.fnal.gov.

LUNA and the Sun

Carlo Broggini (the LUNA Collaboration)

INFN, via Marzolo 8, I-35131 Padova

Abstract. LUNA, Laboratory for Underground Nuclear Astrophysics at Gran Sasso, is studying thermonuclear reactions down to the energy of the stellar nucleosynthesis. The results obtained will be presented and their impact on the determination of the solar neutrino spectrum and on the measurement of the Sun core metallicity will be discussed.

Keywords: proton-proton chain, CNO cycle, solar neutrinos, solar metallicity
PACS: 25.60.Pj; 26.20.+f; 26.65.+t

INTRODUCTION

Nuclear reactions that generate energy and synthesize elements take place inside the stars in a relatively narrow energy window: the Gamow peak. In this region, which is far below the Coulomb energy, the reaction cross-section $\sigma(E)$ drops almost exponentially with decreasing energy E:

$$\sigma(E) = \frac{S(E)}{E} exp(-2\pi\eta) \tag{1}$$

where $S(E)$ is the astrophysical factor and η is given by $2\pi\eta = 31.29 Z_1 Z_2 (\mu/E)^{1/2}$. Z_1 and Z_2 are the nuclear charges of the interacting particles in the entrance channel, μ is the reduced mass (in amu), and E is the center of mass energy (in keV).

The extremely low value of the cross-section, ranging from pico to femto-barn and even below, has always prevented its measurement in a laboratory at the Earth's surface, where the signal to background ratio is too small mainly because of cosmic ray interactions. Instead, the observed energy dependence of the cross-section at high energies is extrapolated to the low energy region, leading to substantial uncertainties.

In order to explore this new energy window we have installed two electrostatic accelerators underground in the Gran Sasso laboratory: a 50 kV accelerator [1] and a 400 kV one [2]. The qualifying features of both the accelerators are a very small beam energy spread and a very high hydrogen and helium current even at low energy. The mountain provides a natural shielding equivalent to at least 3800 meters of water which reduces the muon and neutron fluxes by a factor 10^6 and 10^3, respectively. The γ ray flux is like the surface one, but a detector can be more effectively shielded underground due to the suppression of the cosmic ray induced background.

CP1165, *Nuclear Structure and Dynamics '09*
edited by M. Milin, T. Nikšić, D. Vretenar, and S. Szilner
© 2009 American Institute of Physics 978-0-7354-0702-2/09/$25.00

THE $^3He(^3He,2p)^4He$ REACTION

The initial activity of LUNA has been focused on the ^{3}He(^{3}He,2p)^{4}He cross section measurement within the solar Gamow peak (16-28 keV). Such reaction is a key one of the proton-proton chain. A resonance at the thermal energy of the Sun was suggested long time ago [3] [4] to explain the observed ^{8}B solar neutrino flux: it would decrease the relative contribution of the alternative reaction ^{3}He(α,γ)^{7}Be, which generates the branch responsible for ^{7}Be and ^{8}B neutrino production in the Sun. A narrow resonance with a peak S-factor 10-100 times the value extrapolated from high energy measurements could not be ruled out with the pre-LUNA data (such an enhancement is required to reduce the ^{7}Be and ^{8}B solar neutrinos by a factor 2-3). As a matter of fact, ^{3}He(^{3}He,2p)^{4}He cross section measurements stopped at the center of mass energy of 24.5 keV (σ=7$\pm$2 pb)[5], just at the upper edge of the thermal energy region of the Sun.

Briefly, the LUNA 50 kV accelerator facility consisted of a duoplasmatron ion source, an extraction/acceleration system, a double-focusing 90° analyzing magnet, a window-less gas-target system and a beam calorimeter. The beam energy spread was very small (the source spread was less than 20 eV, acceleration voltage known with an accuracy of better than 10^{-4}), and the beam current was high even at low energy (about 300 μA). Eight thick (1 mm) silicon detectors of 5x5 cm^2 area were placed around the beam inside the target chamber, where there was a constant ^{3}He gas pressure of 0.5 mbar.

The simultaneous detection of 2 protons has been the signature which unambiguously identified a ^{3}He(^{3}He,2p)^{4}He fusion reaction (detection efficiency: 5.3$\pm$ 0.2%, Q-value of the reaction: 12.86 MeV). Its cross section varies by more than two orders of magnitude in the measured energy range. At the lowest energy of 16.5 keV, it has the value of 0.02 pb, which corresponds to a rate of about 2 events/month, rather low even for the "silent" experiments of underground physics.

The LUNA result [6] showed that the ^{3}He(^{3}He,2p)^{4}He cross section does not have any narrow resonance within the Gamow peak of the Sun. Consequently, the astrophysical solution of the ^{8}B and ^{7}Be solar neutrino problem based on its existence has been ruled out. With ^{3}He(^{3}He,2p)^{4}He LUNA provided the first cross section measurement of a key reaction of the proton-proton chain at the thermal energy of the Sun. In this way it also showed that, by going underground and by using the typical techniques of low background physics, it is possible to measure nuclear cross sections down to the energy of the nucleosynthesis inside stars.

THE BERYLLIUM AND BORON NEUTRINOS FROM THE SUN

^{3}He(α,γ)^{7}Be (Q-value=1.586 MeV) is the key reaction for the production of ^{7}Be and ^{8}B neutrinos in the Sun since their flux depends almost linearly on the ^{3}He(α,γ)^{7}Be cross section. The cross section can be determined either from the detection of the prompt γ ray or from the counting of the produced ^{7}Be nuclei. The latter requires the detection of the 478 keV γ due to the excited ^{7}Li populated in the decay of ^{7}Be.

Both methods have been used in the past to determine the cross section in the energy range E$_{c.m.}$ $\geq$ 107 keV but the S$_{3,4}$ extracted from the measurements of the induced ^{7}Be activity are 13% higher than the values obtained from the detection of the prompt γ-rays.

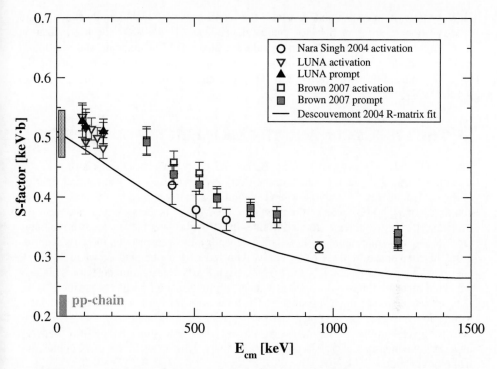

FIGURE 1. Astrophysical S(E)-factor for ^{3}He$(\alpha,\gamma)^7$Be. The results from the modern, high precision experiments are shown with their total error. Horizontal bar: solar Gamow peak.

The underground experiment has been performed with the ^{4}He$^+$ beam from the 400 kV accelerator in conjunction with a windowless gas target filled with ^{3}He at 0.7 mbar. The beam enters the target chamber and it is stopped on a power calorimeter. The ^{7}Be nucleus produced by the reaction inside the ^{3}He gas target are implanted into the calorimeter cap which, after the irradiation, is removed and placed in front of a germanium detector for the measurement of the ^{7}Be activity.

In the first phase of the experiment we obtained the ^{3}He$(\alpha,\gamma)^7$Be cross section from the activation data [7] [8] alone with a total uncertainty of about 4%. In the second phase we performed a new high accuracy measurement using simultaneously prompt and activation methods. The prompt capture γ-ray is detected by a 135% germanium detector heavily shielded and placed in close geometry with the target. Data have been collected at $E_{c.m.} = 170, 106, 93$ keV. In this interval the cross section varies from 10.25 nbarn to 0.23 nbarn, with a total error of about 4%.

The astrophysical factor obtained with the two methods [9] is the same within the quoted experimental error (Fig.1). Similar conclusions have then been reached in a new simultaneous activation and prompt experiment [10] which covers the $E_{c.m.}$ energy range from 330 keV to 1230 keV.

The energy dependence of the cross section seems to be theoretically well determined.

However, results recently obtained with the recoil separator technique between 1 and 3 MeV show a different S-factor energy dependence. If we leave the normalization as the only free parameter we can rescale the fit of [11] to our data and we obtain $S_{3,4}(0)=0.560\pm0.017$ keV barn. Thanks to our small error, the total uncertainty on the ^{8}B solar neutrino flux goes from 12 to 10%, whereas the one on the ^{7}Be flux goes from 9.4 to 5.5% [9].

THE CNO CYCLE AND THE METALLICITY OF THE SUN

The CNO cycle was proposed by H. Bethe and C. von Weizsäcker as a process for hydrogen burning in stars. In our Sun the CNO cycle accounts for just a small fraction of the nuclear energy production, whereas the main part is supplied by the proton-proton chain. ^{14}N(p,γ)^{15}O (Q-value of the reaction: 7.297 MeV) is the slowest reaction of the cycle and it rules its energy production rate. In particular, it is the key reaction to know the ^{13}N and ^{15}O solar neutrino flux, which depends almost linearly on its cross section.

In the first phase of the LUNA study, data have been obtained down to 119 keV energy with solid targets of TiN and a 126% High Purity Germanium detector. This way we could measure the five different radiative capture transitions which contribute to the ^{14}N(p,γ)^{15}O cross section at low energy. The total cross section was then measured down to very low energy in the second phase of the experiment by using a 4π BGO summing detector placed around a windowless gas target (1 mbar pressure). At the lowest energy of 70 keV we measured a cross section of 0.24 pbarn, with an event rate of 11 counts/day from the reaction.

The results we obtained first with the germanium detector data [12][13] and then with the BGO set-up [14] are about a factor two lower than the existing extrapolation at very low energy (Fig. 2). As a consequence the CNO neutrino yield in the Sun is decreased by about a factor two, with respect to the estimates.

The lower cross section is affecting also stars which are more evolved than our Sun. In particular, the age of the Globular Clusters is increased by 0.7-1 Gyr [15] and the dredge-up of carbon to the surface of AGB stars is much more efficient [16].

The main conclusion from the LUNA data has been confirmed by an independent study at higher energy [17]. However, there is a 15% difference between the total S-factor extrapolated by the two experiments at the Gamow peak of the Sun. In particular, this difference arises from the extrapolation of the capture to the ground state in ^{15}O, a transition strongly affected by interference effects between several resonances and the direct capture mechanism.

In order to provide precise data for the ground state capture we performed a third phase of the ^{14}N(p,γ)^{15}O study with a segmented germanium detector to reduce the summing correction and, in order to obtain sufficient statistics, we concentrate on the beam energy region immediately above the 259 keV resonance, where precise data effectively constrain the R-matrix fit for the ground state transition.

With these improvements we could finally reduce to 8% the total error on the S-factor: $S_{1,14}(0)=1.57\pm0.13$ keV barn [18]. This is significant because, finally solved the solar neutrino problem, we are now facing the solar composition problem: the conflict between helioseismology and the new metal abundances that emerged from improved

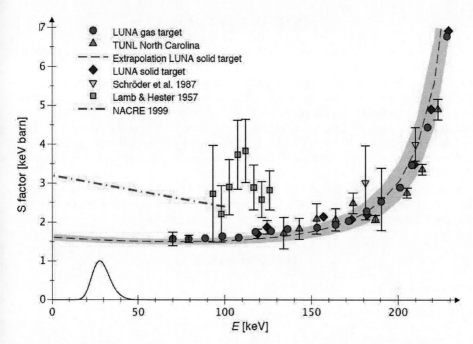

FIGURE 2. Astrophysical S(E)-factor of $^{14}N(p,\gamma)^{15}O$ as function of the center of mass energy E. The errors are statistical only (the systematic ones are similar). The solar Gamow peak is shown in arbitrary units.

modeling of the photosphere [19] [20]. Thanks to the relatively small error we achieved on the $^{14}N(p,\gamma)^{15}O$ cross section it will soon be possible to measure the metallicity of the core of the Sun by comparing the detected CNO neutrino flux with the predicted one. As a matter of fact, the CNO neutrino flux is decreased by about 35% in going from the high to the low metallicity scenario. This way it will be possible to test a key assumption of the standard Solar Model, that the early Sun was chemically homogeneous.

The solar phase of LUNA has almost reached the end and a new and rich program of nuclear astrophysics mainly devoted to the Mg-Al and Ne-Na cycles has already started about 2 years ago with the study of $^{25}Mg(p,\gamma)^{26}Al$. Two are the reasons why this reaction is relevant for astrophysics: the sky map taken by the satellites which see the 1.8 MeV γ from the ^{26}Al decay and the anomalous meteoritic abundance of ^{26}Mg.

We are now measuring $^{15}N(p,\gamma)^{16}O$ with enriched ^{15}N targets. $^{15}N(p,\gamma)^{16}O$ is the leak reaction from the CN cycle to the CNO one. The results we already obtained with nitrogen of natural isotopic composition (0.366% ^{15}N) extend to energies lower than ever measured before and provide a cross section which is about a factor two lower than previously believed at novae energies [21].

CONCLUSIONS

Underground nuclear astrophysics started almost 18 years ago with the goal of exploring the fascinating domain of nuclear astrophysics at very low energy. During these years LUNA has proved that, by going underground and by using the typical techniques of low background physics, it is possible to measure nuclear cross sections down to the energy of the nucleosynthesis inside stars. In particular, the measurements of $^3\text{He}(^3\text{He},2p)^4\text{He}$ has shown that nuclear physics was not the origin of the solar neutrino puzzle.

The cross section of $^3\text{He}(\alpha,\gamma)^7\text{Be}$ has been measured with two different experimental approaches and with a 4% total error. Thanks to this small error, the total uncertainty on the ^7Be solar neutrino flux has been reduced to 5.5%.

Finally, the study of $^{14}\text{N}(p,\gamma)^{15}\text{O}$ has shown that the expected CNO solar neutrino flux has to be decreased by about a factor two, with an error small enough to pave the way for a measurement of the central metallicity of the Sun.

REFERENCES

1. U. Greife, et al., *Nucl. Instr. Meth.* **A 350**, 327 (1994).
2. A. Formicola, et al., *Nucl. Instr. Meth.* **A 507**, 609 (2003).
3. W.A. Fowler, *Nature* **238**, 24 (1972).
4. V.N. Fetysov and Y.S. Kopysov, *Phys. Lett.* **40B**, 602 (1972).
5. A. Krauss, et al., *Nucl. Phys.* **A 467**, 273 (1987).
6. R. Bonetti, et al., *Phys. Rev. Lett.* **82**, 5205 (1999).
7. D. Bemmerer, et al., *Phys. Rev. Lett.* **97**, 122502 (2006).
8. G. Gyurky, et al., *Phys. Rev.* **C 75**, 035805 (2007).
9. F. Confortola, et al., *Phys. Rev.* **C 75**, 065803 (2007).
10. T.A.D. Brown, et al., *Phys. Rev.* **C 76**, 055801 (2007).
11. P. Descouvement, et al., *Data Nucl. Data Tables* **88**, 203 (2004).
12. C. Formicola, et al., *Phys. Lett.* **591B**, 61 (2004).
13. G. Imbriani, et al., *Eur. Phys. J.* **A 25**, 455 (2005).
14. A. Lemut, et al., *Phys. Lett.* **634B**, 483 (2006).
15. G. Imbriani, et al., *Astronomy and Astrophysics* **420**, 625 (2004).
16. F. Herwig and S.M. Austin, *Astrophysical Journal* **613**, L73 (2004) (2004).
17. R.C. Runkle, et al., *Phys. Rev. Lett.* **94**, 082503 (2005).
18. M. Marta, et al., *Phys. Rev.* **C 78**, 022802 (2008).
19. C. Pena-Garay and A.M. Serenelli, *arXiv:*, 0811.2424 (2008).
20. W.C. Haxton and A.M. Serenelli, *arXiv:*, 0902.0036 (2009).
21. D. Bemmerer, et al., *J. Phys. G: Nucl. Part. Phys.* **36**, 045202 (2009).

Prospects for and Status of CUORE – The Cryogenic Underground Observatory for Rare Events

Eric B. Norman[1,2,3]

[1]*Nuclear Engineering Department, University of California, Berkeley, CA 94720 U.S.A.*
[2]*Physics Division, Lawrence Livermore National Laboratory, Livermore, CA 94551 U. S. A.*
[3]*Nuclear Science Division, Lawrence Berkeley National Laboratory, Berkeley, CA 94720 U.S.A.*
On behalf of the CUORE and CUORICINO collaborations

Abstract. The present status of the Cryogenic Underground Observatory for Rare Events (CUORE) is presented along with the latest results from its prototype, CUORICINO.

Keywords double beta decay, neutrino mass, cryogenic detectors
PACS: 14.40.Pq, 23.40.-s, 23.40.Bw, 27.60.+j

INTRODUCTION

CUORE (Cryogenic Underground Observatory for Rare Events) is a next generation experiment designed to search for the neutrinoless DBD of ^{130}Te using a bolometric technique. The source/detector will be composed of 988 5x5x5-cm single crystals of TeO$_2$ housed in a common dilution refrigerator and operated at a temperature of 8-10 mK. The total mass of ^{130}Te contained in CUORE will be approximately 204 kg. Attached to each crystal will be a neutron-transmutation doped (NTD) germanium thermistor that will measure the small temperature rise produced in a crystal when radiation is absorbed. A schematic illustration of the CUORE detector is shown in Fig. 1. Details about the TeO$_2$ cryogenic detector are contained in a NIM A paper [1] and the physics potential of CUORE is described in an article in Astroparticle Physics [2]. A complete description of the CUORE project is also available online [3]. The estimated sensitivity of CUORE illustrated in Fig. 2 is sufficient to cover essentially all of the so-called inverted mass hierarchy region deduced from ν oscillation experiments. There are several compelling reasons to study ^{130}Te DBD. The $\beta\beta$ decay of ^{130}Te has been observed in geo-chemical experiments. Thus, a direct laboratory measurement of the 2ν $\beta\beta$ decay rate will provide an excellent calibration for 0ν-DBD. Second, because of its large decay energy and large expected nuclear matrix element, the half-life of ^{130}Te is predicted to be shorter than that of a number of other candidate isotopes. Third, based on the sensitivity needed to reach the mass scales inferred from the above-mentioned oscillation experiments, the ^{130}Te experiment can be done utilizing the natural abundance of ^{130}Te (34%), without the time and expense of obtaining separated isotopes. Of all the proposed or planned next

CP1165, *Nuclear Structure and Dynamics '09*
edited by M. Milin, T. Nikšić, D. Vretenar, and S. Szilner
© 2009 American Institute of Physics 978-0-7354-0702-2/09/$25.00

generation DBD experiments, only CUORE can reach the needed sensitivity without isotopic enrichment.

SET-UP

One of the great strengths of CUORE is its modular nature and the fact that many isotopes can be incorporated in a bolometer. Once CUORE is constructed, one could remove a number of crystals from the center of the array and replace them with crystals containing other materials of interest. The rest of CUORE could then be used as an anti-coincidence shield, which would provide the lowest background environment in the world in which to search for rare events. Exquisitely sensitive searches could be made for rare nuclear decays or for exotic processes such as the decay of the electron. CUORE is now under construction in the Gran Sasso National Laboratory (LNGS) in Italy.

Array of 988 crystals:
19 towers of 52 crystals/tower.
$M = 0.78$ ton of TeO_2

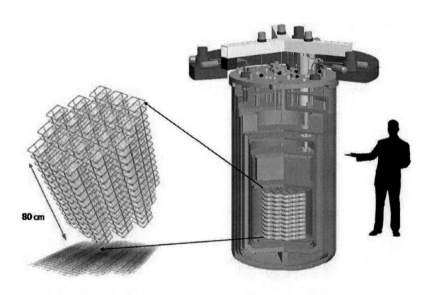

FIGURE 1. Schematic drawing of the CUORE detector

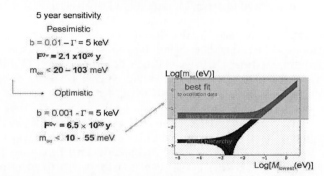

5 year sensitivity

Pessimistic

$b = 0.01 - \Gamma = 5$ keV

$F^{0\nu} = 2.1 \times 10^{26}$ y

$m_{ee} < 20 - 103$ meV

Optimistic

$b = 0.001 - \Gamma = 5$ keV

$F^{0\nu} = 6.5 \times 10^{26}$ y

$m_{ee} < 10 - 55$ meV

FIGURE 2. Estimated sensitivity of CUORE from 5 years of running.

A prototype experiment, CUORICINO, which consisted of 62 TeO$_2$ crystals, collected data for several years at the LNGS and published its physics results [4-6] – limits on neutrino mass only slightly less stringent than those obtained from previous [76]Ge experiments.

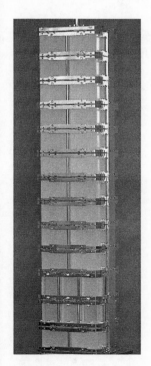

FIGURE 3. CUORICINO

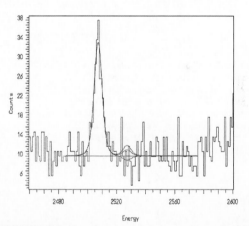

FIGURE 4. Background spectrum obtained from CUORICINO in an exposure of 18 kg-y. The background in the 0νββ region = 0.18±0.01 counts/(keV/kg/y). No peak is observed at the expected position of 0νββ (2527 keV) and a limit of $t_{0\nu1/2} >$ 2.94 x 10^{24} y at 90% C.L. has been obtained. The inferred limit on neutrino mass derived from this results is: $m_\nu < 0.2 - 0.7$ eV, depending on the assumed value of the nuclear matrix element.

During the past few years, we have conducted a number of studies of possible contributions to the background for CUORE. We performed measurements of cross sections for neutron-induced reactions on natural isotopic abundance Te metal at the GEANIE facility at the Los Alamos Neutron Science Center (LANSCE). The results of these experiments are contained in the Ph.D. thesis of Michelle Dolinski [7], and show that neutrons will not contribute significantly to the CUORE background. We also placed an array of large plastic scintillation detectors above CUORICINO to study muon-induced backgrounds. While we did observe a number of coincident events between these scintillators and CUORICINO, all such events involved multiple TeO_2 crystals firing. A double beta decay event, on the other hand, would be fully contained in a single crystal. Thus, muons will also contribute negligibly to the CUORE background. Finally, we performed a high-precision measurement of the $Q_{\beta\beta}$ value using the Canadian Penning Trap at Argonne National Laboratory [8]. The results of this measurement lowered the Q-value from the previously adopted value of 2530 ± 2 keV to 2527.0 ± 0.3 keV.

In conclusion, CUORE is now under construction at the LNGS. It is expected to be completed in 2012 with data taking commencing soon after. If neutrinos turn out to be of Majorana character and the neutrino mass hierarchy is of the inverted type, then CUORE should have the sensitivity to definitively observe the process of neutrinoless double beta decay.

ACKNOWLEDGMENTS

This work was supported in part by the U. S. Department of Energy under contract numbers DE-AC52-07NA27344 at LLNL and DE-AC02-05CH11231 at LBNL.

REFERENCES

1. C. Arnaboldi et al., *Nucl. Instrum. & Meth*. A **518**, 775 (2004).
2. C. Arnaboldi et al., *Astropart. Phys.* **20**, 91(2003).
3. R. Ardito et al., http://xxx.lanl.gov/hep-ex/0501010.
4. C. Arnaboldi et al., *Phys. Lett*. B **584**, 260(2004).
5. C. Arnaboldi et al. *Phys. Rev. Lett*. **95**, 142501 (2005).
6. C. Arnaboldi et al., *Phys. Rev.* C **78**, 035502 (2008).
7. M. J. Dolinski, LLNL-TH-408040 (2008).
8. N. D. Scielzo et al., Phys. Rev. C (in press).

Electron Screening Effects on α-decay

A. Musumarra[c], F. Farinon[b,d], C. Nociforo[b], H. Geissel[b,d],
G. Baur[a], K.-H. Behr[b], A. Bonasera[c], F. Bosch[b], D. Boutin[b], A. Brünle[b],
L. Chen[d], A. Del Zoppo[c], C. Dimopoulou[b], A. Di Pietro[c], T. Faestermann[e],
P. Figuera[c], K. Hagino[f], R. Janik[g], C. Karagiannis[b], P.Kienle[e,h],
S. Kimura[c], R. Knöbel[b,d], I. Kojouharov[b], C. Kozhuharov[b],
T. Kuboki[p], J. Kurcewicz[b], N. Kurz[b], K. Langanke[b], M. Lattuada[c],
S.A. Litvinov[b], Yu.A. Litvinov[b], G.Martinez-Pinedo[b], M. Mazzocco[i],
F. Montes[l], Y. Motizuki[m], F. Nolden[b], T. Ohtsubo[n], Y. Okuma[n],
Z. Patyk[o], M.G. Pellegriti[c], W. Plaß[d], S. Pietri[b], Z. Podolyak[q],
A. Prochazka[b,d], C. Scheidenberger[b,d], V. Scuderi[c], B. Sitar[g], M. Steck[b],
P. Strmen[g], B. Sun[b], T. Suzuki[p], I. Szarka[g], D. Torresi[c], H. Weick[b],
J.S. Winfield[b], M. Winkler[b], H.J. Wollersheim[b] and T. Yamaguchi[p]

[a]FZ Jülich, Germany
[b]GSI, Darmstadt, Germany
[c]INFN-LNS and University of Catania, Catania, Italy
[d]Justus-Liebig Universität, Giessen, Germany
[e]TU München, Germany
[f]Tohoku University, Senday, Japan
[g]Comenius University, Bratislava, Slovakia
[h]SMI, Wien, Austria
[i]INFN and University of Padua, Padua, Italy
[l]Michigan State University, East Lansing, U.S.A.
[m]RIKEN, Wako, Japan
[n]Niigata University, Japan
[o]Soltan Institute for Nuclear Studies, Warsaw, Poland
[p]Saitama University, Japan
[q]University of Surrey, Guildford, United Kingdom

Abstract. An open problem in Nuclear Astrophysics concerns the understanding of electron-screening effects on nuclear reaction rates at stellar energies. In this framework, we have proposed to investigate the influence of the electron cloud on α-decay by measuring Q-values and α-decay half-lives of fully stripped, H-like and He-like ions. These kinds of measurements have been feasible just recently for highly-charged radioactive nuclides by fragmentation of ^{238}U at relativistic energies at the FRS-ESR facility at GSI. In this way it is possible to produce, efficiently separate and store highly-charged α-emitters. Candidates for the proposed investigation were carefully selected and will be studied by using the Schottky Mass Spectroscopy technique.

In order to establish a solid reference data set, lifetimes and Q_α-value measurements of the corresponding neutrals have been performed directly at the FRS, by implanting the separated ions into an active Silicon stopper.

Keywords: Electron Screening, Alpha-Decay, Lifetimes
PACS: 23.60.+e

CP1165, *Nuclear Structure and Dynamics '09*
edited by M. Milin, T. Nikšić, D. Vretenar, and S. Szilner
© 2009 American Institute of Physics 978-0-7354-0702-2/09/$25.00

ELECTRON SCREENING EFFECTS IN CHARGED-PARTICLE REACTIONS AT ASTROPHYSICAL ENERGIES

One important subject in nuclear astrophysics concerns the interpretation of electron screening effects observed in low-energy nuclear reactions with light nuclei [1]. As pointed out by several authors [1,2], screening corrections deeply affect the reaction rates at low relative energies. Screening in the laboratory largely differs from plasma screening in stars, forcing a double step procedure in extracting astrophysical reaction rates. Therefore it is essential to clarify at least the screening effects on the measured cross sections.

The screening enhancement factor in charged-particle reactions at the astrophysical energies is usually written as $f(E) = \dfrac{\sigma_s(E)}{\sigma_b(E)} = \exp\left(\dfrac{\pi \eta U_e}{E}\right)$ [2], where $\sigma_s(E)$ and $\sigma_b(E)$ are the screened and bare cross sections of an arbitrary charged-particle reaction respectively, η is the Sommerfeld parameter and U_e the so called electron screening energy. In order to extract information on U_e, $\sigma_s(E)$ and $\sigma_b(E)$ must be determined. Large experimental sources of uncertainty for $\sigma_s(E)$ are due to the small reaction cross sections involved, the missing knowledge of stopping powers and the high accuracy needed in the knowledge of the relative energy between the interacting ions[3]. Concerning $\sigma_b(E)$, so far it can be evaluated by extrapolation or, at the best, by using R-matrix fit.

In this uncertain scenario comes out the discrepancy between experimental U_e^{exp} and theoretical U_e^{th} values so far deduced. In particular, the U_e^{exp} values mostly exceed the maximum admitted theoretical ones and in some cases also by a factor two [2,4].

ELECTRON SCREENING EFFECTS IN α-DECAY

Since extracting information on U_e by studying nuclear reactions at very low energy is quite ambiguous, we suggest a completely different experimental method by simplifying, as much as possible, the system affected by the electron screening. It is well known that Q_α−values and nuclear α-decay lifetimes should be different for bare nuclei with respect neutral atoms. Q_α−values are simply related to the electron screening energy by the formula:

$$Q_\alpha(bare) - Q_\alpha(screened) = U_e(adiabatic) \qquad (1)$$

In case of alpha-emitters with Z=70-90 U_e(adiabatic)=30-40 keV. Thus the measurement of $Q_\alpha(bare)$ and $Q_\alpha(screened)$ allows direct determination of Electron Screening energy U_e. Furthermore we recently theoretically investigated the effects of electron screening on half-lives of alpha-emitters [5]. It turns out that relative changes in half-lives due to Electron Screening $\Delta T_{1/2}/T_{1/2}$ are expected to be in the order of few per mil.

EXPERIMENTAL METHOD

The FRS-ESR facility at GSI(Darmstadt, Germany) offers a unique opportunity to perform precise half-lives and Q-value measurements for highly charged radioactive ions [6,7,8]. Time-resolved Schottky Mass Spectrometry (SMS) [8] is ideally suited for measuring decay half-lives and Q-values of radioactive ions in the $T_{1/2}$ range from about a few seconds to a few ten of minutes. In our case highly charged alpha decaying species can be produced by fragmentation of relativistic ^{238}U ions, selected by the FRS and injected in the ESR for in-flight half-life and Q_α-value measurements. The ^{213}Fr nucleus has been put forward as a good candidate for such an investigation. Shown in table 1 the main decay properties of such nuclide, also reported expected screening energy and decay constant relative changes.

TABLE 1. ^{213}Fr alpha decay properties, expected screening energy and $\Delta\lambda/\lambda$ relative changes.

	$T_{1/2}$	α-branch(%)	Q_α(MeV)	U^{ad}(keV)	$\Delta\lambda/\lambda$ %
^{213}Fr	34.6 s (3)	99.45	6.905	38.0	0.5

Half-life measurement of bare ^{213}Fr can be performed by using the SMS technique in regime of many particle or single particle decay [9]. For implementing single particle decay and bare Q_α-value measurements, parent (^{213}Fr^{87+}) and daughter (^{209}At^{85+}) nuclei must circulate in the ESR at the same time, in order to record simultaneously Schottky frequencies for both species.

PRELIMINARY $T_{1/2}$ AND Q_α MEASUREMENT OF NEUTRALS ^{214}Ra AND ^{213}Fr

As a first step, in order to establish a solid reference data set, we performed precise half-life and Q_α measurements for ^{214}Ra and ^{213}Fr neutral atoms by using the RISING silicon implantation-decay set-up [10], installed at the FRS-S4 focal plane. It consisted of 6 DSSSD 1 mm thick with 16x16 strips. Fig 1 shows the implantation pattern of ^{213}Fr in the first layer of the stack. The secondary alpha-emitters have been produced by using projectile fragmentation of ^{238}U at 1 GeV·A on a Be target and then separated in the FRS by applying the well known Bρ-ΔE-Bρ method [6].

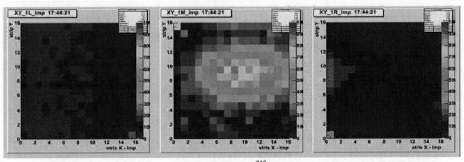

FIGURE 1. Implantation pattern on DSSSD for ^{213}Fr settings during 4 second spill time.

Owing to the high selectivity of the FRS, a clear identification of the implanted ions is shown in Fig. 2 in case of ^{214}Ra: characteristic Q_α-values peaks detected during 12 s interspill time are clearly resolved with very low background. A decay curve shown in the upper left panel of Fig. 1 has been obtained after gating on the ^{214}Ra decay peak. A preliminary half-life of $T_{1/2}=(2.485\pm0.025)$s has been deduced. This measurement is in agreement with the accepted value [11] and is more precise than any previous measurements, confirming the validity of the technique used. The presence of systematic errors will be estimated by performing correlation analysis between the identified implanted ion and α-decay and furthermore by extending the analysis to all the other implanted α-emitters.

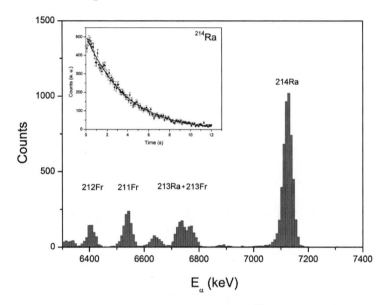

FIGURE 2. Q_α-value spectrum; upper left panel: ^{214}Ra decay curve.

REFERENCES

1. S. Engstler, A. Krauss, K. Neldner, C. Rolfs, U. Schröder, K. Langanke, *Phys. Lett.* **202B,** 179 (1980).
2. F. Strieder, C. Rolfs, C. Spitaleri and P. Corvisiero, *Naturwissenschaften* **88,** 461 (2001).
3. M. Aliotta, et al., *Nucl. Phys.* **A 690**, 790 (2001).
4. T. D. Shoppa, S. E. Koonin, K. Langanke and R. Seki, *Phys. Rev.* C **48**, 837 (1993).
5. Z. Patyk, H. Geissel Yu. A. Litvinov, A. Musumarra, C. Nociforo, *Phys. Rev.* C **78**, 54317 (2008).
6. H. Geissel, et al., *Nucl. Instr. and Meth.* B **70**, 286 (1992).
7. B. Franzke, *Nucl. Instr. and Meth.* B **24/25**, 18 (1987).
8. F. Bosch, *J. Phys.* B **36**, 585 (2003).
9. H. Geissel, Yu.A. Litvinov, B. Pfeiffer, et al., in Proc. FINUSTAR05, Cos, Greece, 2005.
10. R. Kumar, et al., *Nucl. Instr. and Meth.* A **598**, 754 (2009).
11. Y.A. Akovali, *Nucl. Data Sheets* **76,** 127 (1995).

Electron Screening in Reaction Between Protons and Lithium Nuclei

M. Lipoglavšek, I. Čadež, S. Markelj, P. Pelicon, J. Vales and P. Vavpetič

Jožef Stefan Institute, Jamova cesta 39, 1000 Ljubljana, Slovenia

Abstract. We have measured the cross section for the ^{1}H(^{7}Li,α)^{4}He reaction at lithium beam energies from 0.34 to 1.05 MeV. Hydrogen was forced by diffusion into Pd and PdAg alloy foils. A large electron screening effect was not observed at high hydrogen concentrations, while at low H/metal ratio the results are inconclusive. A dependence of the screening potential on Hall coefficient of the metallic host could not be confirmed.

Keywords: Electron screening, low-energy nuclear reactions
PACS: 25.10.+s, 25.40.Ep, 25.90.+k

Due to Coulomb repulsion the cross section σ for charged particle induced nuclear reactions drops rapidly with decreasing beam energy. To separate the strong energy dependence the astrophysical S factor is introduced. The cross section is then written as a function of c.m.s. energy E as

$$\sigma(E) = \frac{S(E)}{E} e^{-2\pi\eta}. \tag{1}$$

It is known that the cross section increases at low energies when the interacting nuclei are not bare, i.e. are in the form of atoms and molecules or in plasma [1]. The enhancement ratio could be written as

$$f(E) = \frac{\sigma(E+U_e)}{\sigma(E)}, \tag{2}$$

where U_e is the screeening potential energy. It was recently observed by two independent groups that the cross section for various light ion reactions increases even more when one of the reactants is implanted into a metal [2-5]. The cross section increase was attributed to metallic valence electrons, which may come closer to the implanted ion and more effectively screen its charge than in an atom. However, the size of the screening effect strongly depends on the host material and the reason for this dependence is not known. Raiola et al. [6] have observed a connection between U_e and the Hall coefficient of the metallic host, while Kasagi [2] suggested that U_e depends on deuterium concentration in the metal. To further investigate electron screening in metals we tried to test both hypotheses by studying the effect in Pd and PdAg alloys at different hydrogen concentrations.

To simplify the experiment we employed the inverse kinematics reaction ^{1}H(^{7}Li,α)^{4}He and measured emitted α particles at a backward angle of 150°. In inverse kinematics the reactions occur on average deeper in the target at the same center of mass energy as in normal kinematics. In this way we hoped to be less sensitive to

CP1165, *Nuclear Structure and Dynamics '09*
edited by M. Milin, T. Nikšić, D. Vretenar, and S. Szilner
© 2009 American Institute of Physics 978-0-7354-0702-2/09/$25.00

surface contamination than in the ^{2}H+^{2}H reaction. The ^{7}Li$^+$ beams with energies between 0.34 and 1.05 MeV were accelerated by the Tandetron accelerator at Jožef Stefan Institute. To detect α particles we used a 100 μm thick silicon detector with an area of 300 mm^2 placed 42 mm from the target. The solid angle Ω embraced by the detector was about 1% of 4π. The detector was covered by a 3 μm thick Al foil to prevent scattered beam particles from hitting the detector. The metallic targets were 100 to 150 μm thick. The summary of all used targets is given in table 1. Before the experiment the targets were soaked in hydrogen gas at 1 bar. Different soaking times resulted in different hydrogen concentrations in the targets. Hydrogen concentrations were determined by elastic recoil detection analysis (ERDA), where the target was tilted by 75° and elastically scattered protons measured at 30° with respect to the ^{7}Li^{2+} beam with an energy of 4.3 MeV [7]. Polymide (Kapton) was used as a reference for both cross section and ERDA measurements [8]. All ERDA measurements showed a constant depth profile of hydrogen down to a depth to which these measurements are sensitive. The only deviation from this behavior was a peak on the surface of the foils. It was present in all foils, but since the number of atoms in this peak was relatively small, the surface peak only became clearly visible when hydrogen concentration in the bulk was low. The width of the peak in ERDA spectra coincides with the measurement resolution and therefore, the peak corresponds to a very thin layer at the surface. The surface peak is most likely due to hydrogen dynamics inside the foils. It disappeared only when there was no hydrogen in the bulk of the foils, i.e. after heating to 900 °C in vacuum, but reappeared with hydrogen loading. The energy loss of lithium in $4.5 \cdot 10^{16}$ hydrogen atoms/cm^2 is 850 eV at 1 MeV beam energy and 500 eV at the lowest beam energy [9]. Since this is less than the uncertainty of the beam energy, the energy loss in the surface peak was neglected in our calculations.

Measured thick target α-particle yields are shown in fig. 1 together with yields calculated as follows. A textbook definition of reaction cross section for a thin target is

$$N_\alpha = 2\Omega W N_{Li} \frac{\rho N_A x}{M} \sigma, \qquad (3)$$

where N_α is the number of detected α particles, W their angular distribution, N_{Li} the number of Li ions, N_A Avogadro's number and ρ, x and M the density, thickness and molar mass of the target. The factor of 2 comes from two equivalent α particles produced in the reaction. $\rho N_A x/M$ represents the number of hydrogen atoms in Kapton or metallic target in an area hit by the beam. For the surface peak this factor is given in column 4 of table 1. The contribution of the surface peak to the α-particle yield was evaluated using the above equation. However, the contribution of the hydrogen distributed below the surface of the metal had to be calculated by transforming eq. 3 into differential form and integrating over energies from the beam energy E_0 to 0.

$$N_\alpha = 2N_{Li} \frac{\rho N_A}{M} \int_{E_0}^{0} \Omega W \frac{\sigma(E)}{dE_{Li}/dx} dE_{Li}. \qquad (4)$$

Stopping power dE_{Li}/dx was calculated using SRIM [9], except for Kapton, where it was taken from ref. [10]. The astrophysical S factor was taken from ref. [4] as

$$S(E) = 0.055 + 0.21E - 0.31E^2 [MeV \cdot b], \qquad (5)$$

where E is in MeV. The α-particle angular distribution was taken from ref. [11]. Electron screening effect was taken into account by replacing $\sigma(E)$ with $\sigma(E+U_e)$.

Therefore, the bare nuclei cross section was taken from ref. [4] and the only free parameter in the fit was U_e. The U_e resulting from one parameter least squares fits to the data are summarized in table 1. As can be seen from table and fig. 1, the measured yields in Kapton as well as high concentration Pd and $Pd_{77}Ag_{23}$ foils can be well described with calculations for bare nuclei.

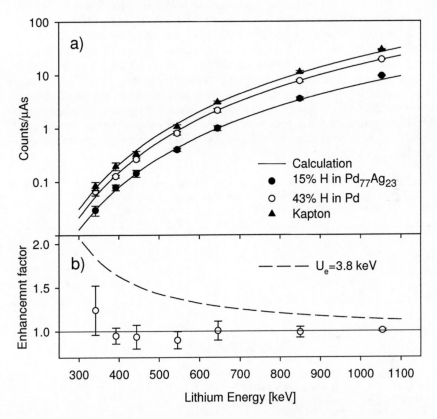

FIGURE 1. Thick target α-particle yields for kapton, high concentration Pd and $Pd_{77}Ag_{23}$ targets. Enhancement factor obtained by dividing measurements and calculation from a) is shown for Pd target in b). Calculation with U_e=3.8keV as reported in ref. [4] is drawn with a dashed line.

The measurements on targets that contained low concentrations of hydrogen were hampered by the presence of a large peak of hydrogen concentration on the metallic surface. The contribution of this peak to the α-particle yield is very similar to the electron screening effect, which enhances low energy yields more than the high energy ones. Surface peak, therefore, masks electron screening. Nevertheless, U_e could be fitted also for low concentration targets. Although the measured points did not follow the predicted energy dependence, for the $Pd_{75}Ag_{25}$ and $Pd_{51}Ag_{49}$ foils U_e differed from zero by more than its fitting error. The values U_e=2.2±1.1 and 4.1±1.4 keV fitted for these two targets, respectively, agree well with the value of 3.8±0.3 keV measured for the $Pd_{99}Li_1$ alloy in ref. [4]. This gives us a hint that the electron screening effect might depend on hydrogen concentration in the metal. The large error bar from the U_e fit of the low concentration Pd target would still allow for large screening. On the other hand, these measurements could not confirm the dependence of electron screening on the Hall coefficient of the metal, since in this case we expected the lowest U_e value in the $Pd_{51}Ag_{49}$ alloy which has the lowest Hall coefficient.

To conclude, we have measured cross sections for the $^1H(^7Li,\alpha)^4He$ reaction with lithium beam energies between 0.34 and 1.05 MeV. Hydrogen was loaded into Pd and PdAg alloy targets. At high hydrogen concentrations in the targets no electron screening was observed. No screening was observed also in hydrogen on the surface of the metal. The measurents at low hydrogen concentrations gave inconclusive results that nevertheless indicate a possible concentration dependence of electron screening. We could not confirm the dependence of electron screening on the Hall coefficient of the metal. Our future work will turn to Ni targets, where we do not expect a prominent hydrogen concentration peak on the surface, and to Pd targets prepared with hydrogen implantation.

TABLE 1. List of targets, their hydrogen contents per metallic atom, Hall coefficients [12], number of hydrogen atoms on the surface and fitted screening potential energies.

Target	H concentration	Hall coeff. $[10^{-11}m^3/As]$	Surface peak $[10^{15} atoms/cm^2]$	U_e [keV]
Kapton	$C_{22}H_{10}N_2O_5$	-	-	<0.6
Pd	0.43(1)	-7.7	-	<0.4
$Pd_{77}Ag_{23}$	0.149(5)	-20	45(5)	<0.7
$Pd_{75}Ag_{25}$	0.023(1)	-21	20(1)	2.2(11)
$Pd_{51}Ag_{49}$	0.0155(5)	-35	23(1)	4.1(14)
Pd	0.0058(3)	-7.7	20(1)	<10

REFERENCES

1. H. J. Assenbaum, K. Langanke and C. Rolfs, *Z. Phys.* A **327**, 461 (1987).
2. J. Kasagi, *Prog. Theor. Phys. Suppl.* **154**, 365 (2004).
3. C. Rolfs, *Prog. Theor. Phys. Suppl.* **154**, 373 (2004).
4. J. Cruz et al., *Phys. Lett.* **B624**, 181 (2005).
5. K. U. Kettner et al., *J. Phys.* G **32**, 489 (2006).
6. F. Raiola et al., *Eur. Phys. J.* A **19**, 283 (2004).
7. S. Markelj et al., *Nucl. Instr. Meth. Phys. Res.*, B **261**, 498 (2007).
8. P. Pelicon et al., *Nucl. Instr. Meth. Phys. Res.*, B **227**, 591 (2005).
9. J. F. Ziegler, J. P. Biersack, and M. D. Ziegler, The Stopping and Range of Ions in Matter, Lulu Press Co.; Morrisville, NC, 27560 USA; http://www.srim.org.
10. F. Munnik et al., *J. Appl. Phys.* **86**, 3934 (1999).
11. C. Rolfs and R. W. Kavanagh, *Nucl. Phys.* **A455**, 179 (1986).
12. W. W. Shulz, P. B. Allen, and N. Trivedi, *Phys. Rev.* B **45**, 10886 (1992).

Effects of Distortion on the Intercluster Motion in Light Nuclei

R.G. Pizzone[a,b], C. Spitaleri[a,b], C. Bertulani[c], A. Mukhamedzhanov[d], L. Blokhintsev[e], B. Irgaziev[f], Đ. Miljanić[g], S. Cherubini[a,b], M. La Cognata[a,b], L. Lamia[a,b], S. Romano[a,b], A. Tumino[h]

[a]INFN Laboratori Nazionali del Sud, Catania, Italy
[b]DMFCI - Università di Catania, Catania, Italy
[c]Texas A&M University, Commerce, USA
[d]Cyclotron Institute, Texas A&M University, College Station, USA
[e]University of Moscow, Moscow, Russia
[f]University of Tashkent, Tashkent, Uzbekhistan
[g]Ruđer Bošković Institute, Zagreb, Croatia
[h]Università Kore, Enna, Italy

Deuteron induced quasi-free scattering and reactions have been extensively investigated in the past few decades as well as ^{6}Li, ^{3}He and ^{9}Be induced ones. This was done not only for nuclear structure and processes study but also for the important astrophysical implication (Trojan Horse Method, THM). In particular the width of the spectator momentum distribution in ^{6}Li and deuterium, which have widely been used as a Trojan Horse nuclei, will be studied as a function of the transferred momentum. Trojan horse method applications will also be discussed in these cases.

Keywords: cluster model, indirect methods in nuclear astrophysics
PACS: 21.60.Gx, 25.10.+s

The study of direct contributions to nuclear reactions involving light nuclei showed a large increase in the last decades due to the development of indirect methods trying to measure the bare nucleus cross sections at astrophysical energies.

Among these methods there are the Coulomb dissociation [1], the ANC [2] and the Trojan Horse Method (THM). The main features of this latter method are extensively discussed elsewhere [3,4].

Basically THM allows us to measure the bare-nucleus two-body cross sections (or equivalently the bare-nucleus astrophysical S(E) factors) by means of quasi-free three-body reactions. The method is therefore an extension of excitation function measurements at energies energies above Coulomb barrier to reactions of astrophysical interest which take place at ultra-low energies.

The present paper can be regarded as part of an experimental as well as theoretical work aimed at analyzing the behavior as a function of the transferred momentum of the momentum distribution width in ^{2}H,^{3}He,^{6}Li and ^{9}Be usually assumed as TH nuclei in several previous experiments. We have reanalyzed our previous data from different experiments and we have compared our results with others present in literature. In

CP1165, *Nuclear Structure and Dynamics '09*
edited by M. Milin, T. Nikšić, D. Vretenar, and S. Szilner
© 2009 American Institute of Physics 978-0-7354-0702-2/09/$25.00

particular the behaivour with transferred momentum of the full width at half maximum (FWHM) of the relative momentum distribution inside the ^{2}H and ^{6}Li is studied.

Since the extraction of the bare-nucleus S(E) factor uses the momentum distribution of the spectator cluster inside the TH nucleus, it is important to evaluate the impact of the uncertainty of the momentum distribution width on the final result.

This study will also help to evaluate the dependence of the THM astrophysical factor on the momentum distribution FWHM, which might introduce additional uncertainties to results obtained by means of the method. This task was already fulfilled for the case of the α-d clusters in ^{6}Li. In that case a recent work [5] shows how errors introduced in THM results by momentum distribution uncertainties are much less than other experimental error sources.

In Plane Wave Impulse Approximation (PWIA) the cross section of the three body reaction can be factorized into two terms and it is given by:

$$\frac{d^3\sigma}{dE_{cm}d\Omega_1 d\Omega_2} \propto KF \cdot |G(P_s)|^2 \cdot \left(\frac{d\sigma^N}{d\Omega}\right)_{cm} \qquad (1)$$

Where $(d\sigma^N/d\Omega)_{cm}$ is the off-energy-shell differential cross section for the two body A(x,c)C reaction, KF is a kinematical factor containing the final state phase-space factor, $|G(P_S)|^2$ is the Fourier transform of the radial wave function for the inter-cluster motion usually described in terms of the appropriate wave-functions depending on the system properties.

Since the extraction of the bare-nucleus S(E) factor uses the momentum distribution of the spectator cluster inside the TH nucleus, it is important to evaluate the impact of the uncertainty of the momentum distribution width on the final result.

In previous works [5,6] it has been shown that the full width at half maximum, $|G(P_S)|^2$ for the α-d momentum distribution in ^{6}Li slowly increases with increasing mean transferred momentum, q_t. The aim of the present work is to extend this study to the deuterium case.

The Galileian invariant mean transferred momentum q_t can be defined or quasi-free processes as:

$$\vec{q}_t = \left(\frac{m_B}{m_A}\right)^{1/2} \vec{p}_A + \left(\frac{m_A}{m_B}\right)^{1/2} \vec{p}_B \qquad (2)$$

where the vectors $\mathbf{p_A}$, $\mathbf{p_B}$ are the momentum of the projectile and ejectiles A and B=c+C, respectively. Thus q_t is defined as the modulus of the mean transferred momentum to c, and C. The standard procedure for the extraction from the experimental data of the momentum distribution is described in details [7].

Experimental momentum distribution results referring to different Trojan Horse nuclei carried out at different energies are reported in pictures 1-2, respectively for the ^{6}Li and deuterium case. It is possible to see as for the same TH nucleus the momentum distribution full width changes as the energy and thus the transferred momentum changes.

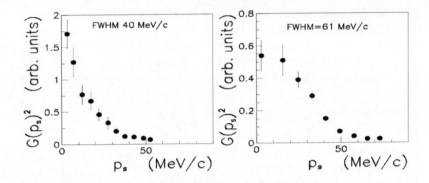

FIGURE 1. momentum distribution for d inside ^{6}Li for the ^{6}Li(^{6}Li,αα)^{4}He reaction at 3,6 and 5,9 MeV [7]

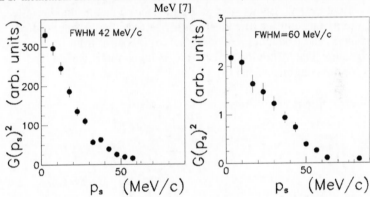

FIGURE 2. : momentum distribution for p in d for the p(d,pp)n reaction at 5 MeV and the ^{11}B(d,α^{8}Be)n at 27 MeV [9]

The above results outline a clear dependence of the full width for the momentum distribution with mean transferred momentum which we will examine in details in the next section. The present paper will also help to point out the distortion effects which arise at low energies in the study of three body processes.

DEPENDENCE OF THE MOMENTUM DISTRIBUTION FROM TRANSFERRED MOMENTUM

Data from several Trojan Horse experiments [9] using the different TH clusters were extracted as shown before. For each data-set the momentum distribution FWHM was measured and the average transferred momentum was calculated. We report in figure 3 the behaviour of the FWHM of the inter-cluster momentum distribution with increasing transferred momentum for They are reported together with the transferred momentum for the different cases. This was already done for the ^{6}Li case in previous works [5,6].

It is evident how, increasing q_t, the FWHM W smoothly increases until the predicted PWIA asymptotic value is reached in the region where q_t is large. These data strongly confirm the behavior already discussed in [5,6] and are with the present paper coherently reproduced also for the other nuclei.

This experimental behaviour was fitted for all cases by using the following function:

$$W(q) = f_0 (1 - \exp(-q_t / q_0)) \qquad (3)$$

where f_0 represents the asymptotic width and q_0 is the fitting parameter value.

What seems clear from this analysis is that as far as the transferred momentum is large compared to K, defined as $K = (2 \mu E_b)^{1/2}$, (and the beam energy is large as well) the momentum distribution shape is not distorted and its width resembles the value expected from theory. But as soon as the transferred momentum become comparable with K the momentum, i.e. the ideal quasi free condition is not verified, distribution shape changes and its width became smaller (as it is clear from fig. 1-2). This seems a clear signature that distortion effects are gaining more weight and this behavior appears for the examined nuclei [9].

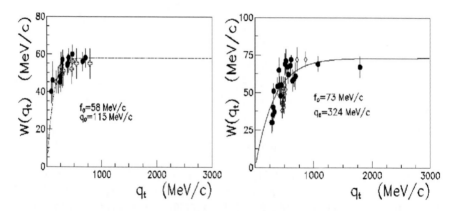

FIGURE 3. : behaviour of the momentum distribution width as a function of the transferred momentum for deuteron (left) and ^{6}Li (right). In both cases the fit reported in the text is plot.

REFERENCES

1. G. Baur, *Phys. Lett.* B **178**, 135 (1986).
2. A.M. Mukhamedzhanov, et al., *Phys. Rev.* C **63**, 024612 (2001).
3. S. Cherubini et al., *Ap. J.* **457**, 855 (1996).
4. C. Spitaleri et al., *Phys. Rev.* C **60**, 055802 (1999).
5. R.G. Pizzone et al., *Phys. Rev.* C **71**, 058801 (2005).
6. S. Barbarino et al., *Phys. Rev.*, C **21**, 1104 (1980).
7. C. Spitaleri et al., *Phys. Rev.* C **63**, 055806 (2004).
8. C. Spitaleri et al., *Phys. Rev.* C **63**, 005801 (2001).
9. R.G. Pizzone et al., *Phys. Rev.* C, to be published (2009).

Intermediate Nuclear Structure for $2\nu2\beta$ Decay of ^{48}Ca Studied by (p,n) and (n,p) Reactions at 300 MeV

H. Sakai and K. Yako, for the ICHOR collaboration

University of Tokyo (Department of Physics, Hongo 7-3-1, Bunkyo, Tokyo 113-0033, Japan

Abstract. Angular distributions of the double differential cross sections for the ^{48}Ca(p,n) and the ^{48}Ti(n,p) reactions were measured at 300 MeV. A multipole decomposition technique was applied to the spectra to extract the Gamow-Teller (GT) transition strengths. In the (n,p) spectrum beyond 8 MeV excitation energy extra $B(\mathrm{GT}^+)$ strengths which are not predicted by the shell model calculation. This extra $B(\mathrm{GT}^+)$ strengths significantly contribute to the nuclear matrix element of the $2\nu2\beta$-decay.

Keywords: double beta decay, Gamow-Teller
PACS: PACS numbers; 25.40Kv,23.40,27.40.+z

INTRODUCTION

The two neutrino double beta $(2\nu2\beta)$ decay proceeds through a sequence of Gamow-Teller (GT) transitions, namely from the parent nucleus to the intermediate nucleus and then from the intermediate nucleus to the final daughter nucleus.

The nuclear matrix element $M^{2\nu} = \sum_m \langle 0^+||O_{GT-}||1_m^+\rangle\langle 1_m^+||O_{GT-}||0^+\rangle/E'_m$ for the $2\nu2\beta^-$ decay thus consists of the β^- decay matrix elements $\langle 1_m^+||O_{GT-}||0^+\rangle$ for the parent nucleus decay and the β^- decay matrix elements $\langle 0^+||O_{GT-}||1_m^+\rangle$ for the intermediate nucleus decay. These GT β^- decay matrix elements can be studied experimentally through the (p,n) reaction for the parent nucleus decay and the (n,p) reaction for the intermediate nucleus decay.

EXPERIMENTS

The $2\nu2\beta$-decay nucleus ^{48}Ca is studied. The charge exchange (p,n) and (n,p) measurements at 300 MeV were performed using the neutron time-of-flight facility and the (n,p) facility, respectively, at RCNP. The results on the ^{48}Ca(p,n) and the ^{48}Ti(n,p) reactions are presented.

The (p,n) data were obtained with a neutron time-of-flight (NTOF) system [1] with a neutron detector NPOL3 system [2] with a resolution of about 420 keV at FWHM. The data were taken at angles every 1 degree from 0 to 16 degrees and every 2 degrees up to 40 degrees for the ^{48}Ca$(p,n)^{48}$Sc reaction.

The (n,p) measurement was performed at the (n,p) facility [3]. The neutron beam was produced by the ^{7}Li(p,n) reaction. A proton beam with a current of 300 nA hit

CP1165, *Nuclear Structure and Dynamics '09*
edited by M. Milin, T. Nikšić, D. Vretenar, and S. Szilner
© 2009 American Institute of Physics 978-0-7354-0702-2/09/$25.00

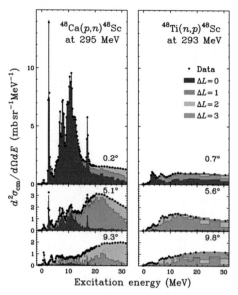

FIGURE 1. Double differential cross sections for the ^{48}Ca$(p,n)^{48}$Sc (left panel) and ^{48}Ti$(n,p)^{48}$Sc (right panel) reactions. The histograms show the results of the MD analyses.

a ^{7}Li target with a thickness of 210 mg/cm^2, producing 1×10^6 s^{-1} neutrons on the target area of $30^W \times 20^H$ mm^2 located 95 cm downstream from the ^{7}Li target. Three ^{48}Ti targets with thicknesses of 298, 296, and 293 mg/cm^2 and a polyethylene (CH$_2$) target with a thickness of 46 mg/cm^2 were mounted for the normalization purpose in a multi-target system [3]. The trajectories of scattered protons are obtained from the multi-wire drift chambers in the multi-target system and those located downstream of the ^{48}Ti targets. The outgoing protons were then momentum-analyzed by Large Acceptance Spectrometer (LAS) and were detected by the focal plane detectors. The overall energy resolution is 1.2 MeV for the (n,p) spectra due mainly to the target thicknesses. The double differential cross sections up to 40 MeV excitation energy were measured over an angular range of 0°–12° in the laboratory frame.

RESULTS

The multipole decomposition (MD) analyses were applied to the angular distributions of the cross section spectra to extract the $\Delta L = 0$ components , which were used to deduce $B(GT^{\pm})$. Figure shows the double differential cross sections for ^{48}Ca$(p,n)^{48}$Sc (left panel) and ^{48}Ti$(n,p)^{48}$Sc (right panel) reactions. The histograms show the results of the multipole decomposition analyses. The details of MD analysis can be found in Refs. [4, 5]. The reliability of the DWIA calculations for 300-MeV (p,n) and (n,p) reactions is well demonstrated in Fig. 2. The peak at 2.5 MeV in the (p,n) spectrum is the lowest 1$^+$ state in ^{48}Sc. The 1.0 MeV-bin (b) contains transitions to 2$^+$(1.14 MeV)

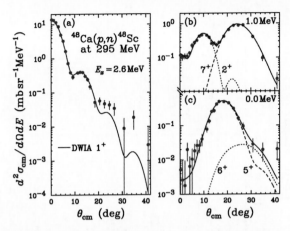

FIGURE 2. Experimental angular distribution of the double differential cross section for the ^{48}Ca$(p,n)^{48}$Sc reaction at (a) $E_x = 2.6$ MeV (2.5 MeV $< E_x <$ 2.7 MeV), (b) $E_x = 1.0$ MeV, and (c) 0.0 MeV. The curves represent DWIA calculations described in the text.

and $7^+(1.10$ MeV), showing a small but finite cross section at $0°$. On the other hand, the angular distribution at 0.0 MeV, which contains $5^+(0.13$ MeV) and $6^+(0$ MeV) transitions, has almost no contribution at $0°$. Such behaviors of the angular distribution are fairly well reproduced by the calculations described above indicating the reliability of the calculated angular distributions based on the DWIA used in the MD analyses.

The results of the MD analyses are shown in Fig. 1. The the $B(GT)$ value is extracted by using the proportionality relation from the $\Delta L = 0$ component of the cross section, $\sigma_{\Delta L=0}(q,\omega)$,

$$\sigma_{\Delta L=0}(q,\omega) = \hat{\sigma}_{GT} F(q,\omega) B(GT), \tag{1}$$

where $\hat{\sigma}_{GT}$ is the GT unit cross section and $F(q,\omega)$ is the kinematical correction factor [6]. The GT unit cross section has been determined from the mass-number dependence studied at 300 MeV [7] and the value is $\hat{\sigma}_{GT} = 4.69 \pm 0.35$ mb/sr.

The strength distributions are shown in Fig. 3. The strength is denoted as $B(GT + IVSM)$ because it contains the isovector spin monopole (IVSM) component [8].

The curves in Fig. 3 show the shell model prediction on $B(GT^\pm)$ distributions by Horoi *et al.* [9] employing the full fp-shell model space and the GXPF1A interaction [10]. Note the quenched GT operator of $\widetilde{\sigma t_\pm} = 0.77\sigma t_\pm$ is used.

The calculation reproduced the gross feature of the experimental $B(GT^-)$ spectrum below 15 MeV including the region of the GTGR, while the $B(GT^+)$ spectrum is roughly described only up to 8 MeV excitation. As is clear, the calculation fails to describe the GT$^+$ strengths above 8 MeV.

The nuclear matrix element $M^{2\nu}$ for the $2\nu2\beta^-$ decay of ^{48}Ca may be inferred from the measured $B(GT^\pm)$ values. Since we are unable to fix the phase of $\langle 1_m^+||O_{GT\pm}||0^+\rangle$, $M^{2\nu}(E_x)$ is replaced by $\sqrt{B(GT^+)/dE}\sqrt{B(GT^-)/dE}/E_x$ and denoted by $\overline{M}^{2\nu}(E_x)$. When we integrate $\overline{M}^{2\nu}(E_x)$ up to $E_x = 30$ MeV, the experimental $\overline{M}^{2\nu}_{exp}$ value amounts

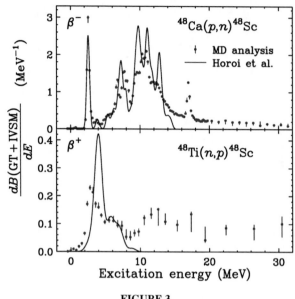

FIGURE 3.

to about twice of the shell model prediction $\overline{M}_{SM}^{2\nu}$ and about 10 times larger than that of $M^{2\nu}$ deduced from the experimental half-life of 4.4×10^{19} yr for the ^{48}Ca $2\beta^-$ decay.

The present results will be published elsewhere [11].

ACKNOWLEDGMENTS

The authors thank the collaborators of the ICHOR Project for their excellent contribution. This work is supported by the Grant-in-Aid of Specially Promoted Research (Grant No. 17002003) of the Ministry of Education, Culture, Sports, Science, and Technology of Japan.

REFERENCES

1. H. Sakai, et al., *Nucl. Instr. and Meth.* A **369**, 120 (1996).
2. T. Wakasa, et al., *Nucl. Instr. and Meth.* A **547**, 560 (2005).
3. K. Yako, et al., *Nucl. Instr. and Meth.* A **592**, 88 (2008).
4. M. Ichimura, H. Sakai, and T. Wakasa, *Prog. in Part. and Nucl. Phys.* **56**, 446 (2006).
5. T. Wakasa, et al., *Phys. Rev.* C **55**, 2909 (1997).
6. T.N. Taddeucci, et al., *Nucl. Phys.* **A 469**, 125 (1987).
7. M. Sasano, et al., *Phys. Rev.* C **79**, 024602 (2009).
8. D.L. Prout, et al., *Phys. Rev.* C **63**, 014603 (2000).
9. M. Horoi, S. Stoica, and B.A. Brown, *Phys. Rev.* C **75**, 034303 (2007).
10. M. Honma, et al., *Eur. Phys. J.* A **25**, Suppl. 499 (2005).
11. K. Yako, et al., to be published in *Phys. Rev. Lett.*

Isospin Corrections for Superallowed β-Decay in Self-consistent Relativistic RPA Approach

Haozhao Liang[*,†], Nguyen Van Giai[†] and Jie Meng[*]

[*]State Key Lab Nucl. Phys. & Tech., School of Physics, Peking University, Beijing 100871, China
[†]Institut de Physique Nucléaire, IN2P3-CNRS and Université Paris-Sud, 91406 Orsay, France

Abstract. Self-consistent random phase approximation based on relativistic Hartree-Fock approach is applied to calculate the isospin symmetry-breaking corrections δ_c for the nuclear $0^+ \to 0^+$ superallowed transitions. The nucleus-independent Ft values and the unitarity of the Cabibbo-Kobayashi-Maskawa matrix are then discussed.

Keywords: CKM matrix, superallowed β decay, isospin symmetry-breaking corrections, relativistic Hartree-Fock, random phase approximation
PACS: 23.40.Bw, 12.15.Hh, 21.60.Jz, 24.10.Jv

There are four experimental methods to determine the element $|V_{ud}|$ of the Cabibbo-Kobayashi-Maskawa (CKM) matrix [1, 2]: nuclear $0^+ \to 0^+$ superallowed β decays [3], neutron decay [4], pion β decay [5] and nuclear mirror transitions [6]. The first method provides the most precise determination of the $|V_{ud}|$ value [7]. In order to determine the $|V_{ud}|$ via the nucleus-independent Ft value, the radiative corrections $\Delta_R^V, \delta_R', \delta_{NS}$ and the isospin symmetry-breaking (ISB) corrections δ_c for the experimental ft values have to be taken into account [3, 8], i.e.,

$$V_{ud}^2 = \frac{K}{2G_F^2(1+\Delta_R^V)Ft} \quad \text{and} \quad Ft = ft(1+\delta_R')(1+\delta_{NS}-\delta_c). \tag{1}$$

In this work, the random phase approximation based on relativistic Hartree-Fock approach (RHF+RPA) [9] with effective interaction PKO3 [10] is applied to calculate the ISB corrections δ_c. The corrected nucleus-independent Ft value can then be obtained by combining the δ_c corrections as well as the experimental ft values in the most recent survey [3] and the improved radiative corrections [11].

In panel (a) of Fig. 1, the corrected Ft values are plotted in full circles as a function of the charge Z of the daughter nucleus. For comparison, the uncorrected experimental ft values [3] and the partially corrected Ft values, only including the radiative corrections, are shown as the open squares and triangles, respectively. One can find the importance of the radiative and ISB corrections by comparing the three sets of data. It can also be seen that the ISB corrections become more important when the charge Z increases. With the present corrections, the average $\overline{FT}$ value is $\overline{FT} = 3081.4(7)$ s with $\chi^2/\nu = 1.1$, which leads to $|V_{ud}| = 0.97273(27)$.

Using the other two top row matrix elements, $|V_{us}| = 0.2255(19)$ and $|V_{ub}| = 0.00393(36)$ [7], one can test the unitarity of the CKM matrix. In panel (b) of Fig. 1, the sum of squared top row elements obtained by RHF+RPA calculations is shown in comparison with those in shell model (H&T) [3] as well as in neutron decay [7], pion β

CP1165, *Nuclear Structure and Dynamics '09*
edited by M. Milin, T. Nikšić, D. Vretenar, and S. Szilner

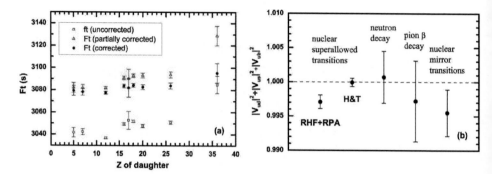

FIGURE 1. (Color online) (a) Corrected Ft values (full circle) as a function of the charge Z for the daughter nucleus. The uncorrected experimental ft values [3] (open square) and partially corrected ($\delta_c = 0$) Ft values (open triangle) are shown for comparison. (b) Sum of squared top row elements of the CKM matrix obtained by RHF+RPA calculations, in comparison with those in shell model (H&T) [3], in neutron decay [7], pion β decay [5] and nuclear mirror transitions [6].

decay [5] and nuclear mirror transitions [6]. It is found that the present value deviates somewhat from the unitarity condition, which is in contradiction with the conclusion of Hardy and Towner [3]. Meanwhile, it is interesting to note that the present result agrees with those obtained by other experimental methods, especially the most recent measurements with nuclear mirror transitions.

ACKNOWLEDGMENTS

This work is partly supported by the Chinese-Croatian project "Nuclear structure far from stability", Major State 973 Program 2007CB815000, the NSFC under Grant Nos. 10435010, 10775004 and 10221003.

REFERENCES

1. N. Cabibbo, *Phys. Rev. Lett.* **10**, 531 (1963).
2. M. Kobayashi, and T. Maskawa, *Prog. Theor. Phys.* **49**, 652 (1973).
3. J. C. Hardy, and I. S. Towner, arXiv:nucl-ex/0812.1202 (2008).
4. D. Thompson, *J. Phys. G* **16**, 1423 (1990).
5. D. Počanić, et al., *Phys. Rev. Lett.* **93**, 181803 (2004).
6. O. Naviliat-Cuncic, and N. Severijns, *Phys. Rev. Lett.* **102**, 142302 (2009).
7. C. Amsler, et al., *Phys. Lett.* **B667**, 1 (2008).
8. H. Liang, N. Van Giai, and J. Meng, arXiv:nucl-th/0904.3673 (2009).
9. H. Liang, N. Van Giai, and J. Meng, *Phys. Rev. Lett.* **101**, 122502 (2008).
10. W. H. Long, H. Sagawa, J. Meng, and N. Van Giai, *Europhys. Lett.* **82**, 12001 (2008).
11. I. S. Towner, and J. C. Hardy, *Phys. Rev. C* **77**, 025501 (2008).

Relativistic QRPA Calculation of β-Decay Rates of r-process Nuclei

T. Marketin*, N. Paar*, T. Nikšić*, D. Vretenar* and P. Ring†

*Physics Department, University of Zagreb, Croatia
†Physik Department der Technischen Universität München, Germany

Abstract. A systematic, fully self-consistent calculation of β-decay rates is presented, based on a microscopic theoretical framework. Analysis is performed on a large number of nuclei from the valley of β stability towards the neutron drip-line. Nuclear ground state is determined using the Relativistic Hartree-Bogoliubov (RHB) model with density-dependent meson-nucleon coupling constants. Transition rates are calculated within the proton-neutron relativistic quasiparticle RPA (pn-RQRPA) using the same interaction that was used in the RHB equations.

Keywords: RHB,pn-RQRPA,r-process,β-decay
PACS: 21.60.Jz;23.40.-s;26.30.Hj

INTRODUCTION

The rapid neutron-capture process is responsible for the creation of many nuclei heavier than iron. However, the exact site and conditions under which the r-process takes place remain unknown. The actual nuclei involved are, at present, beyond experimental reach, and therefore one must turn to nuclear structure models for relevant data used in nucleosynthesis modeling. An important part of any r-process model are the β-decay half-lives of neutron rich nuclei, which determine the speed of the process. As the r-process deals almost exclusively with heavy, neutron-rich nuclei, the natural choice of the theoretical framework is the mean-field formalism.

In previous work concerning β-decay rates only Gamow-Teller transitions were considered [1, 2], but there are indications that forbidden transitions play an important role [3, 4], especially in the heavier regions. We have, therefore, expanded our model via multipole analysis of the weak Hamiltonian, written in the current-current formalism, to systematically include contributions of the forbidden transitions [5] and implemented it within the RHB + pn-RQRPA framework. This approach was previously applied to reaction rates of other weak processes [6, 7], and here we present a preliminary calculation of β-decay half-lives for neutron rich nuclei.

RESULTS AND DISCUSSION

Using previously introduced momentum dependent interaction D3C*, adjusted to nuclear matter properties and bulk properties of finite nuclei, we obtained the effective nucleon mass of $m^* = 0.79m$ [2]. In the pairing channel the pairing part of the Gogny D1S interaction was used [8]. Proton-neutron RQRPA [9] uses the same interactions in

CP1165, *Nuclear Structure and Dynamics '09*
edited by M. Milin, T. Nikšić, D. Vretenar, and S. Szilner
© 2009 American Institute of Physics 978-0-7354-0702-2/09/$25.00

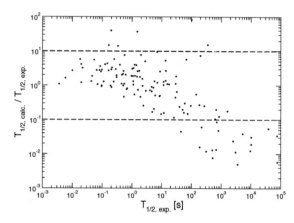

FIGURE 1. Ratio of calculated and experimental half-lives plotted versus the experimental values. Horizontal dashed lines mark a deviation of one order of magnitude

the particle-hole and particle-particle channels as in the RHB equations, making this a fully self-consistent theoretical framework. Calculation of the beta-decay half-lives was performed on more than 100 nuclei, for all multipoles up to $J = 2$, using the renormalized axial-vector coupling constant $g_A = 1$.

In Fig. 1. we plot the ratio of calculated and experimental half-lives versus the experimental half-lives. The major deviation from experiment occurs for nuclei near the valley of stability with long half-lives. The reason for this is overestimation of the difference of the proton and neutron chemical potentials which are used to approximate the Q-values of the transitions. However, for shorter lived nuclei, that are interesting for nucleosynthesis, the obtained values are systematically within an order of magnitude of the experiment.

In conclusion, the results of this preliminary effort show good agreement with experimental values for short-lived nuclei. Combined with previous studies of neutrino-nucleus cross sections and total muon capture rates, the RHB + pn-RQRPA framework allows a quantitative description of all astrophysically relevant weak interaction rates in a fully microscopic, self-cosistent manner.

REFERENCES

1. T. Nikšić, T. Marketin, D. Vretenar, N. Paar and P. Ring, *Phys. Rev.* **C71**, 014308 (2005).
2. T. Marketin, D. Vretenar and P. Ring, *Phys. Rev.* **C75**, 024304 (2006).
3. I. N. Borzov, *Phys. Rev.* **C67**, 025802 (2003).
4. P. Möller, B. Pfeiffer and K-L Kratz, *Phys. Rev.* **C67**, 055802 (2003).
5. J. S. O'Connell, T. W. Donnely and J. D. Walecka, *Phys. Rev.* **C6**, 719 (1972).
6. N. Paar, D. Vretenar, T. Marketin and P. Ring, *Phys. Rev.* **C77**, 024608 (2008).
7. T. Marketin, N. Paar, T. Nikšić and D. Vretenar, *Phys. Rev.* **C79**, 054323 (2009).
8. J. F. Berger, M. Girod and D. Gogny, *Nucl. Phys.* **A428**, 23 (1984).
9. N. Paar, T. Nikšić, D. Vretenar and P. Ring, *Phys. Rev.* **C69**, 054303 (2004).

The MIREDO Facility in Novi Sad

D.Mrda, I.Bikit, M.Veskovic, J.Slivka, M.Krmar, N.Todorovic,
S.Forkapic, N.Jovancevic, G.Soti and J.Papuga

*Department of Physics, Faculty of Sciences, University of Novi Sad,
Trg Dositeja Obradovica 4, 21 000 Novi Sad, Serbia*

Abstract. The MIREDO (Muon Induced Rare Event Dynamic Observatory) spectrometer system will be mainly devoted for the study of cosmic muon induced nuclear reactions. The system will be developed around the 100 % nominal efficiency ultra low-background HPGe spectrometer. Our previous detection system enabled only the detection of secondary muon induced events mostly from the interaction of cosmic muons with the shielding materials [1, 2]. The improved geometry of the MIREDO scintillation detection system will enable the direct measurement of the cross section of the muon induced reactions with the sample material.

Keywords: muons, rare events, spectrometer system, fast slow coincidence circuit,
PACS: 29.30 Kv, 29.40 Wk

The MIREDO spectrometer system will be developed around the 100 % nominal efficiency ultra low-background HPGe spectrometer, made by Canberra. Detector shield is constructed with layered bulk lead. The passive shield has an inner lining to stop the lead K-shell X-rays in the energy range of 75-85 keV. Lining materials are low-background tin with a thickness of 1 mm, and high purity copper with a thicknesses of 1.5 mm. With the addition of two plastic scintillators (Fig.1) and the fast-slow coincidence circuit, the coincident events between the plastic detectors and the HPGe spectrometer will be investigated. The sample material will be placed around the end-cap of Ge detector and covering part of Pb shield will be open in order to minimize influence of secondaries on Ge coincidence spectrum produced by cosmic muons in lead.

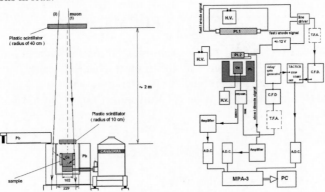

FIGURE 1. The schematic view of: a) the MIREDO facility and b) of coincidence circuit

CP1165, *Nuclear Structure and Dynamics '09*
edited by M. Milin, T. Nikšić, D. Vretenar, and S. Szilner
© 2009 American Institute of Physics 978-0-7354-0702-2/09/$25.00

During the first tests of MIREDO facility, only the bigger plastic detector and Ge detector were operated in the coincident mode, without any sample inside lead shield. The total time of data acquisition was 527314 s. The multi-parameter operating software MPANT was used for acquisition of the following spectra (Fig.2). By analyzing of these spectra it is possible to found which part of muon spectrum is mainly responsible for certain events registered in coincidence spectrum of Ge detector. Also, the time variation of muon flux and spectral lines of Ge detector can be investigated. By changing the distance between detectors, different solid angles covering incident muon paths can be chosen. The muon induced events in different materials, such as CuSO4, NaCl, Pb and Fe will be analyzed.

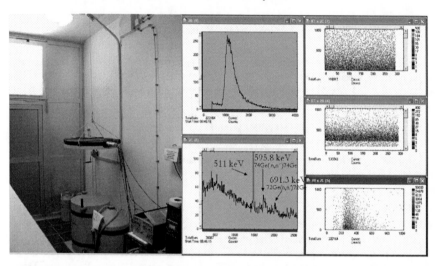

FIGURE 2. a) MIREDO facility during first tests and b) spectra acquired by MPA-3 multi-parameter system: coincidence spectrum of plastic detector, coincidence spectrum of Ge detector, 2D time dependent spectrum of Ge detector, 2D time dependent spectrum of plastic detector and 2D coincidence spectrum of Ge detector and plastic detector

ACKNOWLEDGMENTS

The authors acknowledge the financial support of the Ministry of Science of Republic of Serbia in the frame of the project Nuclear Spectroscopy and Rare Processes (141002) and the Provincial Secretariat for Science and Technological Development of Autonomous Province of Vojvodina (project No 114-451-00631).

REFERENCES

1. D. Mrđa, et al., *Radiat. Meas.* **42**, 1361 (2007).
2. I.Bikit, et al., *Nucl. Instrum. Meth. Phys. Res.* A, doi:10.1016/j.nima.2009.05.153

Absolute Source Activity Measurement with a Single Detector

I.Bikit, T.Nemes, D.Mrdja and S.Forkapic

Department of Physics, Faculty of Sciences, University of Novi Sad,
Trg Dositeja Obradovica 4, 21 000 Novi Sad, Serbia

Abstract. In the present paper the activity of ^{60}Co source was measured using the full absorption, sum and random coincidences (pile up) peaks and the total spectrum area in the gamma spectra. By the exact treatment of the chance coincidence and pile-up events, surprisingly good results were obtained. With the source on the detector end-cap (when the angular correlation effects are negligible), this simple method yields absolute activity values deviating from the reference activity for about 1 percent.

Keywords: Absolute source activity, chance coincidences, angular correlation correction
PACS: 29.30 Kv, 29.40 Wk

The absolute source activity measurement with multi detector systems are commonly used [1]. However the absolute source activity measurements with the single detector are much less recognized. The sum peak method was introduced by a series of papers of Brinkman [2]. The application of the principles of coincidence counting for a single detector was afterwards used only combined with additional fitting of Monte Carlo data [3].

For the measurements of point sources of ^{60}Co, HPGe coaxial detector with relative efficiency of 35% in lead shield was used. Portable workstation "InSpector 2000" (Model 1300-Canberra) was used for digital processing of signals, which improves time and energy resolution. Activities of two point sources of ^{60}Co with reference activities of 246(4) kBq (Areva, No 40885/186) and 5.46(11) kBq (Amersham No 11188) were measured. Measurements with the stronger ^{60}Co source were performed placing the source on the detector end-cap and increasing the distance up to 64.1 mm from detector end-cap. For these measurements the live time was set to be 600 s. Spectrum of the weaker ^{60}Co source was recorded placing the source on the detector cap for 18 hours.

Neglecting angular correlation correction (w=1), we arrive to the final expression for the source activity:

$$A = \frac{N_T}{t}\left[\left[\left(\frac{N_\Sigma}{2\sqrt{N_{C1}N_{C2}}}-1\right)\left(\sqrt{\frac{4N_T\sqrt{N_{C1}N_{C2}}}{N_1N_2}+1}-1\right)\right]^{-1}+1\right] \qquad (1)$$

where: A is the measured source activity calculated from the final expression for source activity; N_1 is the 1173 keV net peak area; N_2 is the 1332 keV net peak area; N_Σ is the 2505 keV sum peak net area; N_{C1} is the 2×1173 keV chance coincidence peak

CP1165, *Nuclear Structure and Dynamics '09*
edited by M. Milin, T. Nikšić, D. Vretenar, and S. Szilner
© 2009 American Institute of Physics 978-0-7354-0702-2/09/$25.00

net area; N_{C2} is the 2×1332 keV chance coincidence peak net area; N_T is total counts in the spectrum, N_C is the calculated number of chance coincidence counts in the sum-peak and t is the time of measurement.

Minimum discrepancy between observed and reference activity of point sources of ^{60}Co (A_0=246.1 kBq) is obtained for measurement when the source is placed on detector end-cap. The increased deviation of A/A_0 from unity plotted in Fig. 1. is in good agreement with the published angular correlation correction factors [3]. The results for the effective resolution time 2τ of the system with the order of magnitude 0.5 µs do not depend on the count rate, and independently confirm the consistency of our method. The measured activity for the weaker source (A_0=5.46 kBq) on the end-cap is also in good agreement with the reference value.

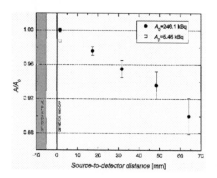

FIGURE 1. Ratio between measured and reference activity (A/A_0) obtained from different distances of point source ^{60}Co to detector end-cap.

The activity of the ^{60}Co source was determined with the relative error of less than 1% what exhibits clearly that with exact treatment of the chance coincidence and pile-up events no additional fitting methods and/or Monte Carlo calculations are needed for accurate source activity measurements.

ACKNOWLEDGMENTS

The authors acknowledge the financial support of the Ministry of Science of Republic of Serbia in the frame of the project Nuclear Spectroscopy and Rare Processes (141002) and the Provincial Secretariat for Science and Technological Development of Autonomous Province of Vojvodina (project No 114-451-00631).

REFERENCES

1. G.F. Knoll, *Radiation Detection and Measurement*, New York, John Wiley & Sons, 1979.
2. G.A. Brinkman, A.H.W. Aten Jr., J.Th. Veenboer, *Int. J. Appl. Radiat. Isotopes* **14**, 153 (1963).
3. I.J. Kim, C.S. Park, H.D. Choi, *Appl. Radiat. Isotopes* **58**, 227 (2003).
4. G.A. Brinkman, A.H.W. Aten Jr., *Int. J. Appl. Radiat. Isotopes* **16**, 177 (1965).
5. I.J. Kim, G.M. Sun, H.D. Choi, Y.D. Bae, *J. Korean. Nucl. Soc.* **34**, 22 (2002).
6. D.M.Scates, J.K.Hartwell , *Appl. Radiat. Isotopes* **63**, 465 (2005).

Study of Neutron Induced Activity in Low-Background Gamma Spectroscopy Systems

N. Jovancevic, M. Krmar, I. Bikit, D. Mrda, M. Veskovic, J. Slivka, N. Todorovic, S. Forkapic, G. Soti, J. Papuga

Department of Physics, Faculty of Sciences, University of Novi Sad,
Trg Dositeja Obradovica 4, 21 000 Novi Sad, Serbia

Abstract. Cosmic-rays induced neutrons can produce measurable activity in carefully selected materials used for construction of low-background gamma spectroscopy systems. In order to improve shielding of low-background systems, flux of neutrons inside the shield and level of induced activity should be measured.

Keywords: low-level gama spectrometry, cosmic-rays induced neutrons, shield.
PACS: 28.41.Qb, 29.30.Hs

Background reduction is sometimes only one possible way to improve sensitivity of experiment [1,2,3]. At the see level natural radioactivity is most important source of background in Ge gamma spectroscopy [4]. The dominant component of background is gamma radiation emitted by natural radionuclides. Another significant source of background is neutrons. There are two sources of neutrons associated with local natural radioactivity: neutrons produced via (α,n) reactions and spontaneous fission of U and Th . Both groups of neutrons have low energy in MeV energy region.

Two different detectors were used to collect evidence about neutron contribution to the background:
1. HPGe detector, 25 % relative efficiency, shielded by 25 cm of pre -WW II iron.
2. HPGe detector, 100 % relative efficiency, Detector was shielded by commercial low-background shield: 15 cm lead, 1 mm tin and 1.5 mm of copper.

Detector in iron shield was surrounded by 9" x 9" NaI Compton suppressed system. Influence of different materials placed in Marinely geometry around detector in commercial shield was studied. Time of measurement was up to 5000 ks.

Two different types of gamma lines were observed at background spectra: standard spectroscopy lines collected after thermal neutron capture in Ge, which provide evidence about thermal neutron flux in detector vicinity, and long - tail gamma lines detected after inelastic scattering of fast neutrons on Ge nuclei, which can be used as a measure of presence of fast neutrons around detector [5]. Thermal neutron flux is 4.3(28) times higher in lead shield than in iron shield. Fast neutron flux is 35(2) times higher in lead shield than in iron shield. The measurements mentioned above were used to check how different materials surrounding detector could change thermal and fast neutron flux. Results are given in Table 1.

NaI Compton suppression system can be source of gamma radiation due to interactions with neutrons. PVC Marinelli container reduces fast neutron number more efficiently than bulk sample located around detectors. Absence of significant reduction

CP1165, *Nuclear Structure and Dynamics '09*
edited by M. Milin, T. Nikšić, D. Vretenar, and S. Szilner
© 2009 American Institute of Physics 978-0-7354-0702-2/09/$25.00

of neutron number by bulk materials located around detector can be explained by production of new neutrons in materials surrounding detector.

TABLE 1. Ratio of intensities of 139.5 keV gamma line detected with surrounding material and without it can be measure of relative changes of thermal neutron flux. Relative changes of fast neutron flux affected by presence of some material around detector can be described by ratio of 691 keV line.

	NaI	NaCl	CuSO$_4$·5H$_2$O	PVC	Paraffin
R_T	1.1(6)	1.04(4)	0.384(12)	3.70(25)	0.323(9)
R_F	0.252(21)	0.748(9)	1.34(2)	4.46(5)	1.260(21)

Intensity of ^{56}Fe (n,n') gamma line changes with known frequency of 28 days (Fig.1). Measured relative intensities of other ^{56}Fe gamma lines differ from relative intensities measured in inelastic scattering experiments.

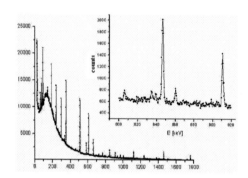

FIGURE 1. ^{56}Fe (n,n') gamma line observed in spectra recorded by iron shielded detector. Intensity of this line was measured during one year to get time dependence.

ACKNOWLEDGMENTS

The authors acknowledge the financial support of the Ministry of Science of Republic of Serbia in the frame of project Nuclear Spectroscopy and Rare Processes (141002) and the Provincial Secretariat for Science and Technological Development of Autonomous Province of Vojvodina (project No 114-451-00631).

REFERENCES

1. R. Wordel, et al., Study of neutron and muon background in low-level germanium gamma-ray spectrometry, *Nucl. Instr. and Meth.* A 369 (1996) 557-562.
2. G. Heusser, Cosmic ray interaction study with low-level Ge-spectrometry, *Nucl. Instr. and Meth.* A 369 (1996) 539-543.
3. G. Heusser, *Annu. Rev.Nucl. Part. Sci.* 1995. 45:543-90.
4. Y. Feige, B. G. Oltman, J. Kastner, Production rates of neutrons in soils due to natural radioactivity, *J. Geophys. Res.* 73 (1968) 3135-3142.
5. G. P. Skoro, I. V. Anicin, A. H. Kukoc, Dj. Krmpotic, P. Adzic, R. Vukanovic and M. Zupancic, *Nucl. Instrum. and Meth. In Phys. Res.* A, 316 (1992) 333.   5.

List of Participants

Yasuhisa Abe
Osaka University
Osaka, Japan
abey@rcnp.osaka-u.ac.jp

Dieter Ackermann
GSI
Darmstadt, Germany
d.ackermann@gsi.de

Luis Acosta
University of Huelva
Huelva, Spain
luis.acosta@dfa.uhu.es

Gurgen Adamian
Joint Institute for Nuclear Research
Dubna, Russia
adamian@theor.jinr.ru

Anatoli Afanasjev
Mississippi State University
Mississippi, USA
aa242@msstate.edu

Martin Alcorta
Instituto de Estructura de la Materia
Madrid, Spain
alcorta@iem.cfmac.csic.es

Hicham Al Falou
Saint Mary's University - TRIUMF
Caen, Canada
alfalou@triumf.ca

Nori Aoi
RIKEN Nishina Center
Hirosawa, Wako, Japan
aoi@riken.jp

Nick Ashwood
University of Birmingham
Edgbaston, Birmingham, UK
n.i.ashwood@bham.ac.uk

Nikolai Antonenko
Joint Institute for Nuclear Research
Dubna, Russia
antonenk@theor.jinr.ru

Thomas Aumann
GSI
Darmstadt, Germany
t.aumann@gsi.de

Zoran Basrak
Ruđer Bošković Institute
Zagreb, Croatia
basrak@irb.hr

Istvan Bikit
University of Novi Sad
Novi Sad, Serbia
bikit@im.ns.ac.yu

Alexey Bogachev
Joint Institute for Nuclear Research
Dubna, Russia
bogachev@nr.jinr.ru

Konstanze Boretzky
GSI
Darmstadt, Germany
k.boretzky@gsi.de

Ismail Boztosun
Erciyes University
Kayseri, Turkey
boztosun@erciyes.edu.tr

Slobodan Brant
University of Zagreb
Zagreb, Croatia
brant@phy.hr

Carlo Broggini
Istituto Nazionale di Fisica Nucleare
Padova, Italy
broggini@pd.infn.it

Giuseppe Cardella
INFN-Sezione CT
Catania, Italy
cardella@ct.infn.it

Gillis Carlsson
University of Jyväskylä
Jyväskylä, Finland
gillis.carlsson@matfys.lth.se

Wilton Catford
University of Surrey
Guildford, UK
w.catford@surrey.ac.uk

William Collis
ISCMNS
The Willows, Hobro, UK
mr.collis@physics.org

Gianluca Colo
University of Milano
Milano, Italy
colo@mi.infn.it

Lorenzo Corradi
Laboratori Nazionali di Legnaro
Legnaro, Italy
corradi@lnl.infn.it

Anthony Cowley
University of Stellenbosch
Stellenbosch, South Africa
aac@sun.ac.za

Raquel Crespo
Instituto Superior Tecnico
Oeiras, Portugal
raquel.crespo@tagus.ist.utl.pt

Jozsef Cseh
Hungarian Academy of Sciences
Debrecen, Hungary
cseh@atomki.hu

Judit Darai
University of Debrecen
Debrecen, Hungary
darai@atomki.hu

Stijn De Baerdemacker
Ghent University
Ghent, Belgium
stijn.debaerdemacker@ugent.be

Alessia Di Pietro
Laboratori Nazionali del Sud
Catania, Italy
dipietro@lns.infn.it

Tomas Dytrych
Louisiana State University
Baton Rouge, USA
tdytrych@phys.lsu.edu

Haris Djapo
TU Darmstadt
Darmstadt, Germany
haris@crunch.ikp.physik.tu-darmstadt.de

Jacek Dobaczewski
University of Warsaw, Poland
University of Jyväskylä, Finland
dobaczew@fuw.edu.pl

Pieter Doornenbal
RIKEN Nishina Center
Hirosawa, Saitama, Wako, Japan
pieter@ribf.riken.jp

Dusan Dragosavac
University of Santiago de Compostela
Santiago de Compostela, Spain
dusan.dragosavac@usc.es

Marianne Dufour
Université de Strasbourg
Strasbourg, France
marianne.dufour@ires.in2p3.fr

Thomas Duguet
CEA-SPhN, Centre de Saclay
Gif-sur-Yvette, France
thomas.duguet@cea.fr

Jean-Paul Ebran
Institut de Physique Nucléaire
Orsay, France
ebran@ipno.in2p3.fr

M. Nizamettin Erduran
Istanbul University
Istanbul, Turkey
erduran@istanbul.edu.tr

Enrico Farnea
INFN Padova
Padova, Italy
farnea@pd.infn.it

Paolo Finelli
University of Bologna
Bologna, Italy
paolo.finelli@bo.infn.it

Enrico Fiori
Institut de Physique Nucléaire
Orsay, France
enrico.fiori@csnsm.in2p3.fr

Sofija Forkapic
University of Novi Sad
Novi Sad, Serbia
sofija@im.ns.ac.yu

Martin Freer
University of Birmingham
Birmingham, UK
M.Freer@bham.ac.uk

Brian Fulton
University of York
York, UK
brf2@york.ac.uk

Hans Fynbo
University of Aarhus
Aarhus, Denmark
fynbo@phys.au.dk

Mitko Gaidarov
Institute of Nuclear Research
and Nuclear Energy, Sofia, Bulgaria
gaidarov@inrne.bas.bg

José Enrique García-Ramos
University of Huelva
Huelva, Spain
enrique.ramos@dfaie.uhu.es

Hans Geissel
GSI and Justus-Liebig Uni.
Giessen, Germany
h.geissel@gsi.de

Paulo Gomes
Universidade Federal Fluminense
Niterói, Brasil
paulogom@if.uff.br

Magdalena Gorska
GSI
Darmstadt, Germany
m.gorska@gsi.de

Marcella Grasso
Institut de Physique Nucléaire
Orsay, France
grasso@ipno.in2p3.fr

Paolo Guazzoni
University of Milan and INFN
Milan, Italy
paolo.guazzoni@mi.infn.it

Florent Haas
Institut Hubert Curien
Strasbourg, France
florent.haas@ires.in2p3.fr

Michael Heil
GSI
Darmstadt, Germany
m.heil@gsi.de

Veerle Hellemans
University of Notre Dame
Notre Dame, USA
vhellema@nd.edu

Kris Heyde
University of Gent
Gent, Belgium
kris.heyde@ugent.be

Francesco Iachello
Yale University
New Haven, USA
francesco.iachello@yale.edu

Yuichi Ichikawa
University of Tokyo, Tokyo 113-0033
RIKEN, Hirosawa, Saitama 351-0198, Japan
ichikawa@ribf.riken.jp

Mikhail Itkis
Joint Institute for Nuclear Research
Dubna, Russia
itkis@jinr.ru

Makoto Ito
Kansai University
Osaka, Japan
itomk@riken.jp

David Jenkins
University of York
York, UK
dj4@york.ac.uk

Jacob Johanson
Department of physics and astronomy
University of Aarhus, Danemark
jacobsj@phys.au.dk

Ari Jokinen
University of Jyväskylä
Jyväskylä, Finland
ari.s.a.jokinen@jyu.fi

Nikola Jovancevic
University of Novi Sad
Novi Sad, Serbia
nikola.jovancevic@if.ns.ac.yu

Elias Khan
Institut de Physique Nucléaire
Orsay, France
khan@ipno.in2p3.fr

Mladen Kiš
Ruđer Bošković Institute
Zagreb, Croatia
mkis@irb.hr

Gokhan Kocak
Erciyes University
Kayseri, Turkey
gkocak@erciyes.edu.tr

Magdalena Kowalska
CERN, Geneva, Switzerland
ISOLDE, Switzerland
kowalska@cern.ch

Eduard Kozulin
Flerov Laboratory of Nuclear Reactions
JINR, Dubna, Russia
 kozulin@dubna.ru

Thorsten Kroell
TU Darmstadt
Darmstadt, Germany
tkroell@ikp.tu-darmstadt.de

Yasemin Kucuk
Erciyes University
Kayseri, Turkey
 ylkucuk@gmail.com

Denis Lacroix
GANIL-Caen
Caen, France
lacroix@ganil.fr

Georgios Lalazissis
Aristotle University of Thessaloniki
Thessaloniki, Greece
glalazis@auth.gr

Edoardo G. Lanza
INFN - Catania
Catania, Italy
lanza@ct.infn.it

Dorothee Lebhertz
Institut Hubert Curien
Strasbourg, France
dorothee.lebhertz@ires.in2p3.fr

Alinka Lepine-Szily
Saõ Paolo University
Saõ Paolo, Brasil
alinka.lepine@dfn.if.usp.br

Geza Levai
ATOMKI
Debrecen, Hungary
levai@atomki.hu

Amiram Leviatan
The Hebrew University
Jerusalem, Israel
ami@phys.huji.ac.il

Zhipan Li
Peking University
Beijing, China
lizhipan1983@126.com

Haozhao Liang
Peking University
Beijing, China
hzliang@pku.edu.cn

Matej Lipoglavsek
Jozef Stefan Institute
Ljubljana, Slovenia
matej.lipoglavsek@ijs.si

Herbert Löhner
KVI, University of Groningen
Groningen, Netherlands
loehner@kvi.nl

Santo Lunardi
University of Padova
Padova, Italy
lunardi@pd.infn.it

David Lunney
CSNSM/IN2P3/CNRS
Paris, France
lunney@csnsm.in2p3.fr

Adam Maj
IFJ PAN Krakow
Krakow, Poland
Adam.Maj@ifj.edu.pl

Victor Mandelzweig
Hebrew University Jerusalem
Jerusalem, Israel
victor@phys.huji.ac.il

Nicholas Manton
University of Cambridge
Cambridge, UK
N.S.Manton@damtp.cam.ac.uk

Tomislav Marketin
University of Zagreb
Zagreb, Croatia
marketin@phy.hr

Masayuki Matsuo
Niigata University
Niigata, Japan
matsuo@nt.sc.niigata-u.ac.jp

Jie Meng
Peking University
Beijing China
mengj@pku.edu.cn

Daniele Mengoni
Università di Padova
Padova, Italy
daniele.mengoni@pd.infn.it

Matko Milin
University of Zagreb
Zagreb, Croatia
matkom@phy.hr

Đuro Miljanić
Ruđer Bošković Institute
Zagreb, Croatia
djuro.miljanic@irb.hr

Daniele Montanari
Universiti degli Studi di Milano
Milano, Italy
daniele.montanari@mi.infn.it

David Morrissey
NSCL, Michigan State University
East Lansing, Michigan, USA
morrissey@nscl.msu.edu

Dusan Mrdja
University of Novi Sad
Novi Sad, Serbia
mrdjad@im.ns.ac.yu

Ivan Mukha
University of Seville
Seville, Spain
mukha@us.es

Agatino Musumarra
University of Catania
Catania, Italy
musumarra@lns.infn.it

Kazuo Muto
Tokyo Institute of Technology
Tokyo, Japan
muto@th.phys.titech.ac.jp

Hitoshi Nakada
Chiba University
Chiba, Japan
nakada@faculty.chiba-u.jp

Takashi Nakatsukasa
RIKEN Nishina Center
Wako, Japan
nakatsukasa@riken.jp

Thomas Neff
GSI
Darmstadt, Germany
t.neff@gsi.de

Tamara Nikšić
University of Zagreb
Zagreb, Croatia
tniksic@phy.hr

Yifei Niu
University of Zagreb
Zagreb, Croatia
nyfster@gmail.com

Chiara Nociforo
GSI
Darmstadt, Germany
c.nociforo@gsi.de

Kosuke Nomura
University of Tokyo
Tokyo, Japan
nomura@nt.phys.s.u-tokyo.ac.jp

Eric Norman
University of California
Berkeley, USA
ebnorman@lbl.gov

Wolfram von Oertzen
Helmholz-zentrum
Berlin, Germany
oertzen@hmi.de

Yuri Oganessian
Flerov Laboratory of Nuclear Reactions
JINR, Dubna, Russia
oganessian@jinr.ru

Takaharu Otsuka
University of Tokyo, Tokyo 113-0033
RIKEN, Hirosawa, Saitama 351-0198, Japan
otsuka@phys.s.u-tokyo.ac.jp

Jasna Papuga
University of Novi Sad
Novi Sad, Serbia
jasna.papuga@if.ns.ac.yu

Nils Paar
University of Zagreb
Zagreb, Croatia
npaar@phy.hr

Daniel Peña Arteaga
Institut de Physique Nucléaire d'Orsay
Orsay, France
pena@ipno.in2p3.fr

Dirceu Pereira
Saõ Paolo University
Saõ Paolo, Brasil
dpereira@dfn.if.usp.br

Sophie Péru
CEA DAM DIF
Arpajon, France
sophie.peru-desenfants@cea.fr

Costel Petrache
IPN Orsay and University Paris Sud
France
petrache@ipno.in2p3.fr

Ernest Piasecki
Warsaw University
Warsaw, Poland
piasecki@fuw.edu.pl

Norbert Pietralla
TU Darmstadt
Darmstadt, Germany
pietralla@ikp.tu-darmstadt.de

Nathalie Pillet
CEA, Bruyeres-le-Chatel
France
nathalie.pillet@cea.fr

Rosario Gianluca Pizzone
Laboratori Nazionali del Sud
Catania, Italy
rgpizzone@lns.infn.it

Zsolt Podolyak
University of Surrey
Guildford, UK
Z.Podolyak@surrey.ac.uk

Giovanni Pollarolo
University of Torino
Torino, Italy
nanni@to.infn.it

Gregory Potel
Universiti degli Studi di Milano
Milano, Italy
gregory.potel@mi.infn.it

Lovro Prepolec
Ruđer Bošković Institute
Zagreb, Croatia
lovro@lnr.irb.hr

Riccardo Raabe
GANIL
Caen, France
raabe@ganil.fr

K. Ernst Rehm
Argonne National Laboratory
Argonne, USA
rehm@anl.gov

Peter Ring
Technische Universität München
München, Germany
ring@ph.tum.de

Roman Sagaydak
Joint Institute for Nuclear Research
Dubna, Russia
sagaidak@nrmail.jinr.ru

Hiroyuki Sagawa
University of Aizu
Aizu, Japan
sagawa@u-aizu.ac.jp

Hideyuki Sakai
University of Tokyo
Tokyo, Japan
sakai@phys.s.u-tokyo.ac.jp

Rodolfo Marcelo Sanchez Alarcon
GSI
Darmstadt, Germany
R.Sanchez@gsi.de

Pedro Sarriguren
Instituto Estructura de la Materia
Madrid, Spain
sarriguren@iem.cfmac.csic.es

Wojciech Satula
University of Warsaw
Warsaw, Poland
satula@fuw.edu.pl

Fernando Scarlassara
Università di Padova
Padova, Italy
scarlassara@pd.infn.it

Peter Schuck
Insitut de Physique Nucléaire
Orsay, France
schuck@ipno.in2p3.fr

Zdravko Siketić
Ruđer Bošković Institute
Zagreb, Croatia
zsiketic@irb.hr

Natko Skukan
Ruđer Bošković Institute
Zagreb, Croatia
nskukan@irb.hr

Neven Soić
Ruđer Bošković Institute
Zagreb, Croatia
soic@lnr.irb.hr

Pietro Sona
Università Firenze
Florence, Italy
sona@fi.infn.it

Olivier Sorlin
GANIL
Caen, France
sorlin@ganil.fr

Paul Stevenson
University of Surrey
Guildford, UK
p.stevenson@surrey.ac.uk

Kazuko Sugawara-Tanabe
Otsuma Women's University
Tokyo, Japan
tanabe@otsuma.ac.jp

Toshio Suzuki
Nihon University
Tokyo, Japan
suzuki@phys.chs.nihon-u.ac.jp

Ivo Šlaus
Ruđer Bošković Institute
Zagreb, Croatia
slaus@irb.hr

Suzana Szilner
Ruđer Bošković Institute
Zagreb, Croatia
szilner@irb.hr

Natasa Todorovic
University of Novi Sad
Novi Sad, Serbia
zikic@im.ns.ac.yu

Jussi Toivanen
University of Jyväskylä
Jyväskylä, Finland
jussi.toivanen@phys.jyu.fi

Dimitar Tonev
INRE, BAS
Sofia, Bulgaria
mitko@lnl.infn.it

Koshiroh Tsukiyama
University of Tokyo
Tokyo, Japan
tsuki@nt.phys.s.u-tokyo.ac.jp

Aurora Tumino
Laboratori Nazionali del Sud
Catania, Italy
tumino@lns.infn.it

Sait Umar
Vanderbilt University
Vanderbilt, USA
umar@compsci.cas.vanderbilt.edu

Milivoj Uroić
Ruđer Bošković Institute
Zagreb, Croatia
muroic@irb.hr

Jose Javier Valiente Dobon
Laboratori Nazionali di Legnaro
Legnaro, Italy
valiente@lnl.infn.it

Andrea Vitturi
Dipartimento di Fisica and INFN
Padova, Italy
vitturi@pd.infn.it

Dario Vretenar
University of Zagreb
Zagreb, Croatia
vretenar@phy.hr

Shaun Wyngaardt
Stellenbosch University
Stellenbosch, South Africa
shaunmw@sun.ac.za

Xinxing Xu
China Institute of Atomic Energy
Beijing, China
xuxinxing@ciae.ac.cn

Nobuaki Yoshida
Kansai University
Takatsuki, Japan
yoshida@res.kutc.kansai-u.ac.jp

Ivana Zamboni Mićanović
Ruđer Bošković Institute
Zagreb, Croatia
izamboni@irb.hr

Luisa Zetta
University of Milan and INFN
Milan, Italy
luisa.zetta@mi.infn.it

Ying Zhang
Peking University
Beijing, China
yzhangjcnp@pku.edu.cn

Shan-Gui Zhou
Chinese Academy of Sciences
Beijing, China
sgzhou@itp.ac.cn

Mikhail Zhukov
Chalmers University of Technology
Göteborg, Sweden
f2bmz@fy.chalmers.se

Butterworth, J., 363

454

Lenske, H., 90
Lenzi, S. M., 386
Leoni, S., 386
Lépine-Szily, A., 33
Lesinski, T., 243
Leske, J., 120, 225
Lévai, G., 211
Leviatan, A., 199
Li, C., 106
Li, K., 82
Li, Z. P., 219
Liang, H. Z., 279
Liang, H., 431
Lin, C. J., 106
Lipoglavšek, M., 419
Lister, K., 225
Litvinov, S. A., 415
Litvinov, Yu. A., 90, 181, 415
Litvinova, E., 145
Liu, Z. H., 106
Lo Bianco, G., 363, 386
Lombardo, I., 309
Lommel, B., 90
Lo Nigro, S., 309
Loria, D., 309
Lozeva, R., 120
Lukić, S., 114
Lukyanov, K. V., 347
Lukyanov, S., 120
Lukyanov, V. K., 347
Lunardi, S., 386
Lunney, D., 94, 118
Lutter, R., 363

M

Mabiala, J., 323
Macchiavelli, A. O., 363
Madurga, M., 27
Mahata, K., 90, 181
Maier, H. K., 86
Maierbeck, P., 90, 181
Maiolino, C., 309
Maira, A., 27, 317
Maj, A., 386
Majer, M., 31
Mann, R., 124
Manton, N. S., 229
Margueron, J., 158

Markelj, S., 419
Marketin, T., 433
Marley, P. L., 331
Martel, I., 317
Martinez-Pinedo, G., 415
Maruhn, J. A., 383
Mason, P., 375, 386
Massen, S. E., 347
Matsuo, M., 263
Maurer, J., 124
Mazzocco, M., 415
McEwan, P., 13
Meczynski, W., 386
Meng, J., 189, 219, 279, 431
Mengon, D., 386
Mertzimekis, T. J., 120
Michalon, A., 331
Michimasa, S., 82
Milin, M., 31, 349
Miljanić, Đ., 31, 349, 423
Million, B., 386
Minaya Ramirez, E., 94, 118
Modamio, V., 120
Montagnoli, G., 375, 386
Montanari, D., 386
Montes, F., 114, 415
Moro, A. M., 317
Morrissey, D. J., 65
Motizuki, Y., 415
Motobayashi, T., 82, 98
Mouginot, B., 120
Mrdja, D., 435, 437, 439
Mukha, I., 317
Mukhamedzhanov, A., 335, 423
Muñoz, A., 27
Musumarra, A., 31, 90, 415
Muto, K., 57

N

Naimi, S., 94
Nakabayashi, T., 98
Nakada, H., 271
Nakamura, T., 82, 98
Nakao, T., 82, 98
Nakatsukasa, T., 173
Namihira, K., 82
Napoli, D. R., 386
Nara Singh, B. S., 363

457

RETURN TO: PHYSICS-ASTRONOMY LIBRARY
351 LeConte Hall   510-642-3122

LOAN PERIOD 1	2	3
1-MONTH		
4		